Polyelectrolyte Solutions

A THEORETICAL INTRODUCTION

MOLECULAR BIOLOGY

An International Series of Monographs and Textbooks

EDITED BY

NATHAN O. KAPLAN

Graduate Department of Biochemistry
Brandeis University,
Waltham, Mass.

HAROLD A. SCHERAGA

Department of Chemistry
Cornell University,
Ithaca, New York

Polyelectrolyte Solutions

A THEORETICAL INTRODUCTION

STUART A. RICE
University of Chicago

and

MITSURU NAGASAWA
Nagoya University

with a contribution by
HERBERT MORAWETZ
Polytechnic Institute of Brooklyn

1961

ACADEMIC PRESS · *London and New York*

to our wives
MARIAN and TERUKO

PREFACE

The past decade has witnessed an extensive development in two directions of the theory of electrolyte solutions. On the one hand, the investigations of Mayer and of Kirkwood and Poirier have finally established from first principles the validity of the Debye-Hückel limiting law. These studies provide a formal framework for extending the theory of electrolytes to concentrations higher than the traditional "slightly polluted water," and they provide insight into the nature of the approximations which make valid the use of the Poisson-Boltzmann equation in dilute solution. Moreover, the recent works of Onsager and Kim, of Fuoss and Onsager, of Falkenhagen and Kelbg, and of Pitts have extended, within the framework of the Poisson-Boltzmann equation, the theory of transport processes to include the effects of finite ion size. While it is doubtful if the validity of the Poisson-Boltzmann equation permits further development, the inclusion of finite ion size does permit the description of dissipative processes, in terms of an ion-size parameter, to be extended to much higher concentrations than heretofore, and it suggests a more satisfactory interpretation of the varying behavior of different strong electrolytes. Further extension of the theory of dissipative processes probably requires the more powerful tools of the Mayer theory combined with the general statistical theory of transport.

On the other hand, there has been an explosive development of the theory of very highly charged macromolecular ions. Whereas, prior to 1940, almost no quantitative theory of colloid solutions existed, and the measurements of Kern represented the only investigations of flexible polyelectrolytes, there are now a number of theories for the equilibrium properties of these substances, and there have been several attempts at the theory of transport. In all instances, the theory of polyelectrolytes has been developed within the framework of the Poisson-Boltzmann equation; in some cases the analysis uses a linearized form, while in others the analysis adopts different approximations equivalent to the assumption that the domain occupied by the polyion has zero net charge. Although there are many difficulties associated with the current theory, and, though qualitative or semiquantitative agreement with experiment is all that can be attained at present, it may be asserted safely that the broad features of the basic physical processes responsible for the observed phenomena have been delineated and that the major problems are quantitative rather than qualitative.

The situation sketched in the preceding two paragraphs suggested to the authors the utility of a unified survey of the theory of polyelectrolyte solutions with emphasis on the basic physical processes. The need for such a survey was further emphasized by the difficulty in providing suitable general reference material in this area for graduate students starting research problems. Once the decision that a review would be valuable was reached only two questions remained: at what level should the material be presented and in what form.

We found it easy to decide on the level of the presentation and have chosen to present the subject matter discussed in a manner we believe suitable for second year graduate students. The nomenclature "second year graduate student" means simply that we assume that the reader is familiar with elementary thermodynamics and, more to the point, elementary statistical mechanics.

On the other hand, we found it much more difficult to decide on the length, selection of material, and amount of detail which would be desirable. Indeed, these pages began in 1957 as a review article dealing with the status of the theory of polyelectrolyte solutions. At the urging of Professor Jui Wang, the original plan for a review article was expanded into the present book. Had we known what slaves we were to become to this project, we would undoubtedly never have been persuaded by his arguments. Be that as it may, we herein present to the reader an introduction to the theory of polyelectrolyte solutions. It will be apparent that we have made no attempt to hide our individual preferences and prejudices. For this we make no excuse, rather trusting to the ultimate averaging produced by different authors over a period of years. It will also be apparent that we have not prepared an exhaustive survey of the literature, but have preferred to give a detailed analysis of various aspects of the theory together with some, but by no means all, of the relevant experimental information.

We have taken as the motto of our presentation the well known statement of Lord Kelvin:

"Where you can measure what you are speaking about and express it in numbers, you know something about it, and when you cannot measure it, when you cannot express it in numbers, your knowledge is of a meagre and unsatisfactory kind. It may be the beginning of knowledge, but you have scarcely in your thought advanced to the stage of a science." Throughout the text we have maintained a commitment to quantitative analysis wherever possible. We have not avoided mathematics; in our opinion all too often such a policy indicates more of an unwillingness to work at the subject matter and to think clearly than an unfamiliarity with specific mathematical techniques. All in all, it will not take the reader long to decide that this is in some senses a peculiar book.

The text has been organized into eleven chapters. Chapter 1 is based upon lectures given by one of us (S.A.R.) in a course dealing with the molecular theory of solutions.* In view of the excellent surveys of the theory of solutions of nonpolar polymers now available, this topic has been completely omitted. We have found that a thorough understanding of the nature of the interactions in solutions of small molecules is of great assistance in understanding the properties of polymer solutions. The utility of this material is confirmed by the fact that the most recent developments in the theory of polymer solutions make extensive use of the McMillan-Mayer theory and diagram techniques. Chapter 2 deals with the nature of solutions of simple electrolytes. Attention is focused throughout on the validity of the Poisson-Boltzmann equation. We are aware that there are more powerful methods for studying the properties of ionic solutions than any we discuss. However, since almost all poly-electrolyte theory is based upon the Poisson-Boltzmann equation we have preferred to confine attention to that aspect of the problem. The inspiration for the approach used comes from a series of lectures by Professor A. S. Coolidge. In Chapters 3 and 4 we study the properties of rigid polyions. Inasmuch as many polyions of this class are inorganic colloids, and the agreement between theory and experiment has been presented in detail in other monographs, we confine our attention to the theory. We have found, as commented above with respect to the content of Chapter 1, that an understanding of the behavior of rigid polyelectrolytes is of great assistance in unraveling the properties of flexible polyelectrolytes from complex experimental observations. Chapter 5 (by Professor H. Morawetz) presents the theory of expansion and counterion distribution in solutions of flexible polyelectrolytes. Chapters 6 and 7 are based extensively on a series of papers by one of us (S.A.R.). We have preferred to develop one theory in great detail and compare it with the predictions of other theories, rather than develop all theories in a sketchy fashion. In Chapters 8, 9, 10, and 11 we make the comparisons alluded to and indicate the current relationship between theory and experiment.

Because of the rapid development of theory and the ever increasing number of studies of the physical chemistry of polyelectrolyte solutions, it is likely that this book will need revision in a very short time. We therefore hope that this presentation will prove to be a suitable introduction to a fascinating and important field of research. If this hope is fulfilled or if any interest in further work is stimulated by this book, our aim will have been accomplished.

* Other lecturers in this course were Professors R. M. Mazo, P. C. Mangelsdorf, and O. J. Kleppa. Some of Professor Mazo's lecture material is also in Chapter 1.

There remains now only the pleasant task of thanking all those who have been of assistance to us. We are indebted to Professors Paul Doty, I. Kagawa, and F. E. Harris for their general interest and encouragement over the years; to our students, postdoctoral research fellows, and colleagues who have in the day-to-day exchange of ideas contributed a great deal to our understanding of the phenomena discussed; to Mrs. Iris Ross for typing the manuscript; to the National Science Foundation, the National Institutes of Health, and the Fulbright Commission for financial assistance; to Oxford University Press, John Wiley and Sons, Cambridge University Press, the *Journal of the American Chemical Society*, the *Journal of Physical Chemistry*, the *Journal of Chemical Physics*, *Zeitschrift für Physikalische Chemie, N.F.*, the *Journal of Polymer Science*, the *Proceedings of the National Academy of Sciences*, the *Proceedings of The Royal Society*, the *Journal of Molecular Physics*, the *Transactions of the Faraday Society*, the *Journal of Colloid Science*, *Recueil des Travaux Chimiques des Pays-Bas*, and to the authors concerned, and Drs. R. Steiner and P. Pierce for permission to reproduce diagrams and figures.

STUART A. RICE
MITSURU NAGASAWA

University of Chicago
Chicago, Illinois
May, 1961

SUGGESTED READING PATTERNS

We have arranged the material in this book in what we believe to be a logical sequence. Nevertheless, it must be recognized that even the most orderly and precise development of a subject is not necessarily the best for all purposes, especially when questions of pedagogy and motivation are involved. With this in mind we recommend below several alternative sequences in which we think this book can be profitably read. The suggestions we make are particularized with reference to possible groups of readers. However, a casual perusal of the text will reveal that many formulae are repeated in different chapters. This policy was consciously adopted in the attempt to render each chapter independent of others with the hope that it was thereby made easier for the reader to select an individual sequence for reading the text.

Theoretical Chemist: 1, 2, 3, 4, 5, 6, 7, 8, 9, 10, 11.

Experimental Scientist: 5, 8, 9, 10, 11, 3, 6, 7, 1, 2, 4
(no background in polymer chemistry).

Experimental Scientist: 8, 9, 10, 11, 2, 3, 6, 7, 5, 4, 1
(working in a general area covered by this text).

Graduate Student: 5, 3, 8, 9, 10, 11, 6, 7, 1, 2, 4.

CONTENTS

1. The Molecular Theory of Solutions

1.0 Introduction

The theory of polyelectrolytes deals in large part with the molecular interpretation of the macroscopic properties of solutions of highly charged ions. Aside from questions of magnitudes, the macroscopic effects observed have at their roots the same kinds of interparticle interactions responsible for the properties of solutions of ordinary molecules and ions. It is therefore hardly necessary to apologize for a detailed treatment of the properties of mixtures of nonpolymeric molecules such as will be given in this chapter. Nevertheless, it is pertinent to point out that, for each of the theories discussed herein, there is an application or an analog in the more approximate polyelectrolyte theories. From the McMillan–Mayer[1] and Kirkwood–Buff[2] theories, Hill[3] has developed a general treatment of protein solutions including the effects of binding; to the perturbation theory of solutions corresponds the Rice and Harris[4] treatment of polyion expansion and the Berkowitz et al.[5] treatment of the Poisson–Boltzmann equation for polyelectrolytes; from the cell theory of mixtures Rice and Harris[6] constructed a theory of ion exchange behavior. It is clear that an understanding of the forms in which interparticle interactions are manifest in simple mixtures is prerequisite to an understanding of the complicated interplay of forces in polyelectrolyte solutions. Of even greater importance is an understanding of the approximations inherent in the models proposed to describe polyelectrolyte behavior. These approximations are best assessed by comparison with the rigorous theory.

1.1 The Grand Ensemble and Fluctuations

It appears that the " royal road " to the statistical mechanical theory of multicomponent systems is by way of the Grand Canonical Ensemble of Gibbs. In this section are discussed the basic formulas and their application to the theory of composition fluctuations necessary for the understanding of the Kirkwood–Buff[2] theory of mixtures. The method used is that of J. G. Kirkwood* and J. E. Mayer.[7]

The ensemble to be considered is prepared as follows. Each member (system) is a volume, V, and the matter contained therein, which is part of an

* Mimeographed Lecture Notes, Princeton, 1947. Unpublished.

infinite system. The boundaries of the system are permeable to the r chemical species present, and to energy so that the composition and energy may fluctuate. The basic formula of the Grand Ensemble theory is the following. Let $N = (N_1, N_2, \cdots N_r)$ denote N_1 particles of type 1, N_2 of type 2, etc. Then the probability of a system picked at random from the ensemble being in a state of energy E_k and having exactly N molecules in the volume V is

$$P(E_k, N) = \omega_k \exp[\beta(\Omega + N \cdot \mu - E_k)] \qquad (1.1.1)$$

$\beta = 1/kT$, ω_k is the number of states of energy E_k (and depends on N), $N \cdot \mu = \sum_{i=1}^{r} N_i \mu_i$, where μ_i is the chemical potential per molecule of species i, and Ω is a normalization factor. In fact $\Omega = -pV$.

Equation (1.1.1) may be derived by maximizing the entropy, holding the energy E and mean composition $N = (N_1 \cdots N_r)$ fixed, the extensive thermodynamic variables being related to the probability defined in Eq. (1.1.1) by:

$$S = -k \sum_{E_k, N} P(E_k, N) \log \frac{P(E_k, N)}{\omega_k}$$

$$E = \sum_{E_k, N} E_k P(E_k, N) \qquad (1.1.2)$$

$$\langle N \rangle = \sum_{E_k, N} N P(E_k, N).$$

A somewhat more perspicuous method of obtaining Eq. (1.1.1) relies more directly on the fundamental subdivision which mechanically defines the grand canonical ensemble. Consider, then, not an infinite volume, but a finite large volume MV. Let the total volume MV be divided into M smaller regions of volume V each. The total energy and the total number of molecules of the ensemble are to be regarded as fixed, while the systems of volume V are separated, in thought, by semipermeable diathermic membranes.

Consider now a number ω of quantum states having energy between E and $E + \Delta E$, available to the N molecules of one region. Let m be the number of such systems in these ω states and let ω^m be the total number of quantum states available to the ensemble of systems with the restriction that exactly m are contained in the ω states. The equilibrium distribution is found by minimizing the total free energy. Under the constraints of constant energy and total number of molecules, this is equivalent to maximizing the entropy of the ensemble.

The number of distinct ways of selecting m systems is $M!/m!(M-m)!$ so that

$$\Omega(m) = \frac{M!}{m!(M-m)!} \omega^m \Omega_r(M-m) \qquad (1.1.3)$$

where $\Omega(M - m)$ is the total number of quantum states available to the remaining $M - m$ systems. Thus,

$$S(m) = k \ln \Omega(m)$$

$$= k \ln M! - k \ln m! - k \ln (M - m)! + km \ln \omega + S_r(M - m) \quad (1.1.4)$$

and

$$S(m - 1) = k \ln M! - k \ln (m - 1)! - k \ln (M - m + 1)$$

$$+ k(m - 1) \ln \omega + S_r(M - m + 1). \quad (1.1.5)$$

The condition of maximization is that

$$\Delta S(m) = S(m) - S(m - 1) = 0$$

$$= k \ln \frac{(M - m + 1)\omega}{m} + S_r(M - m) - S_r(M - m + 1). \quad (1.1.6)$$

The change from $m - 1$ to m of the number of regions in the ω quantum states involves, in the rest of the system, a *decrease* in volume of V, a *decrease* in the number of molecules N, and a *decrease* in energy of magnitude E. Thus,

$$S_r(M - m) - S_r(M - m + 1) = - V\left(\frac{\partial S_r}{\partial V}\right)_{N,E} - N\left(\frac{\partial S_r}{\partial N}\right)_{V,E} - E\left(\frac{\partial S}{\partial E}\right)_{V,N}$$

$$= - \frac{pV}{T} + \frac{N\mu}{T} - \frac{E}{T}. \quad (1.1.7)$$

The limiting process involved in the preceding relation corresponds exactly to the initial definition of the ensemble, i.e., it is understood that $M \to \infty$. Since $M \gg m \gg 1$,

$$\frac{m}{M\omega} = \exp\left(\frac{- pV + N\mu - E_k}{kT}\right) \quad (1.1.8)$$

which is to be interpreted as the probability that a given region that has the chemical potential μ, temperature T, and volume V will occupy a *given* quantum state with N molecules and energy E_k. Clearly, after multiplication by the degeneracy ω, it is also the probability that, in an infinite system in which the temperature and chemical potential are known, a volume V picked at random from the system will contain exactly N molecules and occupy a specified quantum state with energy E_k. Thus m/M is the same as $P(E_k, N)$.

It is convenient to define a Grand Partition Function by

$$\Xi = \sum_{E_k,N} \omega_k \exp\{\beta[N \cdot \mu - E_k]\}. \quad (1.1.9)$$

Clearly $$\Xi = e^{-\beta\Omega} \tag{1.1.10}$$

also $$\Xi = \sum_{N} e^{\beta N \cdot \mu} Q_N \tag{1.1.11}$$

where Q_N is the partition function for a Petit Canonical Ensemble of composition N. Now

$$Q_N = \sum_{E_k} \omega_k e^{-\beta E_k} \tag{1.1.12}$$

$$= e^{-\beta A_N}$$

where A_N is the Helmholtz function of the Petit Ensemble. Thus, by combination,

$$\Xi = \sum_{N} \exp\{\beta[N \cdot \mu - A_N]\}. \tag{1.1.13}$$

The use of the Ξ for the calculation of thermodynamic functions will be reserved for the discussion of the McMillan–Mayer theory. For the present, a more pertinent subject for discussion concerns the possible fluctuations of composition.

The probability density of a composition N, regardless of energy, is

$$P(N) = \sum_{E_k} P(E_k, N) = \exp\{\beta[\Omega + N \cdot \mu]\} Q_N \tag{1.1.14}$$

$$= \exp\{\beta[\Omega + N \cdot \mu - A_N]\}$$

by Eq. (1.1.12). To discuss fluctuations about the mean composition, $\langle N \rangle$, expand A_N in a Taylor's series

$$A_N = A_{\langle N \rangle} + (N - \langle N \rangle) \cdot \tilde{\mu} + \frac{1}{\beta} \sum_{i,j=1}^{r} \xi_i \xi_j B_{ij} + \cdots \tag{1.1.15}$$

$$\left. \begin{array}{l} \xi_i = \left(\dfrac{N_i - \langle N_i \rangle}{\langle N_i \rangle} \right) \\[2ex] \tilde{\mu}_i = \left(\dfrac{\partial A_N}{\partial N_i} \right)_{T,V,N_{j \neq i}} \bigg|_{N = \langle N \rangle} \\[2ex] B_{ij} = \dfrac{\beta}{2} \langle N_i \rangle \langle N_j \rangle \left(\dfrac{\partial^2 A_N}{\partial N_i \, \partial N_j} \right)_{T,V,N_k} \bigg|_{N = \langle N \rangle} \end{array} \right\} \tag{1.1.16}$$

Substitution of the truncated Taylor's series into the definition of the probability density of finding a given composition leads to

$$P(N) = \exp\{\beta[\Omega + \langle N \rangle \cdot \mu - A_{\langle N \rangle}] \tag{1.1.17}$$

$$+ \beta \sum_{i=1}^{r} \xi_i \langle N_i \rangle (\mu_i - \tilde{\mu}_i) - \sum_{i,j=1}^{r} \xi_i \xi_j B_{ij}\}.$$

The ξ_i are relative deviations from the mean composition, and are separated by $1/\langle N_i \rangle$. With the number of molecules of the order of 10^{23}, it is justifiable to treat the ξ_i as continuous variables. To proceed, define a probability density, $P(\xi)$, such that $P(\xi)d\xi$ is the probability of finding relative deviations between ξ_1 and $\xi_1 + d\xi_1$, ξ_2 and $\xi_2 + d\xi_2$, etc. Since there are $\prod_i \langle N_i \rangle$ compositions per unit (hyper-) volume of ξ space

$$P(\xi) = (\prod_i \langle N_i \rangle)P(N). \tag{1.1.18}$$

From (1.1.17)

$$P(\xi) = C \exp\{ \sum_{i=1}^{r} \xi_i \beta \langle N_i \rangle (\mu_i - \tilde{\mu}_i) - \sum_{i,j=1}^{r} \xi_i \xi_j B_{ij}\}$$

$$C = (\prod_{i=1}^{r} \langle N_i \rangle) \exp\{\beta[\Omega + \langle N \rangle \cdot \mu - A_{\langle N \rangle}]\}. \tag{1.1.19}$$

The next task is to evaluate the normalization constant C. From the definition of a probability density, C is defined by requiring that

$$\int_{\xi \text{ space}} P(\xi)d\xi = 1. \tag{1.1.20}$$

Now ξ_i ranges from -1 to ∞. However, the probability of relative deviations even much less than one is minute. Therefore, the lower limit of ξ_i may be set as $-\infty$. Then (1.1.20) becomes

$$C \int_{-\infty}^{\infty} \cdots \int \exp\{ \sum_{i=1}^{r} \xi_i \beta \langle N_i \rangle (\mu_i - \tilde{\mu}_i) - \sum_{i,j=1}^{r} \xi_i \xi_j B_{ij}\}d\xi = 1. \tag{1.1.21}$$

To evaluate the integral in (1.1.21), the quadratic form $\Sigma \xi_i \xi_j B_{ij}$, is diagonalized. The linear term in the exponent vanishes identically. By the theory of thermodynamic stability, the matrix $\mathbf{B}$, of the B_{ij} is positive definite. Thus it is always possible to find a matrix $\mathbf{a}^+$ such that

or

$$\mathbf{a}^+\mathbf{Ba} = 1 \tag{1.1.22}$$

$$\sum_{ij} a_{ik}a_{jm} \cdot B_{ij} = \delta_{km}.$$

If new variables η_j are introduced by

$$\xi_i = \sum_j a_{ij}\eta_j \tag{1.1.23}$$

then

$$\sum_{ij} \xi_i \xi_j B_{ij} = \sum_{i,j,k,l} a_{ik}a_{jl}\eta_k \eta_l B_{ij} = \sum_k \eta_k^2 \tag{1.1.24}$$

by (1.1.22). The Jacobian of the transformation is required to transform the volume elements in the integral. This is

$$\frac{\partial(\xi_1 \cdots \xi_r)}{\partial(\eta_1 \cdots \eta_r)} = |\mathbf{a}| = |\mathbf{B}|^{-\frac{1}{2}} \tag{1.1.25}$$

where $| \;\;|$ denotes determinant. The linear term in the exponent in (1.1.21) can be written

$$\left.\begin{array}{l} \displaystyle\sum_{i=1}^{r} \xi_i \beta \langle N_i \rangle (\mu_i - \tilde{\mu}_i) = 2 \sum_{k=1}^{r} \alpha_k \eta_k \\[3mm] \displaystyle\alpha_k = \frac{1}{2} \sum_i \beta \langle N_i \rangle (\mu_i - \tilde{\mu}_i) a_{ik} \end{array}\right\}. \tag{1.1.26}$$

Thus (1.1.21) becomes

$$\frac{C}{|\mathbf{B}|^{\frac{1}{2}}} \int_{-\infty}^{\infty} \cdots \int \exp\{2\Sigma\alpha_k\eta_k - \Sigma\eta_k^2\}\, d\boldsymbol{\eta} = 1. \tag{1.1.27}$$

Equation (1.1.27) is in a form whereby it may be shown that $\alpha_k \equiv 0$. $\langle N_i \rangle$ is given by

$$\langle N_i \rangle = \int_{-\infty}^{\infty} \cdots \int N_i P(\xi)\, d\xi. \tag{1.1.28}$$

Combining this with (1.1.20)

$$\int_{-\infty}^{\infty} \cdots \int \xi_i P(\xi)\, d\xi = 0. \tag{1.1.29}$$

In terms of the η's

$$\frac{C}{|\mathbf{B}|^{\frac{1}{2}}} \sum_j a_{ij} \int_{-\infty}^{\infty} \eta_j \exp(2\Sigma\alpha_k\eta_k - \Sigma\eta_k^2)\, d\boldsymbol{\eta} = 0. \tag{1.1.30}$$

Equation (1.1.30) is easily transformed to

$$\frac{C}{|\mathbf{B}|^{\frac{1}{2}}} \sum_j a_{ij}\alpha_j \int_{-\infty}^{\infty} P(\boldsymbol{\eta})\, d\boldsymbol{\eta} = 0 \tag{1.1.31}$$

or

$$\sum_{j=1}^{r} a_{ij}\alpha_j = 0. \tag{1.1.32}$$

Since $|\mathbf{a}|$ does not vanish, $\alpha_j = 0$ for all j. Thus

$$\sum_i \langle N_i \rangle (\mu_i - \tilde{\mu}_i) a_{ij} = 0. \tag{1.1.33}$$

By the same argument

$$\langle N_i \rangle (\mu_i - \tilde{\mu}_i) = 0. \tag{1.1.34}$$

Since, in general, $\langle N_i \rangle \neq 0$

$$\mu_i = \tilde{\mu}_i. \tag{1.1.35}$$

Then Eq. (1.1.27) may be rewritten in the form

$$\frac{C}{|\mathbf{B}|^{1/2}} \int_{-\infty}^{\infty} \cdots \int \exp(-\sum_k \eta_k^2) \, d\boldsymbol{\eta} = 1 \tag{1.1.36}$$

$$= C\left(\frac{\pi^r}{|\mathbf{B}|}\right)^{1/2}$$

and finally,

$$C = \left(\frac{|\mathbf{B}|}{\pi^r}\right)^{1/2}. \tag{1.1.37}$$

From the vanishing of the linear term in the exponent it is clear that

$$\langle \xi_i \rangle = \int P(\boldsymbol{\xi}) \xi_i \, d\boldsymbol{\xi} = 0.$$

What about the cross correlations, $\langle \xi_i \xi_j \rangle$? For these,

$$\langle \xi_i \xi_j \rangle = \frac{|\mathbf{B}|^{1/2}}{\pi^{r/2}} \int_{-\infty}^{\infty} \cdots \int \xi_i \xi_j \exp(-\Sigma \xi_i \xi_j B_{ij}) \, d\boldsymbol{\xi}$$

$$= -\left(\frac{\mathbf{B}}{\pi^2}\right)^{1/2} \frac{\partial}{\partial B_{ij}} \int \cdots \int \exp\{-\Sigma \xi_i \xi_j B_{ij}\} \, d\boldsymbol{\xi}$$

$$= -|\mathbf{B}|^{1/2} \frac{\partial}{\partial B_{ij}} |\mathbf{B}|^{-1/2} \tag{1.1.38}$$

$$= \frac{1}{2} \frac{|\mathbf{B}|_{ij}}{|\mathbf{B}|}$$

where, by the rules of differentiation of determinants, $|\mathbf{B}|_{ij} = \partial |\mathbf{B}|/\partial B_{ij}$ is the cofactor of B_{ij} in $|\mathbf{B}|$. Note that

$$\langle \xi_i \xi_j \rangle = \frac{\langle N_i N_j \rangle - \langle N_i \rangle \langle N_j \rangle}{\langle N_i \rangle \langle N_j \rangle}. \tag{1.1.39}$$

The quantities B_{ij} which have been used are defined by derivatives at constant temperature and volume. It is usually more convenient to work at constant temperature and pressure, these conditions corresponding to ordinary laboratory conditions. The necessary transformation is obtained as follows. First note the thermodynamical equation

$$\left(\frac{\partial \mu_i}{\partial \langle N_k \rangle}\right)_{T,V,\langle N_j \rangle} = \left(\frac{\partial \mu_i}{\partial \langle N_k \rangle}\right)_{T,p,\langle N_j \rangle} + \left(\frac{\partial \mu_i}{\partial p}\right)_{T,N} \left(\frac{\partial p}{\partial \langle N_k \rangle}\right)_{T,V,\langle N_j \rangle}. \quad (1.1.40)$$

Now

$$\left(\frac{\partial \mu_i}{\partial p}\right)_{T,N} = \bar{v}_i$$

is the partial molecular volume. Writing

$$\alpha_{ik} = \left(\frac{\partial \mu_i}{\partial \langle N_k \rangle}\right)_{T,p,\langle N_j \rangle} = \alpha_{ki} \quad (1.1.41)$$

and

$$\left(\frac{\partial p}{\partial \langle N_k \rangle}\right)_{T,V,\langle N_j \rangle} = -\frac{\left(\dfrac{\partial V}{\partial \langle N_k \rangle}\right)_{T,p,\langle N_j \rangle}}{\left(\dfrac{\partial V}{\partial p}\right)_{T,N}} \quad (1.1.42)$$

$$= -\frac{\bar{v}_k}{\kappa V}$$

where κ is the compressibility, $(-1/V)(\partial V/\partial p)_{T,N}$, (1.1.40) becomes

$$\left(\frac{\partial \mu_i}{\partial \langle N_k \rangle}\right)_{T,V,\langle N_j \rangle} = \alpha_{ik} + \frac{\bar{v}_i \bar{v}_k}{\kappa V}. \quad (1.1.43)$$

Thus, from Eq. (1.1.16), the definition of B_{ik} is

$$B_{ik} = \frac{1}{2kT} \left(\frac{\bar{v}_i \bar{v}_k \langle N_i \rangle \langle N_k \rangle}{\kappa V} + \alpha_{ik} \langle N_i \rangle \langle N_k \rangle\right). \quad (1.1.44)$$

For later applications it is necessary to discuss one more topic here, namely, composition fluctuations in terms of molecular distribution functions. First define the microscopic number density for particles of type i by

$$v_i^{(1)}(\mathbf{R}_1) = \sum_{\alpha_i = 1}^{N} \delta(\mathbf{R}_{\alpha_i} - \mathbf{R}_1) \quad (1.1.45)$$

and the pair density for species i and j at points $\mathbf{R}_1$ and $\mathbf{R}_2$ respectively by

$$v_{ij}^{(2)} = \sum_{\alpha_i = 1}^{N_i} \sum_{\beta_j = 1}^{N_j} \delta(\mathbf{R}_{\alpha_i} - \mathbf{R}_1)\delta(\mathbf{R}_{\beta_j} - \mathbf{R}_2). \quad (1.1.46)$$

$$\alpha_i \neq \beta_j.$$

These clearly have the property

$$\int v_i^{(1)}(\mathbf{R}_1)\, d\mathbf{R}_1 = N_i$$

$$\iint v_{ij}^{(2)}(\mathbf{R}_1, \mathbf{R}_2)\, d\mathbf{R}_1\, d\mathbf{R}_2 = N_i N_j - N_i \delta_{ij}. \tag{1.1.47}$$

Equation (1.1.1), and the derivation of it, are expressions for the Grand Canonical Ensemble in quantum-mechanical form in an energy representation, although no specific use has been made of quantum mechanics. It is now more convenient to use the classical form in phase space

$$P(\mathbf{N}, \mathbf{p}, \mathbf{R}) = \exp\{[\Omega + \mathbf{N} \cdot \mathbf{\mu} - H_\mathbf{N}(\mathbf{R}, \mathbf{p})]/kT\}P_t \tag{1.1.48}$$
$$P_t = (N!h^{3N})^{-1}$$

where $H_\mathbf{N}$ is the Hamiltonian function for the N particle system, and $\mathbf{R}$ and $\mathbf{p}$ are the position and momentum coordinates, respectively, in phase space. Internal coordinates, which must usually be treated quantum-mechanically, have not been specifically indicated. The average number densities in singlet and pair space are, respectively,

$$\rho_i(\mathbf{R}_1) = \sum_\mathbf{N} \int v_i^{(1)}(\mathbf{R}_1) P(\mathbf{N}, \mathbf{R}, \mathbf{p})\, d\mathbf{R}\, d\mathbf{p} \tag{1.1.49a}$$

$$\rho_{ij}(\mathbf{R}_1, \mathbf{R}_2) = \sum_\mathbf{N} \int v_{ij}^{(2)}(\mathbf{R}_1, \mathbf{R}_2) P(\mathbf{N}, \mathbf{R}, \mathbf{p})\, d\mathbf{R}\, d\mathbf{p}. \tag{1.1.49b}$$

Thus

$$\int \rho_i(\mathbf{R}_1)\, d\mathbf{R}_1 = \langle N_i \rangle$$

$$\iint \rho_{ij}(\mathbf{R}_1, \mathbf{R}_2)\, d\mathbf{R}_1\, d\mathbf{R}_2 = \langle N_i N_j \rangle - \langle N_i \rangle \delta_{ij}$$

$$\iint [\rho_{ij}(\mathbf{R}_1, \mathbf{R}_2) - \rho_i(\mathbf{R}_1)\rho_j(\mathbf{R}_2)]\, d\mathbf{R}_1\, d\mathbf{R}_2 = \tag{1.1.50}$$

$$\langle N_i N_j \rangle - \langle N_i N_j \rangle - \langle N_i \rangle \delta_{ij}.$$

For a fluid system, the average densities have the form

$$\rho_i(\mathbf{R}_1) = \langle N_i \rangle / V = c_i$$
$$\rho_{ij}(\mathbf{R}_1, \mathbf{R}_2) = c_i c_j g_{ij}(R)$$
$$R = |\mathbf{R}_i - \mathbf{R}_j| \tag{1.1.51}$$

assuming, for the last equation, that there are no specific orientational forces. Thus

$$\int (g_{ij} - 1)\, d\mathbf{R}_i = V \frac{\langle N_i N_j \rangle - \langle N_i \rangle \langle N_j \rangle}{\langle N_i \rangle \langle N_j \rangle} - \frac{\delta_{ij}}{c_j}$$

$$= V(\xi_i \xi_j) - \frac{\delta_{ij}}{c_j}. \tag{1.1.52}$$

1.2   The Kirkwood–Buff Theory of Solutions[2]

The preceding development of fluctuation theory has provided most of the tools needed to discuss solution theory in the format of Kirkwood and Buff.[2] First, recall the definition

$$B_{ij} = \frac{\beta}{2} \langle N_i \rangle \langle N_j \rangle \left(\frac{\partial \mu_i}{\partial \langle N_j \rangle} \right)_{T,V,\langle N_k \rangle} \tag{1.1.16}$$

$$\frac{\langle N_i N_j \rangle - \langle N_i \rangle \langle N_j \rangle}{\langle N_i \rangle \langle N_j \rangle} = \frac{1}{2} \frac{|\mathbf{B}|_{ij}}{|\mathbf{B}|}. \tag{1.1.38}$$

The left-hand side of (1.1.38) can be regarded as the ij matrix element of a matrix $\mathbf{A}$. The right-hand side is just the ij element of the matrix $\mathbf{B}^{-1}$ (since $\mathbf{B}$ is symmetric). Therefore,

$$B_{ij} = \frac{1}{2} \frac{|\mathbf{A}|_{ij}}{|\mathbf{A}|}. \tag{1.2.1}$$

It is convenient to define a new matrix $\mathbf{D}$ by

$$D_{ij} = \langle N_i N_j \rangle - \langle N_i \rangle \langle N_j \rangle \tag{1.2.2}$$

and thereby (1.2.1) becomes

$$\frac{1}{kT} \left(\frac{\partial \mu_i}{\partial \langle N_j \rangle} \right)_{T,V,\langle N_k \rangle} = \frac{|\mathbf{D}|_{ij}}{|\mathbf{D}|} \tag{1.2.3}$$

$$= \frac{|\mathbf{R}|_{ij}}{V|\mathbf{R}|}$$

where

$$R_{ij} = c_i \delta_{ij} + c_i c_j G_{ij}$$

$$G_{ij} = \int (g_{ij} - 1)\, d\mathbf{R} \tag{1.2.4}$$

$$c_i = \frac{\langle N_i \rangle}{V}$$

In the following the brackets on the $\langle N \rangle$'s will be dropped, i.e., N_j shall mean $\langle N_j \rangle$, since only average quantities are needed in the further development. It is now possible, using (1.2.3) to derive expressions for the derivatives of the chemical potentials $(\partial \mu_i / \partial N_j)_{T,p,N_k}$, the compressibility κ, and the partial molecular volumes $\bar{v}_i$.

To do this start with the Gibbs–Duhem equation

$$\sum_{i=1}^{r} N_i \left(\frac{\partial \mu_i}{\partial N_j} \right)_{T,p,N_k} = 0 \qquad (1.2.5)$$

and Eq. (1.1.43)

$$\left(\frac{\partial \mu_i}{\partial N_j} \right)_{T,V,N_k} = \left(\frac{\partial \mu_i}{\partial N_j} \right)_{T,p,N_k} + \frac{\bar{v}_i \bar{v}_j}{\kappa V}. \qquad (1.1.43)$$

Multiply (1.1.43) by $N_i N_j$ and sum over i and j. By (1.2.5) and the relations $\Sigma N_i \bar{v}_i = V$, it is found that

$$\sum_{i,j} N_i N_j \left(\frac{\partial \mu_i}{\partial N_j} \right)_{T,V,N_k} = \frac{V}{\kappa}. \qquad (1.2.6)$$

Using (1.2.3)

$$\kappa = \frac{|\mathbf{R}|}{kT} \left(\sum_{i,j} c_i c_j |\mathbf{R}|_{ij} \right)^{-1}. \qquad (1.2.7)$$

In a similar manner, multiply (1.1.43) by N_i and sum.

$$\sum_{i=1}^{r} N_i \left(\frac{\partial \mu_i}{\partial N_j} \right)_{T,V,N_k} = \frac{\bar{v}_j}{\kappa} \qquad (1.2.8)$$

or

$$\bar{v}_j = \sum_{i=1}^{r} c_i |\mathbf{R}|_{ij} \bigg/ \sum_{i,k=1}^{r} c_i c_k |\mathbf{R}|_{ik}. \qquad (1.2.9)$$

By using (1.2.7), (1.2.9), and (1.1.43), it is found that

$$\frac{V}{kT} \left(\frac{\partial \mu_i}{\partial N_j} \right)_{T,p,N_k} = \frac{1}{|\mathbf{R}|} \frac{\sum_{k,l} [c_k c_l (|\mathbf{R}|_{ij} |\mathbf{R}|_{kl} - |\mathbf{R}|_{ik} |\mathbf{R}|_{jl})]}{\sum_{k,l} c_k c_l |\mathbf{R}|_{kl}}. \qquad (1.2.10)$$

For the osmotic pressure of a solution of solvent 1 and solutes $2, \ldots r$, the thermodynamic equation

$$\left(\frac{\partial \pi}{\partial c_i} \right)_{T,\mu_1,c_j} = \sum_{k=2}^{r} c_k \left(\frac{\partial \mu_k}{\partial c_i} \right)_{T,\mu_1,c_j} \qquad (1.2.11)$$

follows directly from the Gibbs–Duhem equation. To proceed further it is necessary to show that the matrix of $(\partial \mu_k / \partial c_i)_{T,\mu_1,c_j} (i, k \neq 1)$ is the inverse of the matrix of $(\partial c_i/\partial \mu_j)_{T,\mu_k}(i, j \neq 1)$. If the concentrations are regarded as functions of T and the μ_i and the volume is held constant,

$$\left(\frac{\partial c_i}{\partial c_l}\right)_{T,\mu_1,c_k} = \delta_{il} = \sum_{j=2}^{r}\left(\frac{\partial c_i}{\partial \mu_j}\right)_{T,\mu_k}\left(\frac{\partial \mu_j}{\partial c_l}\right)_{T,\mu_1,c_k} \tag{1.2.12}$$

which proves our assumption. Thus

$$\left(\frac{\partial \mu_j}{\partial c_l}\right)_{T,\mu_1,c_k} = \frac{|\mathbf{c}|_{jl}}{|\mathbf{c}|} \tag{1.2.13}$$

$$c_{jl} = \left(\frac{\partial c_j}{\partial \mu_l}\right)_{T,\mu_k}$$

To demonstrate the useful relation

$$c_{jl} = c_j\delta_{jl} + c_jc_lG_{jl} \tag{1.2.14}$$

start with

$$c_l = \frac{1}{V} e^{\beta\Omega} \sum_{\mathbf{N}} N_l \exp\{\beta[\mathbf{N}\cdot\boldsymbol{\mu} - A_{\mathbf{N}}]\} \tag{1.2.15}$$

and therefore

$$\begin{aligned}
\frac{1}{\beta} c_{jl} &= \frac{1}{V} e^{\beta\Omega} \sum_{\mathbf{N}} N_j N_l \exp\{\beta[\mathbf{N}\cdot\boldsymbol{\mu} - A_{\mathbf{N}}]\} \\
&\quad + \frac{1}{V} e^{\beta\Omega} \left(\frac{\partial\Omega}{\partial\mu_j}\right)_{T,\mu} \sum_{\mathbf{N}} N_l \exp\{\beta[\mathbf{N}\cdot\boldsymbol{\mu} - A_{\mathbf{N}}]\} \\
&= \frac{\langle N_j N_l\rangle - \langle N_j\rangle\langle N_l\rangle}{V}
\end{aligned} \tag{1.2.16}$$

since $(\partial\Omega/\partial\mu_l)_{T,\mu_k} = -\langle N_l\rangle$. Thus, by combination of the preceding results,

$$\begin{aligned}
c_{jl} &= \frac{1}{kT}\left(\frac{\langle N_j N_l\rangle - \langle N_j\rangle\langle N_l\rangle}{V}\right) \\
&= c_j\delta_{lj} + c_jc_lG_{jl} = R_{jl}, \qquad j, l \neq 1.
\end{aligned} \tag{1.2.17}$$

Cf. Eqs. (1.2.2), (1.2.4).] Returning to the concentration derivative of the osmotic pressure and using Eq. (1.2.17) leads to

$$\left(\frac{\partial\pi}{\partial c_i}\right)_{T,\mu_1,cj} = kT \sum_{k=2}^{r} c_k \frac{|\mathbf{R}'|_{ik}}{|\mathbf{R}'|} \tag{1.2.18}$$

where $\mathbf{R}'$ is the $(n-1) \times (n-1)$ matrix formed by striking out the first row and column of $\mathbf{R}$. Note that no solvent–solvent or solvent–solute matrix elements appear in (1.2.18).

For binary systems some of the formulas are, explicitly,

$$\bar{v}_1 = \frac{1 + (G_{22} - G_{12})c_2}{c_1 + c_2 + c_1 c_2 (G_{11} + G_{22} - 2G_{12})} \tag{1.2.19}$$

$$c_1 \bar{v}_1 + c_2 \bar{v}_2 = 1$$

$$\kappa k T = \frac{1 + G_{11}c_1 + G_{22}c_2 + (G_{11}G_{22} - G_{12})^2 c_1 c_2}{c_1 + c_2 + (G_{11} + G_{22} - 2G_{12})c_1 c_2} \tag{1.2.20}$$

$$\left(\frac{\partial \pi}{\partial c_2}\right)_{T,\mu_1} = \frac{kT}{1 + G_{22}c_2} \tag{1.2.21}$$

$$\frac{1}{kT}\left(\frac{\partial \mu_2}{\partial c_2}\right)_{T,p} = \frac{1}{c_2} + \frac{G_{12} - G_{22}}{1 + c_2(G_{22} - G_{12})} \tag{1.2.22}$$

$$\frac{1}{kT}\left(\frac{\partial \mu_1}{\partial c_1}\right)_{T,p} = \frac{1}{c_1} + \frac{G_{12} - G_{11}}{1 + c_1(G_{11} - G_{12})} \tag{1.2.23}$$

$$\frac{1}{kT}\left(\frac{\partial \mu_2}{\partial x_2}\right)_{T,p} = \frac{1}{x_2} + \frac{c_1(2G_{12} - G_{11} - G_{22})}{1 + c_1 x_2 (G_{11} + G_{22} - 2G_{12})} \tag{1.2.24}$$

This completes the derivation of the Kirkwood–Buff theory. An application to the fluctuation forces operative between protein molecules at their isoelectric point in media of low salt concentration will be considered in a later chapter.[8]

1.3 Strongly Interacting Systems

The theory presented in this section, due to Mayer,[7] is of a different nature than most theories of the condensed state. In the theory of gases (Mayer) or the Kirkwood–Born–Green theory of fluids, it is assumed that the intermolecular potential is known, and the thermodynamic properties of the system deduced from it. In what is to follow, this procedure is largely reversed. The object of the investigation will be to consider the distribution of molecules, characterized by suitable distribution functions, and to calculate the thermodynamic functions corresponding to the change in the system from one fugacity to another.

The fugacity mentioned above is directly related to the thermodynamically defined fugacity. Consider an infinite volume of a given substance which is at the temperature T. The fugacity is defined as

$$z = \lim_{\rho_0 \to 0}\left[\rho_0 \exp\left(\frac{\mu - \mu_0}{kT}\right)\right] \tag{1.3.1}$$

where μ_0 is the chemical potential of the system when the density of the system is ρ_0. Given the fact that the chemical potential is a function of density, it is clear that

$$\frac{z}{\rho} = \lim_{\rho_0 \to 0} \frac{\rho_0 e^{-\mu_0(\rho_0)/kT}}{\rho e^{-\mu(\rho)/kT}} \tag{1.3.2}$$

so that

$$\lim_{\rho \to 0} \frac{z}{\rho} \to 1 \tag{1.3.3}$$

or

$$\lim_{z \to 0} \frac{z}{\rho} \to 1.$$

Thus, when the density approaches zero, the system becomes an ideal gas and the fugacity z approaches the density ρ. The density is measured in molecules per unit volume, $\rho = N/V$. Following Mayer, it is convenient to symbolize the set of coordinates denoting molecule i by (i) and the corresponding volume element by $d(i)$. For n molecules, the corresponding symbols $\{n\}$ and $d\{n\}$ will be used. The distribution function $\rho_n(z, \{n\})$, a function of the coordinates of n molecules and the fugacity, is defined by the statement that in an infinite system of fugacity z the probability that n molecules be at the positions $\{n\}$, and in the element of configuration space $d\{n\}$, is proportional to $\rho_n(z, \{n\}) d\{n\}$. More precisely, the normalization constant is chosen such that

$$\lim_{V \to \infty} \frac{1}{\rho^n} \int \rho_n(z, \{n\}) d\{n\} = 1 \tag{1.3.4}$$

i.e., the average value of the function ρ_n integrated over the internal coordinates of the molecules is unity in the $3n$-dimensional Cartesian space of the centers of mass. As will be seen later, the potential of mean force for a set of $\{n\}$ molecules at fugacity z may be defined by the equation

$$-W_n(z, \{n\}) = kT \ln \rho_n(z, \{n\}). \tag{1.3.5}$$

At zero fugacity, the density is zero and the n molecules are isolated in space. Then, for classical systems

$$W_n(0, \{n\}) = U_N(0, \{n\}) \tag{1.3.6}$$

with U_N the potential energy of the system of molecules. Now, the probability of finding exactly N molecules in the volume V in the *specified* quantum state l is

$$P_{lNV} = \exp[(-pV + N\mu - E_l)/kT]. \tag{1.3.7}$$

Consider, for the moment, the molecules are numbered and collected in the quantum state l. Then the probability that they will occupy the coordinates $\{N\}$ within the volume element $d\{N\}$ is $|\psi_l\{N\}|^2 d\{N\}$ where $\psi_l\{N\}$ is the normalized wave function of the system. If a new probability P_{NV} is defined by

$$P_{NV} = e^{(-pV + N\mu)/kT} \sum e^{-E_l/kT} |\psi_l\{N\}|^2 \qquad (1.3.8)$$

then P_{NV} is the probability that there are exactly N molecules in V and these are located at $\{N\}$ irrespective of the quantum state of the system. If the volume V is known to contain exactly $N + n$ molecules, then the probability density that there be N molecules at $\{N\}$ irrespective of the positions of the remaining n molecules is

$$\int \frac{(N + n)!}{n!} P_{N+n,V}\{N + n\} \, d\{n\} \qquad (1.3.9)$$

where the combinatorial factor arises from the observation that any N of the $N + n$ molecules may fill the $\{N\}$ positions, and having chosen the N molecules, all permutations of the molecules among the positions are allowable. By summation over all n, the probability density of the configuration $\{N\}$ at T, z is

$$\rho_N(z, \{N\}) = e^{-pV/kT} \sum_{n \geqslant 0} e^{(N+n)\mu/kT} \frac{(N + n)!}{n!} \sum e^{-E_l/kT} \qquad (1.3.10)$$

$$\times \int |\psi_l\{N + n\}|^2 \, d\{n\}.$$

With the definition

$$\frac{Q_p}{V} = \lim_{\rho_0 \to 0} \rho_0 e^{-\mu_0/kT} = \left(\frac{2\pi mkT}{h^2}\right)^{3/2} \qquad (1.3.11)$$

the fugacity becomes

$$z = \frac{Q_p}{V} e^{\mu/kT}. \qquad (1.3.12)$$

With the use of this definition of the fugacity, a simple substitution gives

$$z^{-N}\rho_N(z, \{N\}) = e^{-pV/kT} \sum_{n \geqslant 0} \frac{z^n}{n!} \int_V (N + n)! \left(\frac{V}{Q_p}\right)^N \sum_l |\psi_l\{N + n\}|^2 \qquad (1.3.13)$$

$$\times e^{-E_l/kT} \, d\{n\}.$$

In the limit as $z \to 0$, corresponding to a decrease in density until the ideal gas state is reached, the only finite term remaining in the sum on the right-hand side is for $n = 0$. Therefore, defining

$$g_N(0, \{N\}) = N! \left(\frac{V}{Q_p}\right)^N \sum_l |\psi_l\{N\}|^2 e^{-E_l/kT} \tag{1.3.14}$$

$$\rho_n(z, \{N\}) = \rho^n(z) g_n(z, \{n\}) \tag{1.3.15}$$

and by substitution

$$z^{-N} \rho_N(z, \{N\}) = e^{-pV/kT} \sum_{n \geqslant 0} \frac{z^n}{n!} \int_V g_{N+n}(0, \{N + n\}) \, d\{n\} \tag{1.3.16}$$

which states that the distribution function at an arbitrary fugacity z is uniquely related to the distribution function for the same number of molecules at infinite dilution. If an activity coefficient is defined by

$$\gamma(z) = \frac{z}{\rho(z)} \tag{1.3.17}$$

and the convention adopted that when the fugacity is zero, the pressure is zero and the activity coefficient unity, $p(0) = 0$, $\gamma(0) = 1$, then

$$e^{Vp(z)/kT} [\gamma(z)]^{-N} g_N(z, \{N\}) = \sum_{n \geqslant 0} \frac{z^n}{n!} \int_V e^{Vp(0)/kT} \frac{g_{N+n}(0, \{N + n\})}{[\gamma(0)]^{N+n}} \, d\{n\} \tag{1.3.18}$$

By the normalization condition, for the case $N = 0$,

$$g_0(z, \{0\}) = 1 \tag{1.3.19}$$

and

$$e^{Vp(z)/kT} = \sum_{n \geqslant 0} \frac{z^n}{n!} \int_V g_n(0, \{n\}) \, d\{n\}. \tag{1.3.20}$$

Now consider the sum

$$I_N\{N\} = \sum_{m \geqslant 0} \frac{(-z)^m}{m!} e^{Vp(z)/kT} \int_V \frac{g_{N+m}(z, \{N + m\})}{[\gamma(z)]^{N+m}} \, d\{m\} \tag{1.3.21}$$

$$= \sum_{m \geqslant 0} \sum_{n \geqslant 0} (-)^m \frac{z^{m+n}}{m! \, n!} \int_V g_{N+m+n}(0, \{N + m + n\}) \, d\{m + n\}$$

by substitution of Eq. (1.3.16). Regrouping the terms and collecting those for which $m + n = j$ gives†

$$
I_N\{N\} - \sum_{j \geq 0} \left[\sum_{i=0}^{j} \frac{(-)^i j!}{i!(j-i)!} \right] \frac{z^j}{j!} \int_V g_{N+j}(0, \{N + n\})\, d\{j\}
$$

$$
= \sum_{j \geq 0} (1 - 1)^j \frac{z^j}{j!} \int_V g_{N+n}(0, \{N + j\})\, d\{j\}
$$

$$
= g_N(0, \{N\})
$$

(1.3.22)

so that the integral expression I_N may be rewritten as

$$
g_N(0, \{N\}) = \sum_{m \geq 0} \frac{(-z)^m}{m!} e^{V P(z)/kT} \int_V \frac{g_{N+m}(z, \{N + m\})}{[\gamma(z)]^{N+m}}\, d\{m\}.
$$

(1.3.23)

Equation (1.3.18), the basic equation for the distribution function $g_N(z, \{N\})$, may be rewritten for the arbitrary new fugacity $z + y$. Using Eq. (1.3.23) in the integrand then leads to

$$
e^{V P(z+y)/kT} \frac{g_N(z + y, \{N\})}{[\gamma(z + y)]^N} = \sum_{n \geq 0} \sum_{m \geq 0} \frac{(-)^m (z + y)^n z^m}{m!\, n!} e^{V P(z)/kT} \times
$$

$$
\int \frac{g_{N+m+n}(z, \{N + m + n\})}{[\gamma(z)]^{N+m+n}}\, d\{m + n\}.
$$

(1.3.24)

Again, regrouping terms with $n + m = M$ and using the multinomial theorem gives

$$
e^{V P(z+y)/kT} \frac{g_N(z + y, \{N\})}{[\gamma(z + y)]^N} = \sum_{M \geq 0} \frac{y^M}{M!} e^{V P(z)/kT} \int_V \frac{g_{N+M}(z, \{N + M\})}{[\gamma(z)]^{N+M}}\, d\{M\}.
$$

(1.3.25)

This relation allows, in principle, the calculation of the pressure $p(z + y)$, the activity coefficient $\gamma(z + y)$, and the distribution (correlation) functions $g_N(z + y, \{N\})$ at any fugacity $z + y$ in terms of the pressure $p(z)$, activity coefficient $\gamma(z)$, and distribution functions $g_N(z, \{N\})$ at any other value z of the fugacity. The equations derived are not restricted to any particular kind

† $\dfrac{(-z)^m(+z)^n}{m!\, n!} = \dfrac{(m + n)!}{(m + n)!} \dfrac{(-z)^m(z)^n}{m!\, n!} =$ one term in multinomial expansion. If $m + n = j$,

$$
\frac{1}{(m + n)!} \sum_{(m+n=j)} \frac{(m + n)!}{m!\, n!} (-z)^m (z)^n = \frac{1}{(m + n)!} (z - z)^{m+n}.
$$

Only the term for which $m = n = 0$ is nonzero.

of system. All the series converge because of the finite size and low compressibility of molecules. Since the series converge too slowly to be of use, transformations to other forms wherein the convergence is more rapid form a proper subject for discussion. It is this topic which we now pursue.

Most of the systems of interest to us obey classical mechanics to an excellent approximation. In the semiclassical limit, when the spacing of the energy levels of a system of N molecules in a volume V is small relative to kT, the probability density of finding the system in configuration $\{N\}$ is

$$Q^{-1} \sum_i |\psi_i\{N\}|^2 e^{-E_i/kT} = \frac{1}{h^f QN!} \int_V e^{-H(\{N\})/kT} d\{p_N\} d\{N\} \quad (1.3.26)$$

with f the total number of degrees of freedom, $H(\{N\},\{p_N\})$ the Hamiltonian of the system, and $\{p_N\}$ the set of momenta of all the molecules. After integration and use of the distribution functions at zero fugacity

$$g_N(0, \{N\}) = N! \frac{V}{Q_p} \sum_i |\psi_i\{N\}|^2 e^{-E_i/kT} \quad (1.3.27)$$

and therefore, $g_N(0, \{N\})$ is proportional to $\exp(-\beta U_N\{N\})$ for all N. By a proper choice of the zero of energy, the preceding relation may be reduced to

$$g_N(0, \{N\}) = \exp(-\beta U_N\{N\}). \quad (1.3.28)$$

What is the force acting on a given molecule in a selected direction? If there are $N + n$ molecules in the volume V, the force on one of the molecules in the direction q_i is

$$f_{q_i}\{n + N\} = - \left(\frac{\partial U_{N+n}\{n + N\}}{\partial q_i} \right)_{q_j} \quad (1.3.29)$$

by the definition of potential energy. Now, the probability density that there be exactly $N + n$ molecules in V and that of these n be located at $\{n\}$, with the remaining molecules distributed in any manner, is

$$\frac{(N + n)!}{n!} P_{N+n,V}\{n + N\} = e^{-Vp(z)/kT} \frac{z^{n+N}}{n!} g_{n+N}(0, \{N + n\}). \quad (1.3.30)$$

The average force acting on the set $\{N\}$ of molecules is to be understood as the force averaged over all positions of the molecules represented by $\{n\}$, and over all possible values of the number of molecules, n. Thus

$$\langle f_{q_i}\{N\} \rangle = \frac{e^{-Vp(z)/kT} \sum_{n \geqslant 0} z^n/n! \int_V (-\partial U_{N+n}/\partial q_i)_{q_j} \exp(-U_{N+n}\{N + n\}/kT) d\{n\}}{e^{-Vp(z)/kT} \sum_{n \geqslant 0} z^n/n! \int_V \exp(-U_{N+n}\{N + n\}/kT) d\{n\}}$$

$$(1.3.31)$$

or

$$\langle f_{q_i}\{N\}\rangle = \left[\frac{\partial}{\partial q_i} kT \ln g_N(z, \{N\})\right]_{q_j} \tag{1.3.32}$$

which shows that W_N is, to within an arbitrary constant, the potential of mean force. In the limit of zero fugacity

$$\lim_{z \to 0} W_N(0, \{N\}) = U_N\{N\}. \tag{1.3.33}$$

It is usually assumed that the potential energy U_N is decomposable into the sum of pairwise interactions. A more general breakdown will be used for W_N since the pairwise decomposition cannot be justified. For W_N write

$$W_N(z, \{N\}) = \sum_{\{n\}_N} w_n(z, \{n\}_N) \tag{1.3.34}$$

where the symbol $\{n\}_N$ denotes a subset (consisting of the coordinates of n molecules) of the set of coordinates $\{N\}$ of N molecules. The sum is to be taken over all possible subsets and the terms w_n with $n > 2$ may be thought of as corrections for the deviation of W_N from a sum of pair terms only. The inverse of (1.3.34) can be shown to be[1]

$$w_n(z, \{n\}) = \sum_{\{N\}_n} (-)^{n-N} W_N(z, \{N\}_n). \tag{1.3.35}$$

We are now in a position to consider an expansion of the distribution functions. If, of a system of M molecules, m are far away from the remaining $M - m$, the distribution function may be written

$$g_M(z, \{M\}) = g_m(z, \{m\}_M) g_{M-m}(z, \{M - m\}_M). \tag{1.3.36}$$

This arises from the fact that the g_N are probabilities, and when the subset m is far from the subset $M - m$, the two groups of molecules do not interact and are therefore independent. Under these conditions, the probability of this configuration is the product of the probabilities of the configurations of the two independent subsets. Because of this property it will be possible to write g_M as a sum of products of functions, each of a smaller number of coordinates. That is, the distribution function may be regarded as unity (corresponding to no interactions at all) plus a series of correction terms corresponding to the presence of pairs, triples, etc., of molecules close to one another. These correction terms vanish unless the configuration they describe exists. Thus, when there is one pair of molecules close to one another, and all other molecules are distant from each other, we correct the value unity by adding a term for the interacting pair of particles. If there are two pairs of interacting molecules, in addition to the correction for the second pair we must also add a term which accounts for there being *simultaneously* two pairs of interacting particles present. This is obviously the product of the two pair correction terms. If a triplet of interacting molecules were present, there would be a term for the triplet and also for all possible pairs that can be formed from the triplet by selecting two molecules at a time. Thus, the distribution function contains not only the correction for a given aggregate, but also for all smaller aggregates which can be formed by its dissociation. If the initial configuration has all molecules far apart, the distribution function starts with unity plus corrections for pairs, triplets, etc., of interacting molecules until the desired state is reached.

The preceding discussion suggests that g_M be expanded in the form

$$g_M(z, \{M\}) = \sum \{k\{m_i\}_M\}_u \prod_{i=1}^{k} \zeta_m(z, \{m_i\}_M) \qquad (1.3.37)$$

where the sum is over all complete sets of k unconnected (i.e., mutually exclusive) subsets $\{m_i\}_M (1 \leqslant i \leqslant k)$ of the set of M molecules. This is denoted by the symbol $\{k\{m_i\}_M\}_u$ with the restriction

$$\sum_{i=1}^{k} m_i = M. \qquad (1.3.38)$$

Note that two subsets of identical size are different if they do not contain the same particular molecules. $\zeta_m(z, \{m_i\}_M)$ is the correction term at fugacity z for $\{m_i\}_M$ proximate molecules. Some examples of this expansion are

$$g_1(z, (i)) = \zeta_1(z, (i)) \qquad (1.3.39)$$

$$g_2(z, (i)(j)) = \zeta_1(z, (i))\, \zeta_1(z, (j)) + \zeta_2(z, (i)(j))$$

$$g_3(z, (i)(j)(k)) = \zeta_1(z, (i))\, \zeta_1(z, (j))\, \zeta_1(z, (k))$$

$$+ \zeta_2(z, (i)(j))\, \zeta_1(z, (k))$$

$$+ \zeta_2(z, (i)(k))\, \zeta_1(z, (j))$$

$$+ \zeta_2(z, (j)(k))\, \zeta_1(z, (i)) + \zeta_3(z, (i)(j)(k)).$$

Now consider the function g_{N+M}. If the coordinates $\{N + M\}$ are such that they fall naturally into two exclusive groups

$$g_{N+M}(z, \{N + M\}) = g_N(z, \{N\})g_M(z, \{M\}) \qquad (1.3.40)$$

$$= g_N(z, \{N\}) \sum \{k\{m_i\}_M\}_u \prod_{i=1}^{k} \zeta_m(z, \{m_i\}_N).$$

However, if some of the N molecules are close to some of the M molecules, an additional correction is required. This may be written

$$\zeta_m(z, \{m_i\}) + \sum \{n\}_N \zeta_{nm}(z, \{n\}_N\{m_i\}_M) \qquad (1.3.41)$$

and the complete expansion becomes

$$g_{N+M}(z, \{N + M\}) = g_N(z, \{N\}) \sum \{k\{m_i\}_M\}_u \prod_{i=1}^{k} [\zeta_m(z, \{m_i\}_M) \quad (1.3.42)$$
$$+ \sum \{n\}_N \zeta_{nm}(z, \{n\}_N\{m_i\}_N)].$$

The relation (1.3.42) may be integrated over all values of the coordinates $\{M\}$. Two kinds of integrals result:

$$V m! \, b_m(z) = \int \cdots \int \zeta_m (z, \{m_i\}_M) \, d\{m\}_M \quad (1.3.43)$$

$$m! \, b_{nm}(z, \{n\}_N) = \int \cdots \int_V \zeta_{nm}(z, \{n\}_N\{m_i\}_M) \, d\{m\}_M. \quad (1.3.44)$$

Both of these integrals depend only upon the numerical value of the m_i and not upon which of the M molecules constitute the m_i. If the interaction forces drop to zero sufficiently rapidly, the cluster integrals (1.3.43) and (1.3.44) are independent of the volume of the system.

It is convenient to subdivide the coordinates $\{M\}$ into μ_m subsets of size m in $M!/\prod_m (m!)^{\mu_m} \mu_m!$ different ways, all of which appear in the sum (1.3.42). Integration over $\{M\}$ then leads to the equations

$$\int \cdots \int_V g_{N+M}(z, \{N + M\}) \, d\{M\}$$

$$= g_N(z, \{N\}) \sum \{k\{m_i\}_M\}_u \prod_{i=1}^{k} \int \cdots \int_V [\zeta_m(z, \{m_i\}_M)$$

$$+ \sum \{n\}_N \zeta_{nm}(z, \{n\}_N\{m_i\}_M) \, d\{m_i\}_M$$

$$= g_N(z, \{N\}) M! \sum_{\mu_m} \prod_m \frac{[V b_m(z) + \sum \{n\}_N b_{nm}(z, \{n\}_N)]^{\mu_m}}{\mu_m!}$$

$$(1.3.45)$$

subject to the restriction that

$$\sum_m m\mu_m = M. \quad (1.3.46)$$

By substitution of (1.3.45) into (1.3.24) one finds

$$\frac{e^{V p(z+y)/kT} g_N(z + y, \{N\})}{[\gamma(z + y)]^N} = e^{V p(z)/kT} g_N(z, \{N\})[\gamma(z)]^{-N}$$

$$\times \sum_{M \geq 0} \sum_{\mu_m} \prod_m \frac{1}{\mu_m!} \left\{ V b_m(z) \left(\frac{y}{\gamma(z)}\right)^m + \sum_{\{n\}_N} b_{nm}(z) \left(\frac{y}{\gamma(z)}\right)^m \right\}^{\mu_m}. \quad (1.3.47)$$

It is easy to recognize that part of the right-hand side of Eq. (1.3.47) is the expansion of an exponential. Thus,

$$\frac{e^{Vp(z+y)/kT} g_N(z+y, \{N\})}{[\gamma(z+y)]^N} = e^{Vp(z)/kT} g_N(z, \{N\})[\gamma(z)]^{-N} \times$$

$$\exp\left\{\sum_{m\geqslant 0}\left[Vb_m(z)\left(\frac{y}{\gamma(z)}\right)^m + \sum_{\{n\}_N} b_{nm}(z)\left(\frac{y}{\gamma(z)}\right)^m\right]\right\} \quad (1.3.48)$$

for all values of the fugacity y which are smaller than the radius of convergence of the series in (1.3.48). Taking the logarithms of both sides

$$\frac{Vp(z+y)}{kT} - \frac{1}{kT}\sum_{\{n\}_N} w_n(z+y, \{n\}_N) - N \ln \gamma(z+y)$$

$$= \frac{Vp(z)}{kT} - \frac{1}{kT}\sum_{\{n\}_N} w_n(z, \{n\}_N) - N \ln \gamma(z)$$

$$+ \sum_{m\geqslant 1} Vb_m(z)\left(\frac{y}{\gamma(z)}\right)^m + \sum_{\{n\}_N}\sum_{M\geqslant 1} b_{nm}(z, \{n\}_N)\left(\frac{y}{\gamma(z)}\right)^m \quad (1.3.49)$$

whereupon, after equating like powers of V,

$$p(z+y) - p(z) = kT\sum_{m\geqslant 1} b_m(z)\left(\frac{y}{\gamma(z)}\right)^m \quad (1.3.50)$$

and

$$w_1(z+y, (i)) + kT \ln \gamma(z+y) = w_1(z, (i)) + kT \ln \gamma(z)$$

$$- kT\sum_{m\geqslant 1} b_{1m}(z, (i))\left(\frac{y}{\gamma(z)}\right)^m \quad (1.3.51)$$

$$w_n(z+y, \{n\}) - w_n(z, \{n\}) = -kT\sum_{m\geqslant 1} b_{nm}(z, \{n\})\left(\frac{y}{\gamma(z)}\right)^m. \quad (1.3.52)$$

Equation (1.3.50) is an extraordinary result. It expresses the difference in pressure corresponding to a difference in fugacity in terms of cluster integrals. In the particular case when $z = 0$, one finds

$$p(y) = kT\sum_{m\geqslant 1} b_m(0)y^m. \quad (1.3.53)$$

Further, the differentiation of (1.3.50) h times with respect to y, followed by evaluation of the derivative at $y = 0$ yields

$$\left(\frac{\partial^h p(y)}{\partial z^h}\right)_T = kTh!\, b_h(z)\left(\frac{1}{\gamma(z)}\right)^h \quad (1.3.54)$$

thus relating the cluster integrals to derivatives of the pressure. For the particular case $h = 1$ this gives

$$\left(\frac{\partial p(z)}{\partial z}\right)_T = \frac{kT}{\gamma(z)}. \tag{1.3.55}$$

If instead, $y + z$ is substituted for z, one obtains the relation

$$b_n(y + z)\left(\frac{y}{\gamma(z + y)}\right)^h = \sum_{m \geq 1} \frac{m!}{(m - h)! h!} b_m(z)\left(\frac{y}{\gamma(z)}\right)^m \tag{1.3.56}$$

whereupon, for $h = 1$, it is found that

$$\frac{y}{\gamma(z + y)} = \sum_{m \geq 1} m b_m(z)\left(\frac{y}{\gamma(z)}\right)^m. \tag{1.3.57}$$

To implement the result embodied in Eq. (1.3.50) we seek an extension of the virial expansion. To this end, the difference of pressures will be developed in a power series in $y/\gamma(z + y)$. We proceed by defining a simple integral $B_m(z)$ in terms of $b_m(z)$ by the relation

$$b_m(z) = \frac{1}{m^2} \sum \prod_{k \geq 2} \frac{(n_k B_k(z))^{n_k}}{n_k!} \tag{1.3.58}$$

with the restriction

$$\sum_{k \geq 2} (k - 1)n_k = m - 1. \tag{1.3.59}$$

Using the standard Lagrange technique, (1.3.58) can be inverted to give[†]

$$B_m(z) = \sum (-)^{\Sigma n_k - 1}[(m - 2 + \Sigma n_k)!/m!] \prod_{k \geq 2} \frac{(kb_k)^{n_k}}{n_k!}. \tag{1.3.60}$$

[†] The inversion is effected as follows: Define B_k' as

$$\sum_{k \geq 2} k(k - 1)B_k' x^{k-1} = -\frac{1}{2\pi i}\oint \frac{1}{y}\sum_{h \geq 1} x^h[\sum_{m \geq 1} m b_m y^m]^{-h} dy$$

$$= \frac{1}{2\pi i}\oint \frac{x}{y}\frac{dy}{(x - \sum m b_m y^m)}.$$

For $|x|$ sufficiently small,

$$|x| < |\sum m b_m y^m|$$

and if y_0 is defined by

$$x = \sum m b_m y_0^m$$

so that by substitution

$$\sum_{k \geq 2} k(k - 1)B_k x^{k-1} = 1 - \frac{x}{\sum m^2 b_m y_0^m}.$$

Let us now define another function $h_{nm}(z, \{n\})$ by the equation

$$b_{nm}(z, \{n\}) = m b_m(z) h_{n1}(z, \{n\}) + \sum_{k \geqslant 2}^{m} h_{nk}(z, \{n\}) \left[\frac{\partial b_n(z)}{\partial B_k(z)} \right]. \qquad (1.3.61)$$

Now multiply by dx/x, where

$$dx = \left(\sum m^2 b_m y_0^m \right) \frac{dy_0}{y_0}$$

$$y_0 = x \exp\left(- \sum_{k \geqslant 2} k B_k' x^{k-1}\right)$$

and integrate. Then equate coefficients of x^{k-1} in the resultant equation to get

$$k(k - 1) B_k' = -\frac{1}{2\pi i} \oint \frac{1}{y} \left(\frac{1}{\sum\limits_{m \geqslant 1} m b_m y^m} \right)^{k-1} dy .$$

However, the integral b_1 is unity, so that

$$-\frac{1}{2\pi i} \oint \frac{1}{y} \left(\frac{1}{\sum\limits_{m \geqslant 1} m b_m y^m} \right)^{k-1} dy = -\frac{1}{2\pi i} \oint \frac{dy}{y} \left(\frac{1}{1 + \sum\limits_{m \geqslant 2} m b_m y^{m-1}} \right)^{k-1}.$$

Note that $-k(k - 1) B_k'$ is just the coefficient of y^{k-1} in the expansion of

$$\left(1 + \sum_{m \geqslant 2} m b_m y^{m-1}\right)^{-k+1}.$$

Note also that if the coefficent of $(y_1)^{h_1}(y_2)^{h_2} \cdots$ in the expansion of $(1 + y_1 + y_2 + \cdots)^{-h}$ is the number $C(h_1, h_2, \cdots)$, then

$$\left[\frac{\partial^{\Sigma h_i} (1 + y_1 + y_2 + \cdots)}{\partial y_1^{h_1} \partial y_2^{h_2} \cdots} \right]_{y_i = 0} = C(h_1 \cdots) h_1! \, h_2! \cdots$$

Therefore

$$C(h_1, h_2, \ldots) = (-)^{\Sigma h_i} \frac{(h - 1 + \Sigma h_i)!}{(h - 1)! \, h_1! \, h_2! \cdots}$$

This result is now used to evaluate the B_k'. We find

$$\left(1 + \sum_{m \geqslant 2} m b_m y^{m-1}\right)^{-k+1} = \sum_{h_1 \cdots h_i \cdots} (-1)^{\Sigma h_i} \frac{((k - 2) + \Sigma h_i)!}{(k - 2)!} \prod_{i \geqslant 1} \frac{[(i + 1) b_{i+1} y^i]^{h_i}}{h_i!}$$

so that finally

$$k(k - 1) B_k' = \sum_k (-1)^{\Sigma h_i - 1} \frac{(k - 2 + \Sigma h_i)!}{(k - 2)!} \prod_{i \geqslant 1} \frac{[(i + 1) b_{i+1}]^{h_i}}{h_i!}$$

with the restriction

$$\sum_{i \geqslant 1} i h_i = k - 1$$

which proves the result stated.

From

$$x = \sum m b_m y_0^m$$

$$\sum_{k \geq 2} k(k-1) B_k x^{k-1} = 1 - \frac{x}{\sum m^2 b_m y_0^m} \tag{1.3.62}$$

it is readily verified that

$$\sum b_m y_0^m = x - \sum (k-1) B_k x^k . \tag{1.3.63}$$

There is a striking similarity between Eqs. (1.3.63) and (1.3.57). In fact,

$$\sum b_m(z) \left(\frac{y}{\gamma(z)} \right)^m = \frac{y}{\gamma(z+y)} - \sum (k-1) B_k(z) \left(\frac{y}{\gamma(z+y)} \right)^k \tag{1.3.64}$$

and

$$p(z+y) - p(z) = kT \left[\frac{y}{\gamma(z+y)} - \sum (k-1) B_k(z) \left(\frac{y}{\gamma(z+y)} \right)^k \right]. \tag{1.3.65}$$

In the special case $z = 0$, Eq. (1.3.65) reduces to the well-known virial equation of state for gases. By differentiation with respect to B_k,

$$\sum \frac{\partial b_m}{\partial B_k} (y_0)^m = -(k-1) x^k - \sum_m m b_m (y_0)^m \frac{\partial \ln y_0}{\partial B_k} \tag{1.3.66}$$

whereas from (1.3.62)

$$\frac{\partial \ln y_0}{\partial B_k} = -kx^{k-1} \tag{1.3.67}$$

so that by combination,

$$\sum_m \frac{\partial b_m}{\partial B_k} (y_0)^m = x^k. \tag{1.3.68}$$

If the relation (1.3.63) is used to change variables, Eq. (1.3.68) becomes

$$\sum_m \frac{\partial b_m(z)}{\partial B_k(z)} \left(\frac{y}{\gamma(z)} \right)^m = \left(\frac{y}{\gamma(z+y)} \right)^k. \tag{1.3.69}$$

The final step requires the substitution of (1.3.69) into (1.3.61) to yield

$$\sum_m b_{nm}(z, \{n\}) \left(\frac{y}{\gamma(z)} \right)^m = h_{n1}(z, \{n\}) \frac{y}{\gamma(z+y)} + \sum_{k \geq 2} h_{nk}(z, \{n\}) \left(\frac{y}{\gamma(z+y)} \right)^k \tag{1.3.70}$$

and

$$w_1(z + y, (i)) + kT \ln \gamma(z + y) = w_1(z, (i))$$
$$+ kT \ln \gamma(z) - kT \sum h_{1m}(z, (i)) \left(\frac{y}{\gamma(z + y)}\right)^m \quad (1.3.71)$$

$$w_n(z + y, \{n\}) - w_n(z, \{n\}) = - kT \sum_{m \geqslant 1} h_{nm}(z, \{n\}) \left(\frac{y}{\gamma(z + y)}\right)^m. \quad (1.3.72)$$

The relationship of these formal results to the well-known imperfect gas theory is most easily established by noting that

$$g_n(0, \{N\}) = \exp\left[- \frac{U_N\{N\}}{kT} \right] \quad (1.3.73)$$

so that $b_m(0)$ is the ordinary cluster integral and B_k is related to the irreducible integrals by

$$B_k = \frac{1}{k} \beta_{k-1}. \quad (1.3.74)$$

It is also clear that, in the limit of zero fugacity,

$$\zeta_m(\{m\}) = \sum_{m \geqslant i > j \geqslant 1}^{\text{cluster}} \prod (e^{u_{ij}/kT} - 1) \quad (1.3.75)$$

with u_{ij} the intermolecular pair potential.

The formalism of this section, while appearing overly ponderous, will in fact provide a powerful tool for the investigation of interactions in polymeric systems in a systematic manner. Such systematic studies have clearly defined domains of validity and, even more important, clearly defined physical interpretations. They are therefore, of great value in understanding molecular behavior.

1.4 The " McMillan–Mayer " Theory of Solutions[1]

The Mayer–McMillan theory[1] was the first molecular theory of solutions, chronologically. It involves a rather detailed investigation of molecular distribution in multicomponent systems and is quite complicated mathematically. The account which will be given here is based on a simplified version due to T. L. Hill.[9] We simplify Hill's version still more by ignoring all internal degrees of freedom of the molecules. While this lessens the generality of the treatment, it is relatively easy, once the simplified theory is understood, to go back to the original papers and understand the more realistic, more complicated, cases also. The general theory is an extension of the formalism of the last section. One of the major reasons for pursuing a simplified treatment herein is that the general theory follows the one-component case so closely. It is hoped that the simplified treatment will also clarify the formal results for pure systems obtained in the last section. Classical mechanics will be used throughout, though the theory is equally valid quantum-mechanically.

As in the case of condensed phases of one component, it is convenient to work with molecular distribution functions, the first few of which were introduced in Sections 1.2 and 1.3. It will be instructive here to start from scratch. Let us take the phase space probability density for N molecules given by Eq. (1.1.48) and integrate over momenta. Since $H = \sum_{i=1}^{N} p_i^2/2m_i + U_N$ the immediate result is

$$P(N;\{N\}) = \exp\{\beta[\Omega + N \cdot \mu - U_N]\} \prod_i \left(\frac{2\pi m_i k T}{N_i! h^2}\right)^{3N_i/2} \tag{1.4.1}$$

If the fugacity $z_i = e^{\beta \mu_i}(2\pi m_i k T/h^2)^{3/2}$ is introduced, (1.4.1) becomes

$$P(N; \{N\}) = e^{\beta[\Omega - U_N]} \frac{z^N}{N!} \tag{1.4.2}$$

where $z^N/N!$ is shorthand for $\prod z_i^{N_i}/N_i!$. Here, h is Planck's constant, and the factor $(N_i! h^{3N_i})^{-1}$ arises because account must be taken, even classically, of the indistinguishability of particles of the same species. Clearly z_i is the fugacity of component i, and as before is so normalized that $z_i/\rho_i \to 1$ as all $\rho_j \to 0$ (ideal gas state).

Now suppose there are *exactly* N molecules in the volume. Then the probability that molecule 1 be at R_1, molecule 2 at $R_2 \cdots$ molecule r at R_n regardless of the positions of the remaining $N - n$ molecules is given by the formula from canonical ensemble theory

$$P_N^{(n)}(\{n\}) = \int e^{-\beta U_N} d\{N-n\} / \int e^{-\beta U_N} d\{N\} \tag{1.4.3}$$

where $\{n\} = (R_1 \cdots R_n)$; $\{N-n\} = (R_1 \cdots R_{N-n})$, etc. This is the so-called specific distribution function. Since the molecules of each species are, in fact, indistinguishable it is convenient to define a generic distribution function

$$\rho_N^{(n)}(\{n\}) = \frac{N!}{(N - n)!} P_N^{(n)}(\{n\}) \tag{1.4.4}$$

which is clearly the probability density that the N_1 molecules of species 1 are at $R_1 \cdots R_{N_1}$ regardless of which one is at which R vector, etc. Note that

$$\int P_N^{(n)} d\{n\} = 1$$

$$\int \rho_N^{(n)} d\{n\} = \frac{N!}{(N - n)!} . \tag{1.4.5}$$

Now the distribution function for $\mathbf{n}$ particles in an open system is the probability that there be exactly N molecules in the volume times $\rho_N^{(n)}$ for the appropriate N, summed over all N. In symbols

$$\rho_{\mathbf{n}}(\{\mathbf{n}\}) = \sum_{N \geqslant n} P(N) \rho_N^{(n)}(\{\mathbf{n}\}). \qquad (1.4.6)$$

By Eq. (1.4.2) this is

$$\rho_{\mathbf{n}}(\{\mathbf{n}\}) = \sum_{N \geqslant n} e^{\beta \Omega} \frac{z^N}{N!} \int e^{-\beta U_N} d\{N\} \times \frac{N!}{(N-n)!} \frac{\int e^{-\beta U_N} d\{N-n\}}{\int e^{-\beta U_N} d\{N\}} \qquad (1.4.7)$$

$$= e^{\beta \Omega} \sum_{N \geqslant n} \frac{z^N}{(N-n)!} \int e^{-\beta U_N} d\{N-n\}.$$

By writing $N - n = m$ (i.e., $N_i - n_i = m_i$) (1.4.7) can be put in the form

$$e^{-\beta \Omega} \frac{\rho_{\mathbf{n}}(z)}{z^n} = \sum_{m \geqslant 0} \frac{z^m}{m!} \int \exp(-\beta U_{n+m}) d\{m\} \qquad (1.4.8)$$

where the depenaence of $\rho_{\mathbf{n}}$ on z has been emphasized and the spatial dependence has been suppressed for notational convenience. Equation (1.4.8) is one of the basic equations of the Mayer–McMillan theory, and it can be used to find another basic equation, that which relates distribution functions at different fugacities. Now, Taylor's theroem states that, for an analytic function, $f(z)$

$$f(\mathbf{z}) = \sum_{l \geqslant 0} \left[\prod_i \frac{(z_i - z_i^*)^{l_i}}{l_i!} \right] \frac{\partial^{l_1 + \cdots + l_v} f(z)}{\partial z_1^{l_1} \partial z_2^{l_2} \cdots \partial z_v^{l_v}} \bigg|_{z=z^*}$$

$$= \sum_{l \geqslant 0} \frac{(\mathbf{z} - \mathbf{z}^*)}{l!} \frac{\partial^l f(z)}{\partial z^l} \bigg|_{z=z^*}$$

Consider now the right-hand side of Eq. (1.4.8) as $f(z)$ so that, by straightforward differentiation,

$$\frac{\partial^l f}{\partial z^l} = \sum_{q \geqslant 0} \frac{z^q}{q!} \int \exp(-\beta U_{n+l+q}) d\{l+q\}. \qquad (1.4.9)$$

The right-hand side of (1.4.9) can be simplified. For, if (1.4.8) is rewritten with a new dummy variable $\mathbf{q}$ written for $\mathbf{m}$, and $\mathbf{n}$ replaced by $\mathbf{n} + \mathbf{l}$, an integration over $d\{l\}$ can be performed to yield

$$\frac{\partial^l f}{\partial z^l} = \frac{e^{-\beta \Omega}}{z^{n+l}} \int \rho_{\mathbf{n}+l}(\mathbf{z}) d\{l\}. \qquad (1.4.10)$$

Evaluating (1.4.10) for $\mathbf{z} = \mathbf{z}^*$, and using $e^{-\beta\Omega}[\rho_n(\mathbf{z})/\mathbf{z}^n]$ for $f(\mathbf{z})$ leads to

$$\frac{e^{-\beta\Omega(\mathbf{z})}}{\mathbf{z}^n} \rho_n(\{\mathbf{n}\}, \mathbf{z}) = e^{-\beta\Omega(\mathbf{z}^*)} \sum_{m \geqslant 0} \frac{(\mathbf{z} - \mathbf{z}^*)^m}{\mathbf{z}^{n+m}\mathbf{m}!} \int \rho_{n+m}(\mathbf{z}^*) \, d\{\mathbf{m}\}. \quad (1.4.11)$$

This is the second fundamental result. There are several important special cases of (1.4.11) but these will be considered later, as it is more pertinent to now consider from a different point of view the physical significance of the distribution functions ρ_n. Consider the average force acting on a molecule of the set $\mathbf{n}$ (averaged over the positions of the $N - \mathbf{n}$ remaining molecules, summed over all N). It will be shown that (a) this average force is derivable from a potential, and (b) this potential is very simply related to ρ_n.

Let particle j be one of the set $\mathbf{n}$ of particles. If there are N particles in the volume, then the instantaneous force on particle j is

$$\mathbf{F}_j = -\nabla_j U_N$$

where ∇_j is the gradient with respect to $\mathbf{R}_j$. The average force $\langle \mathbf{F}_j \rangle_N^n$, averaged over the positions of the $N - \mathbf{n}$ remaining particles, is then

$$\langle \mathbf{F}_j \rangle_N^n = \frac{-\int e^{-\beta U_N} \nabla_j U_{N_N} \{N-\mathbf{n}\}}{\int e^{-\beta U_N} \, d\{N-\mathbf{n}\}} \quad (1.4.12)$$

or

$$\langle \mathbf{F}_j \rangle_N^n = kT\nabla_j \ln \rho_N^{(n)}. \quad (1.4.13)$$

It is necessary to further average over N to get the average force for the open system under consideration

$$\langle \mathbf{F}_j \rangle^n = \frac{\sum\limits_{N \geqslant n} P(N)\rho_N^{(n)}\langle F_j \rangle_N^n}{\sum\limits_{N \geqslant n} P(N)\rho_N^{(n)}} \quad (1.4.14)$$

since $P(N)\rho_N^n$ is the probability that there are N particles in the volume and $\mathbf{n}$ of them occupy the position set $\{\mathbf{n}\}$, and $\langle \mathbf{F}_j \rangle_N^n$ is the average force under these circumstances.

Comparing (1.4.13) and (1.4.14) it is seen that

$$\langle \mathbf{F}_j \rangle_n = kT\nabla_j \ln \rho_n. \quad (1.4.15)$$

Thus it has been shown that the average force on one of $\mathbf{n}$ particles in an open system is derivable from a potential, which is $-kT \ln \rho_n$ + constant. This constant is conventionally taken as $-\ln \rho_1^n \times kT$ i.e., $\rho_{1(1)}^{n_1} \rho_{1(2)}^{n_2} \cdots$ where $\rho_{1(i)} = \langle N_i \rangle / V$. If the potential of average force be denoted by W_n,

$$\frac{\rho_n}{\rho_1^n} = e^{-W_n/kT}. \quad (1.4.16)$$

Now as z goes to zero (all z_i's, not just some of them) $\rho_n \to \rho_1^n \, e^{-\beta U_n}$ and $W_n \to U_n$. To see this return to Eq. (1.4.8). Let $z \to 0$ on the right-hand side. Only the term $m = 0$ survives in the sum, leaving

$$\lim_{z \to 0} e^{-\beta \Omega} \frac{\rho_n}{z_n} = e^{-\beta U_n}. \tag{1.4.17}$$

But

$$\lim_{z \to 0} e^{-\beta \Omega} = 1,$$

for

$$-\beta \Omega = \frac{pV}{kT} = \frac{pV}{nkT} \times n$$

which approaches n for the perfect gas state, and then $n \to 0$ as $\rho \to 0$; n is the total average number of molecules present. Also z is so normalized that $z/\rho_1^n \to 1$ as $z \to 0$. This proves the statement.

It is now possible to exhibit some special cases of (1.4.11). First, send z to zero. Then (1.4.11) becomes

$$e^{-\beta U_n} = e^{-\beta \Omega(z^*)} \sum_{m \geqslant 0} \frac{(-)^m}{m! z^{*n}} \int \rho_{n+m}(z^*) \, d\{m\}. \tag{1.4.18}$$

This set of equations is the inverse of the set (1.4.7). Secondly, put $n = 0$. Since $\rho_0 = 1$ the result obtained is simply

$$\exp[-\beta(\Omega(z) - \Omega(z^*))] = \sum_{m \geqslant 0} \frac{1}{m!} \frac{(z - z^*)^m}{\gamma^m} \int \exp\left(\frac{-W_m(z^*)}{kT}\right) d\{m\} \tag{1.4.19}$$

where γ is an activity coefficient defined by

$$\gamma_i = \frac{z_i}{\rho_i}. \tag{1.4.20}$$

It would be desirable to have a power series in the concentrations, and not in the fugacities, for ρ_n. It will be sufficient to demonstrate how the first few terms may be obtained. The general treatment is algebraically complicated, but not difficult in principle. An easy way of obtaining the expansion is by applying the theory to a special case, osmotic pressure, and working out the details. The method of calculating other thermodynamic functions will then be clear, although the details may be mathematically difficult and tedious. It is pertinent to remark that such an expansion will enable us to find more explicit representations (in powers of the concentrations) for the functions (G_{ij} integrals) appearing in the Kirkwood–Buff formulation of solution theory.

Consider then a solution composed of a solvent, component 1, and solutes, components $j, j = 2 \cdots r$, in an osmotic cell. An osmotic cell is a vessel with two compartments separated by a membrane permeable only to component 1. See Fig. 1.4.1.

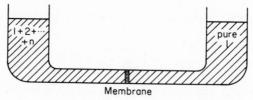

FIG. 1.4.1. Schematic osmotic cell.

Denote quantities referring to the right-hand side of the cell by a star *, with no star for the left-hand side. The theory of thermodynamic equilibrium requires that

$$z_1 = z_1^*$$
$$z_j^* = 0 \qquad j = 2, \cdots r. \tag{1.4.21}$$

Furthermore, $p(\mathbf{z}) - p(\mathbf{z}^*) = \pi$, the osmotic pressure, by definition. Equation (1.4.19) then becomes

$$e^{\pi V/kT} = \sum_{m \geq 0} \frac{\mathbf{z}'^{m'}}{\gamma^{*m'} \mathbf{m}'!} \int \exp\left[\frac{-W_m(z_1^*, 0)}{kT}\right] d\{\mathbf{m}\} \tag{1.4.22}$$

where the prime ($'$) on $\mathbf{z}'$ and $\mathbf{m}'$ corresponds to the restriction that the sum is only over $m_2, m_3 \cdots m_r$ and only $z_2 \cdots z_r$ appear. For, all terms containing z_1 contain a factor $(z_1 - z_1^*)^m$, which vanishes, since $z_1 = z_1^*$. γ^* denotes the activity coefficients of the solute molecules in a solution infinitely dilute with respect to the solutes, but not to the solvent. The usual activity coefficients one meets in thermodynamics are unity in such a case, but γ^* has been defined in such a way that it is unity for an ideal gas.

In the succeeding development attention will be confined to a two-component system, i.e., only one solute. Denote the integral appearing in (1.4.22) by Y_m^*,

$$Y_m^* = \int \exp[-\beta W_m(z_1^*, 0)] d\{\mathbf{m}\} \tag{1.4.23}$$

and then (1.4.23) takes the form

$$e^{\pi V/kT} = \sum_{m \geq 0} \frac{(z/\gamma)^m}{m!} Y_m^*. \tag{1.4.24}$$

This gives $e^{\pi V/kT}$ as a power series in z/γ, since Y_m^* is a function only of z_1^* (z has been written for z_2). What is desired, however, is π/kT as a function of z. Writing

$$\frac{\pi}{kT} = \sum_{j \geq 1} b_j^* \left(\frac{z}{\gamma}\right)^j \tag{1.4.25}$$

the b_j^*'s can be found in terms of the Y_m^*'s by taking the logarithm of (1.4.24).

$$\frac{\pi V}{kT} = \ln\left(\sum_{m \geq 0} \frac{(z/\gamma)^m}{m!} Y_m^*\right). \tag{1.4.26}$$

Now make a Taylor expansion of the logarithm about $z = 0$, remembering $Y_0^* = 1\ddagger$, $Y_1^* = V$

$$\frac{\pi V}{kT} = \sum_{m \geq 1} \frac{(z/\gamma)^m}{m!} Y_m^* - \frac{1}{2}\left(\sum_{m \geq 1} \frac{(z/\gamma)^m}{m!} Y_m^*\right)^2 + \cdots \tag{1.4.27}$$

$$= \frac{z}{\gamma} Y_1^* + \frac{1}{2}\left(\frac{z}{\gamma}\right)^2 Y_2^* - \frac{1}{6}\left(\frac{z}{\gamma}\right)^3 Y_3 + \cdots$$

$$- \frac{1}{2}\left(\frac{z}{\gamma}\right)^2 Y_1^{*2} - \frac{1}{4}\left(\frac{z}{\gamma}\right)^3 Y_1^* Y_2^* + \cdots$$

Comparing (1.4.25) and (1.4.27)

$$1!\, Vb_1 = Y_1^* = V \tag{1.4.28}$$

$$2!\, Vb_2 = Y_2^* - Y_1^{*2}$$

$$3!\, Vb_3 = Y_3^* - 2Y_1^* Y_2^* + 2Y_1^{*3}$$

As the final step, transform from fugacities to concentrations as independent variables. The first step is to note that

$$\frac{z}{\gamma}\left(\frac{\partial(\pi/kT)}{\partial(z/\gamma)}\right)_{T,z^*} = -\frac{1}{V}\left(\frac{\partial\Omega(z)}{\partial\mu_2}\right)_{T,\mu_1} = \frac{\langle m \rangle}{V} \tag{1.4.29}$$

as previously. Thus

$$c = \frac{\langle m \rangle}{V} = \sum_{j \geq 1} jb_j^* \left(\frac{z}{\gamma}\right)j. \tag{1.4.30}$$

What is desired, however, is the inverse of this series, i.e., z/γ as a function of c, and not c as a function of z/γ. The series (1.4.30) must therefore be

‡ Note: This can be seen from Eq. (1.4.10). Simply let $z \to z^*$.

inverted. There are general methods for doing this[†] but a more pedestrian route is perspicuous. The procedure followed is to compute explicitly the first few coefficients in the series

$$\frac{z}{\gamma} = \sum_{n \geqslant 1} a_n c^n. \tag{1.4.31}$$

This is done as follows: replace $z(\gamma)$ in (1.4.30) by the series (1.4.31) and equate like powers of c, i.e.,

$$c = \sum_{j \geqslant 1} j b_j^* \left(\sum_{n \geqslant 1} a_n c^n \right)$$

$$= b_1^*(a_1 c + a_2 c^2 + a_3 c^3 + \cdots) + 2 b_2^*(a_1^2 c^2 + 2 a_1 a_2 c^3 + \cdots)$$

$$+ 3 b_3^*(a_1^3 c^3 + \cdots) + \cdots \tag{1.4.32}$$

to terms of order 3. By (1.4.28) $b_1^* = 1$. Consequently $a_1 = 1$ and

$$a_2 = -2 b_2^*$$

$$a_3 = 8 b_2^{*2} - 3 b_3 \tag{1.4.33}$$

$$\cdots$$

Now use Eq. (1.4.31), with the a's so specified, in Eq. (1.4.27) for $\pi V/kT$, getting

$$\frac{\pi}{kT} = c \left[1 - \sum_{j \geqslant 1} \frac{j}{j+1} \beta_j^* c^j \right] \tag{1.4.34}$$

$$\beta_1^* = 2 b_2$$

$$\beta_2^* = 3(b_3^* - 2 b_2^{*2}).$$

$$\cdots$$

If (1.4.34) is written in virial form

$$\frac{\pi}{kT} = c[1 + B_2^* c + B_3^* c^2 + \cdots] \tag{1.4.35}$$

where B_n^* is the nth osmotic viral coefficient, then clearly

$$B_n^* = -\frac{n-1}{n} \beta_{n-1}^*. \tag{1.4.36}$$

[†] See E. Whittaker and G. Watson, "*Modern Analysis*," 4th ed. pp. 128 ff. Cambridge Univ. Press, London and New York, 1952.

According to Eq. (1.4.35) dilute solutions obey van t' Hoff's law and more concentrated solutions depart from this law in formally the same way as the pressure of a real gas departs from that of an ideal gas at the same temperature and density. The analogy is only formal, however, because the b^* coefficients are integrals of the potentials of *average* force at infinite dilution of the solute species. In the gas case, the b's are integrals of the potentials of *actual* force (potential of average force at infinite dilution of all species).

It is possible to write down the second and third virial coefficients immediately from (1.4.34) and (1.4.36)

$$B_2^* = -\frac{1}{2V} \int \left[\exp\left(\frac{-W_2}{kT}\right) - 1 \right] d\mathbf{R}_1 \, d\mathbf{R}_2 \qquad (1.4.37)$$

$$B_3^* = -\frac{1}{3V} \int \left\{ \exp\left(\frac{-W_3}{kT}\right) - \exp\left[\frac{-W_2(\mathbf{R}_1, \mathbf{R}_2)}{kT}\right] \exp\left[\frac{-W_2(\mathbf{R}_2, \mathbf{R}_3)}{kT}\right] \right.$$

$$- \exp\left[\frac{-W_2(\mathbf{R}_2, \mathbf{R}_3)}{kT}\right] \exp\left[\frac{-W_2(\mathbf{R}_1, \mathbf{R}_2)}{kT}\right]$$

$$- \exp\left[\frac{-W_2(\mathbf{R}_1, \mathbf{R}_3)}{kT}\right] \exp\left[\frac{-W_2(\mathbf{R}_2, \mathbf{R}_3)}{kT}\right] + \exp\left[\frac{-W_2(\mathbf{R}_1, \mathbf{R}_2)}{kT}\right]$$

$$\left. + \exp\left[\frac{-W_2(\mathbf{R}_1, \mathbf{R}_3)}{kT}\right] + \exp\left[\frac{-W_2(\mathbf{R}_2, \mathbf{R}_3)}{kT}\right] - 1 \right\} d\mathbf{R}_1 \, d\mathbf{R}_2 \, d\mathbf{R}_3.$$

$$(1.4.38)$$

Not enough is known about the potentials of average force (except perhaps for electrolytes, Debye–Hückel theory) to make *a priori* calculations of the virial coefficients possible. What is usually done is to measure B_2^* by measuring π, plotting π/c versus c, and measuring the initial slope of the resulting curve. The measured B_2^* as a function of temperature is then used to infer properties of the local environment of the solute molecules.

The case of several solvents and several solutes can be carried through in the same way, but is very complicated. The formal analogy to imperfect gas theory is preserved in this extension. The reader is referred to the paper by McMillan and Mayer[1] for details. Although their notation is somewhat different from that which is used herein, the differences are more apparent than real.

Before closing, one last remark on the formulas (1.4.22) onward is necessary. Note that the properties of the solvent nowhere enter explicitly in the formulas, but lay hidden in the potential of average force at infinite dilution of the solute. This also occurred in the Kirkwood–Buff theory, as will be recalled, cf. Eq. (1.2.18), where the solvent–solute and solvent–solvent matrix elements do not appear in the expression.

It is now possible to show how to get explicit concentration expansions for the osmotic pressure from the Kirkwood–Buff theory. It should be clear what the procedure must be, and therefore the method will only be outlined, leaving explicit verification of the equivalence of the result to (1.4.35) to the reader.

Equation (1.2.21) states

$$\left(\frac{\partial \pi}{\partial c_2}\right)_{T,\mu_1} = \frac{kT}{1 + G_{22}c_2}. \tag{1.2.21}$$

For sufficiently small $c_2 (|G_{22}c_2| < 1)$ it is possible to write

$$\left[\frac{\partial(\pi/kT)}{\partial c_2}\right]_{T,\mu_1} = 1 - G_{22}c_2 + (G_{22}c_2)^2 + \cdots \tag{1.4.39}$$

or

$$\frac{\pi}{kT} = c_2\left[1 - \frac{1}{c_2}\int_0^{c_2} G_{22}c_2\, dc_2 + \frac{1}{c_2}\int_0^{c_2} (G_{22}c_2)^2\, dc_2 + \cdots\right]. \tag{1.4.40}$$

Now G_{22} is, by definition,

$$G_{22} = \frac{1}{V}\int\left[\frac{\rho_2}{c_2^2} - 1\right] d\mathbf{R}_1\, d\mathbf{R}_2$$

where ρ_2 is evaluated at concentrations appropriate to the solution in question, *not* at infinite dilution. Clearly, however, ρ_2 may be expanded about infinite dilution of the solute species, 2, by the use of Eq. (1.4.16), getting

$$G_{22} = G_{22}^0 + \Gamma_{22}^{(1)}c_2 + \Gamma_{22}^{(2)}c_2^2 + \cdots. \tag{1.4.41}$$

Putting (1.4.41) in (1.4.40) the immediate result is

$$\frac{\pi}{kT} = c[1 - \sum \alpha_n c^n]. \tag{1.4.42}$$

It is also clear, almost by inspection, that α_1 is the same as B_2^*, already defined. The verification of the identity of the α_n with B_{n+1}^* takes a bit of work, but is only a question of algebra, with no new principles involved.

This finishes our discussion of exact theories of solutions. The rest of this introductory chapter will be devoted to approximate theories, which, however, are amenable to practical computation. Despite the formal elegance of the theories presented in this section, at present there is little hope of using them for direct exact computations. Their chief utility is in representing an exact form to which all approximate theories must conform to some extent. In other words, the lack of conformity of approximate theories and the Kirk-wood–Buff or McMillan–Mayer formalism can be used to analyze the nature of the approximations. This is the course we shall pursue and is the reason for the discussion of these theories in a book otherwise devoted to polyelectrolytes.

1.5   Perturbation Theories

In recent years, much attention has been focused on the calculation of the properties of a mixture in terms of the properties of suitably chosen pure components from which it differs very little. These perturbation theories all assume the deviations to be small enough to neglect terms of higher order than linear or quadratic. As in all first-order perturbation treatments, the perturbation is itself averaged over the unperturbed system to obtain the first-order correction. All these theories use in fact or in spirit the theorem of corresponding states, which we now proceed to examine.

1.5.1 THE THEOREM OF CORRESPONDING STATES[10]

Consider a dense fluid in which the following conditions are satisfied:[11]

(a) The internal degrees of freedom are independent of the state of aggregation. This assumption enables the partition function to be factored and the internal partition function immediately evaluated.

(b) Classical statistics may be used. With this assumption, H_2, He, and possibly Ne are excluded from consideration.

(c) The potential energy of a pair of molecules is a *universal function* of two parameters, i.e.,

$$u(R) = u^* \phi \left(\frac{R}{R^*} \right). \tag{1.5.1}$$

(d) The total potential energy, U_N, may be represented as the sum of pair potentials,

$$U_N = \sum_{N \geqslant i > j \geqslant 1} \sum u_{ij}(R_{ij}). \tag{1.5.2}$$

Recent quantum-mechanical studies have cast some doubt on the validity of this assumption, but it will be sufficiently accurate for our purposes.

The configurational partition function Q may be written as

$$Q = \frac{1}{N!} \int_V \cdots \int e^{-\beta U_N} d\{N\} \tag{1.5.3}$$

with

$$\beta = \frac{1}{kT}$$

and as usual V is the total volume of the system. Now change variables by introducing the assumptions embodied in Eqs. (1.5.1) and (1.5.2) with the result that

$$Q = \frac{(R^*)^{3N}}{N!} \int \cdots \int \exp\left[-\beta u^* \sum_{i>j} \phi\left(\frac{R_{ij}}{R^*}\right)\right] d\{N/N^*\} \qquad (1.5.4)$$

$$= (R^*)^{3N} Q^*\left(\beta u^*, \frac{V}{(R^*)^3}, N\right) \qquad (1.5.5)$$

where Q^* depends only upon the dimensionless variables βu^*, $V/(R^*)^3$ and of course N. With the assumption (c) that the potential energy is a universal function, Q^* is a universal function. From the configurational Helmholtz free energy,

$$A_{\text{conf}} = -kT \ln Q^* - kT \ln (R^*)^{3N} \qquad (1.5.6)$$

$$= Na(T, v) \qquad (1.5.7)$$

where $a(T, v)$ is an intensive function, it is readily found that

$$Q = (R^*)^{3N} \left\{q^*\left(\frac{kT}{u^*}, \frac{v}{(R^*)^3}\right)\right\}^N \qquad (1.5.8)$$

with v the molar volume. The forms (1.5.7) and (1.5.8) follow from the observation that A is an extensive function and can thus be written as a product of the number of molecules times an intensive function. The logarithmic relation between Q and A then dictates the form (1.5.8). The equation of state,

$$p = kT\left(\frac{\partial \ln Q}{\partial v}\right)_T \qquad (1.5.9)$$

now becomes

$$\tilde{p} = \tilde{p}(\tilde{T}, \tilde{v}) \qquad (1.5.10)$$

where the reduced variables are defined by

$$\tilde{p} = p\,\frac{(R^*)^3}{u^*}$$

$$\tilde{T} = \frac{kT}{u^*} \qquad (1.5.11)$$

$$\tilde{v} = \frac{v}{(R^*)^3}.$$

Equation (1.5.10) is a *universal equation of state*, subject to the four assumptions mentioned. Since, at the critical point,

$$\left(\frac{\partial p}{\partial v}\right)_{T=T_c} = 0$$

$$\left(\frac{\partial^2 p}{\partial v^2}\right)_{T=T_c} = 0 \qquad (1.5.12)$$

$$\tilde{p} = \tilde{p}(\tilde{T}_c, \tilde{v}_c).$$

There are three equations in three unknowns, and it follows immediately that $\tilde{p}_c$, $\tilde{T}_c$, and $\tilde{v}_c$ are universal constants. Also, the reduced triple point data are universal constants. Some examples of this are given in Table 1.5.1.

TABLE 1.5.1

Gas*	$T_c(°K)$	$V_c(cm^3)$	$p_c(atm)$	$\tilde{T}_c$	$\tilde{v}_c$	$\tilde{p}_c$	$\dfrac{\tilde{p}_c\tilde{v}_c}{R\tilde{T}_c}$
He	5.3	57.8	2.26	0.52	5.75	0.027	0.300
H₂	33.3	65.0	12.8	0.90	4.30	0.064	0.304
Ne	44.5	41.7	25.9	1.25	3.33	0.111	0.296
A	151.0	75.2	48.0	1.26	3.16	0.116	0.291
Xe	289.8	120.2	57.9	1.31	2.90	0.132	0.293
N₂	126.1	90.1	33.5	1.33	2.96	0.131	0.292
O₂	154.4	74.4	49.7	1.31	2.69	0.142	0.292
CH₄	190.7	99.0	45.8	1.29	2.96	0.126	0.290

* Reproduced from ref. 10

The data have been reduced using the Lennard–Jones potential

$$u(R) = 4u^* \left[\left(\frac{R^*}{R}\right)^{12} - \left(\frac{R^*}{R}\right)^{6} \right]. \qquad (1.5.13)$$

Note the deviations, due to quantum effects, of the properties of He and H₂ and to a lesser extent of Ne, from the average typical of the last five members of the series.

The way to use this theorem to study solutions was first shown by Longuet-Higgins,[12] whose theory of conformal solutions we discuss after a mathematical interlude.

1.5.2 MATHEMATICAL INTERLUDE. PERTURBATION THEORY[13]

Consider a system for which the configurational partition function is

$$e^{-\beta A_N} = Q = \frac{1}{N!} \int \cdots \int e^{-\beta U_N} d\{N\} \qquad (1.5.14)$$

and assume that the potential energy can be separated as

$$U_N = U_N^{(0)} + U_N^{(1)} \tag{1.5.15}$$

where $U_N^{(0)}$ is the potential energy of the unperturbed system and $U_N^{(1)}$ is the perturbation energy. For the unperturbed system one may write

$$e^{-\beta A_N^{(0)}} = Q^{(0)} = \frac{1}{N!} \int \cdots \int e^{-\beta U_N^{(0)}} d\{N\} \tag{1.5.16}$$

Now, the probability of any particular configuration is

$$\mathscr{P}_0^{(N)} = \frac{e^{-\beta U_N^{(0)}}}{\int \cdots \int e^{-\beta U_N^{(0)}} d\{N\}} \tag{1.5.17}$$

$$= \exp[\beta(A_N^{(0)} - U_N^{(0)})] \tag{1.5.18}$$

and, utilizing the normalization properties of a probability density,

$$\int \cdots \int \mathscr{P}_0^{(N)} d\{N\} = 1. \tag{1.5.19}$$

Operate on Eq. (1.5.14) by multiplying and dividing Q by $Q^{(0)}$ to obtain

$$Q = \frac{1}{N!} \int \cdots \int e^{-\beta U_N} d\{N\} \frac{\int \cdots \int e^{-\beta U_N^{(0)}} d\{N\}}{\int \cdots \int e^{-\beta U_N^{(0)}} d\{N\}} \tag{1.5.20}$$

$$= Q^{(0)} \int \cdots \int \exp[\beta(A_N^{(0)} - U_N^{(0)} - U_N^{(1)})] d\{N\} \tag{1.5.21}$$

$$= Q^{(0)} \int \cdots \int \mathscr{P}_0^{(N)} e^{-\beta U_N^{(1)}} d\{N\} \tag{1.5.22}$$

and therefore

$$e^{-\beta A_N} e^{\beta A_N^0} \equiv e^{-\beta A_N^{(1)}} = \langle e^{-\beta U_N^{(1)}} \rangle_0 \tag{1.5.23}$$

where $A_N^{(1)}$ is the contribution to the free energy from the perturbation and the average value of $e^{-\beta U_N^{(1)}}$ is performed over the unperturbed ensemble,

$$\langle e^{-\beta U_N^{(1)}} \rangle_0 = \int \cdots \int \mathscr{P}_0^{(N)} e^{-\beta U_N^{(1)}} d\{N\} \tag{1.5.24}$$

It is convenient to expand the perturbation free energy as a power series in $\lambda\beta$, where λ is a constant expansion parameter, i.e.,

$$A_N^{(1)} = \sum_{n=1}^{\infty} \frac{\omega_n}{n!} (-\lambda\beta)^{n-1}. \tag{1.5.25}$$

Thus,

$$\exp[-\beta A_N^{(1)}] = \exp\left[\sum_{n=1}^{\infty} \frac{\omega_n}{n!}(-\lambda\beta)^{n-1}\right] \qquad (1.5.26)$$

$$= \langle e^{-\beta V_N^{(1)}\lambda}\rangle_0 \qquad (1.5.27)$$

$$= \sum_{k=0}^{\infty} \frac{(-\lambda\beta)^k}{k!} \langle\alpha^k\rangle_0 \qquad (1.5.28)$$

where λ, the coupling parameter, will eventually be made unity, and

$$\alpha = U_N^{(1)}. \qquad (1.5.29)$$

There is a relationship between the ω_n and the $\langle\alpha^k\rangle_0$. By expanding $e^{-\beta A_N^{(1)}}$,

$$e^{-\beta A_N^{(1)}} = \sum_{k=0}^{\infty} \frac{1}{k!} \left[\sum_{n=1}^{\infty} \frac{\omega_n}{n!}(-\lambda\beta)^n\right]^k \qquad (1.5.30)$$

$$= \sum_{k=0}^{\infty} \frac{1}{k!} \sum_{\substack{n_j \\ \Sigma n_j = k}} k! \prod_{j=1}^{\infty} \frac{1}{n_j!} \left[\frac{(-\lambda\beta)^j \omega_j}{j!}\right]^{n_j} \qquad (1.5.31)$$

where we have used the multinomial theorem

$$(X_1 + X_2 + \cdots + X_k)^n = \sum_{\Sigma n_i = n} \frac{n!}{n_1! n_2! \cdots n_k!} X_1^{n_1} X_2^{n_2} \cdots X_k^{n_k}. \qquad (1.5.32)$$

To reduce the results obtained, collect powers of $\lambda\beta$,

$$e^{-\beta A_N^{(1)}} = \sum_{k=0}^{\infty} \sum_{\substack{n_j \\ \Sigma n_j = k}} (-\lambda\beta)^{\Sigma j n_j} \prod_{j=1}^{\infty} \frac{1}{n_j!} \left(\frac{\omega_j}{j!}\right)^{n_j} \qquad (1.5.33)$$

and continuing

$$e^{-\beta A_N^{(1)}} = \sum_{k=0}^{\infty} \sum_{s=0}^{\infty} (-\lambda\beta)^s \sum_{\substack{n_j \\ \Sigma n_j = k \\ \Sigma j n_j = s}} \prod_{j=1}^{\infty} \frac{1}{n_j!} \left(\frac{\omega_j}{j!}\right)^{n_j} \qquad (1.5.34)$$

$$= \sum_{s=0}^{\infty} (-\lambda\beta)^s \sum_{\substack{n_j \\ \Sigma j n_j = s}} \prod_{j=1}^{\infty} \frac{1}{n_j!} \left(\frac{\omega_j}{j!}\right)^{n_j}. \qquad (1.5.35)$$

To find the relationship between α and ω one need only equate like powers of λ and set $\lambda = 1$

$$\langle\alpha^k\rangle_0 = k! \sum_{\substack{n_j \\ \Sigma j n_j = k}} \prod_{j=1}^{\infty} \frac{1}{n_j!} \left(\frac{\omega_j}{j!}\right)^{n_j}. \qquad (1.5.36)$$

This relation gives for the first few values of k

$$\langle \alpha \rangle_0 = \omega_1$$
$$\langle \alpha^2 \rangle_0 = \omega_2 + \omega_1^2 \tag{1.5.37}$$
$$\vdots$$

and inverting these

$$\omega_1 = \langle \alpha \rangle_0$$
$$\omega_2 = \langle \alpha^2 \rangle_0 - \langle \alpha \rangle_0^2 \tag{1.5.38}$$

the general result being

$$\omega_j = j! \sum_{\substack{n_s \\ \Sigma s n_s = j}} (-)^{\Sigma n_s - 1} (\Sigma n_s - 1)! \prod_{s=1}^{\infty} \frac{1}{n_s!} \left(\frac{\langle \alpha^2 \rangle_0}{s!} \right)^{n_s}. \tag{1.5.39}$$

The Helmholtz free energy is thus of the form

$$A_N = A_N^{(0)} + \omega_1 - \frac{\omega_2}{2kT} \tag{1.5.40}$$

$$\omega_1 = \langle U_N^{(1)} \rangle_0$$
$$\omega_2 = \langle (U_N^{(1)})^2 \rangle_0 - \langle U_N^{(1)} \rangle_0^2. \tag{1.5.41}$$

Let the perturbation potential $U_N^{(1)}$ be pairwise decomposable so that

$$U_N^{(1)} = \frac{1}{2} \sum_{i \neq j} \sum u^{(1)}(i, j) \tag{1.5.42}$$

whereby

$$\omega_1 = \frac{1}{2} \sum_{i \neq j} \sum \langle u^{(1)}(i, j) \rangle_0. \tag{1.5.43}$$

The probability distribution function in the space of two molecules is defined by

$$\mathscr{P}_0^{(2)}(1, 2) = \int \cdots \int \mathscr{P}_0^{(N)} d\{N - 2\} \tag{1.5.44}$$

and if all the molecules are identical

$$\omega_1 = \frac{N(N-1)}{2} \int \int u^{(1)}(1, 2) \mathscr{P}_0^{(2)}(1, 2) d(1) d(2). \tag{1.5.45}$$

Note that

$$\int \int \mathscr{P}_0^{(2)} d(1) d(2) = \int \cdots \int \mathscr{P}_0^{(N)} d\{N\} = 1 \tag{1.5.46}$$
$$\langle u^{(1)}(1, 2) \rangle_0 = \int \cdots \int u^{(1)}(1, 2) \mathscr{P}_0^{(N)} d\{N\}. \tag{1.5.47}$$

To compute ω_2, the square of $U_N^{(1)}$ must be decomposed. To do this write

$$\langle(U_N^{(1)})^2\rangle_0 = \frac{1}{4}\langle\sum_{i\neq j}\sum_{k\neq l} u^{(1)}(i, j)u^{(1)}(k, l)\rangle_0 \tag{1.5.48}$$

$$= \frac{1}{4}\sideset{}{'}\sum_{i,j,k,l}\langle u^{(1)}(i, j)u^{(1)}(k, l)\rangle_0$$

$$+ \sideset{}{'}\sum_{i,j,k}\langle u^{(1)}(i, j)u^{(1)}(j, k)\rangle_0$$

$$+ \frac{1}{2}\sideset{}{'}\sum_{i,j}\langle(u^{(1)}(i, j))^2\rangle_0 \tag{1.5.49}$$

where $\sum'$ means a summation over the indicated indices with no two of the indices allowed to be equal. It is convenient to introduce the integral relations

$$\mathscr{P}_0^{(3)}(1, 2, 3) = \int\cdots\int \mathscr{P}_0^{(N)}d\{N - 3\}$$

$$\mathscr{P}_0^{(4)}(1, 2, 3, 4) = \int\cdots\int \mathscr{P}_0^{(N)}d\{N - 4\} \tag{1.5.50}$$

and once again, if all the molecules are identical,

$$\langle(U_N^{(1)})^2\rangle_0 = \frac{1}{4}\frac{N!}{(N - 4)!}\iiiint \mathscr{P}_0^{(4)}u^{(1)}(1, 2)u^{(1)}(3, 4)d(1)d(2)d(3)d(4)$$

$$+ \frac{N!}{(N - 3)!}\iiint \mathscr{P}_0^{(3)}u^{(1)}(1, 2)u^{(1)}(2, 3)d(1)d(2)d(3) \tag{1.5.51}$$

$$+ \frac{1}{2}\frac{N!}{(N - 2)!}\iint \mathscr{P}_0^{(2)}[u^{(1)}(1, 2)]^2 d(1)d(2)$$

and therefore when N is large

$$\omega_2 = \frac{N^4}{4}\iiiint [\mathscr{P}_0^{(4)}(1, 2, 3, 4) - \mathscr{P}_0^{(2)}(1, 2)\mathscr{P}_0^{(2)}(3, 4)]$$

$$u^{(1)}(1, 2)u^{(1)}(3, 4)d(1)d(2)d(3)d(4)$$

$$+ N^3\iiint\mathscr{P}_0^{(3)}(1, 2, 3)u^{(1)}(1, 2)u^{(1)}(2, 3)d(1)d(2)d(3) \tag{1.5.52}$$

$$+ \frac{N^2}{2}\iint \mathscr{P}_0^{(2)}(1, 2)[u^{(1)}(1, 2)]^2 d(1)d(2).$$

1.5.3 CONFORMAL SOLUTION THEORY

We now turn to a first-order theory of solutions.[13] Consider two similar pure fluids. By similar is meant that the depth of the potential well u^*, and the " size " of the molecule R^* are only slightly different in the two pure fluids. Put

$$f = \frac{u^{*\prime}}{u^*} \qquad \frac{1}{g} = \frac{R^{*\prime}}{R^*}$$

and f and g are *both close to unity*. From Eq. (1.5.8)

$$Q = (R^*)^{3N} \left\{ q^* \left(\frac{kT}{u^*}, \frac{v}{(R^*)^3} \right) \right\}^N \tag{1.5.53}$$

and by the theorem of corresponding states

$$Q' = (R^{*\prime})^{3N} \left\{ q^* \left(\frac{kT}{u^{*\prime}}, \frac{v}{(R^{*\prime})^3} \right) \right\}^N \tag{1.5.54}$$

$$= g^{-3N} \left\{ q^* \left(\frac{T}{f}, g^3 v \right) \right\}^N. \tag{1.5.55}$$

From the relation

$$A = -kT \ln Q \tag{1.5.56}$$

it is found that

$$A' - A = -kT[\ln Q' - \ln Q]. \tag{1.5.57}$$

Since the parameters f and g are close to unity, Q' may be expanded about Q in a Taylor series and all but the leading terms dropped,

$$\ln Q'_1 = \ln Q(T, v) - 3N \ln g + (f - 1)T \frac{\partial \ln Q}{\partial T} + 3(g - 1)v \frac{\partial \ln Q}{\partial v}. \tag{1.5.58}$$

The substitution of Eq. (1.5.58) into Eq. (1.5.57) gives

$$A' - A = -kT \left[-3N \ln g - (f - 1) \frac{U}{kT} + 3(g - 1) \frac{pv}{kT} \right]$$
$$= (f - 1)U + 3(RT - pv)(g - 1). \tag{1.5.59}$$

With the use of the theorem of corresponding states, Eq. (1.5.59) shows how the free energies of the two fluids are related to the perturbation parameters f and g and to the thermodynamic properties of one of the fluids, the reference fluid. On the other hand, to first order, the difference in free energies of the two fluids is simply

$$A' - A = \frac{\int \cdots \int (U'_N - U_N) e^{-\beta U_N} d\{N\}}{\int \cdots \int e^{-\beta U_N} d\{N\}} \tag{1.5.60}$$

as shown in Eq. (1.5.40). From the intermolecular potential

$$u(R) = u^* \phi\left(\frac{R}{R^*}\right) \tag{1.5.61}$$

it is easily shown that

$$U_N = \sum_{i>j} u^* \phi\left(\frac{R_{ij}}{R^*}\right)$$

$$U_N' = \sum_{i>j} f u^* \phi\left(g \frac{R_{ij}}{R^*}\right) = \sum_{i>j} f u(g R_{ij}). \tag{1.5.62}$$

Direct substitution now leads to

$$A' - A = \frac{\int \cdots \int \sum_{i>j} [f u(g R_{ij}) - u(R_{ij})] e^{-\beta U_N} d\{N\}}{\int \cdots \int e^{-\beta U_N} d\{N\}}$$

$$= I(f, g) \tag{1.5.63}$$

where $I(f, g)$ is also given by Eq. (1.5.59).

To extend the formalism to mixtures it is sufficient to assume that the pair potential between a molecule of species r and a molecule of species s is

$$u_{rs}(R) = f_{rs} u(g_{rs} R) \tag{1.5.64}$$

where $u(R)$ is the mutual potential energy of two molecules of some reference species 0, and f_{rs}, g_{rs} are constants depending only on the chemical nature of r and s. By definition, for the reference component,

$$f_{00} = g_{00} = 1. \tag{1.5.65}$$

As before, it is assumed that f_{rs} and g_{rs} are close to unity. Longuet-Higgins also makes the less important approximation of assuming the additivity of molecular radii, i.e.,

$$g_{rs} = \tfrac{1}{2}(g_{rr} + g_{ss}) \tag{1.5.66}$$

When

$$f_{rs} = 1; \qquad g_{rs} = 1 \tag{1.5.67}$$

the solution is a perfect mixture. The potential energy is then

$$U_N = \sum_{i>j} u(R_{ij}). \tag{1.5.68}$$

Consider some one pair of molecules which contributes to U_N the energy $u(R_{ij})$. In the first-order theory, the perturbation is averaged over the unperturbed assembly. In the unperturbed assembly the mode of occupation of the volume elements $d^3 R_i$ and $d^3 R_j$ about R_i and R_j is of no consequence since f_{rs} and g_{rs} are both unity. Thus, since the perturbation energy is averaged over the unperturbed assembly, all modes of occupation of d^3R_i and d^3R_j give the same result as for random mixing. The probability of finding in d^3R_i a molecule of species r at the same time as a molecule of species s in d^3R_j is then easily seen to be just $x_r x_s$ where the x's are the mole fractions of com-molecule of species r at the same time as a molecule of species s in d^3R_j is then easily seen to be just $x_r x_s$ where the x's are the mole fractions of components r and s. The total potential energy in the perturbed solution is then

$$U'_N = \sum_{i>j} \sum_r \sum_s x_r x_s f_{rs}\, u(g_{rs}R_{ij}) \tag{1.5.69}$$

and by substitution one finds

$$A' - A = \sum_r \sum_s x_r x_s \frac{\int \sum_{i>j} [f_{rs}u(g_{rs}R_{ij}) - u(R_{ij})]\, e^{-\beta U_N}\, d\{N\}}{\int e^{-\beta U_N}\, d\{N\}} \tag{1.5.70}$$

$$= \sum_r \sum_s x_r x_s I(f_{rs}, g_{rs}). \tag{1.5.71}$$

The integral has been evaluated previously [Eqs. (1.5.59), (1.5.63)] so that by substitution

$$A' - A = \sum_r \sum_s x_r x_s [U(f_{rs} - 1) + 3(RT - pv)(g_{rs} - 1)] \tag{1.5.72}$$

which again relates the difference of free energy to the perturbation parameters. Since the assumed change in free energy is small, one may write

$$(dA)_{n_j} = -s\, dT - p\, dv + \sum_r \sum_s x_r x_s [U\, df_{rs} - 3(NkT - pv)\, dg_{rs}] \tag{1.5.73}$$

and noting that

$$G = A + pv \tag{1.5.74}$$

leads to the relation

$$(dG)_{n_j} = -s\, dT + v\, dp + \sum_r \sum_s x_r x_s [U\, df_{rs} + 3(NkT - pv)\, dg_{rs}] \tag{1.5.75}$$

or, in integral form

$$G' - G = \sum_r \sum_s x_r x_s [U(f_{rs} - 1) + 3(NkT - pv)(g_{rs} - 1)]. \tag{1.5.76}$$

To compute the excess thermodynamic functions it is convenient to convert to molar units. Then

$$g' - g = \sum_r \sum_s x_r x_s [u_{\mathrm{conf}}(f_{rs} - 1) + 3(RT - pv_m)(g_{rs} - 1)] \tag{1.5.77}$$

and for a single component r

$$g_r - g_r^0 = u_{\mathrm{conf}}^0(f_{rr} - 1) + 3(RT - pv_0)(g_{rr} - 1) \tag{1.5.78}$$

which gives for the excess Gibbs free energy per mole

$$g^E = \sum_r \sum_s [u_{conf}^0(2f_{rs} - f_{rr} - f_{ss}) + 3(RT - pv^0)(2g_{rs} - g_{rr} - g_{ss})].$$

(1.5.79)

If the approximation expressed in Eq. (1.5.66) is valid,

$$g^E = u_{conf}^0 \sum_r \sum_s x_r x_s\, d_{rs}$$

$$d_{rs} = 2f_{rs} - f_{rr} - f_{ss}$$

(1.5.80)

and by standard methods it is immediately found that

$$s^E = -\frac{\partial u_{conf}^0}{\partial T} \sum_r \sum_s x_r x_s\, d_{rs}$$

(1.5.81)

$$h^E = \left[u_{conf}^0 - T\frac{\partial u_{conf}^0}{\partial T} \right] \sum_r \sum_s x_r x_s\, d_{rs}$$

(1.5.82)

$$v^E = \frac{\partial u_{conf}^0}{\partial p} \sum_r \sum_s x_r x_s\, d_{rs}.$$

(1.5.83)

Using the thermodynamic relation

$$\frac{\partial U}{\partial p} = -TV\alpha + pV\kappa$$

$$\alpha = \frac{1}{V}\left(\frac{\partial V}{\partial T}\right)_p; \quad \kappa = -\frac{1}{V}\left(\frac{\partial V}{\partial p}\right)_T$$

it is possible to write

$$v^E = v^0(p\kappa^0 - \alpha^0 T) \sum_{r<s} x_r x_s\, d_{rs}.$$

(1.5.84)

Now u_{conf}^0 is just the difference in energy of the reference component in the condensed phase, and its value as an ideal gas at the same temperature. This is just $-L + RT$, where L is the latent heat (extrapolated to zero pressure). Thus conformal solution theory predicts

$$g^E : s^E : h^E : v^E = (RT - L) : \left(\frac{\partial L}{\partial T} - R\right) : \left(T\frac{\partial L}{\partial T} - L\right) : v^0(p\kappa^0 - T\alpha^0).$$

(1.5.85)

Under ordinary circumstances of temperature and pressure, all the terms on the right have the same sign (negative), and so either all the excess functions are positive, or they are all negative. This is not the case experimentally. For example,[14] the system $CCl_4 - C(CH_3)_4$, which should be quite nearly spherical, has $h^E = 75$ cal/mole, but $v^E/v_{ideal} = -0.5\%$. The theoretical prediction of opposite signs of excess quantities must be looked for in the higher order terms.

Longuet-Higgins was able to fit the experimental data on excess quantities for several binary solutions by equations of the form (1.5.80) — (1.5.85). However, d_{12} was taken as an arbitrary parameter to fit the data. There is some indication [15] that the best d_{12} for fitting the data is not at all concordant with the way d_{12} should be calculated with the theory, Eq. (1.5.80). The quantitative success of conformal solution theory thus seems somewhat fortuitous.

This completes our discussion of conformal solutions. There are other perturbation theories based on the principle of corresponding states.[16] The theoretical basis for these theories can be understood on the considerations of work by J. E. Mayer and L. H. Nosanow.[17] The rest of this section will be devoted to the Mayer–Nosanow theory with, at the end, some indication of how the more detailed results, to which reference is made below, were found.

1.5.4 THE MAYER–NOSANOW THEORY

The Mayer–Nosanow theory is formally independent of any principle of corresponding states. However, as we shall see, some sort of principle must be introduced, in the last analysis, in order to render computations practical. The theory rests on the following assumptions;

a, classical statistics are applicable;

b, the internal partition function is independent of environment;

c, the potential energy of a system may be expressed as the sum of pair potentials.

These assumptions are also part of the earlier theories discussed herein.

For simplicity, our considerations will be restricted to a binary system of molecules of type a and b. There are N_a molecules of type a and N_b of type b. The configurational energy (potential energy) of the system is then

$$U = U(\{N_a\}, \{N_b\}) = \sum_{1 \leqslant i < j \leqslant N_a} u_{aa}(R_{ij})$$

$$+ \sum_{i=1}^{N_a} \sum_{j=1}^{N_b} u_{ab}(R_{ij}) + \sum_{1 \leqslant i < j \leqslant N_b} u_{bb}(R_{ij}) \quad (1.5.86)$$

where $u_{\alpha\beta}(R_{ij})$ is the potential energy of a pair of molecules of type α and β respectively, situated a distance R_{ij} apart. α, β can take on the values a or b.

It is convenient (and is the main trick of the theory) to introduce a fictitious *reference system* characterized by $N(= N_a + N_b)$ identical molecules with a mutual potential energy $u_0(R_{ij})$.

$$U_0 = U_0(\{N\}) = \sum_{1 \leqslant i < j \leqslant N} u_0(R_{ij}) \quad (1.5.87)$$

and a sequence of fictitious systems bridging the gap between the reference system and the actual system. The potential energy of the fictitious system is

$$U_f = U_f(\{N_a\}, \{N_b\}, \lambda) = U_0 + \lambda \, \Delta U$$

$$\Delta U = U - U_0. \qquad 0 \leqslant \lambda \leqslant 1. \tag{1.5.88}$$

Clearly $U_f(\lambda = 0) = U_0; U_f(\lambda = 1) = U$.

Before demonstrating the utility of these definitions it is recommended that the reader review the properties of the Grand Canonical Ensemble and the distribution functions which we developed in earlier sections of this chapter.

The point of introducing the fictitious systems is that one may expand any thermodynamic function of a system characterized by a value of λ in a Taylor's series about $\lambda = 0$. For example, for the Gibbs function

$$G(\lambda) = \sum_{n=0}^{\infty} \frac{G^{(n)}(0)}{n!} \lambda^n$$

$$G^{(n)}(0) = \left(\frac{\partial^n G}{\partial \lambda^n}\right)_{T,p,N_a,N_b,\lambda=0}. \tag{1.5.89}$$

Therefore, the Gibbs function for the actual solution may be written

$$G = \sum_{n=0}^{\infty} \frac{1}{n!} G^{(n)}(0). \tag{1.5.90}$$

The term $(1/n!)G^{(n)}(0)$ will be called the nth order term in the expansion.

It is fairly clear that the expansion is essentially one in terms of ΔU, and so is fairly closely related to conformal solution theory. Note that the idea of corresponding states has not yet been introduced.

The zeroth-order term of (1.5.90) is easy to compute. Since all pair potentials are the same, the reference system is an ideal solution, with labeled molecules. If the Gibbs function for a pure system with unlabeled molecules is G^0, we have

$$G^{(0)}(0) = G^0 + NkT \sum_{\alpha} x_\alpha \ln x_\alpha. \tag{1.5.91}$$

It is convenient to subtract out the trivial ideal free energy of mixing and define

$$G^M = G - NkT \sum_{\alpha} x_\alpha \ln x_\alpha. \tag{1.5.92}$$

The excess Gibbs function is then

$$G^E = G^M - \sum_{\alpha} x_\alpha g^\alpha \tag{1.5.93}$$

where g^α is the Gibbs function per mole for pure α.

Thus, (1.5.90) becomes

$$G^M = G^0 + \sum_{n=1}^{\infty} \frac{1}{n!} G^{(n)}(0). \tag{1.5.94}$$

Now, the grand potential function, $-pV$, also depends on λ, in addition to T, V, z_a, and z_b. This fact can be used to compute the higher terms in (1.5.94).

To calculate $G^{(1)}(0)$ it is convenient to note that

$$\left(\frac{\partial G}{\partial \lambda}\right)_{T,p,N_a,N_b} = -\left(\frac{\partial (pV)}{\partial \lambda}\right)_{T,z,V} \tag{1.5.95}$$

with similar relations for the other thermodynamic functions. To prove (1.5.95) note that

$$-d(pV) = -p\,dV - S\,dT - \sum_\alpha N_\alpha\,d\mu_\alpha + \left(\frac{\partial (pV)}{\partial \lambda}\right)_{T,V,z} d\lambda.$$

But

$$dG = \sum N_\alpha\,d\mu_\alpha + \sum \mu_\alpha\,dN_\alpha$$

$$= -S\,dT + V\,dp + \sum \mu_\alpha\,dN_\alpha + \left(\frac{\partial G}{\partial \lambda}\right)_{T,p,z} d\lambda$$

and thereby

$$-d(pV) = -p\,dV - S\,dT + \sum \mu_\alpha\,dN_\alpha - dG + \left(\frac{\partial (-pV)}{\partial \lambda}\right)_{T,V,z} d\lambda.$$

Substituting the second expression for dG

$$-d(pV) = -p\,dV + V\,dp - \left(\left(\frac{\partial G}{\partial \lambda}\right)_{T,p,N} - \left(\frac{\partial (-pV)}{\partial \lambda}\right)_{T,V,z}\right) d\lambda$$

from which (1.5.95) follows.

From the expression for the grand partition function, whose logarithm multiplied by $-kT$ is pV, one sees immediately that

$$G^{(1)}(0) = \langle \Delta U \rangle_0 \tag{1.5.96}$$

where $\langle \ \rangle_0$ indicates an average with respect to the reference system ($\lambda = 0$). Noting that integration over each $U_{\alpha\beta}$ gives the same result, (1.5.96) may be written

$$G^{(1)}(0) = \langle \tfrac{1}{2}N_a(N_a - 1)\,\Delta U_{aa}(r_{12}) - N_a N_b\,\Delta U_{ab}(r_{12})$$

$$+ \tfrac{1}{2}N_b(N_b - 1)\,\Delta U_{bb}(r_{12})\rangle_0.$$

In terms of the distribution function, $g_2^{(0)}$ (R_{12}), where the [0] refers to the reference solution

$$G^{(1)}(0) = \frac{\rho^{02}}{2} \int_r \sum_{\alpha,\beta} x_\alpha x_\beta \, \Delta U_{\alpha\beta}(R_{12}) g_2^{(0)}(R_{12}) \, d\{2\} \qquad (1.5.97)$$

ρ^0 is the density of the reference liquid at the same T, V, z as the solution.

At this stage there is a pertinent observation to be made. It is, simply, that $G^{(1)}$ (0) is almost a symmetrical function of x_a and x_b. We say almost because the coefficients of $x_a{}^2$ and $x_b{}^2$ will not, in general, be the same.

Returning to the development of G, it is apparent that the next step is the computation of $G^{(2)}(0)$. Mayer and Nosanow have done this, and the resulting formulas are extremely complex. The higher order terms, $G^{(n)}(0)$, $n \geqslant 3$, have not been computed, but are presumably even more complex. We shall here merely indicate the manner in which $G^{(2)}(0)$ is calculated, and then state the result.

By (1.5.95)

$$G^{(2)}(0) = \left(\frac{\partial^2 G}{\partial \lambda^2}\right)_{T,p,N}\bigg|_{\lambda=0} = \left(\frac{\partial \phi_\lambda}{\partial \lambda}\right)_{T,p,N}\bigg|_{\lambda=0} \qquad (1.5.98)$$

$$\phi_\lambda = \left(\frac{\partial(-pV)}{\partial \lambda}\right)_{T,V,z} ; \quad \phi = -pV.$$

Now

$$\left(\frac{\partial \phi_\lambda}{\partial \lambda}\right)_{T,p,N}\bigg|_{\lambda=0} = \phi_{\lambda\lambda}^0 + \phi_{\lambda,V}^0 V_\lambda^0 + \sum_\alpha \phi_{\lambda\alpha}. \qquad (1.5.98')$$

where $V_\lambda = (\partial V/\partial \lambda)_{T,P,N}$ and $\zeta_\alpha = (\partial \ln z_\alpha/\partial \lambda)_{T,P,N}$, and the (0) denotes evaluation at $\lambda = 0$. The subscripts on the ϕ are self-explanatory. $[\phi_{\lambda\alpha} = (\partial^2 \phi/\partial \lambda \, \partial \zeta_\alpha).]$

The $\phi_{\mu\nu}$ functions can be computed directly from the grand partition function, since they are derivatives with respect to the " natural variables " of that function. In accordance with the policy of merely sketching the method, the answers will not be recorded here.

To calculate V_λ^0, ζ_a^0, ζ_b^0, use the facts that $p = -\phi_V^0$, $-kT N_\alpha = \phi_\alpha^0$, and that $(\partial p/\partial \lambda)_{T,P,N} = (\partial N_\alpha/\partial \lambda)_{T,P,N}, = 0$. These lead to the equations

$$-\left(\frac{\partial p}{\partial \lambda}\right)_{T,p,N} = 0 = \phi_{Va}^0 \zeta + \phi_{Vb}^0 \zeta_b^0 + \phi_{V\lambda}^0$$

$$-\frac{kT}{V^0}\left(\frac{\partial N_\alpha}{\partial \lambda}\right)_{T,p,N} = 0 = \phi_{aV}^0 \frac{V_\lambda^0}{V^0} + \frac{\phi_{aa}^0}{V^0}\zeta_a^0 + \frac{\phi_{ab}^0}{V^0}\zeta_b^0 + \frac{\phi_{a\lambda}^0}{V_0} \qquad (1.5.99)$$

$$-\frac{kT}{V^0}\left(\frac{\partial N_b}{\partial \lambda}\right)_{T,p,N} = 0 = \phi_{bV}^0 \frac{V_\lambda^0}{V^0} + \frac{\phi_{ba}^0}{V^0}\zeta_a^0 + \frac{\phi_{bb}^0}{V^0}\zeta_b^0 + \frac{\phi_{b\lambda}^0}{V^0}.$$

Again, the $\phi_{\mu\nu}$ coefficients may be found from the grand partition function, and Eqs. (1.5.99) may be solved for $V_\lambda, \zeta_a, \zeta_b$.

The final result for $G^{(2)}(0)$, may be written

$$G^{(2)}(0) = -\frac{\rho^{02}}{2kT} \int \sum_{\alpha,\beta} x_\alpha x_\beta (\Delta U_{\alpha\beta})^2 g_2^{(0)}(R_{12})\, d\{2\}$$

$$-\frac{\rho^{03}}{kT} \int \sum_{\alpha,\beta,\gamma} x_\alpha\, x_\beta x_\gamma\, \Delta U_{\alpha\beta}\, \Delta U_{\beta\gamma} g_3^{(0)}\, d\{3\}$$

$$-\frac{\rho^{04}}{4kT} \int \sum_{\alpha,\beta,\gamma,\delta} x_\alpha x_\beta x_\gamma x_\delta\, \Delta U_{\alpha\beta}(R_{12})\, \Delta U_{\gamma\delta}(R_{34})[g_4^0 - g_3^0(R_{12})g_2^0(R_{34})]\, d\{4\}$$

$$-\frac{\kappa^0\rho^{04}}{4V} \int \sum_{\alpha,\beta,\gamma,\delta} x_\alpha x_\beta x_\gamma x_\delta\, \Delta U_{\alpha\beta}(R_{12})\, \Delta U_{\gamma\delta}(R_{34}) g_2^0(R_{12})g_2^0(R_{34})\, d\{4\}$$

$$+\frac{\rho^{04}}{2V^0kT} \int \sum_{\alpha,\beta,\gamma,\delta} x_\alpha x_\beta x_\gamma x_\delta\, \Delta U_{\alpha\beta}(R_{12})\, \Delta U_{\gamma\delta}(R_{34}) g_2^0(R_{12}) \times$$

$$[g_3^0(R_3 R_4 R_5) - g_2^0(R_{34})]\, d\{5\} \quad (1.5.100)$$

$$+\frac{\rho^{03}}{V^0kT} \int \sum_{\alpha,\beta,\gamma,\delta} x_\alpha x_\beta x_\gamma x_\delta\, \Delta U_{\alpha\beta}(R_{12})\, \Delta U_{\gamma\delta}(R_{34}) g_2^0(R_{12})g_2^0(R_{34})\, d\{4\}$$

$$+\frac{\rho^{03}x_a x_b}{V^0kT} \int \sum_{\alpha,\beta} x_\alpha x_\beta \vartheta_\alpha(R_{12})\vartheta_\beta(R_{34}) g_2^0(R_{12})g_2^0(R_{34})\, d\{4\}$$

$$\vartheta_\alpha(R) = U_{\alpha a}(R) - U_{\alpha b}(R)$$

$$\kappa = -\frac{1}{V}\left(\frac{\partial V}{\partial p}\right)_T.$$

It is apparent that it would be exceedingly difficult to compute these terms.

One can now, if one wishes, expand the properties of the pure components a and b about those of the reference solution. It may not be immediately obvious how to do this because the reference solution is binary, although ideal, and the technique used above is, on the surface, applicable to expansion of an n-component system about an n-component reference system. The procedure is valid, however, and this may be shown as follows.

As a direct deduction from the grand partition function it is easy to see that

$$\frac{N_a}{z_a} = \frac{N_b}{z_b} \tag{1.5.101}$$

for an ideal binary solution. Since $z_\alpha = \exp(\mu_\alpha - \mu_\alpha^0)/kT$, and, since the system is ideal, $\mu_\alpha = \mu_{\alpha 0}(T, p) + kT \ln x_\alpha$, where $\mu_{\alpha 0}$ is the chemical potential of pure α at the T and p *of the solution*, μ_α^0 is the the chemical potential of α in the ideal gas state. It follows from (1.5.101) that

$$\mu_{a0} - \mu_a^0 = \mu_{b0} - \mu_b^0. \tag{1.5.102}$$

Now, all the distribution functions involve sums of the form

$$\sum_{n_a n_b} \frac{z_a^{n_a} z_b^{n_b}}{n_a! \, n_b!} Q_{n_a + n_b}$$

where $Q_{n_a + n_b}$ is independent of n_a and n_b separately, but only depends on the sum $n_a + n_b$. By (1.5.102) the sum can be written

$$\sum_{n_a, n_b} \frac{x_a^{n_a} x_b^{n_b}}{n_a! \, n_b!} (z_a^0)^{n_a + n_b} Q_{n_a + n_b} \tag{1.5.103}$$

where x_a is the mole fraction of a, z_a is the fugacity of pure a at the T and p *of the solution*. By a simple change of notation (1.5.103) can be written

$$\sum_{N=0}^{\infty} \sum_{n_a=0}^{N} N! \, \frac{x_a^{n_a} x_b^{N-n_a}}{n_a!(n-n_a)!} \frac{(z_a^0)^N}{N!} Q_N. \tag{1.5.104}$$

But

$$\sum_{n_a=0}^{N} \frac{N!}{n_a!(N-n_a)!} x_a^{n_a} x_b^{N-n_a} = (x_a + x_b)^N = 1$$

by the binomial theorem. Therefore, the sum is

$$\sum_{N=0}^{\infty} \frac{(z_a^0)^N}{N!} Q_N$$

which is just the proper sum for a one-component system *at the same temperature and pressure as the ideal binary mixture.*

In the same manner it can easily be seen that the number density of the single component system (with proper T, p) is the same as the number density of the ideal binary mixture. Incidentally, although the proof has been expressed in terms of z_a^0, it is clear that z_b^0 would have done just as well.

The expansions for the pure components about the reference liquid should now be obvious [they are of the same form as (1.5.97) and (1.5.100)] and are not written herein. By forming the appropriate difference, excess quantities may be computed (free energies directly, and other quantities by differentiation).

This completes the discussion of general perturbation theories. The discussion of the physical significance and utility of the results obtained is reserved for a later section.

1.6 The Cell Theory of Mixtures

Preceding sections have dealt with exact theories of mixtures or with attempts to construct perturbation schemes independent of any heuristic model. In this section is discussed a particular model of the liquid state, the cell model. It is not out of place to briefly review the current situation with a view to justifying the utility of crude model calculations.

Since the pioneering work of Guggenheim et al.[18] there have been numerous attempts to construct molecular theories of solutions. The various proposals may be broadly divided into two categories: exact theories based only on the general principles of statistical mechanics discussed in Sections 1.1.–1.4, and approximate theories based on particular models. Aside from the brilliant work of McMillan and Mayer and of Kirkwood and Buff, whose theories in their quantum-mechanical form apply to any type of solution and are formally exact assuming only that the forces between the molecules in solution are known, almost all theories of solution fall into the latter category. Rather than attempting to calculate the absolute free energy, attention has been focused on the simpler problem of calculating the properties of a given mixture from the known properties of the pure components. As seen in Section 1.5, when the deviations from ideal behavior are small, it is possible to treat them as perturbations and, accordingly, to develop a first-order theory of solutions.| These calculations, first made by Longuet-Higgins, make no appeal to any model and are equally applicable to liquids and gases. Though there are many instances when the deviations from ideality are not properly described by first-order relations, the theory remains a valuable guide to which all approximate theories must be reducible if they are to be thermodynamically consistent.

Of the approximate calculations, the best known is the lattice model of Guggenheim.[18] The principal assumptions involved in the calculation are that: (1) the liquid mixture can be represented by a lattice with one molecule per lattice point (or cell); (2) there is no change of volume on mixing; and (3) the partition function may be written

$$Q_N = Q_{internal} \, Q_{vibration} \, Q_{lattice}$$

where $Q_{internal}$ is the partition function for the internal degrees of freedom of the molecule, $Q_{vibration}$ is the partition function that describes the motion of the center of mass of the molecule, and $Q_{lattice}$ is the partition function for the distribution of the molecules over the assumed lattice. It is then assumed that the only changes occurring on mixing are due to changes in the lattice partition function, the internal and vibrational partition functions remaining constant. The theory then predicts that, when there is random mixing, the deviations from ideality are caused by an excess energy of mixing. Inclusion of the effect of nonrandom mixing always leads to an excess entropy that is negative, unless the difference in energy between an AB pair and the $\frac{1}{2}AA$ and $\frac{1}{2}BB$ pairs from which it was made is temperature-dependent. This latter prediction is not in accord with experiment, since there are many instances in which the excess entropy is positive.

The breakdown of the lattice model may be traced to 3 factors corresponding to the approximations listed above. These are (1) the assumption of a regular lattice (applicable only to crystals); (2) the neglect of the change of volume on mixing; and (3) the assumption that the vibrational partition function is constant. These latter are very poor approximations since, in fact, there are volume changes on mixing that lead to changes in density, resulting then in a change in the curvature of the intermolecular potential and, hence, of the vibrational frequency.

The theory that has been most successful in qualitatively correlating the properties of mixtures is an approximate theory based on the cell model of solutions, developed by Prigogine and Mathot as an extension of the Lennard–Jones and Devonshire model of liquids.[14] The cell theory retains only assumption (1) of the preceding paragraph. Rice[19] has shown how to extend the cell model for solutions. A cell-like theory of solutions can be developed that yields, as the first of a series of successive approximations, the Prigogine and Mathot cell model for solutions, and that also converges to the correct configurational probability density. The technique used was first developed by De Boer[20] in a study of the properties of pure liquids.

1.6.1 THE CELL-CLUSTER THEORY[20]

In this section shall be discussed the Lennard–Jones–Prigogine cell model for solutions in a manner designed to emphasize the approximations inherent in the treatment. Consider a system composed of N_1 molecules of species $1, \cdots, N_r$ molecules of species r, with the total number of molecules given by

$$\sum_{s=1}^{r} N_s = N. \tag{1.6.1}$$

The partition function for the system may be written

$$Q_{\mathbf{N}} = \prod_{s=1}^{r} \frac{\lambda_s^{-3N_s}}{N_s!} \int_V \exp(-\beta U_{\mathbf{N}}) \, d\{\mathbf{N}\} \tag{1.6.2}$$

with

$$\lambda_s^2 = \frac{h^2}{2\pi m_s kT}; \qquad \beta = \frac{1}{kT} \tag{1.6.3}$$

and $U_{\mathbf{N}}$ is the potential of intermolecular force. Imagine that the volume V is spanned by a virtual lattice of N cells. Consider a given, arbitrary, distribution of the molecules over the lattice. If all the molecules were at lattice sites, the potential energy would be

$$^cU_N^0 = \frac{1}{2} \sum_{l_i}^N \sum_{l_j}^N (u_{l_il_j}^{st})^0 \tag{1.6.4}$$

with $(u_{l_il_j}^{st})^0$ the pair potential between molecules of species s and t at the sites l_i and l_j respectively, and where the superscript 0 indicates that the molecules are at lattice sites. $^cU_N^0$ is a function of the distribution of the molecules on the lattice. In general, quantities that depend on the distribution of the molecules on the lattice are so indicated by the superscript c to the left of the symbol. If each molecule is constrained to move only within the cell formed by the repulsive potentials of the surrounding molecules, then the potential energy due to a small deviation $\mathbf{r}_1$ from the lattice point is

$$^c\psi_{l_1}^{(s)}(\mathbf{r}_1) = {}^c\chi_{l_1}^{(s)}(\mathbf{r}_1) - {}^c\chi_{l_1}^{0(s)} = \sum_{l_i \neq l_1} u_{l_il_1}^{st}(\mathbf{q}_{l_1l_i}) - \sum_{l_i \neq l_1} (u_{l_il_1}^{st})^0 \tag{1.6.5}$$

with the $\mathbf{q}_{l_1l_i}$ being the vector distances between the molecules in cells l_1 and l_i, and where

$$^c\chi_{l_1}^{0(s)} = \sum_{l_i \neq l_1} (u_{l_il_1}^{st})^0 \tag{1.6.6}$$

is the potential energy when molecule 1 is at the center of cell l_1. The summation is to be carried out over all cells surrounding the cell l_1. For the purpose of calculating the molecular force field, all the neighboring molecules are supposed to be at the centers of their cells. The potential energy for the given distribution of the molecules on the lattice is thus

$$^cU_N(\mathbf{r}_1 \cdots \mathbf{r}_N) = {}^cU_N^0 + \sum_{l_i} {}^c\psi_{l_i}^{(s)}(\mathbf{r}_i). \tag{1.6.7}$$

The probability density of finding the molecule in cell l_1, which is of species s, displaced a distance $\mathbf{r}_1$ from the cell center is

$$^cW_{l_1}^{(s)}(\mathbf{r}_1) = \exp\{-\beta {}^c\psi_{l_1}^{(s)}(\mathbf{r}_1)\} \tag{1.6.8}$$

where the probability density $^cW_{l_1}^{(s)}(\mathbf{r}_1)$ is normalized such that $^cW_{l_1}^{(s)}(0)$ is unity. The normalization of $^cW_{l_1}^{(s)}(\mathbf{r}_1)$ is a consequence of the fact that the potential energy $\{^c\psi_{l_1}^{(s)}(\mathbf{r}_1)\}$ is normalized such that it is zero when the molecule is at the lattice site. Since the probability density defined in Eq. (1.6.8) only partially describes the system of N molecules, we shall refer to it as a partial probability density.

Perhaps the major physical feature of the Lennard–Jones model of a liquid is recognition of the fact that a crystal is much more nearly like a liquid than is an ideal gas. Therefore, the use of a crystal as a reference state, instead of an ideal gas, enhances the utility of any limited series of correction terms designed to describe the liquid in terms of the reference state. We may introduce the lattice as a reference state very simply. If, for each distribution of the molecules over the lattice, the potential energy is written in forms of a deviation from the lattice energy corresponding to the distribution, then the partition function may be written

$$Q_{\mathbf{N}} = \left(\prod_{s=1}^{r} \frac{\lambda_s^{-3N_s}}{N_s!} \right) \exp(-\beta^c U_{\mathbf{N}}^0) \times$$

$$\int^V \exp[-\beta\{U_{\mathbf{N}}(\mathbf{q}_1 \cdots \mathbf{q}_N) - {}^c U_{\mathbf{N}}^0\}] \, d\{\mathbf{N}\} \tag{1.6.9}$$

In Eq. (1.6.9), the intermolecular potential energy $U_{\mathbf{N}}$ is written as a function of the coordinates of all the molecules, $\mathbf{q}_1, \cdots, \mathbf{q}_N$. Since ${}^c U_{\mathbf{N}}^0$ is a constant determined by the arbitrarily chosen reference lattice and is independent of the coordinates of the molecules, Eq. (1.6.9) holds for any and all crystalline reference states. If, however, the motion of a molecule is restricted to such an extent that it must remain within the cell formed by its nearest neighbors, then the coordinates $\mathbf{q}_1, \cdots, \mathbf{q}_N$ may be replaced with the coordinates $\mathbf{r}_1, \cdots, \mathbf{r}_N$, and the partition function is obtained by summing over all possible distributions of the molecules over the lattice. Thus

$$Q_{\mathbf{N}} = \sum_c \left(\prod_{s=1}^{r} \frac{\lambda_s^{-3N_s}}{N_s!} \right) \exp(-\beta^c U_{\mathbf{N}}^0) \times$$

$$\int^\Delta \exp[-\beta\{^c U_{\mathbf{N}}(\mathbf{r}_1 \cdots \mathbf{r}_N) - {}^c U_{\mathbf{N}}^0\}] \, d\{\mathbf{N}\} \tag{1.6.10}$$

with Δ the volume of a cell. If the molecules execute independent motions within their separate cells, the probability density of finding the molecules $1, \cdots, N$ displaced from their lattice points by the distances $\mathbf{r}, \cdots, \mathbf{r}_N$ is

$${}^c W(\mathbf{r}_1 \cdots \mathbf{r}_N) = {}^c W_{l_1}^{(s)}(\mathbf{r}_1) \cdots {}^c W_{l_N}^{(s)}(\mathbf{r}_N) \tag{1.6.11}$$

leading to the relation

$$Q_{\mathbf{N}} = \sum_c \prod_s \frac{\lambda_s^{-3N_s}}{N_s!} {}^c J_s^{N_s} \exp(-\beta^c U_{\mathbf{N}}^0) \tag{1.6.12}$$

with

$${}^c J_s = \int^\Delta \exp\{-\beta^c \psi_{l_i}^{(s)}(\mathbf{r}_i)\} \, d(i). \tag{1.6.13}$$

To reduce Eq. (1.6.12) to the result obtained by Prigogine and Mathot it is necessary to assume that there is random mixing. In that case, the sum over all distributions is replaced by the factor $N!$. Accordingly, the principal assumptions of the usual cell theory of solutions are: (1) the liquid mixture can be divided into a lattice of cells, each containing only 1 molecule; (2) there is random mixing of all components; and (3) the molecules execute independent motions within their respective cells.

The first approximation neglects density fluctuations on the microscopic level. Pople and Prigogine and Jannsens [21] have developed approximate methods to account for multiple cell occupancy in pure fluids and Pople[22] has studied the problem of nonrandom mixing. Both of these problems will be discussed in a later part of this section. The last approximation is responsible for the Lennard–Jones model of a liquid being a " one-particle " model with insufficient correlation in motion between neighboring cells. The chief contribution of the cell cluster theory is the removal of approximation 3.

Before proceeding, it is convenient to separate the partition function by the relation

$$Q_N = Q_N^{(random)} \left(\frac{Q_N}{Q_N^{(random)}} \right)$$

where $Q_N^{(random)}$ is the partition function for a system that has a random distribution of the molecules over the assumed lattice: that is, there is random mixing. The utility of this separation depends upon our ability to evaluate $Q_N^{(random)}$ precisely and to show that the ratio $Q_N/Q_N^{(random)}$ is close to unity.

1.6.2 THE EVALUATION OF $Q_N^{(random)}$

As noted, Eq. (1.6.11) corresponds to the approximation that each molecule executes motions within its cell that are independent of the motions of the molecules in all neighboring cells. This therefore corresponds, in the theory of imperfect gases, to the perfect gas. In this instance, as in the theory of gases, the theory can be made exact by calculating the effect of interacting pairs of molecules, triplets of molecules, etc. For this purpose we introduce cell clusters analogous to the clusters considered in the imperfect gas theory. A cell cluster consists of a group of, say, l neighboring cells that are considered to form one larger cell. The total volume of this larger cell is available for and commonly shared by the l molecules. Within the large cell, the molecules move under the influence of two force fields. The first is due to the intermolecular potential arising from their mutual interaction

$$U_{l_1 \cdots l_l}^{(s \cdots z)}(\mathbf{r}_1 \cdots \mathbf{r}_l) = \tfrac{1}{2} \sum_{s,t} \left\{ \sum_{l_i}^{n_s} \sum_{l_j}^{n_t} u_{l_i l_j}^{st}(\mathbf{q}) \right\} \tag{1.6.14}$$

where n_s is the number of molecules of species s in the cluster and $\sum_s n_s = l$. The pair interaction potential is $u_{l_i l_j}^{st}(\mathbf{q})$ between particles of species s and t in cells l_i and l_j when separated by a distance $\mathbf{q}$. The second force is due to the interaction of the molecules in the cell with the $N - l$ surrounding molecules, which are taken to be at rest at their respective lattice points.

$$\chi_{l_i \cdots l_l}^{(s \cdots z)} = \sum_{l_i \neq l_1 \cdots l_l} \{ u_{l_i l_1}^{st}(\mathbf{q}_{l_i l_1}) + \cdots + u_{l_i l_1}^{st}(\mathbf{q}_{l_i l_1}) \}. \tag{1.6.15}$$

The potential energy when all the molecules are at their lattice sites is

$$\chi_{l_1\cdots l_l}^{0(s\cdots z)} + U_l^{0(s\cdots z)}_{1\cdots l_l} = \sum_{l_i \neq l_1 \cdots l_l} \{(u_{l_i l_i}^{st})^0 + \cdots + (u_{l_i l_i}^{st})^0\} \tag{1.6.16}$$

$$+ \frac{1}{2} \sum_{l_i}^{l} \sum_{l_j}^{l} (u_{l_i l_j}^{st})^0.$$

Once again it is convenient to define a set of partial probability densities for the molecules in a cell cluster of l cells to be displaced from their lattice points by the distances $r_1, \cdots, r_l$

$$W_{l_1}^{(s)}(r_1) = \exp[-\beta\{\chi_{l_1}^{(s)}(r_1) - \chi_{l_1}^{0(s)}\}]$$

.

.

.

$$W_{l_1\cdots l_l}^{(s\cdots z)}(r_1 \cdots r_l) = \exp[-\beta\{(U_{l_1\cdots l_l}^{(s\cdots z)} + \chi_{l_1\cdots l_l}^{(s\cdots z)}) - (U_{l_1\cdots l_l}^{0(s\cdots z)} + \chi_{l_1\cdots l_l}^{0(s\cdots z)})\}].$$
$$\tag{1.6.17}$$

These probability densities are normalized in the same way as those defined in Eq. (1.6.8). Note that as l approaches N, the cell cluster approaches one encompassing the entire fluid. From Eqs. (1.6.16) and (1.6.17) it is seen that the partial probability densities approach the limiting value

$$W_N^{(s)}(r_1 \cdots r_N) = W_{l_1\cdots l_l}^{(s\cdots z)}(r_1 \cdots r_N) = \exp[-\beta\{U_N(q_1 \cdots q_N) - U_N^0\}]$$
$$\tag{1.6.18}$$

which is the correct configurational probability density for the system. It should be recalled that Eqs. (1.6.14) and (1.6.18) refer to a system in which the molecules are *randomly mixed*.

As in the theory of imperfect gases, the probability density shown in Eq. (1.6.18) may be expressed as the sums and differences of products of the partial probability densities. That is, a series of functions may be defined by the relations

$$W_{l_1}^{(s)}(r_1) = \hat{U}_{l_1}^{n_1\cdots n_r}(r_1)$$

$$W_{l_1 l_2}^{(s,t)}(r_1, r_2) = \hat{U}_{l_1 l_2}^{n_1\cdots n_r}(r_1, r_2) + \sum_P \hat{U}_l^{n_1\cdots n_r}(r_1) \times \hat{U}_l^{n_1\cdots n_r}(r_2) \tag{1.6.19}$$

.

.

.

$$W_{l_1\cdots l_N}^{(s\cdots z)}(r_1 \cdots r_N) = \sum_P \prod \hat{U}_{l_1\cdots l_N}^{n_1\cdots n_r}$$

where the superscripts $n_1, \cdots, n_r$ refer to the number of molecules of species $1, \cdots, r$ in the cell cluster, and the subscripts $l_1, \cdots, l_l$ describe the shape of the cluster. It will be convenient at a later stage to describe the shape of the cluster by a single subscript, k, and suppress the elaborate subscript rotation

$l_1, \cdots, l_l$. Similarly, the composition of the cluster will be denoted by $\{n_i\}$, the number of molecules of species $1, \cdots, r$ present in the cluster. In Eq. (1.6.19), the summation $\sum_P$ is to be carried out over all permutations of the molecules over the different cell clusters. The general procedure is to divide the molecules 1 to N into a number of groups, take the product of the corresponding $\hat{U}$ functions, and sum over all products obtained by dividing the molecules in groups in all possible ways.

Equation (1.6.19) may be successively solved to yield

$$\hat{U}_{l_1}^{n_1 \cdots n_r}(\mathbf{r}_1) = W_{l_1}^{(s)}(\mathbf{r}_1)$$

$$\hat{U}_{l_1 l_2}^{n_1 \cdots n_r}(\mathbf{r}_1, \mathbf{r}_2) = W_{l_1 l_2}^{(s,t)}(\mathbf{r}_1, \mathbf{r}_2) - \sum_P W_{l_1}^{(s)}(\mathbf{r}_1) W_{l_2}^{(t)}(\mathbf{r}_2). \tag{1.6.20}$$

Examination of Eq. (1.6.20) shows that the $\hat{U}$ functions, as defined here, are zero when any one of the molecules concerned is at a lattice point. The probability density for such a cluster reduces to the product of the probability densities for 2 clusters, each of smaller size. This is entirely analogous to the observation that in the gas phase, the function

$$f_{ij} = \exp(-\beta u_{ij}) - 1$$

approaches zero as the distance between the interacting molecules increases to infinity. The summations of Eq. (1.6.18) must be made subject to the conservation condition

$$\sum_{k,n_s} n_s \zeta_{k,n_1 \cdots n_r} = N_s \tag{1.6.21}$$

where $\zeta_{k,n_1, \cdots n_r}$ is the number of clusters of type k containing n_1 molecules of species 1, n_2 of species 2, etc. Now, cluster integrals $J_{k,n_1 \cdots n_r}$ for a cluster containing n_1 molecules of species 1, n_2 of 2, etc., may be defined by the relation

$$J_{k,n_1 \cdots n_r} = \frac{1}{n_1! \cdots n_r!} \int \hat{U}_k^{\{n_i\}} \, d\{\sum n_s\} \, d\{\sum n_s\} \tag{1.6.22}$$

The total contribution to the partition function of the set of clusters character-ized by the numbers $\zeta_{k,n_1 \cdots n_r}$ is just the number of ways in which N_1 numbered molecules of species 1, N_2 numbered molecules of species 2, etc., can be distributed among the clusters. Since terms resulting from the permutations of the molecules over the cell cluster give the same results, the total partition function becomes

$$Q_N^{(\text{random})} = \prod_{s=1}^{r} \lambda_s^{-3N_s} \exp(-\beta U_N^0) \times \sum_{\{\zeta_{k,n_1 \cdots n_r}\}} g(\zeta_{k,n_1 \cdots n_r}) \times \tag{1.6.23}$$

$$\prod_{k,n_1 \cdots n_r} J_{k,n_1 \cdots n_r}^{\zeta_{k,n_1 \cdots n_r}}$$

where $g(\zeta_{k,n_1}, \cdots {}_{n_r})$ is the number of ways of distributing the given number of cell clusters of given size and shape over the reference lattice. The excess free energy is thus, *for the random configuration* (the superscript ^{0} denotes pure components)

$$\frac{A^{E \, (\text{random})}}{NkT} = \frac{A^{(\text{random})}}{NkT} - \sum_s \frac{A_s^{(\text{random})}}{NkT} - \sum_s x_s \ln x_s$$

$$= \frac{U_N^0 - \sum U_{N_s}^{00}}{NkT} - \sum_s x_s \ln x_s - \ln \left\{ \sum g(\zeta_{k,n_1} \cdots {}_{n_r}) \times \right. \qquad (1.6.24)$$

$$\left. \prod_{k,\{n_s\}} J_{k,n_1 \cdots n_r}^{\zeta_{k,n_1} \cdots n_r} \right\} + \sum_s \ln \left\{ \sum g(\zeta_{k,n_s}) \prod_{k,n_s} J_{k,n_s}^{0 \zeta_{k,n_s}} \right\}.$$

If only clusters of single molecules are considered,

$$\sum g(\zeta_{k,n_1} \cdots {}_{n_r}) \prod_{k,\{n_s\}} J_{k,n_1 \cdots n_r}^{\zeta_{k,n_1} \cdots n_r} = \frac{N!}{\prod_s N_s!} \prod_s J_s^{N_s} \qquad (1.6.25)$$

since only the terms $\zeta_{1,1000}\cdots, \zeta_{1,01000}\cdots$, etc., are nonzero, and there is only one shape for a cluster of size unity.

Before turning to the implementation of Eqs. (1.6.23) and (1.6.24), it is pertinent to reiterate the three principal assumptions involved in the theory. The first approximation is that only one configuration of the system has been considered, that is, the random configuration. The second approximation is the assumption that the molecules are sufficiently alike in size and shape to fit on the same lattice. The effect of differences in the sizes of the component molecules of a solution has been discussed by Prigogine and his co-workers.[14] The third approximation is one of convenience. That is, the force acting on the l molecules of a cell cluster has been calculated as if the remaining molecules were at the centers of their cells. In actuality, this is not the case, and there will be a correlation energy that represents the difference in energy between the assumed lattice configuration and the real configuration of the molecules in the system. It is possible to introduce this correlation energy for clusters of all sizes. Clearly, as the size of the cluster increases, the importance of the correlation energy decreases and, in the limit of 1 cluster containing N molecules, it vanishes identically. The introduction of the correlation energy would, however, serve the purpose of making the total potential energy of the system identically correct for clusters of all sizes. With the theory formulated in this section, the total potential energy, as well as the configurational probability density, both approach the correct values in the limiting case when the cluster size is N. Thus the potential energy at each stage is slightly inaccurate, but converges to the proper value.

1.6.3 APPLICATION TO A BINARY SOLUTION

In this section an approximate calculation will be made of the excess free energy of a binary solution with random mixing. The difficulty is, of course, the evaluation of the combinatorial factor $g(\zeta_{k,n_1 \cdots n_r})$. The evaluation of the combinatorial factor has been accomplished for the related problem of mixtures of open-chain polymers and monomers and for mixtures of monomers with triangular trimers and tetrahedral tetramers. For illustrative purposes, confine attention to mixtures of dimers and monomers; that is, in the random mixing approximation, the series is truncated after the terms corresponding to cell clusters of 2 molecules, and the solution may be considered to be one of dimers and monomers. To this end the combinatorial factors may be evaluated as if the solution were composed of the 5 components: monomers 1 and 2, and dimers 11, 22, 12. The combinatorial factor is thus[18]

$$g(\zeta_{k,n_1 \cdots n_r}) = \left\{ \prod_s \left(\frac{\rho_s}{\sigma_s} \right)^{\eta^{(s)}} \right\} \left\{ \frac{(\sum_s r_s \eta^{(s)})!}{\prod_s \eta^{(s)}!} \right\} \left\{ \frac{(\sum_s q_s \eta^{(s)})!}{(\sum_s r_s \eta)!^{(s)}} \right\}^{z/2} \tag{1.6.26}$$

with z the number of nearest neighbors in the lattice, ρ_s the number of alternative arrangements of a " molecule " of type s when one of its elements is fixed ($\rho_s = z$ for a dimer), and σ_s is the symmetry number for the " molecule ". Each of the $\eta^{(s)}$ " molecules " of species s occupies r_s sites. The number of pairs of sites, of which one is occupied by a " molecule " of type s and the other is not occupied by an element of the same " molecule " is $z q_s$. The q_s and r_s are related by the equation

$$q_s = r_s - \frac{2}{z}(r_s - 1). \tag{1.6.27}$$

After the use of Stirling's approximation and some algebraic rearrangement, the excess free energy assumes the form

$$\frac{A^{E \text{ (random)}}}{NkT} = x_1 \ln \frac{J_{10}^0}{J_{10}} + x_2 \ln \frac{J_{01}^0}{J_{01}} + \frac{z}{2} x_1 \ln \frac{J_{20}^0}{J_{20}} + \frac{z}{2} x_2 \ln \frac{J_{02}^0}{J_{02}}$$

$$+ \frac{z}{2} x_1 x_2 \ln \frac{J_{02} J_{02}}{J_{11}^2} + \frac{U_N^0 - U_{N_1}^{00} - U_{N_2}^{00}}{NkT}. \tag{1.6.28}$$

Note that Eq. (1.6.27) reduces to the well-known result,[17]

$$\frac{A^{E \text{ (random)}}}{NkT} = x_1 \ln \frac{J_{10}^0}{J_{10}} + x_2 \ln \frac{J_{01}^0}{J_{01}} + \frac{U_N^0 - U_{N_1}^{00} - U_{N_2}^{00}}{NkT} \tag{1.6.29}$$

when only clusters of one molecule are considered. The excess entropy and excess enthalpy can be obtained from Eq. (1.6.28) by straightforward methods.†

The effects of nonrandom mixing must now be studied. The relation between $Q_N^{(random)}$ and Q_N, it will be recalled, is

$$Q_N = Q_N^{(random)} \quad \left(\frac{Q_N}{Q_N^{(random)}}\right)$$

$$= Q_N^{(random)} \quad \exp \beta(A^{(random)} - A). \tag{1.6.30}$$

The systematic investigation of $Q_N/Q_N^{(random)}$ is a problem that will not be treated in this book. It is of interest, however, to investigate the order of magnitude of this correct term, an idea of which may be obtained from the lattice model for solutions. The most undersirable features of this model should largely be avoided, since only a ratio of partition functions is required. Using the quasi-chemical approximation and Bragg–Williams approximation to estimate the ratio $Q_N/Q_N^{(random)}$ one easily obtains the relation (written for a binary solution)

$$\frac{A^{(random)} - A}{kT} = \frac{2w}{kT} \frac{N_1 N_2}{N} + N_1 \ln x_1 + N_2 \ln x_2$$

$$- \frac{N_1 z}{2} \left[\ln \frac{\{4N_1 N_2 e^{2w/kT} + (N_1 - N_2)^2\}^{1/2} + N_1 - N_2}{\{4N_1 N_2 e^{2w/kT} + (N_1 - N_2)^2\}^{1/2} + N_1 + N_2} - \frac{z - 2}{z} \ln x_1 \right]$$

$$- \frac{N_2 z}{2} \left[\ln \frac{\{4N_1 N_2 e^{2w/kT} + (N_1 - N_2)^2\}^{1/2} + N_2 - N_1}{\{4N_1 N_2 e^{2w/kT} + (N_1 - N_2)^2\}^{1/2} + N_2 + N_1} - \frac{z - 2}{2} \ln x_2 \right]$$

$$\tag{1.6.31}$$

where w is defined as $w = (u_{l_1,l_2}^{12})^0 - \frac{1}{2}\{(u_{l_1,l_2}^{11})^0 + (u_{l_1,l_2}^{22})^0\}$. Now it is possible to estimate the order of magnitude of $Q_N/Q_N^{(random)}$ for $N_1 = N_2$, in which case

$$\frac{A^{(random)} - A}{kT} = -\frac{Nz}{z} \ln \left(\frac{2}{e^{-w/2kT} + e^{w/2kT}}\right) \tag{1.6.32}$$

$$\frac{A^{(random)} - A}{NkT} = 0.0024$$

for the typical value $w/kT = 0.05$.

The effect of nonrandom mixing is thus estimated to be of the order of 0.2% of the total partition function. If the excess free energy is of the order of

† The reader should note that the notation U^{00} refers to a static lattice energy for a pure component. The notation J^0 also refers to a pure component, since, in the case of the cell partition function, a superscript zero could not possibly refer to the static pseudo-lattice.

300 cal/mole of solution at 300°K, then the nonrandom contribution would be about 0.5%. Of course, the derivatives of the free energy will have a somewhat greater contribution from the nonrandom mixing. It should be noted that our conclusions are in accord with the perturbation calculation of Pople.[22]

1.6.4 CELL CLUSTER THEORY AND MULTIPLE CELL OCCUPANCY

Salzburg and Kirkwood[23] and Kirkwood[24] have shown, by a different technique than that used here, that the cell theory of liquids and mixtures can be derived from the general principles of statistical mechanics by well-defined approximations. In this method, the volume V of the system is also spanned by a virtual lattice of N cells. If each of the virtual cells has a volume Δ, the phase integral over the total volume may be replaced by a sum of integrals over the cells. This leads to the relation

$$\frac{Q'_{N_1 \cdots N_r}}{\prod_s N_s!} = \sum_{\xi_1{}^1 \cdots \xi_1{}^N = 0}^{N_1} \cdots \sum_{\xi_r{}^1 \cdots \xi_r{}^N = 0}^{N_r} \frac{Q_{N_1 \cdots N_r}^{(\xi_1{}^1 \cdots \xi_r{}^N)}}{\prod_{s=1}^{r} \prod_{m=1}^{n} \xi_m^s!} \tag{1.6.33}$$

where $Q'_{N_1 N_2 \cdots N_r}$ is defined as usual by

$$Q'_{N_1 \cdots N_r} = \int_V \exp(-\beta U_{\mathbf{N}}) \, d\{\mathbf{N}\} \tag{1.6.34}$$

and ξ_m^s is the number of molecules of species s occupying the cell m. The subsidiary conservation conditions require that

$$\sum_s \xi_m^s = \text{number of molecules in cell } m$$

$$\sum_m \xi_m^s = N_s. \tag{1.6.35}$$

If $Q'^{(1)}$ denotes the sum containing integrals for all distinguishable configurations corresponding to single occupancy of each cell, then a parameter σ may be defined formally by the relation

$$\sigma^N = \sum_{\{\xi_1{}^i\}}^{N_1} \cdots \sum_{\{\xi_r{}^i\}}^{N_r} \frac{1}{\prod_{s=1}^{r} \prod_{m=1}^{N} \xi_m^s!} \frac{Q'^{(\xi_1{}^1 \cdots \xi_r{}^N)}_{N_1 \cdots N_r}}{Q^{(1)}}$$

$$\sum_{s=1}^{N} \xi_m^s = N_s \tag{1.6.36}$$

and the partition function may be written

$$Q_{\mathbf{N}} = \prod_{s=1}^{r} \lambda_s^{-3N_s} \sigma^N Q'^{(1)}. \tag{1.6.37}$$

The partition function $Q'^{(1)}$ corresponds to the Lennard–Jones and Devonshire model of liquids and the parameter σ to the correction that is necessary to make the arbitrary division of the phase space correct. It should be noted that there is an important difference between the cell-cluster theory, in which terms higher than those corresponding to the Lennard–Jones and Devonshire model refer to the cooperative motions of molecules moving in larger and larger cells, and the multiple-occupation theory in which the correction terms involved in the parameter σ refer to the multiple occupation of 1 cell, of the same volume Δ, by several molecules. The evaluation of σ may be made by a method first used by Pople for pure liquids.[21]

It is convenient to assume that

$$\sigma^N = \sum_{\{\xi_1^i\}}^{N} \cdots \sum_{\{\xi_r^i\}}^{N_r} \prod_{s=1}^{r} \prod_{m=1}^{N} \frac{\omega_{\zeta m}^{\zeta s}}{\zeta_m^{\zeta s}!} \tag{1.6.38}$$

which is equivalent to the approximation that there is no interference between multiply occupied cells. There is an ω for each different type of occupancy. We shall discuss the meaning of the ω's subsequently. Let a given distribution contain $x_{n_1 n_2 \cdots n_r}$ cells which have n_1 molecules of species 1, n_2 of species 2, etc. The $x_{\{n_s\}}$ must satisfy the relations

$$\sum_{n_1, n_2, \cdots = 0}^{N} n_s x_{n_1 \cdots n_r} = N_s \tag{1.6.39}$$

$$\sum_{n_1} \cdots \sum_{n_r} x_{n_1 \cdots n_r} = N.$$

The number of ways of placing N particles such that there are $x_{n_1 n_2 \cdots n_r}$ cells containing n_1 molecules of 1, n_2 of 2, etc., is

$$\frac{N!}{\prod\limits_{n_1} \cdots \prod\limits_{n_r} (x_{n_1 \cdots n_r})!} \tag{1.6.40}$$

so that

$$\sigma^N = \sum_{\{x_{\{n_s\}}\}} \frac{N!}{\prod\limits_{n_1} \cdots \prod\limits_{n_r} (x_{\{n_s\}})!} \prod_s \prod_m \left(\frac{\omega_{\zeta m}^{\zeta s}}{\zeta_m^{\zeta s}!} \right). \tag{1.6.41}$$

Equation (1.6.41) is readily transformed to

$$\sigma^N = \sum_{\{x_{\{n_s\}}\}} \frac{N!}{\prod\limits_{n_1} \cdots \prod\limits_{n_r} (x_{\{n_s\}})!} \prod_{n_1 \cdots n_r} \left(\frac{\omega_{n_1 \cdots n_r}}{\prod n_s!} \right)^{x_{\{n_s\}}}. \tag{1.6.42}$$

The equilibrium values of the $x_{\{n_s\}}$ may be determined by minimizing the free energy of the system. Using Stirling's approximation and differentiating with respect to $x_{\{n_s\}}$ gives

$$\ln x_{\{n_s\}} = \ln \frac{\omega_{n_1 \cdots n_r}}{\Pi_{n_s}!} + \ln N\lambda + \ln K_1^{n_1} K_2^{n_2} \cdots K_r^{n_r} \qquad (1.6.43)$$

where the K_s and λ result from the subsidiary conditions

$$N\lambda\delta \sum x_{\{n_s\}} = 0$$
$$K_s\delta \sum n_s x_{\{n_s\}} = 0. \qquad (1.6.44)$$

The equilibrium distribution of the $x_{\{n_s\}}$ is thus

$$x_{\{n_s\}} = N\lambda \frac{\omega_{n_1 \cdots n_r}}{\Pi \, n_s!} \prod K_s^{n_s} \qquad (1.6.45)$$

and the subsidiary conditions become

$$\sum N\lambda \frac{\omega_{n_1 \cdots n_r}}{\Pi n_s!} \prod K_s^{n_s} = N$$

$$\sum N\lambda n_s \frac{\omega_{n_1 \cdots n_r}}{\Pi n_s!} \prod K_s^{n_s} = N_s \qquad (1.6.46)$$

which suffice to determine the K_s and λ.

Specialize now to the case of a binary solution. If a function $f(K_1, K_2)$ is defined by the relation

$$f(K_1, K_2) = \sum \sum \omega_{n_1} \omega_{n_2} K_1^{n_1} K_2^{n_2} \qquad (1.6.47)$$

then, after substitution, summation and the use of Stirling's approximation, the relation

$$\sigma = \frac{1}{\lambda K_1^{x_1} K_2^{x_2}} = \frac{K_1(\partial f/\partial K_1) - K_2(\partial f/\partial K_2)}{K_1^{x_1} K_2^{x_2}} \qquad (1.6.48)$$

is obtained. Including only terms up to double occupancy, the parameters K_1, K_2, and λ are determined by the conditions

$$\lambda(1 + K_1^2 + \frac{\omega_{20}}{2} K_1^2 + K_2 + \frac{\omega_{02}}{2} K_2^2 + \omega_{11} K_1 K_2) = 1$$

$$\lambda(K_1 + \omega_{20} K_1^2 + \omega_{11} K_1 K_2) = x_1$$

$$\lambda(K_2 + \omega_{02} K + \omega_{11} K_1 K_2) = x_2 \qquad (1.6.49)$$

$$\omega_{10} = \omega_{01} = \omega_{00} = 1.$$

If the approximation

$$\omega_{11} = (\omega_{02}\omega_{20})^{1/2} \qquad (1.6.50)$$

is made, Eq. (1.6.49) can be solved very easily, leading to the relations

$$K_1 = \frac{x_1(2\sqrt{\omega_{02}} + \sqrt{2})}{x_1(\sqrt{2\omega_{02}\omega_{20}} + \sqrt{\omega_{20}}) + x_2(\sqrt{2\omega_{02}\omega_{20}} + \sqrt{\omega_{02}})} \qquad (1.6.51)$$

$$K_2 = \frac{x_2(2\sqrt{\omega_{20}} + \sqrt{2})}{x_2(\sqrt{2\omega_{02}\omega_{20}} + \sqrt{\omega_{02}}) + x_1(\sqrt{2\omega_{20}\omega_{02}} + \sqrt{\omega_{20}})}$$

which reduce to

$$K_1^0 = \sqrt{\frac{2}{\omega_{20}}} \qquad (1.6.52)$$

$$K_2^0 = \sqrt{\frac{2}{\omega_{02}}}$$

in the limits of zero mole fraction of species 1 or 2. The contribution to the excess free energy is then

$$\begin{aligned}
\frac{\Delta A}{NkT} \text{ (double occupancy)} &= -\ln \sigma + x_1 \ln \sigma_1 + x_2 \ln \sigma_2 \\
&= x_1 \ln \left(\frac{2 + K_1^0}{K_1^0}\right)\left(\frac{K_1^{x_1} K_2^{x_2}}{2 + K_1 + K_2}\right) \qquad (1.6.53) \\
&\quad + x_2 \ln \left(\frac{2 + K_2^0}{K_2^0}\right)\left(\frac{K_1^{x_1} K_2^{x_2}}{2 + K_1 + K_2}\right)
\end{aligned}$$

where the superscript ^{0} refers, as before, to the pure components.

Now, the $\omega_{n_1 n_2}$ were defined as the ratio by which the restricted phase integral is changed when the configuration corresponding to 1 molecule per cell is changed to the configuration in question. For instance,

$$\omega_{20}^0 = \frac{\displaystyle\int^{\Delta_1}\int^{\Delta_1} \exp[-\beta\{\psi_{l_1}(\mathbf{r}_1) + \psi_{l_1}(\mathbf{r}_2) + u_{l_1 l_1}(\mathbf{q})\}]\, d(1)d(2)}{\left[\displaystyle\int^{\Delta_1} \exp\{-\beta\psi_{l_1}(\mathbf{r}_1)\}\, d(1)\right]^2} \cdot \qquad (1.6.54)$$

On the other hand, the cell-cluster integral J_{20}^0 is defined as

$$\begin{aligned}
2J_{20}^0 &= \int \hat{U}_{20}(\mathbf{r}_1, \mathbf{r}_2)\, d(1)\, d(2) \\
&= \int^{2\Delta} \exp[-\beta\{\chi_{l_1 l_2}(\mathbf{r}_1, \mathbf{r}_2) + u_{l_1 l_2}(\mathbf{q}) - \chi_{l_1 l_2}^0 \qquad (1.6.55) \\
&\quad - (u_{l_1 l_2})^0\}]\, d(1)\, d(2) - 2\int^\Delta \exp\{-\beta\psi_{l_1}(\mathbf{r}_1)\}\, d(1)\int^\Delta \exp\{-\beta\psi_{l_2}(\mathbf{r}_2)\}\, d(2).
\end{aligned}$$

It is quite clear that multiple cell occupancy cannot contribute appreciably to the thermodynamic properties at liquid densities, since most fluids are about 80% close packed. This indicates that if the cell volume be taken as V/N, there is not enough room for 2 molecules in a single cell. Pople's calculations verify this conclusion. On the other hand, the cell-cluster integrals should be much larger since the 2 molecules share a double cell. Quantitative comparisons cannot yet be made since the values of the cell-cluster integrals have not been computed.

1.6.5 THE ADDITIVITY HYPOTHESIS

Recently it has been proposed that the free energy of a mixture can be considered, aside from the free energy of mixing, to be the sum of the free energies that the components would have as pure liquids at the same reduced temperature and reduced volume that characterizes their environment in the mixture.[16] That is, the total free energy may be written

$$A = N_1 A_1^0(\tilde{T}_1, \tilde{v}_1) + N_2 A_2^0(\tilde{T}_2, \tilde{v}_2) + A_{\text{ideal mixing}} \qquad (1.6.56)$$

where $\tilde{T}_1$, $\tilde{T}_2$, $\tilde{v}_1$, $\tilde{v}_2$ are the reduced temperature and volume characteristic of the surroundings of a molecule of species 1 or 2 in the mixture. It can be shown that Eq. (1.6.56) is exact, if the ordinary cell model for solutions is used (Eq. (1.6.29)).[25] Note that the deviations from Eq. (1.6.56) will be described by the terms

$$\sum_{\zeta_{k,\{n\}}} g(\zeta_{k,n_1} \cdots {}_{n_r}) \prod_{k,n_s} J_{k,n_1}^{\zeta_{k,\{n\}}} \cdots {}_{n_r} - \frac{N!}{\prod_s N_s!} \prod_s J_s^{N_s} \qquad (1.6.57)$$

in the cell-cluster expansion. A similar relation can be written in terms of the multiple occupation expansion. It is clear from Eq. (1.6.55) that the use of Eq. (1.6.56) will reduce the magnitude of the excess free energy, but this energy will not vanish identically. This deviation from additivity is due to the fact that the reduced temperatures are defined for single cells, whereas there are double-cell interaction terms, etc., to be accounted for. A quantitative estimate of the validity of Eq. (1.6.56) must await further computation, though calculations based on the hole model for solutions[25] suggest that it is reasonably accurate numerically, even though it is physically incorrect (see following). The structure of the cell-cluster integral suggests that, in principle, the additivity hypothesis could be extended to eliminate identically the discrepancy in Eqs. (1.6.55) and (1.6.57) by the definition of a series of reduced temperatures and reduced volumes corresponding to single-cell clusters, double-cell clusters, etc. That is, for a pure liquid we may write

$$A_s^0 = A_s^{(\text{L.J.})} + A_s^{(2)} + A_s^{(3)} + \cdots + A_s^{(N)} \qquad (1.6.56')$$

where the first term is the free energy computed for the Lennard–Jones model and the successive terms are the corrections necessary to account for the correlations between pairs, triplets, . . . , of molecules.

To establish an extended additivity theorem requires a careful investigation of the theorem of corresponding states as applied to mixtures. The following analysis is due to Rice.[26]

It is pertinent, at the very outset, to point out that in contrast to the development of Section 1.5, attention is focused on the question of the applicability of the theorem of corresponding states in a form intuitively plausible for solutions. Therefore a development very different in form from the usual treatment is used. In particular, the form developed will refer not so much to the correspondence of thermodynamic functions of the mixture and components as to certain well-defined molecular contributions to the thermodynamic functions.

1.6.6 PURE FLUIDS

In this section shall be demonstrated the theorem of corresponding states for a pure fluid in the notation of the cell cluster theory of liquids. This preliminary proof of a well-known result is introduced to facilitate later arguments.

As before, at the outset of the analysis imagine the fluid to be spanned by a virtual lattice which divides it into cells. Let the number of cells be equal to the number of molecules. A cell cluster is a group of l neighboring cells which are considered to form one large cell, the total volume of which is available for and shared commonly by all the l molecules in the cell cluster. Within this cell the molecules move under the influence of their mutual interaction and their interaction with the surrounding N molecules. The precise method of computation used to evaluate the interactions mentioned is, for the moment, immaterial. For the pure fluid, in analogy with Eq. (1.6.19), it is convenient to define a set of partial probability densities for the l molecules in a cellcluster to be a specified set of displacements from their virtual lattice points. Thus

$$W_{l_1}(\mathbf{r}_1) = \exp - \left(\frac{1}{kT}\right)[\chi_1(\mathbf{r}_1) - \chi_1^0]$$

$$W_{l_1,l_2}(\mathbf{r}_1, \mathbf{r}_2) = \exp - \left(\frac{1}{kT}\right)[\chi_{12}(\mathbf{r}_1, \mathbf{r}_2) + U_{12}(\mathbf{r}_{12}) - \chi_{12}^0 - U_{12}^0]$$

$$\cdot$$
$$\cdot \qquad\qquad (1.6.58)$$
$$\cdot$$

$$W_{l_1\cdots l_l}(\mathbf{r}_1 \cdots \mathbf{r}_l) = \exp - \left(\frac{1}{kT}\right) \times$$
$$[\chi_1 \ldots {}_l(\mathbf{r}_1 \cdots \mathbf{r}_l) + U_1 \ldots {}_l(\mathbf{r}_1 \cdots \mathbf{r}_l) - \chi_1^0 \ldots {}_l - U_1^0 \ldots {}_l]$$

where the intermolecular potential due to the surrounding molecules is χ and that due to intracell interactions is U, and the superscripts (0) refer to the state when all molecules are at rest on their lattice points. Now introduce $\hat{U}$ functions defined by the recursive relations

$$W_{l_1}(\mathbf{r}_1) = \hat{U}_{l_1}(\mathbf{r}_1)$$

$$W_{l_1,l_2}(\mathbf{r}_1, \mathbf{r}_2) = \hat{U}_{l_1}(\mathbf{r}_1)\hat{U}_{l_2}(\mathbf{r}_2) - \hat{U}_{l_1,l_2}(\mathbf{r}_{12})$$

$$\cdot$$
$$\cdot \qquad\qquad\qquad\qquad\qquad (1.6.59)$$
$$\cdot$$

$$W_{l_1\cdots l_N}(\mathbf{r}_1\cdots\mathbf{r}_N) = \sum \prod \hat{U}_{\{l_i\}}$$

where the $\hat{U}$ functions are obtained by dividing the set of cells in all possible ways into cell clusters. This subdivision will be different for different geometric arrangements of the cells, and all possible geometries must be accounted for in the complete set. In the same manner as in the preceding development, cell cluster integrals for cells of l molecules of shape k are defined as

$$J_l^k = \frac{1}{l!}\int \hat{U}_l^k(\mathbf{r}_1\cdots\mathbf{r}_l)\,d\{l\}. \qquad (1.6.60)$$

If there are ζ_{lk} cell clusters of size l and shape k, then the partition function may be written

$$Q_N = \lambda^{-3N}\exp(-\beta U_N^0)\sum_{\{\zeta_{lk}\}} g(\zeta_{lk})\prod_{l,k}(J_l^k)^{\zeta_{lk}} \qquad (1.6.61)$$

$$\lambda^3 = \left(\frac{h^2}{2\pi mkT}\right) \qquad (1.6.62)$$

where $g(\zeta_{lk})$ is the total number of ways in which the N cells can be divided into ζ_1 cell clusters of size one, ζ_2 of size two, $\cdots$, m is the mass of a molecule and U_N^0 is the lattice energy of the arbitrarily chosen reference lattice. To study corresponding states, define the reduced distance, reduced temperature, reduced partial probability density, and reduced $\hat{U}$ function by

$$\tilde{r} = \frac{r}{\sigma}; \qquad \tilde{T} = \frac{kT}{u^*}; \qquad \tilde{\lambda} = \frac{\lambda}{\sigma}$$

$$\tilde{W}_l^k = \lambda^{3l}W_l^k = \lambda^{3l}\exp\left\{\left(-\frac{1}{\tilde{T}}\right)\right.$$

$$\times \left[\sum_1^l\sum_{l+1}^N \phi(r_{ij}) + \tfrac{1}{2}\sum_1^l\sum_1^l \phi(r_{ij}) - \sum_1^l\sum_{l+1}^N \phi(r_{ij}^0) - \tfrac{1}{2}\sum_1^l\sum_1^l \phi(r_{ij}^0)\right]\right\}$$

$$\tilde{U}_l^k = \lambda^{3l}\hat{U}_l^k \qquad (1.6.63)$$

the superscript k again referring to the cluster shape. By substitution of the reduced variables into Eq. (1.6.60) it is seen that

$$J_l^k = \frac{1}{l!} \left(\frac{\sigma}{\lambda}\right)^{3l} \int \tilde{U}_l^k(\tilde{r}_1 \cdots \tilde{r}_l) \, d^3\tilde{r}_1 \cdots d^3\tilde{r}_l$$

$$= \frac{1}{l!} \tilde{\lambda}^{-3l} \int \sum \prod \tilde{W}_i^k \, d^3\tilde{r}_1 \cdots d^3\tilde{r}_l \qquad (1.6.64)$$

and by reference to Eq. (1.6.63) it is readily seen that the $\tilde{W}_l^k$ are functions of the reduced variables only and hence

$$J_l^k = \tilde{\lambda}^{-3l} \mathscr{J}_l^k(\tilde{T}, \tilde{v}) \qquad (1.6.65)$$

where the reduced volume, $\tilde{v} = V/N\sigma^3$, enters through the dependence of J_l^k on the cell volume. The substitution of Eq. (1.6.65) into Eq. (1.6.61) leads to a reduced partition function, one of the several possible forms which represent the theorem of corresponding states.

1.6.7 EXTENSION TO MIXTURES

The first question to be asked when considering mixtures is whether or not a theorem of corresponding states indeed exists. The configurational partition function of a mixture is a function of the temperature, volume, numbers of molecules, and the parameters of the pair potential (assuming pairwise additive forces). If the pair potential for unlike molecules can also be expressed in the form of Eq. (1.5.1) and if there exists some functional relationship between the parameters of the pair potential for unlike molecules and those for like molecules, $\sigma_{st} = f(\{u_{ss}^*\}, \{\sigma_{ss}\})$ and $u_{st}^* = g(\{u_{ss}^*\})$, then we may establish a reduced equation of state. The precise functional form of the combining rules for the $\{u_{st}^*\}$ and the $\{\sigma_{st}\}$ is immaterial, but they must exist for the following to be true. For, by utilizing the combining rules we may eliminate the mixture pair potential constants in terms of the component pair potential constants. Thus

$$Q_N(T, V, \{N\}, \{u_{st}^*\}, \{\sigma_{st}\}) = F(T, V, \{N\}, \{u_{ss}^*\}, \{\sigma_{ss}\}) \qquad (1.6.66)$$

and there must exist a reduced equation of state for the mixture obtainable by defining suitable reduced variables. From the purely macroscopic point of view, reduced temperatures and volumes may be conveniently defined in terms of, say, the critical mixing temperature and the volume at the point of critical mixing. The thermodynamic criteria for phase stability together with the equation of state provide, in principle, the necessary simultaneous equations which uniquely define the plait point as a corresponding point. As usual, the plait point constants of the pure components may be related to thermodynamic functions of the pure components. This procedure is equivalent to choosing some one substance as a reference point and measuring all thermo-dynamic functions from this common zero. The pair potential constants then clearly determine the thermodynamic functions and may be expressed in terms of them. The use of experimental data then permits, in principle, the tabulation of reduced equations of state as a function of concentration. From these equations of state are deducible all the thermodynamic properties of the mixture.

The difficulty with such a program is that it requires an enormous amount of data covering wide ranges of composition, temperature, and pair potential parameters in order to establish the empirical reduced equation of state over a wide range of the reduced variables. Further, there is no indication of the nature of the molecular behavior responsible for the properties observed. For these reasons, though the macroscopic approach can be implemented, a molecular development appears preferable.

Wojtowicz, Salzburg, and Kirkwood and Brown have independently shown that in the random mixing approximation there does indeed exist a correspondence between the properties of a mixture and the properties of the pure components.[16] The argument, which is very general, proceeds by showing that if the total potential energy of interaction is replaced by a randomized potential energy defined by

$$\langle U \rangle = \sum_{i>k} \sum \sum_s \sum_t x_s x_t u_{st}(R_{ik}) \tag{1.6.67}$$

then the configurational partition function for the mixture is of identical analytic form with that of a pure component. Kirkwood *et al.* and Brown concentrated attention on different aspects of the development of this result. However, for purposes of actual evaluation, both used expansions. Kirkwood and co-workers examined the Margules expansion for the excess chemical potential whereas Brown developed what is essentially a second-order con-formal solution theory. As in the more general case mentioned previously, these theories represent the properties of the mixture in terms of the properties of the components without direct and explicit discussion of the molecular behavior responsible for the observed phenomena. Also, although it is in principle not necessary to resort to expansions for numerical evaluation if enough data are available, the paucity of information makes such a develop-ment a practical necessity. In view of these two observations, it is of interest to examine corresponding states in a manner that refers explicitly to the molecular contributions to the thermodynamic properties and, hopefully, somewhat clarifies the relationship between molecular dynamics and macro-scopic properties. Finally, so long as it is practically necessary to utilize expansions it is worthwhile to examine different expansions to ascertain their relative utility.

The preceding sections dealt with a development of the cell cluster theory of solutions for random mixtures, culminating in Eqs. (1.6.22) and (1.6.23). The assumption of random configurations on the lattice sites implies that the sizes of the molecules are the same, or very nearly so, for all species in the mixture. This is a simple geometric consequence of the spatial requirements for the packing of spheres. If the molecules differed markedly in size it would in general be impossible geometrically to construct a shell of neighbors with the same composition as the bulk mixture. It is only for spheres of equal or nearly equal size that random mixing can be expected to obtain.

An obvious method of introducing a reduced temperature and reduced volume for the mixture consistent with our assumption of random mixing is to utilize for the pair potential the randomized pair interaction energy appearing under the sum in Eq. (1.6.67). This has the result that all clusters containing an equal number of molecules and of the same shape give the same contribution independent of the composition of the cluster. To show that this definition of reduced temperature and volume is consistent with the assumption of random mixing we merely note that the configurational probability density for a single cluster containing all N molecules, W_N, may be developed into sums and products of $\hat{U}$ functions, just as in the last line of Eq. (1.6.58). Write

$$W_N = W_{l_1 \cdots l_N s} \tag{1.6.68}$$

and observe that if the randomized potential is used as described, then the development in Eq. (1.6.68) corresponds to the expansion of the configurational probability density for a pure system with a potential equal to the randomized potential. In this method of establishing correspondence, each cluster of the mixture *is made the same as each corresponding cluster in the pure fluid*. Note that the peculiarity of having all clusters of a given size and shape equal, independent of composition, is equivalent to the fact that the clusters in a pure fluid depend only upon size and shape and are, of course, independent of composition.

The method of introducing reduced variables discussed previously is analogous to the method used by Brown and Kirkwood and co-workers. For, it is clear that Eq. (1.6.68) represents part of the formal cluster expansion of the configurational partition function. With the choice of randomized potential energy given in Eq. (1.6.67) this method and that of Brown and Kirkwood *et al.* become formally indentical. It will be seen in the following that an advantage of the cluster development rests in the possible alterations one may introduce into the method of identifying corresponding states.

Although the method of introducing reduced variables described in the foregoing is formally exact and provides a direct correspondence between clusters in the mixture and clusters in the pure component, *it suffers from the intuitive disadvantage of making similar clusters of differing composition equal.* On the other hand, one would intuitively expect the unlike pair inter-action to differ from the like pair interactions, which in turn differ from one another. We therefore hopefully anticipate that a more suitable reduced variable scheme could be found which would account for the cluster compo-sition. Such a scheme ought to improve the representation of the thermo-dynamic properties of the fluid by a truncated cluster expansion.

One may proceed to develop a consistent sequence of approximations by expanding the exponential factor of each cluster integral in a power series in $(kT)^{-1}$ and considering the nth partial sums. This is just the moment expansion used in perturbation theory, the theory of order-disorder, and else-where. Let the total potential energy of a given cluster be Υ_l^k which we expand as follows:

$$\exp\left(-\frac{\Upsilon_l^k}{kT}\right) = \sum_{\xi=0}^{\infty} \frac{M_\xi}{\xi!}\left(-\frac{1}{kT}\right)^\xi \tag{1.6.69}$$

where the ξth moment, M_ξ, is given by

$$M_\xi = \prod_{s=1}^{r} \frac{n_s!}{n!} \sum (\Upsilon_i^k)^\xi. \tag{1.6.70}$$

By expanding Υ_l^k in a power series in $(kT)^{-1}$,

$$\Upsilon_l^k = \sum_{p=1}^{\infty} \frac{\Lambda_p}{p!}\left(-\frac{1}{kT}\right)^{p-1} \tag{1.6.71}$$

where the various Λ_p can be expressed in terms of the moments M_ξ. The results of such a procedure are well known. For example, M_1 and M_2 are given by

$$\Lambda_1 = M_1$$
$$\Lambda_2 = M_2 - M_1^2. \tag{1.6.72}$$

The moment expansion is to be made for each cluster, and all terms of equal order in (kT) collected. This then gives an ordered cluster expansion in which the leading term refers to infinite temperature. It is easy to show that when all terms of order $(kT)^{-1}$ are neglected,

$$\Upsilon_l^k = \sum_{s=1}^{r} \sum_{t=1}^{r} \sum_{i<j} u_{st}(R_{ij})x_s^k x_t^k \tag{1.6.73}$$

with x_s^k and x_t^k the mole fractions of s and t in the cluster of shape k. This potential corresponds to what we shall call local internal random mixing. It is important to point out that Eq. (1.6.73) is correct for clusters containing more than say three molecules, but due to the use of Stirling's approximation for factorials, the result for a double cluster is slightly inaccurate. On the other hand, the computation of the cluster integrals for two molecules is sufficiently easy to be performed exactly, and Eq. (1.6.73) introduces no great inaccuracy. We shall use Eq. (1.6.73) for all clusters bearing in mind the reservations mentioned when an actual numerical computation is to be made.

The leading term in the ordered cluster development gives *composition dependent* cluster integrals with a simple randomized potential. In the next approximation when terms of order $(kT)^{-1}$ are retained but those of order $(kT)^{-2}$ neglected, no simple randomized potential is found. It has perhaps not been sufficiently emphasized that random mixing is a necessity in the formulation. The $\hat{U}$ functions are defined in a manner such that they vanish when any or all of the particles are situated at their virtual lattice sites. In a mixture, this requirement must be met for all possible internal permutations of the molecules composing the cell cluster. Clearly this is only possible if some mean potential is utilized for both external and internal interactions. If the direct interactions are considered, clusters of size two remain uniquely defined (with external randomization) but larger clusters can be made to differ from one another by, say, placing all molecules of one species together in one instance and uniformly dispersing them in the other instance. The mean potential to which we are led by the moment expansion eliminates this problem. It is in this sense that random mixing is necessary. The composition dependent cluster integrals enable us to modify the theorem of corresponding states as follows.

Consider in detail the structure of the potential energy of a cell cluster. In Eq. (1.6.58) the field in which a molecule moves is separated into two parts, the external and internal fields. In general, in a mixture, the composition of an arbitrarily chosen cluster is not identical with the bulk composition, although the average over all clusters has this property. Thus, in a mixture it is necessary to introduce at least two reduced temperatures for each cluster, one for the internal field and one for the external field. For small clusters, the composition of the surrounding medium is essentially independent of the cluster composition, and therefore so is the corresponding reduced temperature. When the cluster becomes large enough to encompass macroscopic regions of the fluid this is no longer true.

Consider now how reduced variables are defined. In general, the method of definining a reduced temperature or distance consists of forcing the average interaction potential to have the same form as that of a pure component. For a pure component this may be done for any two parameter pair potential, but for a mixture this is only possible for restricted forms of the pair potential. In any event, the net result of the reduction is to be the resolution of the interaction energy into a product of a reduced temperature and a universal function of the reduced distance. In the case of a mixture, in order to factor the interaction energy into the form desired, it is necessary to describe the method of computation of the cell field. For our purposes it will be most convenient to compute the interactions between the l molecules of a cell cluster and the $N - l$ external molecules by situating all the molecules at the sites of the virtual lattice when l is two or greater. Note that this is not a restriction on the generality of the complete theory since if all terms are computed one still obtains the exact configurational partition function. From the practical point of view, even for small clusters the effects of neglecting corelations in motion between internal and external molecules are quite small when the energy is considered, and only slightly larger when the entropy is considered. The major contribution to the correlations in motion come from interactions within the cell, to which no dynamical approximations are made. One molecule clusters are, of course, treated on an entirely different basis since in this case there are no internal interactions and the excess energy of the molecule over that corresponding to no motion at the center of the cell is of dominant importance. By computing the energy for molecules fixed at virtual lattice sites, the term U_{N-l}^0 cancels an identical term in the partial probability density, as can be seen from Eq. (1.6.58). The two reduced temperatures of importance in this scheme then correspond to the external interactions of single cell clusters, $\tilde{T}_1$, and the internal interactions of larger cell clusters, $\tilde{T}_m$.

Consider first the case when all molecules are of equal size. Then, the pair potential may be expressed as

$$u_{st}(R) = u_{st}^* \phi \left(\frac{R}{\sigma} \right) \tag{1.6.74}$$

and the reduced partial probability density expressed in the form

$$\tilde{W}_{\{n_i\}}^k = \prod_i \lambda_i^{3n_i} W_{\{n_i\}}^k$$

$$= \prod_i \lambda_i^{3n_i} \exp - \frac{1}{\tilde{T}_m} \left[\frac{1}{2} \sum_1^l \sum_1^l (\phi(\tilde{r}_{ij}) - \phi(\tilde{r}_{ij}^0)) \right]; \qquad l \geq 2$$

$$\tilde{W}_{(1)} = \lambda_i^3 \exp - \frac{1}{\tilde{T}_1} \left[\sum_2^N (\phi(\tilde{r}_{ij}) - \phi(\tilde{r}_{ij}^0)) \right]; \qquad l = 1 \tag{1.6.75}$$

and likewise for the $\hat{U}$ functions,

$$\tilde{U}_{\{n_i\}}^k = \prod_i \lambda_i^{3n_i} \hat{U}_{\{n_i\}}^k$$

$$\tilde{\lambda}_i = \frac{\lambda_i}{\sigma}; \qquad \sum n_i = n \tag{1.6.76}$$

and with the reduced temperatures $\tilde{T}_1$ and $\tilde{T}_m$ defined by the relations

$$\tilde{T}_1 = \frac{kT}{\sum \sum x_s x_t u_{st}^*}; \qquad l = 1$$

$$\tilde{T}_m = \frac{kT}{\sum\sum x_s^k x_t^k u_{st}^*}; \qquad l \geq 2$$

(1.6.77)

with x_s and x_t the mole fractions of s and t in the mixture. By substitution,

$$J_{\{n_i\}}^k = \prod_i \frac{1}{n_i!} \left(\frac{\sigma}{\lambda_i}\right)^{3n_i} \int \sum \prod \tilde{W}_{\{n_i\}}^k \, d^3\tilde{r}_1 \cdots d^3\tilde{r}_n$$

$$= \prod_i \tilde{\lambda}^{-3n_i} \mathscr{I}_{\{n_i\}}^k (\tilde{T}_m, \tilde{T}_1, \tilde{v})$$

(1.6.78)

in analogy with Eq. (1.6.65). The argument $\tilde{T}_1$ refers only to single cell clusters.

Thus far two assumptions of dominant importance have been made. The first assumption, that there was a random distribution of molecular pairs, appears extremely difficult to remove. It was introduced originally to permit a logical choice of reference lattice and allow a partial summation of the configurational partition function. It has not been possible to formulate any corresponding states treatment for nonrandom mixtures via molecular considerations. Of course the very general argument given at the beginning of this section is independent of any assumption about the character of the solution, but it does not rely on any molecular interpretation. The second major assumption, that the molecules are all of equal size, is removed by the following device. The pair potential expressed in Eq. (1.6.74), generalized to different molecular sizes, is substituted into Eqs. (1.6.67) and (1.6.73). By forcing the average interaction potential into the same functional form as that of a pair of molecules it is readily found that

$$\langle V \rangle = \langle u^* \rangle \phi \left(\frac{R}{\langle \sigma \rangle}\right) = \sum_s \sum_t x_s x_t u_{st}^* \phi \left(\frac{R}{\sigma_{st}}\right)$$

$$\langle u^* \rangle = \frac{[x_1^2 u_{11}^* \sigma_{11}^6 + 2x_1 x_2 u_{12}^* \sigma_{12}^6 + x_2^2 u_{22}^* \sigma_{22}^6]^2}{x_1^2 u_{11}^* \sigma_{11}^{12} + 2x_1 x_2 u_{12}^* \sigma_{12}^{12} + x_2^2 u_{22}^* \sigma_{22}^{12}}$$

$$\langle \sigma \rangle = \left[\frac{x_1^2 u_{11}^* \sigma_{11}^{12} + 2x_1 x_2 u_{12}^* \sigma_{12}^{12} + x_2^2 u_{22}^* \sigma_{22}^{12}}{x_1^2 u_{11}^* \sigma_{11}^6 + 2x_1 x_2 u_{12}^* \sigma_{12}^6 + x_2^2 u_{22}^* \sigma_{22}^6}\right]^{1/6}$$

(1.6.79)

The composition-dependent energy and size parameters will vary somewhat with the choice of the function $\phi(R/\sigma)$. The results given in Eq. (1.6.79) are for the well-known Lennard–Jones 6–12 potential and the parameters can be evaluated for any inverse power pair potential with analogous expressions obtainable. By procedures completely equivalent to those already discussed, it is readily seen that the cell cluster integrals may be expressed in the reduced form

$$J^k_{\{n_i\}} = \prod_i \tilde{\lambda}^{-3n_i} \mathscr{J}^k_{\{n_i\}}(\tilde{T}_1, \tilde{T}_m, \tilde{v}_1, \tilde{v}_m) \qquad (1.6.80)$$

where the reduced temperatures and volumes are defined as

$$\tilde{T}_1 = \frac{kT}{\langle u^* \rangle_1} \; ; \qquad \tilde{T}_m = \frac{kT}{\langle u^* \rangle_m}$$

$$\tilde{v}_1 = \frac{V}{N \langle \sigma \rangle_1^3} \; ; \qquad \tilde{v}_m = \frac{V}{N \langle \sigma \rangle_m^3} \qquad (1.6.81)$$

and where $\tilde{T}_1$ and $\tilde{v}_1$ refer only to the case of single cell clusters. Clusters containing two or more molecules depend only on $\tilde{T}_m$ and $\tilde{v}_m$.

1.6.8 DISCUSSION

The preceding development shows that, within the approximation of random mixing, reduced temperatures and volumes may be formulated which permit the cluster integrals for the mixture to be adjusted to be equal to the corresponding integrals of the same shape k and number of molecules by adjusting $\tilde{T}_m$ and $\tilde{v}_m$ to be equal to $\tilde{T}$ and $\tilde{v}$, and for one molecule clusters adjusting $\tilde{T}_1$ and $\tilde{v}_1$ analogously. Since for each cluster the reduced variables depend upon the cluster composition they may differ appreciably from cluster to cluster. In the limit as the cell cluster size increases to encompass all the molecules in the fluid, the condition of local internal random mixing becomes identical with the random mixing condition which leads to the randomized potential expressed in Eq. (1.6.67). In this limit, the definitions of the reduced variables coincide with that of Kirkwood and co-workers.

Equation (1.6.81) represents the theorem of corresponding states in terms of cell clusters, not thermodynamic functions. In principle, if the properties of the pure components are known in terms of the cell cluster integrals, then the properties of the mixture, with random mixing in the sense defined, can be expressed in terms of the properties of the pure components at suitable reduced temperatures and volumes, not necessarily the same for all clusters. This is the format of the extended additivity hypothesis.

The term random mixing has been used many times; the random mixing approximation appears in several places. If we consider first the molecules inside a cell cluster, it has been shown that there are two possible kinds of random mixing. In the first, consistency with the over-all assumption of a random distribution requires that each cell cluster be computed with an averaged potential energy randomized with respect to the over-all composition of the mixture and taken to be the same for all clusters independent of composition. This corresponds to the development of Kirkwood and Brown. On the other hand, by using a moment expansion one can express the effective potential within a cell cluster by randomizing it with respect to the cluster composition. This method has the advantage of preserving differences between clusters of different composition and still being consistent with over-all random mixing. For, when any cell cluster becomes large enough to encompass macroscopic regions of the fluid, the contributions to the thermodynamic functions from clusters which depart markedly from the over-all mixture composition is very small. This situation arises simply because to construct a large cluster deviating markedly from the bulk composition one would need to have a region of space covering many contiguous cells which also deviated from the mean composition. The probability that such a macroscopic region has a given concentration may be calculated from fluctuation theory and can be seen to be very small if the concentration is not the mean composition. Thus, as the cell clusters become larger, only those with the mean composition contribute significantly to the thermodynamic properties of the fluid. This is obviously consistent with over-all random mixing.

A second place in which the random mixing approximation is encountered is in the computation of the mean molecular field due to the rest of the fluid external to a cell cluster. In this computation, random mixing implies that the distribution of molecules amongst the virtual lattice sites is random. The use of the combination of the second type of cluster potential with the external distribution described above is the type of random mixing envisaged.

Finally, we made an approximate evaluation of the combinatorial factor $g(\zeta_{k,\{n_i\}})$ which consisted of truncating the series after terms corresponding to clusters of two cells and computing the combinatory factor as that for the random distribution on a lattice of sites of a hypothetical solution of dimers and monomers. It is not necessary to make this approximation to define reduced temperatures or volumes. For practical use of these results some approximation to the combinatorial factor must be made, but this approximation has nothing to do with the molecular fields.

It is pertinent to re-emphasize that the theorem of corresponding states as presented herein differs from the usual formulation. It is common to use the theorem of corresponding states to relate the thermodynamic properties of one system to those of another via well-defined thermodynamic derivatives. However, our intention has been to relate the properties of one substance to those of another via a well-defined theoretical form of the equation of state, in this case the cell cluster formulation. To adjust the reduced ptemeratures and volumes of pure components to be equal to those in the mixture, we expand (compress) and heat (cool) each pure component until successively each cell cluster contribution to the free energy in the pure component is identical with the same cell cluster contribution in the mixture. This does not imply that all cell cluster integrals have the same functional form, but only that the use of a reduced temperature and volume suffices to make a given cell cluster in the pure component have the same functional form as the same cell cluster in the liquid. The requirement for the use of this form of the theorem of corresponding states is the knowledge of the contributions to the thermodynamic properties of each of the cell clusters for some one pure fluid. A common analytic form for all cell clusters would only be required if it were insisted that the same physical process produce a similar effect on all clusters i.e., an increase in volume cause them all to decrease. Nowhere has such an assumption been made.

It should be pointed out that the contribution to cell clusters containing more than one molecule corresponds to the inclusion of correlations between the molecules in the fluid. It is not at all obvious that the correlations in the mixture will be sufficiently like those in the components to make these terms unimportant. In particular, it is to be anticipated that the poor results obtained for the entropies of mixing in both the cell models (single cells only) and in the second-order treatments of Brown and Kirkwood, will be somewhat improved by accounting for these differential correlation effects.

REFERENCES

1. W. G. McMillan and J. E. Mayer, *J. Chem. Phys.* **13**, 276 (1945).
2. J. G. Kirkwood and F. P. Buff, *J. Chem. Phys.* **19**, 774 (1951).
3. T. L. Hill, *J. Chem. Phys.* **23**, 623, 2270 (1955).
4. F. E Harris and S. A. Rice, *J. Phys. Chem.* **58**, 725, 733 (1954); S. A. Rice and F. E. Harris, *J. Chem. Phys.* **24**, 326, 336 (1956).
5. F. T. Wall and J. Berkowitz, *J. Chem. Phys.* **26**, 114 (1957); M. Nagasawa and I. Kagawa, *Bull. Chem. Soc. Japan* **30**, 961 (1957); S. Lifson, *J. Chem. Phys.* **27**, 700 (1957).
6. S. A. Rice and F. E. Harris, *Z. physik. Chem.* **8**, 207 (1956).
7. J. E. Mayer, *J. Chem. Phys.* **10**, 629 (1942).
8. But see F. P. Buff and R. Brout, *J. Chem. Phys.* **23**, 774 (1955).
9. T. L. Hill, *Lecture and Review Series*, No. 54–2, Naval Medical Research Institute, National Naval Medical Center, Bethesda, Maryland. See also T. L. Hill, " Statistical Mechanics." McGraw Hill, New York, 1956.
10. See, for example, J. Hirschfelder, C. Curtiss, and R. Byrd, " Molecular Theory of Gases and Liquids." Wiley, New York, 1954.
11. K. S. Pitzer, *J. Chem. Phys.* **7**, 583 (1939).
12. R. W. Zwanzig, *J. Chem. Phys.* **22**, 1420 (1954).
13. H. C. Longuet-Higgins, *Proc. Roy. Soc.* **A205**, 247 (1951).
14. See I. Prigogine, " Molecular Theory of Solutions." North Holland Publishing Co., Amsterdam, 1958.
15. F. Wallbrook, *J. Chem. Phys.* **23**, 749 (1955).
16. R. L. Scott, *J. Chem. Phys.* **15**, 193 (1956); I. Prigogine, R. Bellemans, and A. Englert-Chowles, *J. Chem. Phys.* **24**, 518 (1956); J. G. Kirkwood, Z. Salsburg, and P. Wojtowicz, *J. Chem. Phys.* **26**, 1553 (1957); *Ibid.* **27**, 505 (1957); W. B. Brown, *Proc. Roy. Soc.* **A240**, 561 (1951); *Phil. Trans. Roy. Soc.* **A250**, 175 (1957).

17. L. H. NOSANOW, *J. Chem. Phys.* **30**, 1596 (1959).
18. See E. A. GUGGENHEIM, "Mixtures." Oxford Univ. Press, London and New York, 1952.
19. S. A. RICE, *Ann. N. Y. Acad. Sci.* **65**, 33 (1956); *J. Chem. Phys.* **24**, 1283 (1956).
20. J. DE BOER, *Physica* **20**, 655 (1954).
21. J. POPLE, *Phil. Mag.* **41**, 459 (1951); I. PRIGOGINE AND P. JANNSENS, *Physica* **16**, 895 (1950).
22. J. POPLE, *Trans. Faraday Soc.* **49**, 591 (1953).
23. Z. SALSBURG AND J. G. KIRKWOOD, *J. Chem. Phys.* **20**, 1538 (1952).
24. J. G. KIRKWOOD, *J. Chem. Phys.* **18**, 380 (1950).
25. S. A. RICE, *J. Chem. Phys.* **24**, 357 (1956).
26. S. A. RICE, *J. Chem. Phys.* **29**, 141 (1958).

2. The Equilibrium Properties of Dilute Electrolyte Solutions

2.0 Introduction

A major goal of the statistical theory of solutions is the prediction of the macroscopic properties of the solution from the known properties of the component molecules. The degree to which this goal can be achieved depends, of course, upon the complexity of the problem and the possibility of making simple approximations to the often unattainable exact answer. In this chapter we shall specialize some of the results of the general molecular theory discussed previously and be concerned with the properties of solutions which contain charged molecular species, but which are nevertheless electrically neutral as a whole. Such charged solutes interact very strongly with both the solvent molecules and with other solute molecules. Moreover, the very long range of the solute–solute interactions, which are Coulombic in origin, prevents the use of approximations which would have the physical effect of limiting the range of interaction between charged molecules to some small number of other molecules in the immediate neighborhood of a given ion. It is immediately apparent that the mathematical complexity of the problem is greatly increased by the inapplicability of this approximation since it is always necessary to consider the simultaneous interaction of many particles rather than only the interaction of a few particles.

The theory of electrolytes, at least as usually presented, limits itself to a discussion of the properties of the solution attributable to the electrostatic forces operative between the ions. Thus, though the heat of mixing of an ionic compound and a dielectric medium may be very large, this effect is not considered. A typical effect which is discussed is the heat of dilution, i.e., the change in enthalpy on altering the concentration of an already existent electrolyte solution. In the sense that many gross effects which are due to the interaction of the charged ions with the solvent are neglected, the theory of electrolyte solutions is very incomplete. In some measure, the neglect of these ion–solvent interactions tends to limit the validity of the usual theory to the range where the solvent is in such great excess that there is little change in solute–solvent interactions with changes in solute concentration. Though some attention will be devoted to the problem of ion–solvent interactions, in the main we shall follow the classical outlines of the theory of electrolytes and discuss the effect of ion–ion interactions for small ions, colloids, and polyelectrolytes.

2.1 Some Historical Comments

In this section will be given a brief quasi-historical discussion of the development of our current views of the structures of electrolyte solutions. The fact that solutions of certain inorganic compounds will permit the passage of an electric current has been known since about 1800. At first, attention was focused on the remarkable chemical effects produced by the passage of the current but little progress could be made in unraveling the apparently mysterious chemical reactions proceeding under the influence of the current until the charge-carrying species could be identified, even if only in theory. The fact that solutions of electrolytes will conduct an electric current at all led to early speculations that the solute molecules were charged. Grotthuss[1] postulated that electrolysis was due to a successive decomposition and recombination among the molecules of the electrolyte. He proposed that the electrical field oriented the salt molecules and then forced away the ends of the chain with a resultant shift of partners all along the line. An alternative point of view, due to Berzelius,[1] asserted that the passage of the electric current was due to the electrical field separating a salt into its positive and negative components. However, if it is necessary for the applied field to induce the originally uncharged solute molecules to dissociate into ions, then there would be a minimum electrical field required for conduction, the solution being nonconducting for applied fields smaller than this critical value. Since this implication directly contradicts the observed behavior, the hypothesis of field induced ionization had to be discarded. The disagreement with experiment cited above was first pointed out by Clausius[2] who replaced the original notion with the hypothesis that the solute in an electrolyte solution was charged even in the absence of any external field. However, Clausius' contention was that the fraction of solute molecules present as ions was always small.

Parallel with these speculations, important empirical contributions were made by Hittorf,[1] who showed that the velocities with which the ions in an electrolyte move are not equal, and by Kohlrausch[1] who showed that the low concentration mobility, or relative migration velocity of an ion, was independent of the salt in which it was contained. These facts are in harmony only with the assumption that the ions are (relatively) free during conduction. But since the ions are presumed to exist in the absence of any applied field, and it would be extremely unusual if all electrolytes dissociated to exactly the same (small) fractional extent, these data point to complete or nearly complete dissociation of the solute into ions. Kohlrausch did not draw this conclusion and it remained for Arrhenius[3] to propose an extention of Clausius' hypothesis that electrolytes are largely dissociated in solution. The experimental evidence which led to Arrhenius' proposal consisted mainly of conductivity data previously cited and of measurements of the abnormal colligative properties of electrolyte solutions: freezing point depression, boiling point elevation, and osmotic pressure. It is important to note that, at the time that Arrhenius' proposal was made, the theory of solutions was dominated by the superficial resemblance between the properties of dilute solutions and the behavior of an ideal gas. It is only with the advent of detailed molecular theories, such as that of McMillan and Mayer, that the analogy could be explored in detail and the discrepancies noted. In modern terminology, the consequence of the assumption that the solute behaves as an ideal gas, and the solvent merely as a container is that the properties of the solution are considered to be determined by the translational degrees of freedom only, and that the interactions between solute molecules and between solute and solvent molecules need not be considered. Though considerations based on the identification of a solute solution and an ideal gas cannot be justified in detail from either the thermodynamic or the molecular point of view, in the hands of Van't Hoff[1] this analogy led to the establishment of the laws describing the colligative properties of a dilute ideal solution. The neglect of all intermolecular forces, of course, leads to the conclusion that the properties of the solute when in solution depend only upon the number of solute molecules per unit volume. In terms of this theory, then, measurements of the colligative properties of a solution determine the number of free particles in the solution.

The mathematical formulation of the Arrhenius theory is quite simple since it is based solely on the postulated existence of a dissociation equilibrium between ideal solutes. The dissociation constant of the salt $A_{\nu_+}B_{\nu_-}$ is simply characterized as

$$K_x(T, p, \text{solvent}) = \frac{x_+^{\nu_+} x_-^{\nu_-}}{x_0} \tag{2.1.1}$$

with the x_i the mole fraction of species i. In terms of α, the fraction of solute dissociated, the mass action expression (2.1.1) becomes

$$K_x(T, p, \text{solvent}) = \frac{\nu_+^{\nu_+} \nu_-^{\nu_-} \alpha^{\nu_+ + \nu_-}}{1 - \alpha} \left[\frac{n}{n_1 + n(1 + (\nu_+ + \nu_- - 1)\alpha)} \right] \tag{2.1.2}$$

with n the number of moles of solute $A_{\nu_+}B_{\nu_-}$ and n_1 the number of moles of solvent. In the interpretation of the behavior of strong electrolytes, it is fruitful to adopt as a standard of reference the properties to be expected from an ideal solution, in which the electrolyte is fully dissociated. In that case the sum of the mole fractions of all solute species would be $(v_+ + v_-)n/n_1$, where $n_1 \gg n$. The extent to which, say, the observed colligative properties fall short of the value corresponding to these conditions may then be characterized by an osmotic coefficient g, defined in terms of the chemical potential by

$$\mu_i = \mu_i^0 + g_i RT \ln x_i \tag{2.1.3}$$

so that e.g. the osmotic pressure and the freezing point depression become

$$\pi \bar{v}_1 = g_1 \frac{(v_+ + v_-)n}{n_1} RT$$

$$\Delta T = g_1 \frac{RT_0^2}{L_f} (v_+ + v_-) \frac{n}{n_1}. \tag{2.1.4}$$

It is to be noted that, although the solution has been assumed to be ideal, g_1 differs from unity since the total number of particles in solution is not constant, but changes on dilution due to dissociation. In fact,

$$1 - g_1 = \frac{v_+ + v_- - 1}{v_+ + v_-} (1 - \alpha) \tag{2.1.5}$$

or

$$1 - g_1 = \frac{v_+ + v_- - 1}{v_+ + v_-} \cdot \frac{\alpha^{v_+ + v_-} v_+^{v_+} v_-^{v_-}}{K_x(T, p)} \frac{n}{n_1} \tag{2.1.6}$$

where $n_1 \gg n$.

To what extent does this simple model conform to reality? If the conductance of electrolyte solutions is explained in terms of free ions and undissociated molecules, then the conductance must be proportional to the degree of dissociation,

$$\Lambda = \alpha \Lambda_0 \tag{2.1.7}$$

with Λ the molal conductance. Values of α calculated from conductance data and from thermodynamic data are in fair accord as can be seen from Table 2.11.

Although the qualitative agreement is heartening, it is instructive to proceed with the argument and pursue the deficiencies of the classical theory of Arrhenius. If the concentration is expressed in terms of moles per liter of solution, c_i, then $c_i = n_i/V$ where V is the volume of the solution. In terms of this new concentration unit, Eq. (2.1.2) assumes the form

$$\frac{c_+^{v_+} c_-^{v_-}}{c_0} = v_+^{v_+} v_-^{v_-} \frac{\alpha^v}{1 - \alpha} c^{v-1}; \qquad v = v_+ + v_-$$

$$K_c(T, p, n_i) = K_x(T, p) \left[\Sigma \frac{n_i}{V} \right]^{v-1}. \tag{2.1.8}$$

TABLE 2.1.1[a]

DEGREE OF DISSOCIATION FOR TYPICAL ELECTROLYTES (18°C)

	Concentration (gm-eq/l)									
Electrolyte	5×10^{-3}		10^{-2}		2×10^{-2}		5×10^{-2}		10^{-1}	
	a	*b*	*a*	*b*	*a*	*b*	*a*	*b*	*a*	*b*
KCl	0.956;	0.963	0.941;	0.943	0.922;	0.918	0.889;	0.885	0.860;	0.851
NaCl	0.952;	0.953	0.936;	0.938	0.914;	0.922	0.878;	0.892	0.852;	0.875
LiCl	0.949;	0.944	0.931;	0.937	0.908;	0.928	0.870;	0.912	0.846;	0.901
HCl	0.981;	0.991	0.973;	0.972	0.962;	0.957	0.944;	0.933	0.923;	0.912
$MgSO_4$	0.739;	0.694	0.666;	0.618	0.592;	0.536	0.498;	0.420	0.434;	0.324

[a]Reproduced from ref. 4.
a: From conductance measurements.
b: From freezing point depression measurements.

When α is close to unity,

$$1 - g_1 = \frac{\nu - 1}{\nu} c^{\nu-1} \frac{\nu_+^{\nu_+} \nu_-^{\nu_-}}{K_x(T, p)} \left[\Sigma \frac{n_i}{V} \right]^{1-\nu}$$

$$= \frac{\nu - 1}{\nu} c^{\nu-1} \frac{\nu_+^{\nu_+} \nu_-^{\nu_-}}{K_c}. \tag{2.1.9}$$

Note that K_c is not, from the rigorous point of view, a true equilibrium constant since it depends on the concentration through the factor $(\Sigma n_i/V)^{\nu-1}$ which is the total number of moles and is not a constant. However, in dilute solutions, since $n_1 \gg n_i$ ($i \neq 1$), this factor is effectively constant. If now, $(1 - g_1)$ is plotted versus the concentration, $c = n/V$, for a binary electrolyte, the slope at the origin should be $1/2K_c$, i.e., the curve is linear in c. For electrolytes which dissociate into more than two ions, $\nu = \nu_+ + \nu_- > 2$, and the curve should approach the origin as a tangent to the c axis. Experimentally, for electrolytes for which $\alpha \approx 1$, it is found that[3]

$$1 - g_1 = \text{constant} \times c^{\frac{1}{2}} \tag{2.1.10}$$

although for electrolytes for which $\alpha \ll 1$, Eq. (2.1.9) is approximately followed. The observed behavior of strong electrolytes is thus not reconcilable with the requirements of Eq. (2.1.9). In a similar manner, combining the relations for conductance and degree of dissociation with that for the dissociation constant,

$$\frac{\Lambda_0 - \Lambda}{\Lambda_0} = \nu_+^{\nu_+} \nu_-^{\nu_-} \frac{\alpha^\nu}{K_c} c^{\nu-1} \tag{2.1.11}$$

which predicts that $(\Lambda_0 - \Lambda/\Lambda_0)$ is proportional to an integral power of the concentration. This is generally true only if α is sufficiently close to unity, so that the relative variation of α with c is small compared with the relative variation of $(1 - \alpha)$. Here again it is well known that the observed conductance of an electrolyte solution is consistent with the relation[5]

$$\Lambda = \Lambda_0 - \text{constant} \times c^{\frac{1}{2}} \qquad (2.1.12)$$

and therefore irreconcilable with the requirements of Eq. (2.1.11). A slight rearrangement of Eq. (2.1.11) leads to the well-known Ostwald dilution law for an electrolyte,[6]

$$\frac{v_+^{v_+} v_-^{v_-} (\Lambda/\Lambda_0)^v}{1 - (\Lambda/\Lambda_0)} c^{v-1} = K_c. \qquad (2.1.13)$$

It has previously been mentioned that when $\alpha \ll 1$, the weak electrolyte seems to conform to the Arrhenius theory approximately, whereas when $\alpha \approx 1$, this is not so. In Table 2.1.2 are listed dissociation constants computed from the conductance data listed in Table 2.1.1 using Eq. (2.1.13). In Table 2.1.3 are similar data for the weak electrolyte acetic acid.

TABLE 2.1.2

APPARENT DISSOCIATION CONSTANTS FOR STRONG ELECTROLYTES

Electrolyte	Apparent dissociation constants				
	0.005 gm-eq/1	0.01 gm-eq/1	0.02 gm-eq/1	0.05 gm-eq/1	0.1 gm-eq/1
KCl	0.104	0.150	0.218	0.356	0.529
NaCl	0.0945	0.137	0.195	0.317	0.490
LiCl	0.0884	0.126	0.179	0.292	0.465
HCl	0.253	0.350	0.487	0.795	1.11
MgSO$_4$	0.0105	0.0133	0.0172	0.0247	0.0332

It is immediately noted that the dissociation constant K_c is not at all constant for strong electrolytes. It is now apparent that though there is ample evidence that strong electrolytes are completely dissociated into ions, these ions do not behave as independent particles even at relatively high dilution and this latter aspect of the classical dissociation theory must be discarded. On the other hand, for weak electrolytes, where the number of ions per unit volume is very small and $\alpha \ll 1$, Eq. (2.1.13) is closely followed as the data in Table 2.1.3 show.

TABLE 2.1.3[a]

APPARENT DISSOCIATION CONSTANT OF ACETIC ACID (25°C)

c (moles/liter)	$K_c \times 10^5$
0.07369	1.835
0.03685	1.825
0.01842	1.818
0.009208	1.810
0.004606	1.811
0.002303	1.807
0.001151	1.806
0.0005757	1.807
0.0002879	1.807

[a]Reproduced from ref. 4.

It is, of course, now well-known from what origin the deviations from the dissociation theory arise. However it should be recalled that it was not until the advent of X-ray crystallography that it was discovered that the compounds which were strong electrolytes in solution were also ionic in the crystalline state. Since there was no trace of molecule formation even in the crystal, the assignment of the concentration dependence of the thermodynamic properties of electrolyte solutions to incomplete dissociation had to be modified. In recent years some evidence has accumulated for the existence of ion pairs in solution so that there does exist, in a certain sense, an equilibrium between dissociated and ion-paired species. However, these ion pairs are not compounds characteristic of the salt in the crystalline state, but rather represent transient species formed due to the competition between attractive electrostatic forces and the entropic forces due to bombardment by the solvent molecules. In dilute solutions of strong electrolytes in solvents of high dielectric constant, the number of ions pairs is usually vanishingly small.

In general, in sufficiently dilute solutions of strong electrolytes, the concentration dependence of all the reduced colligative properties as well as the conductance and reduced viscosity is given by[5]

$$F(c) - F(0) = \text{constant} \times c^{1/2} \qquad (2.1.14)$$

where $F(0)$ is the value of F at zero solute concentration and the constant depends upon the property being measured. It is further observed that the proportionality constant is a function of the dielectric constant of the medium and of the charge on the ions. For example, $(1 - g_1)$ decreases with increasing dielectric constant and increases with increasing charge on the ion. The deviations from ideality in solutions of nonelectrolytes are never linear in the square root of the solute concentration.[7] This fact, coupled with the observed dependence of the properties of the solution on the dielectric constant and the ionic charge, suggests that the unique properties of solutions of electrolytes are in some way related to the long-range electrostatic interactions of the ions.

A possible model for solutions of strong electrolytes, then, represents the solute as completely dissociated and treats the concentration dependence of the solution properties as entirely due to the interactions between the ions. Amongst the earliest investigators to advocate that the anomalies peculiar to solutions of electrolytes aere due to Coulombic interactions between the ions, are Van Laar,[8] Sutherland,[9] Hertz,[10] Milner,[11] and Bjerrum.[12] In particular, Hertz computed the conductivity of an electrolyte solution by carrying over the formalism developed in the kinetic theory of gases taking cognizance of the interionic electrical forces. The applicability of the considerations developed for a dilute gas to a system in which the ion is in essentially continuous interaction with other molecules and ions is subject to serious objections. In fact, it is only recently that the statistical mechanics of transport has been sufficiently developed to make possible, even in principle, a calculation of the absolute value of the mobility of an ion. Nevertheless, the recognition of the importance of interionic forces was a major contribution of this work. Milner was the first to attempt to calculate the effects of the interionic forces on the thermodynamic properties of the solution. His procedure consisted of weighting all configurations of the ions about a central ion in accordance with the appropriate Boltzmann factor. Due to mathematical difficulties, Milner evaluated the configurational partition function by a graphical procedure. Aside from trivial errors, these calculations are in complete agreement with the results obtained from Debye–Hückel theory.

The most prominent theory of electrolyte solutions in the period between 1910 and 1923 was due to Ghosh.[13] This theory assumed that the solute was completely dissociated and that the deviations were due to electrical forces between the ions. To calculate the effects of the electrostatic forces, Ghosh used a cell model and erroneously presumed that the ions were sufficiently localized to form a quasi-lattice spanning the available volume. The electrical energy of such a quasi-crystal was then approximately computed in terms of the dielectric constant of the medium. If it is assumed that the role of the solvent is merely to provide a medium in which the solute particles move, then the effects of the electrostatic forces may be estimated from the viral theorem.[14] This procedure leads to the prediction that $(1 - g_1)$ should be proportional to the one-third power of the solute concentration, in disagreement with experiment. Despite the gross errors in Ghosh's theory, it does show clearly that the introduction of interionic forces could modify the properties of the solution markedly.

The most significant advance in the theory of electrolytes to follow the original dissociation equilibrium theory of Arrhenius was the development of a simple analytical treatment of the effects of interionic forces. This development, due to Debye and Hückel,[15] makes use of a continuum approximation in which it is assumed that the real discrete charge distribution is replaceable by a continuous charge distribution. The distribution of charge about some central ion is then governed by the competition between the electrostatic forces which would tend to cause the charge to neutralize itself, and the thermal force due to the bombardment by the solvent molecules which tends to drive the ions apart and to increase the entropy of the solution. If the work done in separating a pair of ions is assumed proportional to the mean electrostatic potential, and the charge distribution is assumed to be continuous, then the mean electrostatic potential must obey Poissons equation, since it will be recalled that Poisson's equation relates to electrostatic potential to the charge density in a continuous dielectric medium. The solution of the resultant differential equation with appropriate boundary conditions then provides an approximation to the electrostatic potential and thus to the distribution of the ions. The justification for the unorthodox use of the continuum approximation as well as several other mathematical approximations can only be investigated from the point of view of statistical mechanics.

Though the spectacular successes of the Debye–Hückel theory have tended to overshadow questions of its validity, we shall conern ourselves in part with a detailed investigation of just this aspect of the theory. Such an investigation is of importance for our purposes since a thorough understanding of the inherent limitations and approximations of the theory will be necessary in our later discussion of more complex systems containing colloids or polyelectrolytes.

Following the original Debye–Hückel theory, significant advances have been made by Onsager,[16] Falkenhagen,[17] Kirkwood,[18] Mayer,[19] and several others. In recent years attention has been focused on the properties of concentrated electrolyte solutions. An approximate theory of such solutions due to Eigen and Wicke[20] has received much attention. In view of the excellent reviews of modern developments now available[20a, 20b] we shall confine our attention throughout to the basic validity of the Poisson–Boltzmann equation and to its approximate solution.*

2.2 Energy Relations in Dielectric Media[21-23]

In the preceding section we have treated the properties of solutions of electrolytes without any specific discussion of the energetic changes required to bring together the system of charges. In this section we briefly consider the

*We wish to remind the reader that the modern theory of simple electrolytes has advanced far beyond this stage. However, since almost the entire body of theory connected with polyelectrolytes is based on the Poisson-Boltzmann equation, we have restricted attention to this type of analysis.

relationship between the electrical properties of the medium and the energy of assembly of the solution, reserving for a later section a discussion of the local interaction between an ion and the solvent.

From the outset all processes shall be considered to be isothermal. Consider a set of charges $\{z_i q\}$ in free space. If these charges are intially infinitely distant from one another, the change in energy in introducing them into the volume V is easily seen to be

$$W = \tfrac{1}{2} \sum_i \sum_j \frac{z_i z_j q^2}{R_{ij}} \tag{2.2.1}$$

where R_{ij} is the distance between ions i and j. If the electrostatic potential at the point $\mathbf{R}_i$ (where the charge $z_i q$ is located) due to all other charges in the system is denoted by ψ_i^*, then the energy required for the process described above may be rewritten in the form

$$W = \tfrac{1}{2} \sum z_i q \psi_i^*. \tag{2.2.2}$$

The potential due to the charge on ion i is omitted from ψ_i^*. Since the electrostatic forces are conservative the energy change due to bringing the charges together in the volume V is equal to the work expended. It will prove convenient to find the analog of Eq. (2.2.2) for a continuous charge distribution. Consider then the work required to remove the set of charges discussed above to infinite mutual separation and also to completely disperse the charge into infinitesimal charges. By reversing the process previously described the work for complete dispersal is clearly

$$W = \frac{1}{2} \int_V \int_V \frac{\rho_i \rho_j}{R_{ij}} \, dV_i \, dV_j \tag{2.2.3}$$

where ρ_i and ρ_j are the charge densities in the infinitesimal volume elements dV_i and dV_j separated by the distance R_{ij}. Using the definition of the electrostatic potential and introducing Poisson's equation,

$$\nabla \cdot \mathscr{E}_i = -\nabla^2 \psi_i = \frac{4\pi \rho_i}{D} \tag{2.2.4}$$

Eq. (2.2.3) becomes

$$W = \frac{1}{2} \int_V \rho_i \psi_i \, dV_i = \frac{1}{8\pi} \int_V \nabla \cdot \mathscr{E}_i \psi_i \, dV_i \tag{2.2.5}$$

where $\mathscr{E}_i$ is the electrostatic field strength. In the following, the subscript i denoting that the quantitities are due to all other charge but that located in the volume element dV_i may be dropped. Relation (2.2.5) may be rewritten as

$$\frac{1}{8\pi} \int_V [\nabla \cdot (\mathscr{E}\psi) - \mathscr{E} \cdot \nabla\psi] \, dV. \tag{2.2.6}$$

The first term vanishes by an application of Gauss' theroem since $\mathscr{E}$ vanishes on the surface of a volume element and $\nabla\psi = -\mathscr{E}$. Thus,

$$W = \frac{1}{8\pi} \int_V \mathscr{E}^2 \, dV \tag{2.2.7}$$

and it is clear that $\mathscr{E}^2/8\pi$ is the energy density due to the field.

It is now convenient to draw up a balance sheet for the energy changes involved in the assemblage of the set of charges. These charges are presumed to be located at the points $\{\mathbf{R}_i\}$ and are not subject to movement by secondary forces, such as interaction with a solvent medium. We refer to this system as a set of *controlled* charges. In a real medium, due to the molecular structure of the solvent, each ion is subject to continual bombardment by solvent molecules. Due to this Brownian movement a set of real ions in a real solvent will be referred to as a set of *uncontrolled* charges. In the instance studied above, namely the assembly of a set of controlled charges in a vacuum, the work done is exactly equal to the field energy stored up in space. It will be seen in the following that when a set of controlled charges is assembled in a dielectric medium, in addition to the energy stored in the electro-static field, energy is also stored as potential energy in the dielectric, and some free energy is used up in reorganizing the dipoles of the dielectric. That is, due to the reorganization of the dipolar structure of the medium, there is a decrease in the entropy of the dielectric and some heat is "squeezed out."

Consider now that a one component dielectric is introduced everywhere. As before, temporarily consider only isothermal processes. Under these conditions the work required to assemble the charge is no longer given by Eq. (2.2.1) but is instead

$$W = \frac{1}{2D} \sum_i \sum_j \frac{z_i z_j q^2}{R_{ij}} \tag{2.2.8}$$

and the work of assemblage in a dielectric medium by the relation

$$W = \frac{W(\text{free space})}{D} = \frac{D}{8\pi} \int_V \mathscr{E} \cdot \mathscr{E} \, dV \tag{2.2.9}$$

for an isotropic medium.

To demonstrate that Eq. (2.2.9) is correct, consider an infinitesimal incre-ment of charge $\delta\rho$ to be added to the system, the process to occur at constant volume. Under these conditions, $pdV = 0$, and no mechanical work is done.

The work done in adding the charge is thus only the additional electric work which is given by

$$\delta W = \int_V \psi \delta \rho \, dV$$

$$= \frac{1}{4\pi} \int_V \psi \delta (\nabla \cdot \mathscr{D}) \, dV \qquad (2.2.10)$$

$$= \frac{1}{4\pi} \int_V \psi \nabla \cdot (\delta \mathscr{D}) \, dV$$

where we have used the fundamental equation

$$\mathscr{D} = D\mathscr{E} \qquad (2.2.11)$$

$$\nabla \cdot \mathscr{D} = 4\pi \rho$$

and $\mathscr{D}$ is the electric displacement. As before, ρ is the charge density in the system; $\mathscr{D}$ represents the electric field whose sources are the true charges in the medium. The relationship between $\mathscr{D}$ and $\mathscr{E}$ is

$$\mathscr{D} = \mathscr{E} + \frac{4\pi}{V} \mathscr{P} \qquad (2.2.12)$$

where $\mathscr{P}$ is the total polarization of the medium. The difference between $\mathscr{D}$ and $\mathscr{E}$ is just the field arising from the fictitious charges due to the polarization of the dielectric medium. Equation (2.2.10) may be written in the form

$$\delta W = \int_S \delta(\mathscr{D}\psi) \cdot d\mathbf{S} - \int_V \delta \mathscr{D} \cdot \nabla \psi \, dV \qquad (2.2.13)$$

$$\psi \nabla \cdot \delta \mathscr{D} = \nabla \cdot \delta(\mathscr{D}\psi) - \delta \mathscr{D} \cdot \nabla \psi$$

where Gauss' theorem has been used to transform the first integral over the volume to an integral over a surface S. If the surface is taken to be infinitely distant, the first term of the right-hand side of Eq. (2.2.13) is zero. The infinitesimal amount of work done is therefore given by the relation

$$\delta W = \frac{1}{4\pi} \int_V \mathscr{E} \cdot \delta \mathscr{D} \, dV \qquad (2.2.14)$$

using the definition of the electric field as $\mathscr{E} = -\nabla \psi$. To integrate Eq. (2.2.14) $\mathscr{E}$ must be specified as a function of $\mathscr{D}$. If the fluid is isotropic with a scalar dielectric constant D, which may be a function of position but not of field strength $\mathscr{E}$, the total work done is simply

$$W = \int_0^{\mathscr{D}} \delta W = \frac{1}{4\pi} \int_0^{\mathscr{D}} \int_V \mathscr{E} \cdot \delta \mathscr{D} \, dV$$

$$= \frac{1}{4\pi} \int_V \int_0^{\mathscr{E}} \frac{D\delta(\mathscr{E} \cdot \mathscr{E})}{2} \, dV \qquad (2.2.15)$$

$$= \frac{D}{8\pi} \int_V \mathscr{E}^2 \, dV$$

and this result is valid only when there is no dielectric saturation.

A simple comparison with the free space case can now be made by using the identity

$$\frac{D\mathscr{E}^2}{8\pi} = \frac{\mathscr{E}^2}{8\pi} + \frac{(D-1)\mathscr{E}^2}{8\pi}. \qquad (2.2.16)$$

The first term on the right-hand side of Eq. (2.2.16) is just the energy that would be stored in the electrostatic field in free space and the second term is the potential energy stored in the dielectric medium. This potential energy is due to the partial alignment of the dipoles of the dielectric along the direction of the electric field. As usual the extent of the orientation is determined by the change in free energy for the process. The gain in potential energy described above is in opposition to the change in entropy due to the increased orientation of the dipoles. Due to the increased orientation of the dipoles and the decrease in entropy of the dielectric, there is a release of heat at constant temperature. It is instructive to pursue this point somewhat further.

As explicitly stated at the outset, all the considerations of this section have been for processes occurring at constant temperature. Since the temperature was assumed constant, in the derivation of Eq. (2.2.15) it was possible to assume that the dielectric constant was independent of time and a function of position only. In general, D is a function of the temperature. To conduct the process isothermally then requires that the system of charges under discussion be kept in contact with a heat bath and that heat be exchanged between the thermostat and the system. Thermodynamically, the maximum work obtainable in an isothermal process is equal to the change in free energy. Dependent upon the path of the given isothermal process (i.e., at constant volume or constant pressure) the appropriate free energy is the Helmholtz or Gibbs free energy. The use of the condition expressed before, $p \, dV = 0$, restricts the preceding discussion to an isothermal process at constant total volume. The change in Helmholtz free energy in an isothermal, isochoric process at constant composition is

$$(\delta A)_T = \delta U - T\delta S \qquad (2.2.17)$$

where U is the internal energy and S the entropy of the system. Since the isothermal-isochoric work done is equal to the change in Helmholtz free energy, from Eq. (2.2.14) is obtained

$$(\delta A)_T = \frac{1}{8\pi} \int_V \mathscr{E} \cdot \delta \mathscr{D} \, dV. \qquad (2.2.18)$$

From Eq. (2.2.17) it is readily verified that the entropy change accompanying the assemblage of the charges is

$$\Delta S = -\left(\frac{\partial \Delta A}{\partial T}\right)_V = \frac{1}{8\pi} \int_V \mathscr{E}^2 \frac{dD}{dT} \, dV \qquad (2.2.19)$$

$$= \frac{1}{8\pi} \int_V \mathscr{E} \cdot \mathscr{D} \frac{d \ln D}{dT} \, dV.$$

Again, with the use of Eq. (2.2.17) the change in internal energy is seen to be

$$\Delta U = \frac{1}{8\pi} \int_V \mathscr{E} \cdot \mathscr{D} \frac{d(TD)}{dT} \, dV. \qquad (2.2.20)$$

Finally, the heat absorbed due to the change in entropy during the application of the field is

$$\delta Q = T\delta S = \frac{1}{8\pi} \int_V \mathscr{E} \cdot \delta \mathscr{D} \frac{d \ln D}{d \ln T} \, dV \qquad (2.2.21)$$

and it is this heat which is "squeezed out" in the reorganization of the dipoles of the dielectric medium.

It is possible to define an ideal dielectric as one for which the temperature dependence of the dielectric constant is given by

$$D = a + b / T \qquad (2.2.22)$$

where a and b are constants, independent of the temperature. Note that, in this case, $D - a$ is inversely proportional to T, and from Eq. (2.2.20) the internal energy density becomes just $(a\mathscr{E}^2/8\pi)$,* which differs from the corresponding result in matter free (empty) space by the constant factor a. Physically, a is simply the difference between the electric susceptibility of a given medium and of vacuum.

---

*These results also follow simply from the Gibbs-Helmholtz equation:

$$\Delta U = \Delta A - T\left(\frac{\partial \Delta A}{\partial T}\right)_V \qquad (2.2.22A)$$

whence, once it is recognised that ΔA is proportional to the reciprocal of the dielectric constant,

It is of interest to generalize the results of the preceding arguments to the case where the dielectric medium is composed of several components. The first law of thermodynamics becomes, in this case,

$$dU = T\,dS - p\,dV + \sum_{i=1}^{r} \mu_i\,dn_i + \mathscr{E}\cdot d\mathscr{P} \tag{2.2.23}$$

where, as before, $\mathscr{P}$ is the total polarization of the dielectric medium, μ_i is the chemical potential of component i, and dn_i is the variation in the number of moles of component i accompanying the change in energy dU. The volume V is entirely within the electrostatic field and we may imagine, for convenience, that the dielectric sample is contained between the plates of a condenser. In the usual manner, Gibbs and Helmholtz free energies may be defined by the relations

$$\begin{aligned} G &= U + pV - TS \\ A &= U - TS \end{aligned} \tag{2.2.24}$$

whence

$$dG = -S\,dT + V\,dp + \sum_{1}^{r} \mu_i\,dn_i + \mathscr{E}\cdot d\mathscr{P}$$

$$dA = -S\,dT - p\,dV + \sum_{1}^{r} \mu_i\,dn_i + \mathscr{E}\cdot d\mathscr{P}. \tag{2.2.25}$$

It is convenient to define two modified free energies G^* and A^* by

$$\begin{aligned} dG^* &= d(G - \mathscr{E}\cdot\mathscr{P}) = -S\,dT + V\,dp + \sum \mu_i\,dn_i - \mathscr{P}\cdot d\mathscr{E} \\ dA^* &= d(A - \mathscr{E}\cdot\mathscr{P}) = -S\,dT - p\,dV + \sum \mu_i\,dn_i - \mathscr{P}\cdot d\mathscr{E} \end{aligned} \tag{2.2.26}$$

and work with these rather than the more usual free energies defined in Eq. (2.2.24).

$$\frac{\partial \Delta A}{\partial T} = \frac{\partial \Delta A}{\partial D}\frac{\partial D}{\partial T} = -\frac{\Delta A}{D}\frac{\partial D}{\partial T} \tag{2.2.22B}$$

and from Eq. (2.2.22A) the change in internal energy becomes

$$\Delta U = \frac{\Delta A}{D}\left(D + T\frac{\partial D}{\partial T}\right). \tag{2.2.22C}$$

For the ideal dielectric defined by Eq. (2.2.22),

$$T\frac{\partial(D-a)}{\partial T} = -(D-a) \tag{2.2.22D}$$

and we find, as above

$$\Delta U = a\frac{\Delta A}{D} \tag{2.2.22E}$$

Consider, for simplicity, a binary mixture containing n_1 moles of component 1 and n_2 moles of component 2. Imagine that the condenser mentioned above is dipped into a very large volume of the mixture. In the region between the plates of the condenser, the electrostatic field will cause changes in the chemical potentials of the components and these changes will be accompanied by changes in composition and density of the material between the plates. If the volume V, enclosed between the plates is small relative to the total volume of the solution, then the solution outside the condenser may be regarded as an infinite reservoir and the small complementary changes in composition and density induced by the changes within volume V may be neglected. For this limiting case, the changes in composition in the small region between the plates may be considered to occur at constant chemical potential of the components. For only two components, Eq. (2.2.26) reduces to

$$dG^* = -S\,dT + V\,dp - \mathscr{P}\cdot d\mathscr{E} + \mu_1\,dn_1 + \mu_2\,dn_2. \qquad (2.2.27)$$

Since dG^* is an exact differential, application of the cross-differentiation identity results in

$$\left(\frac{\partial \mu_1}{\partial |\mathscr{E}|}\right)_{p,T,n_1,n_2} = \left(\frac{\partial |\mathscr{P}|}{\partial n_1}\right)_{p,T,\mathscr{E},n_2}. \qquad (2.2.28)$$

Now, the polarization is related to χ_V, the electric susceptibility of the dielectric per unit volume, by

$$\mathscr{P} = \chi_V \mathscr{E} V. \qquad (2.2.29)$$

Using Eqs. (2.2.12) and (2.2.11), by substitution and rearrangement it is found that

$$\chi_V = \frac{D-1}{4\pi}. \qquad (2.2.30)$$

Substitution of Eq. (2.2.29) into Eq. (2.2.28) and subsequent differentiation leads to*

$$\left(\frac{\partial \mu_1}{\partial \mathscr{E}}\right)_{p,T,n_1,n_2} = -\bar{v}_1 \mathscr{E} \chi_V - V\mathscr{E}\left(\frac{\partial \chi_V}{\partial n_1}\right)_{p,T,\mathscr{E},n_2} \qquad (2.2.31)$$

$$\mathscr{E} = |\mathscr{E}|$$

where $\bar{v}_1$ is the partial molal volume of component 1. Equation (2.2.31) may be rewritten in terms of the mole fractions of the two components as

$$\left(\frac{\partial \mu_1}{\partial \mathscr{E}}\right)_{p,T,n_1,n_2} = -\bar{v}_1 \mathscr{E} \chi_V - v_m \mathscr{E} x_2 \left(\frac{\partial \chi_V}{\partial x_1}\right)_{p,T,\mathscr{E}} \qquad (2.2.32)$$

*We now use the notation $\mathscr{E}$ for the magnitude of the vector $\mathscr{E}$ and drop the modulus bars.

where v_m is the mean molar volume of the solution. It is now easy to determine the change in concentration of component 1 (by the conservation of mass this also determines the change in concentration of component 2) under various conditions. Thus, if the field is applied with the constraints of constant temperature, pressure, and chemical potential of component 1,

$$\left(\frac{\partial x_1}{\partial \mathscr{E}}\right)_{p,T,\mu_1} = -\left(\frac{\partial \mu_1}{\partial \mathscr{E}}\right)_{p,T,x_1} \left(\frac{\partial x_1}{\partial \mu_1}\right)_{p,T,\mathscr{E}} \tag{2.2.33}$$

and the use of Eq. (2.2.32) leads to

$$\left(\frac{\partial x_1}{\partial \mathscr{E}}\right)_{p,T,\mu_1} = \left(\frac{\partial x_1}{\partial \mu_1}\right)_{p,T,\mathscr{E}} \left[\bar{v}_1 \mathscr{E} \chi_V + v_m \mathscr{E} x_2 \left(\frac{\partial \chi_V}{\partial x_1}\right)_{p,T,\mathscr{E}}\right]. \tag{2.2.34}$$

To transform Eq. (2.2.34) write the change in composition at constant temperature $(dx_1)_T$, as a function of the field strength the pressure p, and the chemical potential of component 1, μ_1,

$$(dx_1)_T = \left(\frac{\partial x_1}{\partial \mathscr{E}}\right)_{T,p,\mu_1} d\mathscr{E} + \left(\frac{\partial x_1}{\partial p}\right)_{T,\mu_1,\mathscr{E}} dp + \left(\frac{\partial x_1}{\partial \mu_1}\right)_{T,p,\mathscr{E}} d\mu_1. \tag{2.2.35}$$

Consider now a change in x at constant temperature and chemical potential of components 1 and 2,

$$\left(\frac{\partial x_1}{\partial \mathscr{E}}\right)_{T,\mu_1,\mu_2} = \left(\frac{\partial x_1}{\partial \mathscr{E}}\right)_{T,p,\mu_1} + \left(\frac{\partial x_1}{\partial p}\right)_{T,\mu_1,\mathscr{E}} \left(\frac{\partial p}{\partial \mathscr{E}}\right)_{\mu_1,\mu_2,T}. \tag{2.2.36}$$

In the second terms on the right-hand side of Eq. (2.2.36) both derivatives are easily evaluated. For the first term of the product we use the identity

$$\left(\frac{\partial x_1}{\partial p}\right)_{T,\mu_1,\mathscr{E}} = -\left(\frac{\partial \mu_1}{\partial p}\right)_{T,\mathscr{E},x} \left(\frac{\partial x_1}{\partial \mu_1}\right)_{T,p,\mathscr{E}} \tag{2.2.37}$$

$$= -\bar{v}_1 \left(\frac{\partial x_1}{\partial \mu_1}\right)_{T,p,\mathscr{E}}.$$

For the second term of the product, we return to the definition of dA^* given in Eq. (2.2.26) and after substituting for $\mathscr{P}$ from Eq. (2.2.29) and using the cross-differentiation identity

$$\left(\frac{\partial p}{\partial \mathscr{E}}\right)_{T,\mu_1,\mu_2} = \chi_V \mathscr{E} \tag{2.2.38}$$

the insertion of Eqs. (2.2.37) and (2.2.30) into (2.2.36) leads to the desired result, namely

$$\left(\frac{\partial x_1}{\partial \mathscr{E}}\right)_{T,\mu_1,\mu_2} = \frac{\mathscr{E} v_m x_2}{4\pi} \left(\frac{\partial x_1}{\partial \mu_1}\right)_{T,p,\mathscr{E}} \left(\frac{\partial D}{\partial x_1}\right)_{T,p,\mathscr{E}}. \tag{2.2.39}$$

A special case of intrinsically great interest is that of an ideal solution with the additional constraints that $(\partial D/\partial x_1)_{T,p,\mathscr{E}}$ and v_m both be independent of the field strength. For an ideal solution

$$\frac{\partial x_1}{\partial \mu_1} = \frac{x_1}{RT}.$$

(2.2.40)

Performing the differentiation indicated in Eq. (2.2.39) and integrating, it is found that

$$\ln \frac{x_1}{x_2} - \ln \frac{x_1^0}{x_2^0} = \frac{v_m \mathscr{E}^2}{8\pi RT} \left(\frac{\partial D}{\partial x_1}\right)_{p,T,\mathscr{E}}$$

(2.2.41)

where x_1^0 and x_2^0 are the mole fraction of components one and two in the bulk fluid outside the condenser plates where $\mathscr{E} = 0$. This relation clearly indicates the ability of ions to sort the components of a multicomponent dielectric medium.

There are some general conclusions which can be immediately drawn from the preceding development. For simplicity consider a one-component system. First note that since the dielectric constant D is a function of the mass density ρ_m, and the density changes with the electric field strength due to electrostriction, D is an implicit function of the field strength and is not a constant if $\mathscr{E}$ changes and ρ_m changes. However, if the constraint of constant volume is imposed on the system, ρ_m is a constant independent of $\mathscr{E}$ and hence D is a constant independent of $\mathscr{E}$. (This argument, as always, neglects the possibility of dielectric saturation.) Second, when an electric field is applied to a fluid, its activity is lowered. If constant chemical potential is to be maintained, more fluid must enter the volume V which is in the field until the increase in the vapor pressure of the fluid due to its increased density will just counterbalance the vapor pressure decrease due to the effect of the electric field. In a phenomenological sense, the constant activity is maintained by the balance between the increase in density and the decrease in " activity coefficient." The chemical potential of the fluid in the field may thus be written

$$\mu = \mu^0(T, p, \mathscr{E} = 0) + RT \ln p + \mu^E(\mathscr{E})$$
$$= \mu^0(T, p, \mathscr{E} = 0) + RT \ln \gamma(\mathscr{E})p.$$

(2.2.42)

To reiterate, if the density of the fluid were increased with no other change occurring, the vapor pressure would increase due to the increased repulsive pressure in the fluid. On the other hand, if there were no change in density, due to the presence of the electric field, $\gamma(\mathscr{E})$ which may be considered to be an activity coefficient, decreases. The decrease in $\gamma(\mathscr{E})$ is just sufficient to balance the increase of the vapor pressure and maintain the chemical potential constant. From the molecular point of view, $\gamma(\mathscr{E})$ decreases because accompanying an increase in the field strength $\mathscr{E}$ is an increase in the induced dipole moment in each molecule. This increased dipole moment results in an increased attraction between the molecules and hence would, of itself, lower the vapor pressure and therefore, $\gamma(\mathscr{E})$. The balance between these two tendencies then results in constant activity.

Finally, we note that the presence of an electric field can markedly alter the composition of the fluid. In the vicinity of an ion the electric field strength is huge and we may anticipate rather complete sorting of the solvent. The result of this sorting is that there is a concentration in the neighbourhood of the ion of the more polarizable component. The necessity for considering this effect when discussing the properties of ions in mixed solvents is obvious.

2.3 The Poisson–Boltzmann Equation[18,24,25]

At the very outset of this discussion of the Poisson–Boltzmann equation we shall advance two hypotheses:

(a) the solute in a solution of a strong electrolyte is completely dissociated into ions, and

(b) all deviations from the properties of an ideal solution are due to the electrostatic forces which exist between the ions. That is, if all the ions were discharged, the ghost ions would not interact and the solution would become an ideal solution.

The first hypothesis is an extension of Arrhenius' proposal which only considered the equilibrium between dissociated and undissociated molecules and ions in solution. The statement (a) was first made by Bjerrum who studied the absorption spectra of electrolyte solutions and concluded that for strong electrolytes no equilibrium between dissociated and undissociated species can be noticed, and that this situation persists to rather high solute concentrations. Bjerrum's conclusion has recently been questioned with respect to the question of the existence of ion pairs, but these species differ from those envisaged in the Arrhenius theory in any event.[26] The second hypothesis (b) permits the identification of the experimentally observed excess free energy with the electrical free energy of the solution. The assumption that the solution would be ideal if the ions were uncharged cannot in principle be defended. However, the deviations from ideality of very dilute non-electrolyte solutions are sufficiently small (since the intermolecular forces have .a very short range) that in this case the approximation of ideality will not introduce appreciable error.

In contrast with the situation in solutions of nonelectrolytes, a most important feature of electrolyte solutions is the nonrandom distribution of ions. Because of the strong electrostatic forces and the requirements of electrical neutrality, a positive ion is surrounded, on the average, by an excess of negative charge equal in magnitude to the positive charge. This excess charge is not localized in a lattice but rather forms a fluctuating, diffuse cloud. Consider some positive ion as the center of a coordinate system. At a point P in the neighbourhood of this ion the average electrostatic potential is ψ. Therefore the work required to transport a positive ion from infinity (where the ions are noninteracting) to the point P is $q\psi$ *if the test charge does not perturb the ion atmosphere;* to transport a negative ion to the point P the work is $-q\psi$, where q is the magnitude of the charge on the ion. If the solution contains $N_1, \cdots, N_r$ different ions with valencies $z_1, \cdots, z_r$, then a volume element dV centered at the point P will have an average excess concentration

$$c_i^0 \, e^{-z_i q\psi/kT} \tag{2.3.1}$$

of the ionic species i, and where c_i^0 is the bulk concentration, N_i/V, of the ionic species i. The total charge density in the volume element dV is the sum of the charge densities due to each ionic species, so that

$$\rho = \sum_1^r c_i z_i q = \sum_1^r z_i q c_i^0 \, e^{-z_i q\psi/kT} \tag{2.3.2}$$

This equation is an expression of the assumption that the electrostatic field is simply the linear superposition of the fields due to all the ions, an approximation to be examined shortly. The total charge density and the mean electrostatic potential ψ are related by Poisson's equation

$$\nabla^2 \psi = -\frac{4\pi}{D} \rho \tag{2.2.4}$$

in a medium of dielectric constant D, *if the charge is assumed to be continuously distributed with density* ρ. The nature of this approximation also will be discussed later. The substitution of Eq. (2.3.2) into Eq. (2.2.4) leads to the relation

$$\nabla^2 \psi = -\frac{4\pi}{D} \sum_1^r z_i q c_i^0 \, e^{-z_i q\psi/kT} \tag{2.3.3}$$

with $\nabla\psi = \psi = 0$ at ∞. This is the celebrated Poisson–Boltzmann equation, the basis of modern electrolyte theory. Equation (2.3.3) cannot be solved in closed form. Debye and Hückel linearized the differential equation by expanding the exponential term on the right-hand side. The first term in the expansion vanishes when summed over all species due to the electroneutrality condition

$$\sum_1^r c_i^0 z_i q = 0. \tag{2.3.4}$$

If the quantity $(zq\psi/kT)$ is very small relative to unity, terms higher than the square in the expansion may be neglected, and Eq. (2.3.3) assumes the form

$$\nabla^2\psi = \kappa^2\psi \qquad (2.3.5)$$

$$\kappa^2 = \frac{4\pi}{DkT} \sum_1^r c_i^0 z_i^2 q^2. \qquad (2.3.6)$$

This linearized equation may be solved explicitly.

It is not our purpose to discuss in detail the solutions of Eq. (2.3.5). Rather, our interest is to investigate the difficulties inherent in the statistical-mechanical theory of electrolyte solutions and then to proceed to a justification of the approximations inherent in Eq. (2.3.5).

It is easy to see that a straighforward calculation of the configurational partition function for the model solution composed of point ions in a continuous dielectric medium is fraught with difficulties associated with the nature of the assumed interaction between ions. Since the unmodified Coulomb potential falls off only as R^{-1}, whereas the volume element in space increases as R^3, the configurational integral is divergent. This divergence has its physical origin in the neglect of the presence of the other charges in the system which ensure electroneutrality. Accounting for the presence of these charges introduces a screening factor into the potential of interionic interaction and this screened Coulomb potential falls off sufficiently rapidly with distance that the configurational partition function converges with respect to its upper limit. In the language of Chapter 1, the force acting on an ion is derivable from the potential of mean force and not from the intermolecular potential. In the present case, the existence of other ions coupled with the condition of electroneutrality implies that a given ion cannot be moved without other changes occurring which are not controllable. In particular, the constraint of electroneutrality forces positive ions to follow the motion of a negative ion to some extent. With a negative ion surrounded by positive ions, an external test charge sees an effective charge on the central negative ion which is less than the true charge. This coupling of ionic motions then is expressible as a modified potential of interaction and this is, in the case considered, a screened Coulomb potential.

A second divergence of the configurational partition function arises because of the assumption of point ions. For, with point ions, there is an infinite gain in free energy if positive and negative charges coalesce to form a neutral point. It is clear then that the short-range repulsive forces which have been neglected in the preceding analysis must play an important role in the determination of the free energy. These forces may be considered to determine an excluded volume for the ions and will be expected to modify the considerations leading to Eq. (2.3.3). In particular, the assumption that the work expended in bringing a charge from infinite separation to a point where the potential is ψ, is $q\psi$, is rigorously valid only for point infinitesimal charges. Further, it is not immediately obvious that the Poisson equation (2.2.4), in which ρ is the average charge density, is rigorously applicable. It is necessary, therefore, to examine in greater detail the approximations involved in Eq. (2.3.3).

The procedure to be followed at first will be to examine the detailed balance between the electrostatic forces between the ions and the thermal forces due to Brownian motion.[25] By considering the average force per unit volume, and passing to the limit of an infinitesimal volume element, a force density can be defined. Now, the rigorous form of the Boltzmann distribution is

$$c_i = c_i^0 \, e^{-W_i/kT} \tag{2.3.7}$$

where W_i is the ensemble average work required to place the particle in the given configuration, and the distribution is achieved as a result of the balance between the thermal motion of the particles and the work expended in bringing them to a given distance of separation. Clearly, the average force density arising from the intermolecular interactions is $-c_i \nabla W_i$. It is easily shown (see Chapter 1) that if the momentum coordinates of a system of N particles do not depend upon the position coordinates, then the Hamiltonian of the set of N particles is separable and the momentum coordinates may be integrated over immediately. This is true whenever there are only scalar potentials acting between the molecules. The physical consequence of the separation of variables is that the average kinetic energy is independent of the intermolecular forces. If now we consider somewhere in the solution a plane of unit area, the average kinetic energy of the ions carries an average momentum kTc_i across the plane in unit time. This is equivalent to a pressure of kTc_i, and therefore, a thermal force density of $kT\nabla c_i$. At equilibrium then, by the laws of mechanics,

$$-c_i \nabla W_i = kT\nabla c_i. \tag{2.3.8}$$

If it is assumed that the work required to establish a given configuration is related to the average electrostatic potential by

$$W_i = z_i q \psi \tag{2.3.9}$$

it is immediately evident that all short-range contributions to W_i of nonelectrostatic character have been neglected. By substitution, Eq. (2.3.8) now becomes

$$c_i z_i q \nabla \psi = -kT\nabla c_i \tag{2.3.10}$$

which may be integrated to

$$c_i = c_i^0 \, e^{-z_i q \psi/kT} \tag{2.3.11}$$

The total charge density is the sum of the charge densities due to each ionic species, and the use of Poisson's equation therefore leads to the relation

$$\frac{D}{4\pi} \nabla^2 \psi \nabla \psi = - \sum_1^r c_i^0 z_i q \, e^{-z_i q \psi / kT} \nabla \psi \tag{2.3.12}$$

where we have used the equation

$$\nabla c_i = - \frac{c_i^0 z_i q}{kT} \, e^{-z_i q \psi / kT} \nabla \psi. \tag{2.3.13}$$

By definition, the average field intensity $\langle \mathscr{E} \rangle$ is the negative gradient of the average electrostatic potential,

$$\langle \mathscr{E} \rangle = -\nabla \psi \tag{2.3.14}$$

so that Eq. (2.3.12) becomes

$$\frac{D}{4\pi} \langle \mathscr{E} \rangle \nabla \cdot \langle \mathscr{E} \rangle = \rho \langle \mathscr{E} \rangle = \nabla \cdot \mathbf{\Phi} = \nabla p \tag{2.3.15}$$

$$\mathbf{\Phi} = \frac{D}{8\pi} \left[2 \langle \mathscr{E} \rangle \langle \mathscr{E} \rangle - \langle \mathscr{E} \rangle \cdot \langle \mathscr{E} \rangle \mathbf{1} \right] \tag{2.3.16}$$

$$p = kT \sum_1^r c_i(R) \tag{2.3.17}$$

where $\mathbf{1}$ is the unit tensor.

In any continuous static charge distribution, the forces exerted in the medium are describable in terms of the Maxwell stress tensor

$$\mathbf{\Phi}' = \frac{D}{8\pi} \left[2 \mathscr{E} \mathscr{E} - \mathscr{E}^2 \mathbf{1} \right]. \tag{2.3.18}$$

In the static case the forces on the charge itself, as distinguished from those acting on the medium, produce everywhere a force density

$$\nabla \cdot \mathbf{\Phi}' = \rho \mathscr{E} \tag{2.3.19}$$

and the average force density is therefore,

$$\langle \nabla \cdot \mathbf{\Phi}' \rangle = \langle \rho \mathscr{E} \rangle. \tag{2.3.20}$$

Equation (2.3.20) is not identical with Eq. (2.3.15) since, in general,

$$\rho \langle \mathscr{E} \rangle \neq \langle \rho \mathscr{E} \rangle \tag{2.3.21}$$

since, unless the variables are independent, the product of two averages is not equal to the average of the product. The difference between the two averages arises from the self-interaction of charge on the same ion which contributes to Eq. (2.3.19) but not to Eq. (2.3.15). Under what conditions can we expect the Poisson–Boltzmann equation, Eq. (2.3.3) to be accurate? To answer this question assume that the ions are spheres of radius a, with their charges uniformly distributed throughout their volumes. For simplicity, the dielectric constant is assumed to be the same inside an ion sphere as in the solution. Finally, let the distribution function c_i be expanded about the point P in the form

$$c_i = c_i^0 + c_i^{(1)}\xi + \cdots \tag{2.3.22}$$

where ξ is the distance from the point P. The instantaneous field strength at a point not within an ion is denoted by $\mathscr{E}_0$, and is due to the superposed electric fields of all the neighboring ions. Inside an ion, at a distance r from its center, there is an additional field of magnitude

$$|\mathscr{E}_i| = \frac{z_i q r}{Da^3}. \tag{2.3.23}$$

Consider an ion located at the point $\mathbf{r}$, with $\mathbf{r} \cdot \mathbf{r} < a^2$, so that the origin of the coordinate system is inside the ion, but not coincident with the center of mass of the ion. The average field at the origin due to the charge on the ion is then found to be

$$\langle \mathscr{E}_{ix} \rangle = \frac{z_i q}{Da^3} \int_V (c_i^0 x + c_i^{(1)} x^2 + \cdots) \, dV \tag{2.3.24}$$

where the x axis has been taken along ξ for convenience. The average values of the field components in the y and z directions are zero due to the assumption (2.3.31) and the choice of the coordinate system. Clearly

$$\lim_{a \to 0} \langle \mathscr{E}_{ix} \rangle = 0. \tag{2.3.25}$$

Similarly, the instantaneous value of the charge density, ρ_i, is

$$\rho_i = \frac{3z_i q}{4\pi a^3} \tag{2.3.26}$$

and its average is

$$\langle \rho_i \rangle = \frac{3z_i q}{4\pi a^3} \int_V (c_i^0 + c_i^{(1)} x + \cdots) \, dV \tag{2.3.27}$$

so that in the limit as the radius of the ion tends to zero

$$\lim_{a \to 0} \langle \rho_i \rangle = c_i^0 z_i q = z_i q c_i \tag{2.3.28}$$

in agreement with Eq. (2.3.3). Now, the internal field contributes to the force density an instantaneous value

$$\rho_i \mathscr{E}_{ix} = \frac{3z_i^2 q^2 x}{4\pi Da^6} \tag{2.3.29}$$

and the corresponding average is seen to be

$$\langle \rho_i \mathscr{E}_{ix} \rangle = \frac{3z_i^2 q^2}{4\pi D a^6} \int_V (c_i^0 x + c_i^{(1)} x^2 + \cdots) \, dV \qquad (2.3.30)$$

so that, in contrast to the limiting behavior of (2.3.25) and (2.3.28), as the radius of the ions tends to zero

$$\lim_{a \to 0} \langle \rho_i \mathscr{E}_{ix} \rangle = \infty \qquad (2.3.31)$$

whereas, by taking the product of Eqs. (2.1.25) and (2.3.28)

$$\lim_{a \to 0} \langle \rho_i \rangle \langle \mathscr{E}_{ix} \rangle = 0. \qquad (2.3.32)$$

Physical considerations require that, since the ions are assumed to be stable, there must be some internal cohesive force which balances the repulsion calculated above. The force density in equilibrium with the thermal pressure gradient is then

$$\langle \rho \mathscr{E} \rangle - \sum_i \langle \rho_i \mathscr{E}_i \rangle = \langle \rho \mathscr{E}_0 \rangle \qquad (2.3.33)$$

and to obtain Eq. (2.3.3) it is necessary to assume that

$$\langle \rho \mathscr{E}_0 \rangle = \langle \rho \rangle \langle \mathscr{E}_0 \rangle \qquad (2.3.34)$$

which implies that there are no correlations between the fluctuations in charge density and the field strength. With the use of this assumption

$$\langle \rho \rangle \langle \mathscr{E}_0 \rangle = \langle \rho \rangle [\langle \mathscr{E} \rangle - \sum_i \langle \mathscr{E}_i \rangle] = \langle \rho \rangle \langle \mathscr{E} \rangle \qquad (2.3.35)$$

for point charges. The neglect of the self-energy of the ion by assuming that there is a balancing cohesive force has analogs in many theories. For instance, in the theory of atomic spectra it is assumed that the nucleus is a massive point charge but the electrostatic self-energy of the nucleus is neglected. In general, when studying the properties of stable charged particles, the electrostatic self-energy is omitted from consideration. The Poisson–Boltzmann equation is thus seen to be valid in the *limit as the ion size tends to zero and the correlations between the charge density and the electric field strength can be neglected.* It is pertinent to point out that though this treatment shows that the Poisson–Boltzmann equation is valid in the limit of zero ion size, the neglect of the finite dimensions of the ion leads to a divergence in the phase integral. In the simple derivation of the Poisson–Boltzmann equation given at the outset of this section it was tacitly assumed that a distribution of the ions did exist, but the phase integral divergence indicates that such a distribution is thermodynamically unstable. It is thus clear that the approximation embodied in Eq. (2.3.9) cannot be rigorously correct for ions with finite dimensions. It is probable, however, that the approximation (2.3.9) may be useful over a certain region. A better classification of the various contributions to the free energy is needed to resolve this difficulty. To complete our investigation of the Poisson–Boltzmann equation it is necessary to ascertain to what extent Eq. (2.3.9) is accurate for real ions. We shall therefore proceed to investigate the Poisson–Boltzmann equation from the point of view of classical statistical mechanics[18,24] with proper attention to the short-range repulsive forces which must exist. Such an analysis will supplement the considerations based on the Maxwell stress tensor.

Let A_0 be the Helmholtz free energy of the solution when all the ions are discharged. From the outset we shall omit from consideration the self-energy of the ions since they are internally stable and this self-energy is unaffected by changes in concentration of the solute, etc. If A is the Helmholtz free energy of the real electrolyte solution, then

$$e^{-(A-A_0)/kT} = \frac{Q_N}{Q_N^0} = \frac{\int_V e^{-U_N/kT}\,d\{N\}}{\int_V e^{-U_{N^0}/kT}\,d\{N\}} \tag{2.3.36}$$

where U_N is the mutual potential energy of the ions when charged, and U_N^0 is the mutual potential energy of the discharged ions. These latter forces are of short range, generally repulsive, and cause the phase integral to converge. For simplicity we shall continue to represent the ions as spherically symmetric and with no internal degrees of freedom. The potential energy, U_N may be expressed as the sum of the Coulombic terms and the potential of short-range forces, viz.,

$$U_N = \tfrac{1}{2}\sum_1^N \left[z_i q\psi_i(\mathbf{R}_i) + \phi_i(\mathbf{R}_i)\right] \tag{2.3.37}$$

where the Coulombic term is

$$\psi_i(\mathbf{R}_i) = \sum_{\substack{1 \\ i \neq k}}^N \frac{z_k q}{D|\mathbf{R}_i - \mathbf{R}_k|} \tag{2.3.38}$$

and is simply the instantaneous electrostatic potential at the point $\mathbf{R}_i$ due to all the other ions in the solution. The second term of Eq. (2.3.37)

$$\phi_i(\mathbf{R}_i) = \sum_{\substack{1 \\ k \neq i}}^N u_{ki}(\mathbf{R}_k - \mathbf{R}_i) \tag{2.3.39}$$

is the potential of short-range forces. It should be noted at the very outset that the solvent is being treated as a dielectric continuum, a space in which the ions can move. This is the origin of the factor D in Eq. (2.3.38). A mean electrostatic potential may be formally defined, using the definition of mean value, by the relation

$$\langle \psi_i \rangle = \frac{\int_V \psi_i(\mathbf{R}_i)\, e^{-U_N/kT}\, d\{N\}}{\int_V e^{-U_N/kT}\, d\{N\}} \qquad (2.3.40)$$

from which it is easily seen that

$$\frac{\partial A}{\partial(z_i q)} = \langle \psi_i \rangle. \qquad (2.3.41)$$

This relation is obtained by differentiation of Eq. (2.3.36) with the use of Eq. (2.3.37). A similar differentiation of Eq. (2.3.46) with respect to $z_j q$ leads to the cross differentiation identity

$$\frac{\partial \langle \psi_i \rangle}{\partial(z_j q)} = \frac{\partial \langle \psi_j \rangle}{\partial(z_i q)}. \qquad (2.3.42)$$

The relations (2.3.41) and (2.3.42) show that a variation of the Helmholtz free energy with change in charge can be written as an exact linear differential form,

$$\delta A = \sum_1^N \frac{\partial A}{\partial(z_i q)}\, \delta(z_i q) = \sum_1^N \langle \psi_i \rangle \delta(z_i q). \qquad (2.3.43)$$

Since Eq. (2.3.43) is an exact linear differential form, the integral of δA is independent of the path of integration. This is, of course, merely an expression of the fact that A is a state function. Consider a hypothetical process in which the charges on all the ions are varied continuously and simultaneously from zero to the full charge. Let the instantaneous charge on the ion i be $\lambda z_i q$, where λ is a parameter ranging continuously from zero to unity. Then the excess Helmholtz free energy due to the electrostatic interactions is, in terms of λ,

$$A - A_0 = \sum_1^N \int_0^1 z_i q \langle \psi_i(\lambda z_1 q \cdots \lambda z_N q) \rangle\, d\lambda. \qquad (2.3.44)$$

This relation will be used extensively in our later work.

The instantaneous electrostatic potential at an arbitrary point in the solution may be simply written as

$$\psi(\mathbf{R}) = \sum_1^N \frac{z_i q}{D|\mathbf{R} - \mathbf{R}_i|}. \qquad (2.3.45)$$

If some arbitrary ion is selected, say ion k and fixed in position at the point, $\mathbf{R}_k$, then the mean value of the instantaneous potential at the point $\mathbf{R}$ is

$$^k\langle\psi(\mathbf{R})\rangle = \frac{\displaystyle\int_V \psi(\mathbf{R})\, e^{-U_N/kT}\, d\{N-1\}}{\displaystyle\int_V e^{-U_N/kT}\, d\{N-1\}} \tag{2.3.46}$$

where the coordinates of ion k are omitted from the integration. In general, if the coordinates of one or a set of ions are omitted from integration in the phase integral this will be indicated by superscript letters to the left of the quantity being discussed. If $\psi_k(\mathbf{R})$ is defined to be the instantaneous potential at the point $\mathbf{R}$ due to all ions but the ion k, it is easily verified that

$$\psi_k(\mathbf{R}) - \psi(\mathbf{R}) = -\frac{z_k q}{D|\mathbf{R}-\mathbf{R}_k|}. \tag{2.3.47}$$

So long as the ion k remains fixed at the point $\mathbf{R}_k$, the term $z_k q/(D|\mathbf{R}-\mathbf{R}_k|)$ is a constant. It is therefore unaffected by any averaging operation in which the coordinates of ion k are kept fixed, i.e., an operation of the type indicated in Eq. (2.3.46). Now, a new mean potential can be defined by

$$^k\langle\psi_k(\mathbf{R}_k)\rangle = \frac{\displaystyle\int_V \psi(\mathbf{R}_k)\, e^{-U_N/kT}\, d\{N-1\}}{\displaystyle\int_V e^{-U_N/kT}\, d\{N-1\}} \tag{2.3.48}$$

and $^k\langle\psi_k(\mathbf{R}_k)\rangle$ is seen to be the value of $^k\langle\psi(\mathbf{R})\rangle$ in the interior of ion k due to all the other ions in the solution. From Eq. (2.3.40) and (2.3.46) it is readily verified that

$$\langle\psi_k\rangle = \frac{\displaystyle\int_V {}^k\langle\psi_k(\mathbf{R}_k)\rangle \int_V e^{-U_N/kT}\, d\{N-1\}\, d(k)}{\displaystyle\int_V e^{-U_N/kT}\, d\{N\}}. \tag{2.3.49}$$

If $^k\langle\psi_k(\mathbf{R}_k)\rangle$ is independent of $\mathbf{R}_k$, the two potentials defined by Eqs. (2.3.48) and (2.3.49) are related by

$$^k\langle\psi_k(\mathbf{R}_k)\rangle = \langle\psi_k\rangle \tag{2.3.50}$$

and this will be true everywhere except at the surfaces of the volume. The relation (2.3.50) is essentially a statement to the effect that the system is fluid and isotropic. It is now possible to rewrite Eq. (2.3.46) in the form

$$^k\langle\psi(\mathbf{R})\rangle = \sum_{\substack{1 \\ l \neq k}}^{N} \int \frac{^k\langle\rho_l(\mathbf{R}_l)\rangle}{D|\mathbf{R}_l - \mathbf{R}|} d(l) + \frac{z_k q}{D|\mathbf{R}_k - \mathbf{R}|} \tag{2.3.51}$$

where the charge density is defined by the following:

$$^k\langle\rho_l(\mathbf{R}_l)\rangle = z_l q \frac{\displaystyle\int_V \exp\left\{-\frac{1}{kT}[U_{N-1} + U_l(\mathbf{R}_l)]\right\} d\{N-2\}}{\displaystyle\int_V e^{-U_N/kT} d\{N-1\}}. \tag{2.3.52}$$

In Eq. (2.3.52) the coordinates of both ions k and l remain fixed in the numerator while only the coordinates of ion k are fixed in the denominator. Further, the total potential energy U_N has been separated into a sum of two terms: the potential energy of $N-1$ particles, U_{N-1}, and the potential energy of the lth particle, U_l. The sum of these two potential energies is obviously the total potential energy. It is well known from classical electrostatic theory that the potential given in Eq. (2.3.51) is the solution of the following Poisson equation:

$$\nabla^2(^k\langle\psi(\mathbf{R})\rangle) = -\frac{4\pi}{D}\,^k\langle\rho(\mathbf{R})\rangle \tag{2.3.53}$$

with the total charge density $^k\langle\rho(\mathbf{R})\rangle$ defined by

$$^k\langle\rho(\mathbf{R})\rangle = \sum_{1}^{N} z_l q \frac{\displaystyle\int_V \exp\left\{-\frac{1}{kT}[U_{N-1} + U_l]\right\} d\{N-2\}}{\displaystyle\int_V e^{-U_N/kT} d\{N-1\}}. \tag{2.3.54}$$

Thus far our considerations based on the discussion from Eq. (2.3.36) and onward have been for a single electrolyte and may be easily generalized. If the solution contains $N_1, \cdots N_r$ ions of species $1, \cdots, r$, with the bulk concentration defined by $c_i^0 = N_i/V$, then a trivial generalization of Eq. (2.3.54) leads to the relations

$$^k\langle\rho(\mathbf{R})\rangle = \sum_{1}^{r} c_i^0 z_i q g_2(\mathbf{R}_i, \mathbf{R}) \tag{2.3.55}$$

where $g_2(\mathbf{R}_k, \mathbf{R}_l)$ is the pair correlation function, defined by

$$g_2(\mathbf{R}_k, \mathbf{R}_l) = \frac{V \displaystyle\int_V \exp\left\{-\frac{1}{kT}[U_{N-1} + U_l]\right\} d\{N-2\}}{\displaystyle\int_V e^{-U_N/kT} d\{N-1\}} \tag{2.3.56}$$

$$= \frac{V\,^k\langle\rho_l(\mathbf{R})\rangle}{z_l q} \tag{2.3.57}$$

and the Poisson equation assumes the form

$$\nabla^2({}^k\langle\psi(\mathbf{R})\rangle) = -\frac{4\pi}{D}\sum_i c_i^0 z_i q g_2(\mathbf{R}_i, \mathbf{R}). \tag{2.3.58}$$

It will be recalled that the potential of mean force is related to the corresponding correlation function by

$$g_2 = e^{-W_2/kT}. \tag{2.3.59}$$

By differentiation of Eqs. (2.3.56) and (2.3.59) a differential equation for W_2 may be obtained in terms of the potential U_N. This equation is

$$\nabla^2 W_2 - {}^{ik}\langle\nabla^2 U_i\rangle = -\frac{1}{kT}[{}^{ik}\langle(\nabla U_i)^2\rangle - \langle\nabla W_2\rangle^2] \tag{2.3.60}$$

where, as before, the mean values are taken with ions i and k fixed in their respective positions. In principle, the solution of Eq. (2.3.60) affords the necessary information to obtain ${}^k\langle\psi(\mathbf{R})\rangle$ with appropriate boundary conditions. The actual computation is, unfortunately, not practicable and approximations must be employed.

To obtain the familiar Poisson-Boltzmann equation, Eq. (2.3.3), it will be seen that the same approximation as introduced previously is necessary,

$$z_i q^k\langle\psi(\mathbf{R})\rangle = W_2. \tag{2.3.61}$$

We shall return to this approximation after further development of the exact Poisson–Boltzmann equation. Before Eq. (2.3.58) can be utilized the dependence of g_2 on the mean potential ${}^k\langle\psi(\mathbf{R})\rangle$ must be determined. By differentiation of g_2 with respect to $z_i q$,

$$\frac{\partial g_2}{\partial(z_i q)} = -\frac{1}{kT}g_2[{}^{ik}\langle\psi_i(\mathbf{R})\rangle - {}^k\langle\psi_i(\mathbf{R}_i)\rangle] \tag{2.3.62}$$

where the quantities appearing in Eq. (2.3.62) are defined by

$$^k\langle\psi_i(\mathbf{R}_i)\rangle = \frac{\displaystyle\int_V \psi_i(\mathbf{R}_i)\, e^{-U_N/kT}\, d\{N-1\}}{\displaystyle\int_V e^{-U_N/kT}\, d\{N-1\}} \tag{2.3.63}$$

and

$$^{ik}\langle\psi_i(\mathbf{R})\rangle = \frac{\displaystyle\int_V \psi_i(\mathbf{R})e^{-U_N/kT}\, d\{N-2\}}{\displaystyle\int_V e^{-U_N/kT}\, d\{N-2\}} \tag{2.3.64}$$

and where the subscript i on $^{ik}\langle\,\psi_i(\mathbf{R})\rangle$ again indicates that the term $z_i q/(D\,|\mathbf{R}_i - \mathbf{R}\,|)$ had been subtracted. Physically, $^{ik}(\,\psi_i(\mathbf{R}))$ is the mean value of the electrostatic potential at the point $\mathbf{R}$ when the ions i and k are held fixed at the points $\mathbf{R}$ and $\mathbf{R}_k$ respectively. Now, the mean potential at any arbitrary point $\mathbf{R}'$ when the ions i and k are held fixed at the points $\mathbf{R}_i$ and $\mathbf{R}_k$ is

$$^{ik}\langle\psi_i(\mathbf{R}', \mathbf{R}_i, \mathbf{R}_k)\rangle = \frac{\displaystyle\int_V \psi_i(\mathbf{R}')e^{-U_N/kT}\,d\{N-2\}}{\displaystyle\int_V e^{-U_N/kT}\,d\{N-2\}} \tag{2.3.65}$$

so that the obvious relation

$$^{ik}\langle\psi_i(\mathbf{R}')\rangle = {}^{ik}\langle\psi_i(\mathbf{R}', \mathbf{R}_i, \mathbf{R}_k)\rangle \tag{2.3.66}$$

is immediately verified. Except in the vicinity of a phase boundary the mean potential depends only upon the magnitude of the separations of the charges, since the system is fluid and isotropic. The potential defined by Eq. (2.3.65) is therefore independent of the (temporary) labeling of the ions, and any interchange of co-ordinates or translation of the pair of ions at constant pair separation leaves $^{ik}\langle\,\psi_i(\mathbf{R}', \mathbf{R}_i, \mathbf{R}_k)\rangle$ unaltered. Thus

$$^{ik}\langle\psi_i(\mathbf{R}, \mathbf{R}, \mathbf{R}_k)\rangle = {}^{ik}\langle\psi_i(\mathbf{R}_i, \mathbf{R}_i, \mathbf{R}')\rangle \tag{2.3.67}$$

with

$$|\mathbf{R} - \mathbf{R}_k| = |\mathbf{R}' - \mathbf{R}_i| \tag{2.3.68}$$

and therefore

$$^{ik}\langle\psi_i(\mathbf{R})\rangle = \frac{\displaystyle\int_V \psi_i(\mathbf{R}_i)\exp\left\{-\frac{1}{kT}[U_{N-1} + U_k(\mathbf{R}')]\right\}d\{N-2\}}{\displaystyle\int_V \exp\left\{-\frac{1}{kT}[U_{N-1} + U_k(\mathbf{R}')]\right\}d\{N-2\}} \tag{2.3.69}$$

subject to the condition that $|\mathbf{R}' - \mathbf{R}_i| = |\mathbf{R} - \mathbf{R}_k|$. The integration of Eq. (2.3.62) may now be performed to give

$$g_2 = g_2^0 \exp\left\{-\frac{1}{kT}\int_0^{z_i q} [{}^{ik}\langle\psi_i(\mathbf{R})\rangle - {}^k\langle\psi_i(\mathbf{R}_i)\rangle]\,d(z_i q)\right\} \tag{2.3.70}$$

where the factor g_2^0 is the constant of integration and is equal to

$$g_2^0 = \frac{V\displaystyle\int_V \exp\left\{-\frac{1}{kT}[U_{N-1} + \phi_i(\mathbf{R}_i)]\right\}d\{N-2\}}{\displaystyle\int_V \exp\left\{-\frac{1}{kT}[U_{N-1} + \phi_i(\mathbf{R}_i)]\right\}d\{N-1\}} \tag{2.3.71}$$

with $\phi_i(\mathbf{R}_i)$ the potential of short-range forces acting on the ion i. g_2^0 is clearly the value of the pair correlation function g_2 when the ion i is discharged. g_2^0 therefore represents the distribution of the ions as modified by the discharge of ion i, but contains the contribution due to the short-range repulsive forces originating in ion i. By discharge is of course meant that the electric charge is diminished to zero. Define

$$g_1^0(\mathbf{R}) = \exp\left\{-\frac{1}{kT}[\phi_i(\mathbf{R}) - \phi_i(\mathbf{R}_i)]\right\}\qquad(2.3.72)$$

and when ion i is uncharged

$$g_2^0 = {}^k\langle g_1^0(\mathbf{R}_i)\rangle_{z_iq=0}.\qquad(2.3.73)$$

The substitution of Eq. (2.3.71) into Eq. (2.3.58) gives the *exact* Poisson-Boltzmann equation

$$\nabla^2({}^k\langle\psi(\mathbf{R})\rangle) = -\frac{4\pi}{D}\sum_1^r c_i^0 z_i q g_2^0 \exp\left\{-\frac{1}{kT}\int_0^{z_iq} [{}^{ik}\langle\psi_i(\mathbf{R})\rangle - {}^k\langle\psi_i(\mathbf{R}_i)\rangle]\,d(z_iq)\right\}$$

$$(2.3.74)$$

which is the relation sought. It is important to emphasize that this is an *exact* form of the Poisson equation for the mean potential ${}^k\langle\psi(\mathbf{R})\rangle$. Since each of the terms g_2^0 contains a factor of the form $\exp[-u_{ki}(|\mathbf{R}_k - \mathbf{R}_i|)/kT]$, when $|\mathbf{R} - \mathbf{R}_k|$ becomes small the short-range repulsive forces make $u_{ki}(\mathbf{R}_{ki})$ positively infinite and each g_2^0 vanishes. The region of exclusion of other ions, that is the region wherein $u_{ki}(\mathbf{R}_{ki})$ is positive infinite, may be considered to be an effective rigid sphere defining the " size " of the ion. Thus, inside the ion k the right-hand side of Eq. (2.3.74) vanishes, and Eq. (2.3.74) becomes Laplace's equation. Outside the ion k the g_2^0 are correction terms resulting from the necessarily nonrandom distribution of the discharged ions. The g_2^0 may be expanded as a series in the concentration in the same manner as the virial expansion for the gaseous phase. This type of expansion was detailed in the considerations of Chapter 1. If the solution is dilute, only the first terms of the expansion will be necessary for an adequate approximation to the thermodynamic properties of the solution. The expansion of the distribution function therefore assumes the form

$$g_2^0 = 1 + 2\sum_{j=1}^r b_{ij}c_j^0 + \cdots\qquad(2.3.75)$$

to terms of order c^0, and where

$$b_{ij} = \tfrac{1}{2}\int (1 - e^{-u_{ij}/kT})\,d\mathbf{R}_{ij}.\qquad(2.3.76)$$

The potentials ${}^{ik}\langle\psi_i(\mathbf{R})\rangle$ and ${}^k\langle\psi_i(\mathbf{R}_i)\rangle$ can be related to their limiting values when the charges on ions i and k are allowed to tend to zero by a charging process. Thus, in the spirit of Eq. (2.3.44), the potentials can be written in the alternative form

$$^{ik}\langle\psi_i(\mathbf{R})\rangle = {}^{ik}\langle\psi_i(\mathbf{R})\rangle_{z_iq=0} + {}^{ik}\langle\psi_i(\mathbf{R})\rangle_{z_kq=0} - {}^{ik}\langle\psi_i(\mathbf{R})\rangle_{\substack{z_iq=0\\z_kq=0}}$$

$$+ \int_0^{z_iq}\int_0^{z_kq} \frac{\partial^2}{\partial(z_iq)\,\partial(z_kq)} {}^{ik}\langle\psi_i(\mathbf{R})\rangle\, d(z_iq)\, d(z_kq). \quad (2.3.77)$$

But when the charge z_iq is zero, from Eq. (2.3.65)

$$^{ik}\langle\psi_i(\mathbf{R})\rangle_{z_iq=0} = \frac{\displaystyle\int_V \psi_i(\mathbf{R})\exp\left\{-\frac{1}{kT}[U_{N-1}+\phi_i]\right\}d\{N-2\}}{\displaystyle\int_V \exp\left\{-\frac{1}{kT}[U_{N-1}+\phi_i]\right\}d\{N-2\}} \quad (2.3.78)$$

and if both numerator and denominator of this latter expression are multiplied by the volume V, and this factor of V is taken into the integrals as an integration over the coordinates of ion i, then

$$^{ik}\langle\psi_i(\mathbf{R})\rangle_{z_iq=0} = \frac{\displaystyle\int_V \psi_i(\mathbf{R})\exp\left\{-\frac{1}{kT}[U_{N-1}+\phi_i]\right\}d\{N-1\}}{\displaystyle\int_V \exp\left\{-\frac{1}{kT}[U_{N-1}+\phi_i]\right\}d\{N-1\}}. \quad (2.3.79)$$

The transition from Eq. (2.3.78) to Eq. (2.3.79) is made possible by the omission of the coordinates of ion i from the integrand of both numerator and denominator, since it is the potential at the point $\mathbf{R}$ which is being specified. Recalling the definition of the distribution function g_1^0 given in Eq. (2.3.72), it is easily seen that

$$^{ik}\langle\psi_i(\mathbf{R})\rangle_{z_iq=0} = \frac{{}^k\langle\psi_i(\mathbf{R})g_i^0\rangle_{z_iq=0}}{g_2^0}$$

$$= {}^k\langle\psi_i(\mathbf{R})\rangle_{z_iq=0} + \frac{{}^k\langle\psi_i(\mathbf{R})g_i^0\rangle_{z_iq=0} - {}^k\langle\psi_i(\mathbf{R})\rangle_{z_iq=0}\langle g_i^0\rangle_{z_iq=0}}{{}^k\langle g_i^0\rangle_{z_iq=0}}.$$

$$(2.3.80)$$

Note that the second term on the right-hand side of Eq. (2.3.80) is to be evaluated when the charge z_iq is zero. This term therefore depends primarily on the short-range repulsive forces, and is approximately equal to

$$-2\sum_{j=1}^r b_{ij}c_j^0 g_2\left[\frac{3z_jq}{2Da_{ij}} + {}^{jk}\langle\psi_{ij}(\mathbf{R})\rangle_{z_iq=0} - {}^k\langle\psi_i(\mathbf{R})\rangle_{z_iq=0}\right] \quad (2.3.81)$$

where a_{ij} is the distance of closest approach of ions i and j. By starting with Eq. (2.3.69) and following a similar procedure, it is found that

$$^{ik}\langle\psi_i(\mathbf{R})\rangle_{z_kq=0} = {}^{i}\langle\psi_i(\mathbf{R}_i)\rangle_{z_kq=0} +$$

$$\frac{{}^{i}\langle\psi_i(\mathbf{R}_i)g_k^0(\mathbf{R}')\rangle_{z_kq=0} - {}^{i}\langle\psi_i(\mathbf{R}_i)\rangle_{z_kq=0}{}^{i}\langle g_k^0(\mathbf{R}')\rangle_{z_kq=0}}{{}^{i}\langle g_k^0(\mathbf{R}')\rangle_{z_kq=0}} \qquad (2.3.82)$$

subject to the restriction

$$|\mathbf{R}' - \mathbf{R}_i| = |\mathbf{R} - \mathbf{R}_k|. \qquad (2.3.83)$$

Once again, the second term on the right-hand side of Eq. (2.3.82) is small at low concentration and depends primarily on the short-range forces. If the charges on both ions i and k, z_iq and z_kq, are set equal to zero, then the correlation between the distribution function $g_k^0(\mathbf{R}')$ and the potential $^{ik}\langle\psi_i(\mathbf{R})\rangle$ is zero so that the potential $^{ik}\langle\psi_i(\mathbf{R})\rangle_{z_iq=z_kq=0}$ is zero by reference to Eq. (2.3.82), providing that the distance of closest approach is the same, i.e., if all ions are of equal size. This latter condition ensures that the $g_k^0(\mathbf{R}')$ will be the same. If the sizes of the ions are not the same, this potential will consist of a small term due to the statistical double layer about ion i and will be proportional to the concentration.

Differentiation of Eq. (2.3.64) permits the evaluation of the second derivative appearing in Eq. (2.3.77) in the form

$$\left(\frac{1}{kT}\right)^2 {}^{ik}\langle[\psi_k(\mathbf{R}_k) - {}^{ik}\langle\psi_k(\mathbf{R}_k)\rangle][\psi_i(\mathbf{R}) - {}^{ik}\langle\psi_i(\mathbf{R})\rangle]^2\rangle \qquad (2.3.84)$$

which is a fluctuation of the third order. Combining the various terms yields the final relation

$$\int_0^{z_iq} [{}^{ik}\langle\psi_i(\mathbf{R})\rangle - {}^{k}\langle\psi_i(\mathbf{R}_i)\rangle]\, d(z_iq) = z_iq^k\langle\psi_i(\mathbf{R})\rangle_{z_iq=0}$$

$$+ \int_0^{z_iq} [{}^{i}\langle\psi_i(\mathbf{R}_i)\rangle_{z_kq=0} - {}^{k}\langle\psi_i(\mathbf{R}_i)\rangle]\, d(z_iq) + \Theta_{ki}(\mathbf{R}) \qquad (2.3.85)$$

where the last term on the right-hand side of Eq. (2.3.85) is defined by

$$\Theta_{ki}(\mathbf{R}) = \int_0^{z_iq}\int_0^{z_iq}\int_0^{z_kq} \left(\frac{1}{kT}\right)^2 \frac{\partial^2}{\partial(z_iq)\,\partial(z_kq)}\, {}^{ik}\langle\psi_i(\mathbf{R})\rangle\, d(z_iq)\, d(z_iq)\, d(z_kq)$$

$$+ \int_0^{z_iq} [{}^{ik}\langle\psi_i(\mathbf{R})\rangle_{z_kq=0} - {}^{i}\langle\psi_i(\mathbf{R}_i)\rangle]\, d(z_iq) \qquad (2.3.86)$$

$$+ 2z_iq \left[\frac{{}^{k}\langle\psi_i(\mathbf{R})g_i^0\rangle_{z_iq=0} - {}^{k}\langle\psi_i(\mathbf{R})\rangle_{z_iq=0}\langle g_i^0(\mathbf{R}')\rangle_{z_iq=0}}{{}^{k}\langle g_i^0(\mathbf{R})\rangle_{z_iq=0}}\right.$$

$$\left. - {}^{ik}\langle\psi_i(\mathbf{R})\rangle_{\substack{z_iq=0 \\ z_kq=0}}\right].$$

Thus, the mean potential $^k\langle \psi_i(\mathbf{R})\rangle$ differs from the mean potential $^k\langle \psi(\mathbf{R})\rangle$ by a quantity of order of magnitude $^k\langle \psi(\mathbf{R})\rangle/N_i$. In any averaging process in which ion k is held fixed at the point $\mathbf{R}_k$ but all the other ions are allowed to move from point to point in the solution, the $N-1$ free ions are all equivalent since any one of them can be moved over the entire volume V. The discharge of one of these ions, say of species i, will effect the potential only to the order of magnitude N_i^{-1}. For the same reason

$$^i\langle \psi_i(\mathbf{R}_i)\rangle_{z_kq=0} = {}^i\langle \psi_i(\mathbf{R}_i)\rangle + 0\left(\frac{1}{N_i}\right). \tag{2.3.87}$$

Since, by Eq. (2.3.50)

$$^i\langle \psi_i(\mathbf{R}_i)\rangle = \langle \psi_i \rangle \tag{2.3.88}$$

and we may evidently combine Eqs. (2.3.87) and (2.3.88) to yield

$$\langle \psi_i(\mathbf{R}_i)\rangle = {}^i\langle \psi_i(\mathbf{R}_i)\rangle_{z_kq=0}. \tag{2.3.89}$$

Further, $^k\langle \psi_i(\mathbf{R}_i)\rangle$ differs from $\langle \psi_i(\mathbf{R}_i)\rangle$ by a quantity of negligible order, since the fixed ion k can only influence the potential distribution in the neighborhood of some other ion i if that ion is in a region about the ion k of order of magnitude ω. Since the ratio of probabilities of finding an ion in the volume element ω and anywhere in the total volume V is of the order of magnitude ω/V, and ω is of molecular dimensions, this effect is negligible. To continue, since $^i\langle \psi_i(\mathbf{R}_i)\rangle_{z_kq=0} = \langle \psi_i(\mathbf{R}_i)\rangle$ and $^k\langle \psi_i(\mathbf{R}_i)\rangle = \langle \psi_i(\mathbf{R}_i)\rangle$ to order of magnitude N_i^{-1}, the second term on the right-hand side of Eq. (2.3.87) vanishes to order N^{-1}, and Eq. (2.3.85) is reduced to the form

$$\int_0^{z_iq} [{}^{ik}\langle \psi_i(\mathbf{R})\rangle - {}^k\langle \psi_i(\mathbf{R})\rangle] \, d(z_iq) = z_iq^k\langle \psi(\mathbf{R})\rangle + \Theta_{ki}(\mathbf{R}). \tag{2.3.90}$$

The substitution of (2.3.90) into Eq. (2.3.74) now leads to the mean Poisson equation

$$\nabla^2(^k\langle \psi(\mathbf{R})\rangle) = -\frac{4\pi}{D}\sum_1^r c_i^0 z_iqg_2^0 \exp\left\{-\frac{1}{kT}[z_iq^k\langle \psi(\mathbf{R})\rangle + \Theta_{ki}(\mathbf{R})]\right\} \tag{2.3.91}$$

and Eq. (2.3.91) is exact to terms of the order of magnitude N^{-1}.

We are finally in a position to examine the validity of the assumption expressed in Eq. (2.3.61) and briefly discussed earlier, namely that the potential of mean force is the product of the ionic charge and the mean potential. For simplicity assume that the ions are rigid spheres with no attractive field except for the effects of their charges. For dilute solutions g_2^0 will be set equal to unity, with the result that

$$\nabla^2 {}^k\langle \psi(\mathbf{R})\rangle = 0 \qquad |\mathbf{R} - \mathbf{R}_k| < a_k \tag{2.3.92}$$

$$\nabla^{2\,k}\langle\psi(\mathbf{R})\rangle = -\frac{4\pi}{D}\sum_1^r c_i^0 z_i q \, \exp\left\{-\frac{1}{kT}[z_i q^{\,k}\langle\psi(\mathbf{R})\rangle + \Theta_{ki}(\mathbf{R})]\right\} \quad (2.3.93)$$

$$|\mathbf{R} - \mathbf{R}_k| > a_k$$

while the fluctuation term becomes, finally,

$$\Theta_{ki} = \left(\frac{1}{kT}\right)^2 \int_0^{zq} {}^{ik}\langle[\psi_k(\mathbf{R}_k) - {}^{ik}\langle\psi_k(\mathbf{R}_k)\rangle][\psi_i(\mathbf{R}) - {}^{ik}\langle\psi_i(\mathbf{R})\rangle]^2\rangle] \times$$

$$d(z_i q)\,d(z_i q)\,d(z_k q). \quad (2.3.94)$$

If the fluctuation terms are neglected, Eq. (2.3.94) reduces to Eq. (2.3.3), the fundamental equation of the Debye–Hückel theory, which in the more elaborate notation used here is

$$\nabla^{2\,k}\langle\psi(\mathbf{R})\rangle = -\frac{4\pi}{D}\sum_1^r c_i^0 z_i q \, \exp\left(-\frac{1}{kT}z_i q^{\,k}\langle\psi(\mathbf{R})\rangle\right)$$

$$|\mathbf{R} - \mathbf{R}_k| > a_k \qquad\qquad (2.3.95)$$

$$\nabla^{2\,k}\langle\psi(\mathbf{R})\rangle = 0$$

$$|\mathbf{R} - \mathbf{R}_k| < a_k$$

The nature of this approximation becomes clearer if the fluctuation term is evaluated. An exact determination is impossible but some approximations suffice to indicate the order of magnitude of the error. Recall that

$$^{ik}\langle\psi_k(\mathbf{R}_k)\rangle = \frac{\displaystyle\int_V \psi_k(\mathbf{R}_k)\exp\left\{-\frac{1}{kT}[U_{N-1} + \phi_i]\right\} d\{N\text{-}2\}}{\displaystyle\int_V \exp\left\{-\frac{1}{kT}[U_{N-1} + \phi_i]\right\} d\{N\text{-}2\}}. \quad (2.3.96)$$

By differentiation of Eq. (2.3.96) it is found that

$$^{ik}\langle[\psi_k(\mathbf{R}_k) - {}^{ik}\langle\psi_k(\mathbf{R}_k)\rangle][\psi_i(\mathbf{R}) - {}^{ik}\langle\psi_i(\mathbf{R})\rangle]^2\rangle = kT^2 \frac{\partial^2\,{}^{ik}\langle\psi_k(\mathbf{R}_k)\rangle}{\partial(z_i q)^2} \quad (2.3.97)$$

The mean potential $^{ik}\langle\psi_k(\mathbf{R}_k)\rangle$ may be expanded in a Taylor's series about the value of the mean potential when the charge on ion i, $z_i q$, is zero, yielding

$$^{ik}\langle\psi_k(\mathbf{R}_k)\rangle = {}^{ik}\langle\psi_k(\mathbf{R}_k)\rangle_{z_i q=0} + z_i q\left[\frac{\partial}{\partial(z_i q)}\,{}^{ik}\langle\psi_k(\mathbf{R}_k)\rangle\right]_{z_i q=0}$$

$$+ \sum_2^\infty \frac{(z_i q)^n}{n!}\left[\frac{\partial^n}{\partial(z_i q)^n}\,{}^{ik}\langle\psi_k(\mathbf{R}_k)\rangle\right]_{z_i q=0}. \qquad (2.3.98)$$

If all but the first two terms in the expansion can be neglected, then both the fluctuation expressed in Eq. (2.3.97) and Θ_{ki} vanish. *The neglect of the fluctuation term is therefore equivalent to assuming that $^{ik}\langle\psi_k(\mathbf{R}_k)\rangle$ can be approximated by a linear function of the charge, $z_i q$, on the ion i.* When the field point $\mathbf{R}$ at which the ion i is placed is very far from the point $\mathbf{R}_k$ at which the ion k is located, this ought to be a fair approximation. Aside from the neglect of the secondary (in dilute solution) effects of the short-range forces, this is the only approximation involved in the derivation of Eq. (2.3.95). The effects of the short-range forces will be discussed in a later section. Their neglect corresponds to the previously emphasized assumption that the solution of discharged ions is an ideal solution.

Now the fluctuation terms, Θ_{ki} depend entirely upon the screening influence of the space charge due to all the other ions in the solution on the potential of interaction between ions i and k. Therefore, the quantities Θ_{ki} must tend to zero as the concentration tends to zero. From classical electrostatics it is found that

$$\lim_{c_i{}^0 \to 0} {}^{ik}\langle \psi_k(\mathbf{R}_k) \rangle = \frac{z_i q}{D|\mathbf{R} - \mathbf{R}_k|} \tag{2.3.99}$$

$$\lim_{c_i{}^0 \to 0} \frac{\partial^2}{\partial(z_i q)^2} {}^{ik}\langle \psi_k(\mathbf{R}_k) \rangle = 0 \tag{2.3.100}$$

and therefore that

$$\lim_{c_i{}^0 \to 0} \Theta_{ki} = 0 \tag{2.3.101}$$

by Eq. (2.3.97). Since the potential ${}^k\langle \psi(\mathbf{R}) \rangle$ approaches $z_k q / D|\mathbf{R} - \mathbf{R}_k|$, it is found that

$$\lim_{c_i{}^0 \to 0} \frac{\Theta_{ki}(\mathbf{R})}{{}^k\langle \psi(\mathbf{R}) \rangle} = 0 \tag{2.3.102}$$

and therefore Eq. (2.3.95) is *exact in the limit of zero concentration.* It is only in this limit of zero concentration that the potential of mean force can be rigorously written as the product of the ionic charge and the mean potential.

After this rather extended analysis it is pertinent to summarize our findings. (a) The phase integral for a system of point charges will diverge if the short-range forces responsible for determining the " size " of an ion are neglected. It is just these short-range repulsive forces which prevent the solutions from collapsing into $N/2$ ideal gas points. (b) The apparent divergence at large distance is avoided by the condition of electrical neutrality and the resultant screening effect of the statistical space charge on the interaction between any pair of charges. (c) A closed differential equation for the mean electrostatic potential may be derived from the phase integral with the use of the approximation that the potential of mean force is the product of the ionic charge and the mean potential. This approximation is found to be exact only in the limit of infinite dilution, but it is an accurate approximation in the range where secondary effects of short-range van der Waals forces may be neglected. The primary effect of these latter forces in dilute solution, namely the determination of the finite volume occupied by an ion, cannot be neglected at any time.

2.4   Some Solutions of the Poisson–Boltzmann Equation

Before turning to the study of long-range interactions in solutions of polyelectrolytes, it is convenient to examine the solution of the Poisson–Boltzmann equation,

$$\nabla^2 \psi = -\frac{4\pi}{D} \sum_1^r c_i^0 z_i q e^{-z_i q \psi / kT}. \tag{2.4.1}$$

In general it will not be possible to obtain a solution to Eq. 2.4.1) and various approximations will have to be used. For this reason it is instructive to discuss first the solution of (2.4.1) for a one-dimensional problem: an electrolyte solution in contact with an infinite plane maintained at a potential ψ_w. For this case an analytic solution to Eq. (2.4.1) may be obtained and this compared with the various approximate solutions.

Let x be the distance from the charged wall. Since the plane is supposed to be infinite, it is only the variation of potential with the distance x from the plane that is of interest. The Laplacian operator ∇^2 then reduces to the derivative d^2/dx^2. Using the identity

$$\frac{d\psi}{dx}\frac{d^2\psi}{dx^2} = \frac{1}{2}\frac{d}{dx}\left(\frac{d\psi}{dx}\right)^2 \tag{2.4.2}$$

in Eq. (2.4.1) and integrating leads to

$$\left(\frac{d\psi}{dx}\right)^2 = \frac{4\pi kT}{D}\sum_1^r c_i^0(e^{-z_i q \psi/kT} - 1) \tag{2.4.3}$$

where the constant of integration has been evaluated through the use of the boundary condition

$$\frac{d\psi}{dx} = 0; \qquad x = \infty. \tag{2.4.4}$$

For a single z–z valent electrolyte, the right-hand side of Eq. (2.4.3) may be summed directly. Since there are only two ions, each present at the same concentration,

$$\left(\frac{d\psi}{dx}\right)^2 = \frac{8\pi kT}{D}c^0[e^{zq\psi/kT} + e^{-zq\psi/kT} - 2] \tag{2.4.5}$$

and after extracting the square root,

$$\frac{d\psi}{dx} = -\left(\frac{8\pi kT c^0}{D}\right)^{1/2} 2 \sinh \frac{zq\psi}{2kT} \tag{2.4.6}$$

where the negative sign has been chosen because the change in electrical potential $d\psi/dx$ is negative when the surface charge density is positive and vice versa. Equation (2.4.6) may be considerably simplified by use of two dimensionless variables, a reduced potential η and a reduced distance ξ, defined by

$$\eta = \frac{zq\psi}{kT}; \quad \xi = \left(\frac{8\pi c^0 z^2 q^2}{DkT}\right)^{1/2} x = \kappa x \qquad (2.4.7)$$

so that (2.4.6) assumes the form

$$\frac{d\eta}{e^{\eta/2} - e^{-\eta/2}} = -d\xi = \frac{1}{2}\left[\frac{e^{\eta/2}\, d\eta}{e^{\eta/2} - 1} - \frac{e^{\eta/2}\, d\eta}{e^{\eta/2} + 1}\right]. \qquad (2.4.8)$$

On integration one obtains

$$e^{\eta/2} = \frac{e^{\eta_w/2} + 1 + (e^{\eta_w/2} - 1)e^{-\xi}}{e^{\eta_w/2} + 1 - (e^{\eta_w/2} - 1)e^{-\xi}} \qquad (2.4.9)$$

where the boundary condition

$$\psi = \psi_w; \quad x = 0, \quad \eta_w = \frac{zq\psi_w}{kT} \qquad (2.4.10)$$

has been used and η_w stands for the reduced potential at the surface of the charged wall. Equation (2.4.9) describes the electric potential in the solution as a function of the distance from the plane surface, and has a roughly exponential form, ψ decreasing with increasing x from $\psi = \psi_w$ at $x = 0$ to $\psi = 0$ at $x = \infty$. Alternatively, Eq. (2.4.8) can be written as[†]

$$\eta = 4\tanh^{-1}\left[e^{-\xi}\tanh\frac{\eta_w}{4}\right]. \qquad (2.4.11)$$

As stated previously, it will in general be impossible to find a solution to Eq. (2.4.1). Under certain conditions however, the Poisson–Boltzmann

[†]Equation (2.4.11) can be easily obtained by writing Eq. (2.4.6) in the form

$$dx = -\left(\frac{D}{8\pi kT\, c^0}\right)^{1/2}\operatorname{csch}\frac{\eta}{2}\, d\psi \qquad (2.4.11.1)$$

and using the identity

$$\int\operatorname{csch}\theta\, d\theta = \ln\tanh\left|\frac{\theta}{2}\right| + \text{constant} \qquad (2.4.11.2)$$

so that, when the boundary condition at $x = 0$ is used to evaluate the constant of integration Eq. (2.4.11.1) integrates to

$$x = -\frac{1}{\kappa}\ln\frac{\tanh\left|\dfrac{zq\,\psi}{4kT}\right|}{\tanh\left|\dfrac{zq\,\psi_w}{4kT}\right|} \qquad (2.4.11.3)$$

which may be rearranged to give Eq. (2.4.11).

equation may be linearized. If the quantity $z_i q \psi / kT$ is very small relative to unity, then the right-hand side of Eq. (2.4.1) may be expanded in a power series in $(z_i q \psi / kT)$ and only the first few (two) terms retained. The use of the electroneutrality condition for the solution as a whole then leads to Eq. (2.3.5). Now, Eq. (2.3.5) is particularly easy to solve. By direct quadrature and use of the boundary conditions (2.4.4) and 2.4.10) the result obtained is

$$\eta = \eta_w e^{-\xi} \tag{2.4.12}$$

for the same single z–z valent electrolyte discussed before.

Before discussing the comparison of Eqs. (2.4.11) and (2.4.12) it is interesting to investigate what forces are set up by the presence of the electric field. To do this, consider again the Maxwell stress tensor discussed in Section 2.3. In the special case of a unidirectional field, the components $\mathscr{E}_y$ and $\mathscr{E}_z$ of the electric field strength are zero. From the relations

$$\nabla \cdot \boldsymbol{\Phi} = \nabla p$$

$$p = kT \sum_1^r c_i \tag{2.4.13}$$

is obtained

$$\frac{d\Phi_{xx}}{dx} = \frac{dp}{dx} \tag{2.4.14}$$

since the gradient exists only in the x direction. Moreover, since the off-diagonal components of the stress tensor contain products of the form $\mathscr{E}_y \mathscr{E}_z$, $\mathscr{E}_x \mathscr{E}_y$, only the diagonal elements are nonvanishing. On substituting the value $D\mathscr{E}_x^2 / 8\pi$ for Φ_{xx} and integrating, Eq. (2.4.14) becomes

$$p = \frac{D}{8\pi} \mathscr{E}_x^2 + p_0 \tag{2.4.15}$$

where p_0 is the thermal pressure at any point where the electric field is zero. Due to the presence of the electric field there is an additional pressure, additive to the normal fluid pressure, induced in the direction perpendicular to the field. The direction of the pressure induced is perpendicular to the field because the pressure is equivalent to the electrostatic tension acting across an equipotential surface. If the electric field strength is large there will be a variation of density along the x axis large enough to cause a variation of dielectric constant along x. Thus, even neglecting the effects of dielectric saturation, if the field is large enough D is not constant and Poisson's equation does not apply in the form used. The effects of dielectric saturation, of course, magnify this effect. In practice, since there is at present no means of accounting for the effect, any variation of D with x will be neglected. It is likely that the variation of D with x due solely to electrostriction is sufficiently small not to modify greatly our conclusions.

We are now ready to examine the solution (2.4.11) for several cases. When the potential on the infinite plane is sufficiently small that $(zq\,\psi_w/4kT)$ is very small compared to unity, Eq. (2.4.11) reduces to

$$\eta = \eta_w e^{-\xi}$$
$$\psi = \psi_w e^{-\kappa x}$$

$$(2.4.16)$$

which is identical with the solution (2.4.12) of Eq. (2.3.5). The quantity κ^{-1} is now seen to have the properties of a screening length. It is, in fact, that distance in which the potential falls to $1/e$ of its value at the wall. On the other hand, when $(zq\,\psi_w/4kT)$ is much greater than unity, Eq. (2.4.11) reduces to

$$\eta = \eta_w - e^{\eta_w/2}\xi; \qquad \xi \text{ small}$$
$$\eta = 4e^{-\xi}; \qquad \xi \text{ large}$$

$$(2.4.17)$$

and it is seen that the solutions depart from the simple solution (2.4.12). The discrepancies between the two solutions are tabulated for several selected points in Table 2.4.1.

TABLE 2.4.1

POTENTIAL AS A FUNCTION OF DISTANCE FROM THE WALL AND/OR
ELECTROLYTE CONCENTRATION

		$zq\,\psi_w/kT$			
		1	2	5	10
	1	0.376[a]	0.688	1.30	1.52
		0.368[b]	0.736	1.84	3.68
	2	0.133	0.250	0.460	0.540
		0.135	0.271	0.677	1.353
κx	5	0.00660	0.0125	0.0229	0.0266
		0.00674	0.0135	0.0337	0.0674
	10	0.00044	0.00083	0.000153	0.000178
		0.00045	0.00090	0.00230	0.000450

[a] Solution of exact equation (2.4.11).
[b] Solution of linearized equation (2.4.12).

It is immediately noted that there is a marked discrepancy between the numerical values computed from the solutions, Eqs. (2.4.11) and (2.4.12). This discrepancy amounts to a factor of two when $\xi = 1$ and $\eta_w = 10$. Further it should be noted that the rate of decay of the electric potential is determined by the screening parameter κ, and that κ increases with increasing charge on the ions and with increasing salt concentration. The magnitude of the rate of decay of the potential is thereby related to the magnitude of the statistical space charge in the vicinity of the infinite plane.

The condition which must be fulfilled if Eq. (2.3.5) is to be an adequate approximation to the Poisson–Boltzmann equation is that $(zq\,\psi/kT)$ must be small relative to unity at all points. The success or failure of Eq. (2.3.5) is directly related to the nature of the distribution of charges in the vicinity of the infinite plane. Now, in general, the charge and electric potential are related by

$$\rho = -2c^0zq\,\sinh\frac{zq\psi}{kT} \qquad (2.4.18)$$

for a single z–z electrolyte. On the other hand, if $zq\,\psi/kT$ is small, the potential and charge density are linearly dependent,

$$\rho = -2c^0z^2q^2\,\frac{\psi}{kT} \qquad (2.4.19)$$

where Eq. (2.4.19) is an approximation to Eq. (2.4.18) by the usual power series expansion in which terms of order ψ^2 are neglected. However, since the hyperbolic sine is an exponentially increasing function of η for positive η, the charge density falls off even more rapidly than the electric potential. A computation of the local ionic concentrations, c_+ and c_- shows that for a positive charge density on the infinite plane, even for rather low values of $zq\,\psi_w/kT$ (say of order unity), the statistical space charge is almost entirely composed of negative ions. In the case of the linear approximation, Eq. (2.4.19), the excess of negative charge would be exactly equal to the deficiency of positive charge in the immediate neighborhood of the plane. Thus, even for moderate values of the potential ψ_w, Eqs. (2.4.19) cannot account for the great difference in local ionic concentrations. Only when the potential ψ_w is very small is Eq. (2.4.19) an adequate approximation. These comments are, of course, only a simple amplification of the stated conditions for the validity of Eq. (2.4.1) namely: $zq\,\psi/kT \ll 1$.

We are now ready to attempt to solve Eq. (2.4.1) under more general conditions. It will again be necessary to make the approximation of linearizing the equation. We shall therefore be concerned with solution of Eq. (2.3.5) in some region of space. Let us consider the case when there is spherical symmetry. We choose for our model of the ion a hard sphere of radius a, so that Eq. (2.3.5) is correct when the distance from the center of the sphere

$r > a$, and Laplace's equation holds for distances $r < a$. For simplicity, assume the charge to be uniformly distributed within the rigid sphere. Since the electrical potential must tend to zero as r increases to infinity the solution of the linearized Poisson–Boltzmann equation (2.3.5) is of the form

$$\psi = \frac{Ae^{-\kappa r}}{Dr} \tag{2.4.20}$$

and if the boundary condition

$$\left(\frac{d\psi}{dr}\right)_{r=a} = -\frac{zq}{Da^2} \tag{2.4.21}$$

is used, then the integration constant A can be evaluated and the potential becomes

$$\psi = \frac{zq}{Dr} \frac{e^{\kappa(a-r)}}{1 + \kappa a}. \tag{2.4.22}$$

The screening parameter κ is once again seen to regulate the rate of decay of the electric potential. Let ψ^* be the potential produced by the statistical charge surrounding an ion. We then have, by simple subtraction,

$$\psi^*(r) = \psi(r) - \frac{zq}{Dr}$$

$$= \frac{zq}{Dr} \left(\frac{e^{\kappa(a-r)}}{1 + \kappa a} - 1\right). \tag{2.4.23}$$

If κa is much greater than unity, the variation of the electric potential ψ^* with κ can be neglected. This is due to the fact that, when a is very large, the charge is spread so thinly around the sphere that there is no screening. Thus the simple Coulomb potential is approached. In the other limit, as κa tends to zero,

$$\left.\begin{array}{c} \psi^*(r) = \dfrac{zq}{Dr}(e^{-\kappa r} - 1) \\[2em] \psi^*(0) = -\dfrac{zq\kappa}{D} \end{array}\right\} \kappa a \to 0 \tag{2.4.24}$$

so that $\psi^*(0)$ is independent of the individual characteristics of the ion except for its charge. It is convenient to interpret $\psi^*(0)$ as the electric potential produced at the origin by an electrical atmosphere the mean distance of which from the origin is $1/\kappa$. Now, the potential is largest at the surface of the ion, at which point,

$$\frac{zq\psi}{kT} \approx \frac{z^2q^2}{DkTa} \tag{2.4.25}$$

so that when the condition $zq\psi/kT \ll 1$ is fulfilled everywhere in space, it must be true that

$$a \gg \frac{z^2q^2}{DkT} \approx 7A(H_2O, 25^0 \text{ C}). \tag{2.4.26}$$

In general, the ionic radius a will be less than this quantity and therefore it is to be expected that there will be deviations in the composition of the space charge about an ion from that predicted by the linearized Poisson–Boltzmann equation. Equation (2.4.24), which is the major physical content of the famous Debye–Hückel limiting law, may be expected to be valid only under conditions such that the region of space over which the condition $zq\psi/kT \ll 1$ fails is small relative to the total region of space between the ions. A more precise statement of the condition which must be met is that $\int \rho dV$ over the region where η is not small compared to unity must be small compared to zq. Thus, Eq. (2.4.24) is only expected to be valid at vanishingly small concentrations. In the sense that $zq\psi/kT \ll 1$ is never fulfilled over the entire region of space, Eq. (2.4.24) can never be rigorously justified by the method used in this section. To justify the limiting law theoretically, it is necessary to study the statistical mechanical theory of ionic solutions, as in Section 2.3. From such a study it is found that Eq. (2.4.24) is a reasonable approximation in the same range of concentration for which the potential of mean force is given by $zq\psi$.

It is pertinent at this point to discuss in some further detail the physical nature of the approximations to the solution of the Poisson–Boltzmann equation. We restrict our attention to solutions of binary salts for simplicity.

The zeroth approximation to the electrostatic potential in an electrolyte solution may be considered to be the unmodified Coulomb interaction. It is certainly true that in the limit of zero solute concentration, the potential of mean force acting between two ions must approach the unmodified Coulomb potential. The unmodified Coulomb potential $\psi = zq/Dr$ is the solution of the equation

$$\nabla^2 \eta = 0 \tag{2.4.27}$$

corresponding to zero space charge. The unmodified Coulomb potential will approximate the real potential only at vanishingly small ion concentrations. Compared to conditions in electrolyte solutions of finite concentration, the potential zq/Dr tends to zero much too slowly as r tends to infinity. The neglect of the effect of the space charge appears clearly in Eq. (2.4.27) which may be obtained from Poisson's equation by setting the charge density equal to zero.

Building upon the zero concentration limit, the limiting law starts from the linear relation

$$\nabla^2 \eta = \kappa^2 \eta \qquad (2.3.5)$$

with the solution

$$\eta = \frac{2Q}{r} e^{-\kappa r}; \qquad Q = \frac{q^2}{2DkT}. \qquad (2.4.28)$$

The use of Eq. (2.3.5) is equivalent to the assumption that the effects of all the ions in solution are additive so that the charge density is a linear function of the potential. The use of the linearized equation then is equivalent to assuming that it is possible to linearly superpose the fields of all the ions. This is clearly possible only at very low concentrations where ion–ion interferences are negligible. Now, the space charge, that is the ion atmosphere, about any given ion produces at the origin the potential given by Eq. (2.4.24). This solution to the problem neglects the exclusion of ions due to the finite radius of the central ion. As a result, the electrical potential increases too slowly with decreasing distance at the origin. In fact, the simple zeroth-order approximation, namely the unmodified Coulomb potential, has the correct initial slope of potential versus distance from the ion at the origin.

The second Debye–Hückel approximation also starts from the relation (2.3.5) which assumes a linear relationship between the charge density and the potential. However, the model is now changed in that the ion is regarded as a rigid sphere of radius a. The solution of Eq. (2.3.5) is then, in reduced notation

$$\eta = \frac{2Q}{r} \frac{e^{\kappa a} e^{-\kappa r}}{1 + \kappa a} \qquad (2.4.29)$$

and the potential produced by the ion atmosphere at the origin is now

$$\eta^* = -\frac{2Q\kappa}{1 + \kappa a}. \qquad (2.4.30)$$

Though this solution accounts for the finite size of the central ion, it neglects the breakdown of the linear approximation. The full Poisson–Boltzmann equation has never been solved, except in the one dimensional case, and in any event neglects the fundamental errors discussed previously.

It is of interest to examine the nature of the space charge predicted by the approximate solutions discussed above.

Consider now the space charge contained between concentric spheres centered on ion i with radii a and d respectively. If the radius d is chosen such that $d \gg Q$, then the potential outside the sphere with radius d must satisfy the inequality $\eta \ll 2Q/r$. The charge remaining outside of the sphere with radius d then is subject to the restriction

$$\tilde{Q} < \int_a^d \rho \, dV = \int_a^d c^0 \sinh \frac{2Q}{r} 4\pi r^2 \, dr. \qquad (2.4.31)$$

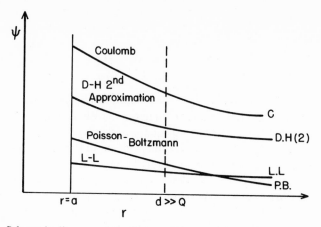

FIG. 2.4.1. Schematic diagram of the electrostatic potential as a function of distance from the ion.

From Eq. (2.4.22) is seen that the total charge tends to zero as κ^2 when κ tends to zero. However, the curves in Fig. 2.4.1 approach curve C (the Coulomb interaction) as κ while the charge decreases as κ^2. As the concentration of solute tends to zero, κ tends to zero and consequently κ^{-1} tends to infinity. When κ^{-1} is sufficiently great relative to the distance $|0 - d|$, then only an insignificant part of the charge is in this region ($4\pi d^3/3$). Therefore the true curve of electrostatic potential versus the distance approaches that characteristic of the limiting law faster than the limiting law approaches the curve C characteristic of the unmodified Coulomb interaction.

Define the radial charge density R^* by the relation.

$$R^* = -4\pi r^2 \rho = D r^2 \nabla^2 \psi \qquad (2.4.32)$$

and typical schematic radial charge distributions are shown in Fig. 2.4.2.

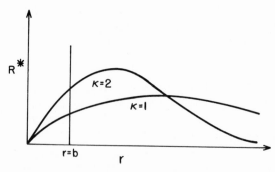

FIG. 2.4.2. Schematic radial charge distributions for two values of the screening length, κ.

As mentioned previously, the maximum in the radial charge density occurs at the distance $r = \kappa^{-1}$. The curves shown in Fig. 2.4.2 have been normalized to unity and represent the radial charge density only when the limiting law is valid. Indeed, when the limiting law is applicable, for a univalent ion,

$$R_{LL}^* = Dr^2\nabla^2\psi = q\kappa^2re^{-\kappa r} \tag{2.4.33}$$

and the potential at the origin due to this charged atmosphere is given by Eq. (2.4.24). Consider now a sphere of radius b centered about the central ion. The charge inside of this sphere arising from the ion atmosphere is

$$-\rho_{inside}^* = \int_0^b R^*\, dr = q[1 - e^{-b\kappa}(1 + b\kappa)]$$
$$= \tfrac{1}{2}qb^2\kappa^2 \tag{2.4.34}$$

when the limiting law is applicable. The potential at the origin due to the space charge may now be divided into two parts, ψ_0^* due to charges outside the sphere with radius b and ψ_i^* due to the space charge inside this sphere:

$$\psi_0^* = \int_b^\infty \frac{R_{LL}^*}{Dr}\, dr = \frac{zq\kappa}{D}\, e^{-\kappa b} \tag{2.4.35}$$

$$\psi_i^* = \int_0^b \frac{R_{LL}^*}{Dr}\, dr = \frac{zq\kappa}{D}(1 - e^{-\kappa b}). \tag{2.4.36}$$

Let us now examine the consequences of a limited class of changes in the model of the ion. If all charges inside the sphere of radius b are excluded, ψ_i^* will vanish and ψ^* will be reduced to the value of ψ_0^*. If the excluded charge is now restored and placed at a distance κ^{-1} from the origin, the potential at the origin becomes

$$\psi^* = \frac{zq\kappa}{D}\, e^{-\kappa b} + \frac{\rho_{inside}^*\kappa}{D} = \frac{zq\kappa}{D}(1 - \kappa be^{-\kappa b}). \tag{2.4.37}$$

On expanding (2.4.37) as a power series, it is found that the first three terms are identical with those obtained from the second Debye–Huckel approximation.

Finally consider the case of a real ion. That is, consider that the radial charge density is similar to that depicted in Fig. 2.4.3.

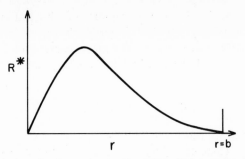

FIG. 2.4.3. Atomic charge density for a Fermi-Thomas statistical ion.

This nonuniform charge density would represent an approximation to the real charge density in an ion, as calculated from quantum theory. For this distribution

$$-\eta_i^* = 2Q\kappa(\kappa\tilde{c}_1 - \tfrac{1}{2}\kappa^2\tilde{d}^2 + \cdots)$$
$$-\eta_0^* = 2Q\kappa(1 - \kappa b + \tfrac{1}{2}\kappa^2(b^2 + b\tilde{c}_2) + \cdots) \qquad (2.4.38)$$

and the total atmosphere contribution becomes

$$-\eta^* = 2Q\kappa(1 - \kappa(b - \tilde{c}_1) + \cdots) \qquad (2.4.39)$$

where the quantities $\tilde{c}_1$, $\tilde{d}$, and $\tilde{c}_2$ are integrals related to the charge removed from the various regions of space.

What conclusions can be drawn from this discussion? First, note that the quantity $(b - \tilde{c}_1)$ appearing in the last and most realistic model is *not* the ionic radius. Second, note that the deviation from the limiting law due to the exclusion of charge from a region of space (finite ionic volume) is in the same direction as that caused by introducing a rigid sphere of radius a as in the Debye–Hückel second approximation. Finally, note that it is always possible to write the atmosphere contribution to the potential in the form

$$-\eta^* = 2Q\kappa(1 - \kappa a^* + \kappa^2 b^* + \cdots) \qquad (2.4.40)$$

where a^* and b^* are empirical coefficients, but that this series cannot be summed exactly, so that

$$-\eta^* \neq \frac{2Q\kappa}{1 + \kappa a^*}. \qquad (2.4.41)$$

Physically this means that it is not possible to consistently describe the atmosphere potential with a simple parameter a^* for the last case discussed. Therefore, the introduction of more realistic repulsive forces than hard sphere repulsions leads to very great difficulties in calculating the atmosphere potential. These difficulties are, of course, further reflected in the calculation of all quantities dependent upon the potential. It is clear that refinements of the model require more powerful techniques than any presented thus far. The tools required are in fact available in the form of cluster expansions as developed by Mayer, a review of which may be found in the text by O'M. Bockris[20a].

REFERENCES

1. C. J. T. DI GROTTHUSS, *Ann. chim. (et phys.)* **58**, 54 (1806). For an interesting account of the history of the theory of electrolytes, see F. J. MOORE, " History of Chemistry," 3rd Ed. McGraw-Hill, New York, 1939.
2. R. CLAUSIUS, *Pogg. Ann.* **101**, 338 (1857).
3. S. ARRHENIUS, *Z. physik. Chem.* **1**, 631 (1887).
4. This table is taken from H. FALKENHAGEN, " Electrolytes." Oxford Univ. Press, London and New York, 1934.
5. For a summary of much pertinent data see H. S. HARNED AND B. B. OWEN, " The Physical Chemistry of Electrolyte Solutions," 2nd ed. Reinhold, New York, 1950.
6. W. OSTWALD, *Z. physik. Chem.* **2**, 36 (1888).
7. See, for example, J. H. HILDEBRAND, AND R. L. SCOTT, " The Solubility of Non-electrolytes," 3rd ed. Reinhold, New York, 1950.
8. J. J. VAN LAAR, *Z. physik. Chem.* **18**, 274 (1895.)
9. W. SUTHERLAND, *Phil. Mag.* (6) **14**, 1 (1907).
10. P. HERTZ, *Ann. Physik* **37**, 1 (1912).
11. S. R. MILNER, *Phil. Mag.* (6) **23**, 551 (1912).
12. N. BJERRUM, *Kgl. Danske Videnskab. Selskabs, Skrifter* (7) **4**, 1 (1906).
13. C. GHOSH, *J. Chem. Soc.* **113**, 449, 627, 707, 790 (1918.)
14. See, for example, R. B. LINDSAY, " Physical Statistics." Wiley, New York, 1940.
15. P. DEBYE AND E. HÜCKEL, *Physik. Z.* **24**, 185, 305 (1923).
16. L. ONSAGER, *Physik. Z.* **28**, 277 (1927); L. ONSAGER AND R. M. FUOSS, *J. Phys. Chem.* **36**, 2689 (1932).
17. H. FALKENHAGEN, *Physik. Z.* **32**, 365, 745 (1931).
18. J. G. KIRKWOOD, *J. Chem. Phys.* **2**, 767 (1934); J. G. KIRKWOOD AND J. C. POIRIER, *J. Phys. Chem.* **58**, 591 (1954).
19. J. E. MAYER, *J. Chem. Phys.* **18**, 1426 (1950).
20. E. WICKE AND M. EIGEN, *Naturwissenschaften* **38**, 453 (1951); E. WICKE AND M. EIGEN, *Z. Elektrochem.* **56**, 551 (1952); **57**, 219 (1953); *Z. Naturforsch.* **8a**, 161 (1953).
20a. J. O'M. BOCKRIS, " Modern Aspects of Electrochemistry." Academic Press, New York, 1959.
20b. R. M. FUOSS AND F. ACCASCINA, " Conductance of Electrolytes," Academic Press, New York, 1960.
21. F. O. KOENIG, *J. Phys. Chem.* **41**, 597 (1937).
22. H. P. FRANK, *J. Chem. Phys.* **23**, 2023 (1955).
23. E. A. GUGGENHEIM, " Thermodynamics." North Holland Publ., Amsterdam, 1951.
24. L. ONSAGER, *Chem. Revs.* **13**, 73 (1933).
25. A. S. COOLIDGE AND W. JUDA, *J. Am. Chem. Soc.* **68**, 608 (1946).
26. See, for example, Interaction in ionic solutions, *Discussions Faraday Soc.* **24** (1957).

3. Some Equilibrium Properties of Solutions of Rigid Polyelectrolytes

3.0 Introduction

The fundamental theory of solutions of rigid impenetrable macroions was published by Verwey and Overbeek in 1948 in their well-known monograph,[1] and has been continuously refined by Levine,[2] Booth,[3] and Kirkwood,[4] and co-workers. Briefly, the model adopted for the interaction between particles represents the polymer as a rigid sphere with a specified surface charge density. The high surface potential tends to force the counterions to cluster close to the macroion, creating a thin double layer which shields the surface charges from similar charges on other colloidal particles. The net repulsive force between two macroions is usually obtained from a suitable solution of the Poisson–Boltzmann equation though more sophisticated methods have also been used. The fact that the macroion is rigid and impenetrable simplifies the problem to the extent that almost no features are intrinsically different from the low molecular weight strong electrolyte case. There are, however, qualitative differences in the relative importance of the various contributions.[1]

For colloidal ions with one type of charge we may imagine the free energy of the solution to consist of four contributions. First, if the charge is due to adsorption or desorption of ions from the solution, there is an intrinsic chemical potential change due to the dissociation reaction. To this must be added the sum of the chemical potentials of all species in solution, the difference in self-energy of an adsorbed ion from a free ion, and the electrical free energy of charging up and assembling the double layer. When considering the interaction of two macroions it is important to note that the overlap of double layers causes a redistribution of small ions in the solution with the net result that the mutual interactions of the small ions are changed. Thus in addition to the expected screened Coulomb force between macroions, there is a force arising from the change in interaction energy of the small ions as they redistribute under the influence of the overlapping double layers.

When consideration is given to the possibility that charges of both sign may exist on the surface of the molecule a new mechanism of interaction between macroions becomes apparent. If there is a net difference in the numbers of positive and negative charges, at the isoelectric point there will be a number of uncharged groups otherwise identical in chemical character with some of the charged groups. For one of the types of charge, there exist under these conditions a large number of different arrangements of the charges on the surface of the molecule. A number of these arrangements will be of approximately equal electrostatic energy and we may expect the charges to fluctuate in occupation of first one and then another set of sites. If two such molecules are brought close to one another, the charges of one ion will tend to polarize the charges on the other ion in a manner such as to separate like charges. This is possible if, as postulated, there exist surface charge distributions of roughly equal energy and the decreased repulsion is, of course, equivalent to a net attraction since the net charge of each particle is zero. It has been proposed that this fluctuation force is operative in enzyme reactions.

One of the fundamental differences between macroions and small ions is that in the former case many experiments are concerned explicitly with the self-energy of the ion. The magnitude of this interaction depends upon the geometric arrangement of the charges, the dielectric constant, and the salt concentration, and may be readily ascertained from the titration curve. Early attempts to characterize this electrostatic energy made the crude approximation of smearing out the discrete charge structure and using Debye–Hückel theory. Refinements such as allowing the ion to be partially penetrable by solvent alter the model only slightly. On the other hand, direct calculation of the electrostatic energy for discrete charge distributions reveals a marked dependence of the titration curve on the details of the distribution of charges, as was to be expected. Physically, this corresponds to the fact that neighboring acid and base groups tend to become zwitterions, whereas neighboring groups of the same kind resist ionization because of the large repulsive interactions created by such ionizations. Consequently, the amount of work required to ionize a group depends strongly on the group environment and a smeared-out charge distribution gives an erroneous although at times fortuitous numerically accurate picture of the molecule.

We now turn to a quantitative formulation of the models discussed in this section.

3.1   Interactions between Rigid Polyions

Consider an isolated rigid polyion of arbitrary shape in a volume V of an aqueous electrolyte solution. The charge on the polyion will be assumed due to the preferential adsorption of n ions of type one and charge q_1. Equivalently, with no change in the treatment, the charge may be considered due to the dissociation of some groups. For the present we restrict attention to the case when only a single species of charge is present on the polyion. We shall follow closely the method introduced by Levine[2] which permits the simultaneous determination of the adsorption isotherm for ions on the particle surface and the mutual free energy of the particles.

To compute the electrostatic free energy we proceed via a hypothetical charging process in which all the ions are charged from zero to their full charge at the same rate. It may be shown that the change in Helmholtz free energy due to a change $d\lambda$ in the fraction of full charge, λ, is

$$\delta A = \left(\frac{\partial A}{\partial \lambda}\right) d\lambda = \left(\frac{\partial A_c}{\partial \lambda}\right) d\lambda + \left(\frac{\partial A_s}{\partial \lambda}\right) d\lambda \qquad (3.1.1)$$

where A_c is the Helmholtz free energy due to interparticle interactions and A_s is the Helmholtz free energy due to the electrical self-energy of all the ions. Let the surface density of adsorbed ions be v, subject to the condition

$$n = \int_S v \, dS = n \int_S v_0(\mathbf{s}) \, dS \qquad (3.1.2)$$

where n is the number of adsorbed ions, $\mathbf{s}$ is a vector specifying position on the polyion surface, and S is the area of the surface of a polyion. The change in interparticle electrostatic energy due to a change $d\lambda$ in charge arising from the interaction of the ions in any small element of volume and from the interaction of ions with the polyion may be written in the form

$$\sum_{i=1}^{r} c_i(\lambda, \mathbf{r}) q_i \Psi_i(\lambda, \mathbf{r}) \, d\lambda \, dv + v(\mathbf{s}) q_1 \Psi_1(\lambda, \mathbf{s}) \, d\lambda \, dS \qquad (3.1.3)$$

where $c_i(\lambda, \mathbf{r})$ is the average volume density of ions of species i and $\Psi_i(\lambda, \mathbf{r})$ and $\Psi_1(\lambda, \mathbf{s})$ are the average electrostatic potentials at the point $\mathbf{r}$ occupied by an ion of species i due to surrounding ions, and the average electrostatic potential of an adsorbed ion due to the surrounding ions, respectively. The change in self-energy of the ions in the same volume and surface elements due to a change $d\lambda$ in charge is

$$\sum_{i=1}^{r} c_i(\lambda, \mathbf{r}) \left(\frac{\partial \chi_i}{\partial \lambda}\right) d\lambda \, dv + v(\mathbf{s}) \left(\frac{\partial \chi_s}{\partial \lambda}\right) d\lambda \, dS \qquad (3.1.4)$$

where $\chi_i(\lambda)$ is the electrical self-energy of an ion of species i which has charge λq_i and $\chi_s(\lambda)$ is the corresponding self-energy of an adsorbed ion carrying charge λq_1.

The existence of adsorbed ions implies the existence of a dissociation equilibrium which may, of course, be characterized by the change in chemical potential per ion adsorbed. If all the ions are discharged, the Helmholtz free energy may be written

$$A_0(n) = \sum_{i=0}^{r} N_i \mu_i(0) + a_0(n) \qquad (3.1.5)$$

$$a_0(n) = \int_S a_0^*(v) \, dS \qquad (3.1.6)$$

where $a_0(n)$ is the change in free energy when n discharged ions are adsorbed on the polyion and $\mu_i(0)$ is the chemical potential of a discharged ion of species i when there are no adsorbed ions. From Eqs. (3.1.1.) and (3.1.5) it is seen that

$$A(n) = A_0(n) + \int_0^1 d\lambda \left(\frac{\partial A_c}{\partial \lambda}\right) + \int_0^1 \left(\frac{\partial A_s}{\partial \lambda}\right) d\lambda. \tag{3.1.7}$$

As usual, the equilibrium conditions will be determined by minimizing the free energy with respect to the relevant parameters.

The average potential $\Psi_i(\mathbf{r})$ may be decomposed into two parts, the potential in the bulk of the solution, $\phi_i(n)$, and the change in potential, $\psi_i(\mathbf{r})$, experienced by the ion i in being brought from the bulk of the medium to the point $\mathbf{r}$. Thus write

$$\Psi_i(\lambda, \mathbf{r}) = \psi_i(\lambda, \mathbf{r}) + \phi_i(\lambda, n) \tag{3.1.8}$$

$$\Psi_1(\lambda, \mathbf{s}) = \psi_1(\lambda, \mathbf{s}) + \phi_1(\lambda, n) \tag{3.1.9}$$

where the average surface potential Ψ_1 has been decomposed in the same manner. By substitution of (3.1.8) and (3.1.9) into (3.1.3) and performing the indicated integrations it is found that (3.1.3) becomes

$$\sum_{i=0}^{r} q_i \left[\int_0^1 \phi_i(\lambda, n) \, d\lambda \int_V c_i(\lambda, \mathbf{r}) \, dv + \int_0^1 d\lambda \int_V c_i(\lambda, \mathbf{r}) \psi_i(\lambda, \mathbf{r}) \, dv \right]$$
$$+ nq_1 \left[\int_0^1 \phi_1(\lambda, n) \, d\lambda + \int_0^1 d\lambda \int_S v_0(\mathbf{s}) \psi_1(\lambda, \mathbf{s}) \, dS \right]. \tag{3.1.10}$$

The average excess electrostatic energy (the excess Helmholtz free energy) associated with the ions of the double layer arises only from the potential differences ψ_i and ψ_1 and is therefore seen to be

$$\delta A_e(\lambda, n) = \frac{\lambda}{2} \sum_0^r q_i \int_V c_i \psi_i \, dv + \frac{\lambda q_1}{2} \int_S v(\mathbf{s}) \psi_1 \, dS \tag{3.1.11}$$

and the total change in Helmholtz free energy due to the excess charge of the double layer is

$$A_e = \int_{\lambda=0}^{\lambda=1} \delta A_e(\lambda, n) \, d\lambda. \tag{3.1.12}$$

Integration of Eq. (3.1.4) gives

$$\int_0^1 \left(\frac{\partial A_s}{\partial \lambda}\right) d\lambda = \sum_1^r \int_0^1 \left(\frac{\partial \chi_i}{\partial \lambda}\right) d\lambda \int_V c_i \, dv + n\chi_s(1) \tag{3.1.13}$$

as the contribution from the self-energy term. Let

$$\tau_i = \mu_i(0) + \chi_i(1) + q_i \int_0^1 \phi_i(\lambda, n) \, d\lambda \qquad (3.1.14)$$

and

$$\Delta\chi_1(\lambda) = \chi_1(\lambda) - \chi_s(\lambda) \qquad (3.1.15)$$

so that the total Helmholtz free energy becomes

$$A(n) = N_0\mu_0(0) + \sum N_i\tau_i + a_0(n) - n\Delta\chi_1(1) + A_e(n) \qquad (3.1.16)$$

where the subscript zero refers to the solvent. Equations (3.1.16) represents a division of the total free energy into several parts. The first part of the free energy is that arising from the chemical potential of the solvent present, $N_0\mu_0(0)$, and from the chemical potentials of the ions, $N_i\tau_i$, including the self-energy of the ions. A second portion of the free energy describes the intrinsic chemical potential change due to adsorption, $a_0(n)$, and a third part describes the difference in self-energy of the adsorbed ion between the bulk and surface phases. Finally, the last contribution to the free energy arises from the excess electrostatic interaction in the double layer over that in the uniform bulk medium, $A_e(n)$,

If there were no ions adsorbed on the polyion surface, the average electrostatic potential in the bulk of the solution would differ from $\phi_i(\lambda, n)$ due to a redistribution of the ions in the solution when the double layer is dispersed. We may therefore conveniently define

$$\phi_i(\lambda, n) = \phi_i^0(\lambda, n = 0) + \Delta\phi_i(\lambda, n) \qquad (3.1.17)$$

and rewrite (3.1.14) in the form

$$\tau_i(n) = \tau_i(0) + A_c^*(n) \qquad (3.1.18)$$

where

$$\tau_i(0) = \mu_i(0) + \chi_i(1) + q_i \int_0^1 \phi_i^0(\lambda, 0) \, d\lambda \qquad (3.1.19)$$

$$A_c^*(n) = q_i \int_0^1 \Delta\phi_i(\lambda, n) \, d\lambda. \qquad (3.1.20)$$

That part of the electrical free energy dependent upon n may be minimized with respect to n to determine the equilibrium ionic adsorption. If we separate the constant terms (with respect to variations of n),

$$W(n) = A(n) - N_0\mu_0(0) - \sum N_i\tau_i(0)$$

$$= a_0(n) - n\,\Delta\chi_1(1) + A_e(n) + A_c^*(n) \qquad (3.1.21)$$

which leads to

$$\frac{\partial W(n)}{\partial n} = 0$$

$$\Delta\chi_1(1) = \frac{\partial a_0(n)}{\partial n} + \frac{\partial A_e}{\partial n} + \frac{\partial A_c^*}{\partial n}. \tag{3.1.22}$$

The most convenient procedure to pursue is to evaluate $W(n)$ in the Debye-Hückel approximation. When $n = 0$, for a uni-univalent salt

$$\kappa_0^2 = \frac{4\pi}{DkT} \sum_1^r c_i^0 q_i^2 \tag{3.1.23}$$

and

$$\phi_i(\lambda, 0) = -\frac{\lambda^2 q_i \kappa_0}{D} \tag{3.1.24}$$

whereas when $n \neq 0$, put

$$\kappa^2 = \frac{4\pi\lambda^2}{DkT} \sum_1^r c_i^0 q_i^2 \tag{3.1.25}$$

$$\phi_i(\lambda, n) = -\frac{\lambda^2 q_i \kappa}{D} \tag{3.1.26}$$

where we have assumed that the net charge density $\sum c_i^0 q_i$ vanishes. This is valid if the volume is very large since then the excess charges balancing the n adsorbed charges are dispersed in an essentially infinite medium and contribute only infinitesimally to the potential at all points in the volume. Let $\hat{c}_i(\mathbf{r})$ be the difference in concentration between the double layer of the colloidal ion and the bulk medium. Then,

$$c_i(\lambda, \mathbf{r}) = c_i^0(\lambda, n) + \hat{c}_i(\lambda, \mathbf{r}) \tag{3.1.27}$$

and

$$\kappa^2(\lambda, n) = \frac{4\pi\lambda^2}{DkTV} \left[\sum_1^r N_i q_i^2 - nq_1^2 - \sum_1^r q_i^2 \int_V \hat{c}_i \, dv \right] \tag{3.1.28}$$

which becomes, by expansion and retention of terms in V^{-1},

$$\kappa(\lambda, n) - \kappa_0\lambda = \Delta\kappa(\lambda, n)$$

$$= -\frac{2\pi\lambda}{DkTV\kappa_0} \left[nq_1^2 + \sum_1^r q_i^2 \int_V \hat{c}_i \, dv \right]. \tag{3.1.29}$$

Similarly, expansion of $\Delta\phi_i(\lambda, n)$ about $\lambda\kappa_0$ yields

$$\Delta\phi_i(\lambda, n) = - \frac{\lambda q_i}{D} \Delta\kappa(\lambda, n) \qquad (3.1.30)$$

and by substitution,

$$A_c^* = 2 \int_0^1 \frac{\lambda^2 \kappa_0}{4D} \sum_1^r q_i^2 \int_V c_i \, dv \, d\lambda + \frac{n q_1^2 \kappa_0}{6D}. \qquad (3.1.31)$$

It is instructive to note that if two potential functions $\psi_c(n)$ and $\psi_c^*(n)$ are defined by

$$\frac{\partial A_c}{\partial n} = q_1 \psi_c \, ; \qquad \frac{\partial A_c^*}{\partial n} = q_1 \psi_c^* \qquad (3.1.32)$$

then $W(n)$ becomes

$$W(n) = a_0(n) - n \Delta\chi_1(1) + q_1 \int_0^n [\psi_c(n) + \psi_c^*(n)] \, dn \qquad (3.1.21')$$

and the equilibrium adsorption is determined by

$$\Delta\chi_1(1) = \frac{\partial a_0}{\partial n} + q_1(\psi_c + \psi_c^*). \qquad (3.1.22')$$

The integral term in (3.1.21') is the work done against the electrical forces when n ions of species 1 are transferred to the surface of the polyion from the bulk medium. The potential of mean force has been divided into two parts, one of which, $q_i\psi_c$, arises from the excess charge of the double layer, and the other, $q_1\psi_c^*$, from the change in mutual energy of the ions in the bulk solution when the n ions of type one are removed and placed on the polyion surface.

Having discussed the electrostatic free energy of an isolated colloidal particle, consider now the interaction of two such polyions. We may immediately write*

$$W(n, r) = 2a_0(n) - 2n \Delta\chi_1(1) + A_e(n, r) + A_c^*(n, r) \qquad (3.1.33)$$

where $A_e(n, r)$ is the electrical free energy due to the excess charge in the two double layers and $A_c^*(n, r)$ is the correction arising from the mutual energy of the ions in the solution. The relations (3.1.12) and (3.1.31) are still valid, except that the integrations are now to be understood as performed at fixed interparticle separation r. The force between the two particles is

$$\left(\frac{\partial W(\kappa, r)}{\partial r} \right)_{n=n_{eq}} = \frac{\partial}{\partial r} [A_e + A_c^*]_{n=n_{eq}}. \qquad (3.1.34)$$

* In anticipation of the calculations for the case of spherical symmetry we shall now drop the vector notation for the variable r.

To implement this general result, the free energy contributions $A_e(n, r)$ and $A_c^*(n, r)$ must be computed. Accordingly we now turn to a consideration of the force acting between two colloidal ions when the potential is calculated from the linearized Poisson–Boltzmann equation. From the Poisson–Boltzmann equation in the form

$$\nabla^2 \psi = -\frac{4\pi\rho}{D} = \lambda f(\lambda\psi) \tag{3.1.35}$$

$$f(\lambda\psi) = -\frac{4\pi}{D} \sum_1^s c_i^0 q_i e^{-\lambda q_i \psi/kT}, \tag{3.1.36}$$

the substitution of (3.1.35) and (3.1.36) into (3.1.11) and differentiating with respect to n gives

$$\frac{\partial(\delta A_e)}{\partial n} = \frac{\lambda q_1}{2} \int_S (\psi_n v + \psi v_n)\, dS - \frac{D\lambda}{8\pi} \int_V (\psi_n f + \lambda\psi\psi_n f')\, dv \tag{3.1.37}$$

where $\psi_n = (\partial\psi/\partial n)$, $v_n = (\partial v/\partial n)$ and $f' = (\partial f/\partial\lambda\psi)$. But from Eq. (3.1.25)

$$\nabla^2 \psi_\lambda = f + \lambda(\psi + \lambda\psi_\lambda)f'$$
$$\nabla^2 \psi_n = \lambda^2 \psi_n f'. \tag{3.1.38}$$

The boundary conditions to be used are as follows. Let the dielectric constant of the colloidal particle be D_0 and let $\psi^{(1)}$ be the potential inside the particle. Then if D and ψ are the corresponding quantities for the solution, in the interior of the particle

$$\nabla^2 \psi^{(1)} = 0$$
$$\nabla^2 \psi_n^{(1)} = 0 ; \qquad \nabla^2 \psi_\lambda^{(1)} = 0 \tag{3.1.39}$$

whereas at the surface

$$\psi^{(1)} = \psi$$
$$D\nabla_n\psi - D_0\nabla_n\psi^{(1)} = -4\pi v\lambda q_1 \tag{3.1.40}$$

where the gradient is taken along the outward normal from the surface. The integrals in Eq. (3.1.37) may be evaluated by the use of Green's theorem, the boundary conditions (3.1.39) and (3.1.40) and several integrations by parts. The result is, for one particle,

$$\frac{\partial(\delta A_e)}{\partial n} = \lambda q_1 \int_S \psi(\lambda) \frac{\partial v}{\partial n}\, dS. \tag{3.1.41}$$

This free energy change may be interpreted as the work done in bringing up δn ions of charge λq_1 to the surface with the result that the surface charge density is increased by $\delta v = (\partial v / \partial n)\delta n$. The electrical work is easily seen to be $\lambda q_1 \psi \delta v dS$, which is of the familiar form charge multiplied by potential. The result for several ions charged simultaneously is obtained from (3.1.41) by summation, but where the potential must be obtained from a solution of the Poisson–Boltzmann equation for the relevant number of particles. For two colloidal ions, after integration with respect to n,

$$A_e(n, r) = 2q_1 \int_0^n \psi(\lambda)v_0(\mathbf{s}) \, dS \, dn. \tag{3.1.42}$$

By exactly the same techniques, the electrical free energy due to the Coulomb interactions of the small ions may be shown to be

$$A_c^*(n) = -\int_0^n \frac{\kappa_0 q_1 q_2}{6D} \, dn$$

$$-2 \int_0^n \int_0^1 \frac{\lambda^3 \kappa_0 q_1^2 q_2}{8DkT} (2n\tilde{\psi}_n - \tilde{\psi} - \lambda\tilde{\psi}_\lambda) \, d\lambda \, dn \tag{3.1.43}$$

where

$$\tilde{\psi} = \int_S \psi(\lambda)v_0 \, dS. \tag{3.1.44}$$

To evaluate A_e and A_c^* explicitly consider the case when $q_i \psi \ll kT$ so that Eq. (3.1.36) may be linearized with the result that

$$\nabla^2 \psi = \lambda^2 \kappa_0^2 \psi. \tag{3.1.45}$$

Now put

$$\tilde{\psi}(\lambda, n) = n\tilde{\psi}^0(\lambda)$$

$$2n\tilde{\psi}_n - \tilde{\psi} - \lambda\tilde{\psi}_\lambda = -\lambda^2 \frac{\partial(\tilde{\psi}/\lambda)}{\partial \lambda} \tag{3.1.46}$$

$$= -\lambda^2 n \frac{\partial(\tilde{\psi}^0/\lambda)}{\partial \lambda}$$

and

$$\psi_c^*(n) = \psi_c^*(0) + n\psi_c^{*0}. \tag{3.1.47}$$

The substitution of (3.1.46) into (3.1.43) and the use of (3.1.32) gives

$$\psi_c^*(0) = -\frac{\kappa_0 q_2}{6D}$$

$$\psi_c^{*0} = \frac{\kappa_0 q_1 q_2}{4DkT} \int_0^1 \lambda^5 \frac{\partial(\tilde{\psi}_0/\lambda)}{\partial\lambda} \, d\lambda \tag{3.1.48}$$

and

$$A_c^*(n) = q_1 \psi_c^*(0)n + \tfrac{1}{2}q_1\psi_c^{*0}n^2. \tag{3.1.49}$$

The first term in Eq. (3.1.49) results from the removal of an ion from its ion atmosphere in the bulk of the solution and placing on the surface of the colloid, all with $n = 0$. That is, even when no work is done against the field of the polyion, it still requires finite work to remove an ion from solution and adsorb it on the macromolecule due to changes in the ionic environment.

Consider now two particles. From Eq. (3.1.42)

$$A_e(n, r) = q_1\tilde{\psi}^0(1, r)n^2 \tag{3.1.50}$$

and from Eq. (3.1.32)

$$\psi_c^*(n, r) = \frac{1}{2q_1} \frac{\partial A_c^*}{\partial n} = \psi_c^*(0) + n\psi_c^{*0}(r) \tag{3.1.51}$$

with

$$\psi_c^{*0} = \frac{\kappa_0 q_1 q_2}{4DkT} \int_0^1 \lambda^5 \frac{\partial(\tilde{\psi}^0/\lambda)}{\partial\lambda} \, d\lambda \tag{3.1.52}$$

and thus

$$A_c(n, r) = 2q_1\psi_c^*(0)n + q_1\psi_c^{*0}(r)n^2. \tag{3.1.53}$$

By substitution into Eq. (3.1.34),

$$\left(\frac{\partial W(n, r)}{\partial r}\right)_{n=n_{eq}} = q_1 n_{eq}^2(r) \frac{\partial}{\partial r} [\tilde{\psi}_0 + \psi_c^{*0}] \tag{3.1.54}$$

or, when $n = n_{eq}$ is independent of r,

$$\delta W(n, r) = q_1 n^2 [\delta\tilde{\psi}_0 + \delta\psi_c^{*0}] \tag{3.1.55}$$

and where

$$\delta\psi_c^{*0} = \frac{\kappa_0 q_1 q_2}{4DkT} \left[\delta\tilde{\psi}^0(1, r) - 5\int_0^1 \lambda^3 \delta\tilde{\psi}^0(\lambda, r) \, d\lambda\right]. \tag{3.1.56}$$

When the potential is constant over the surface of a spherical particle of radius b,

$$n\tilde{\psi}^0(1, \lambda) = \frac{\lambda q_1 n}{Db(1 + \kappa b)} \qquad (3.1.57)$$

and for a single particle

$$A_e(n) = \frac{(q_1 n)^2}{2Db(1 + \kappa b)} \qquad (3.1.58)$$

$$A_c^*(n) = q_1 \psi_c^*(0)n + \tfrac{1}{2} q_1 \psi_c^{0*} n^2$$

$$= -\frac{\kappa_0 q_1 q_2 n}{6D} \left[1 + \frac{6 q_1^2 n}{8 D b k T} \left\{ \frac{1}{1 + \kappa b} - \frac{5}{(\kappa b^5)} \left(\frac{(\kappa b)^4}{4} \right. \right. \right.$$

$$\left. \left. \left. - \frac{(\kappa b)^3}{3} + \frac{(\kappa b)^2}{2} - \kappa b + \ln (1 + \kappa b) \right) \right\} \right]. \qquad (3.1.59)$$

For two interacting spherical ions, the potential at a point P is, to first order,

$$\psi = \text{constant} \left[\frac{e^{-\kappa r_1}}{r_1} + \frac{e^{-\kappa r_2}}{r_2} \right] \qquad (3.1.60)$$

and the average potential of each particle may be shown to be

$$\langle \Psi_1 \rangle = \frac{n q_1}{Db(1 + \kappa b)} \left\{ 1 + \frac{\exp[-\kappa b(r/b - 2)]}{(r/b)(1 + \kappa b)} \right\} \qquad (3.1.61)$$

and thereby

$$\delta \tilde{\psi}^0(\lambda, r) = \frac{\lambda q_1}{Db(1 + \lambda \kappa b)^2} \frac{\exp[-\lambda \kappa b(r/b - 2)]}{(r/b)}. \qquad (3.1.62)$$

After substitution of (3.1.62) into (3.1.56) we obtain

$$\delta \psi_c^{*0}(r) = -\frac{5 \kappa_0 q_1^2 q_2}{4 D^2 b k T} \frac{b}{r(\kappa b)^5} \left[\frac{1}{(r/b - 2)} \left\{ \frac{r}{b} + 1 - \frac{2}{r/b - 2} + \frac{2}{(r/b - 2)^2} \right\} \right.$$

$$\left. - \left(\frac{r}{b} + 2 \right) e^{r/b - 2} \left[Ei^* \left(\frac{r}{b} - 2 \right) - Ei^* \left\{ \left(\frac{r}{b} - 2 \right) (1 + \kappa b) \right\} \right] \right.$$

$$- \frac{e^{-\kappa b(r/b - 2)}}{r/b - 2} \left[\frac{2}{(r/b - 2)^2} + \frac{2(\kappa b - 1)}{r/b - 2} \right. \qquad (3.1.63)$$

$$\left. + \frac{r/b - 2}{1 + \kappa b} + (\kappa b - 1)^2 + 2 + \frac{(\kappa b)^5(r/b - 2)}{5(1 + \kappa b)^5} \right]$$

where

$$Ei^*(x) = \int_x^\infty \frac{e^{-u}}{u} \, du. \qquad (3.1.64)$$

For very large r,

$$\delta\psi_c^{*0}(r) \sim -\frac{30\kappa_0 q_1^2 q_2}{D^2 bkT(\kappa a)^5}\left(\frac{b}{r}\right)^6 \tag{3.1.65}$$

and both forces are repulsive.*

At most distances, $A_e \gg A_c^*$, so that the dominant repulsive potential is given by (3.1.62), but at very large distances the term (3.1.65) is dominant. The potential (3.1.62) was considered by Verwey and Overbeek and shown to lead to quantitative description of the stability of lyophobic colloids with respect to changes in ionic strength, coagulation, etc. For further details the reader is referred to their book.[1]

3.2   Higher Order Approximations

The preceding section has been devoted to a study of the theory of rigid polyion solutions in the region where the surface potential is small relative to kT. In this section we consider the solution to the Poisson–Boltzmann equation when the restriction of small surface potential is removed.

A general solution to the Poisson–Boltzmann equation valid for all values of the electric potential ψ has never been found. Numerical integration on a high-speed electronic computer has been used by Hoskin[5] and a series expansion method by Pierce.[6] The agreement between the two different solutions is excellent when the surface potential is less than 100 mv in water at 25°C, and will be illustrated in detail later in this section.

The series solution starts from the observation that the Poisson–Boltzmann equation may be expanded as follows

$$\nabla_\xi^2 \eta = \eta + \frac{\eta^3}{3!} + \frac{\eta^5}{5!} + \frac{\eta^7}{7!} + \cdots \tag{3.2.1}$$

where

$$\eta = \frac{zq\psi}{kT}, \qquad \xi = \kappa r. \tag{3.2.2}$$

Assume that η may be expressed in a series of the following form

$$\eta = A\eta_1 + A^3\eta_3 + A^5\eta_5 + \cdots \tag{3.2.3}$$

with η_n functions of ξ and A a constant to be determined by the boundary conditions. To insure that η and $d\eta/d\xi$ vanish as $\xi \to \infty$, we require that η_n and $d\eta_n/d\xi$ vanish in the same limit. By substitution of (3.2.3) into (3.2.1),

---

* Note that $q_1\delta\psi$ is the force.

collecting and equating like powers of A, it is found that

$$\nabla^2 \eta_1 - \eta_1 = 0$$

$$\nabla^2 \eta_3 - \eta_3 = \frac{1}{6} \eta_1^3$$

$$\nabla^2 \eta_5 - \eta_5 = \frac{1}{2} \eta_1^2 \eta_3 + \frac{1}{120} \eta_1^5 \tag{3.2.4}$$

$$\nabla^2 \eta_7 - \eta_7 = \frac{1}{2} \eta_1 \eta_3^2 + \frac{1}{2} \eta_1^2 \eta_5 + \frac{1}{24} \eta_1^4 \eta_3 + \frac{1}{5040} \eta_1^7.$$

The equation for η_1 is seen to be the Debye–Hückel approximation with solution $\eta_1 = e^{-\xi}/\xi$. The remaining inhomogeneous equations may all be recast into the form

$$\frac{d^2}{d\xi^2} [\xi \eta_n] - \xi \eta_n = \xi \hat{g}_n(\xi) \tag{3.2.5}$$

or, in operator notation

$$(D + 1)(D - 1)(\xi \eta_n) = \xi \hat{g}_n(\xi). \tag{3.2.6}$$

Equation (3.2.6) may be formally integrated to give

$$(D + 1)\xi \eta_n = e^{\xi} \int_{\infty}^{\xi} t\, e^{-t} \hat{g}_n(t)\, dt = e^{\xi} h_n^-(\xi)$$

$$(D - 1)\xi \eta_n = e^{-\xi} \int_{\infty}^{\xi} t\, e^{t} \hat{g}_n(t)\, dt = e^{-\xi} h_n^+(\xi). \tag{3.2.7}$$

A simple subtraction eliminates the differential operator and provides as the solution for η_n,

$$\eta_n = \frac{1}{2\xi} [e^{\xi} h_n^-(\xi) - e^{-\xi} h_n^+(\xi)]. \tag{3.2.8}$$

As an example of the procedure, consider the evaluation of η_3. From Eq. (3.2.4) and the solution $\eta_1 = e^{-\xi}/\xi$, the inhomogeneous term $\hat{g}_3(\xi)$ is found to be

$$\hat{g}_3(\xi) = \frac{1}{6} \frac{e^{-3\xi}}{\xi^3} \tag{3.2.9}$$

and thereby

$$\eta_3(\xi) = \frac{1}{12\xi} \left[e^{\xi} \int_{\infty}^{\xi} \frac{e^{-4t}}{t^2}\, dt - e^{-\xi} \int_{\infty}^{\xi} \frac{e^{-2t}}{t^2}\, dt \right]. \tag{3.2.10}$$

Using the definition of the exponential integral,

$$Ei(-\alpha\xi) = \int_{\infty}^{\xi} \frac{e^{-\alpha t}}{t}\, dt. \tag{3.2.11}$$

The relation (3.2.10) may be rewritten as

$$\eta_3(\xi) = \frac{e^{-\xi}Ei(-2\xi)}{6\xi} - \frac{e^{\xi}Ei(-4\xi)}{3\xi}. \tag{3.2.12}$$

This result for η_3 may now be used to find the function η_5, and the recursive procedure repeated. The use of Eq. (3.2.7) is not convenient, however, since untabulated functions are obtained from the integration. Pierce proceeds by using the asymptotic expansion of $Ei(-x)$,

$$Ei(-x) = -\frac{e^{-x}}{x}\left[1 - \frac{1}{x} + \frac{2!}{x^2} - \frac{3!}{x^3} + \cdots\right] \tag{3.2.13}$$

which, when combined with Eq. (3.2.12), yields for the asymptotic form of η_3,

$$\eta_3(\xi) = \frac{e^{-3\xi}}{48\xi^3}\left[1 - \frac{3}{2\xi} + \frac{21}{8\xi^2} - \frac{45}{8\xi^3} + \cdots\right]. \tag{3.2.14}$$

Using Eq. (3.2.14) $\hat{g}_5$ may be written in the form of an asymptotic expansion which, after term by term integration by parts, yields

$$\eta_5(\xi) = \frac{e^{-5\xi}}{768\xi^5}\left[\frac{3}{5} - \frac{3}{2\xi} + \frac{57}{2\xi^2} - \frac{543}{8\xi^3} + \cdots\right] \tag{3.2.15}$$

and analogously

$$\eta_7(\xi) = \frac{e^{-7\xi}}{36864\xi^7}\left[\frac{9}{7} - \frac{9}{2\xi} + \frac{1845}{144\xi^2} - \frac{863}{8\xi^3} + \cdots\right]. \tag{3.2.16}$$

From Eqs. (3.2.14)–(3.2.16) derivatives may easily be obtained. The results are valid only for large ξ and for the purposes of evaluation for $\xi < 5$, Pierce used numerical integration.

To complete the formal aspects of the solution, the boundary conditions must be satisfied and thereby the constant A evaluated. The boundary conditions are taken to be

$$\lim_{\xi \to \kappa a} \eta = \eta_0$$
$$\lim_{\xi \to \kappa a} \frac{d\eta}{d\xi} = \eta_0' \tag{3.2.17}$$

either one of which is sufficient to determine the constant A. To determine A, Eq. (3.2.3) is evaluated at the surface of the polyion,

$$\eta_0 = A\eta_1(\kappa b) + A^3\eta_3(\kappa b) + A^5\eta_5(\kappa b) + A^7\eta_7(\kappa b) + \cdots. \tag{3.2.18}$$

Now, $\eta_1(\kappa b) = e^{-\kappa b}/\kappa b$. If Eq. (3.2.1) is multiplied by $\kappa b\, e^{\kappa b}$ and then inverted to obtain A as a power series in $\kappa b\eta_0\, e^{\kappa b}$, the resultant series is

$$A = \kappa b\, e^{\kappa b}\eta_0 + \beta_3(\kappa b)^3\, e^{3\kappa b}\eta_0^3 + \beta_5(\kappa b)^5\, e^{5\kappa b}\eta_0^5 + \cdots \tag{3.2.19}$$

where the coefficients β_n are found to be

$$\beta_3 = -\kappa b\, e^{\kappa b} \eta_3(\kappa b)$$

$$\beta_5 = -3(\kappa b)^2\, e^{2\kappa b} \eta_3^2(\kappa b) - \kappa b\, e^{\kappa b} \eta_5(\kappa b)$$

$$\beta_7 = 8(\kappa b)^2 e^{2\kappa b} \eta_3(\kappa b)\eta_5(\kappa b) \qquad (3.2.20)$$

$$- 12(\kappa b)^3 e^{3\kappa b} \eta_3^3(\kappa b) - \kappa b e^{\kappa b} \eta_7(\kappa b).$$

An analogous procedure may be set up in terms of the derivative of the potential. With the convenient definition

$$f_n(\xi) = \xi^n e^{n\xi} \eta_n(\xi) \qquad (3.2.21)$$

the potential about the polyion may finally be written

$$\eta = A\frac{e^{-\xi}}{\xi} + A^3 \frac{e^{-3\xi}}{\xi^3} f_3(\xi) + A^5 \frac{e^{-5\xi}}{\xi^5} f_5(\xi) + \cdots \qquad (3.2.22)$$

which form gives an indication of the rapidity with which the higher terms fall to zero. Values of $f_n(\xi)$ are tabulated in the Appendix to this Chapter.

TABLE 3.2.1

COMPARISON OF NUMERICAL AND SERIES EXPANSION SOLUTIONS
OF THE POISSON–BOLTZMANN EQUATION[a]

κb	$\eta_0 = 1$		$\eta_0 = 2$		$\eta_0 = 4$		$\eta_0 = 6$	
	Series	Numer.	Series	Numer.	Series	Numer.	Series	Numer.
1	0.9918	0.9915	0.9663	0.9664	0.8732	0.8734	0.7566	0.7489
3	0.9861	0.9861	0.9465	0.9466	0.8159	0.8160	0.6655	0.6692
5	0.9841	0.9841	0.9395	0.9395	0.7958	0.7980	0.6000	0.6470
15	0.9814	0.9814	0.9299	0.9299	0.7670	0.7752	0.4541	0.6205

[a] Reproduced from ref. 6.

It is pertinent to compare the results of Pierce with the numerical integration performed by Hoskin. Hoskin expresses his results in the form

$$\lim_{\xi \to \infty} \eta = \kappa b\, e^{\kappa b} \eta_0 c(\kappa b, \eta_0) \frac{e^{-\xi}}{\xi} \qquad (3.2.23)$$

for large distances, whereupon

$$A = \kappa b\, e^{\kappa b} \eta_0 c(\kappa b, \eta_0). \qquad (3.2.24)$$

The numerical agreement is shown in Table 3.2.1.

It is seen that very satisfactory results are obtained up to values of $\eta_0 = 4$ for values of κb up to about 15. Of course, the enormous advantage of the series method over numerical integration is that it presents the solution in terms of analytic functions which can be manipulated, and not in the form of tabular entries which can only be examined.

3.3   The Force between Two Rigid Polyions

To compute the force acting between two rigid polyions it is convenient to proceed by a different method than that used in Section 3.1. First, for our present purposes it is sufficient to assume that the charge on the polyion is uninfluenced by bringing another polyion up close, and therefore no further consideration need be given to the adsorption isotherm. Second, it is apparent from the preceding section that the solution of the Poisson–Boltzmann equation with any symmetry other than spherical is exceedingly difficult. Accordingly, we shall not seek an iterative solution of the type discussed for the case of the isolated polyion. We consider, instead, that information which can be obtained from the Maxwell stress tensor introduced in Chapter 2.

If W is the electrostatic energy associated with the double layers of two particles, and $\Delta\pi$ the difference between the ideal osmotic pressure of the ions at a point in the double layers and the corresponding pressure at infinity then the free energy of the double layers is

$$A = -W - \int_V \Delta\pi \, dv. \tag{3.3.1}$$

For the special case of a symmetrical electrolyte Eq. (3.3.1) can be rewritten as

$$A = -\frac{D}{8\pi} \int_V |\nabla\psi|^2 \, dv - 2nkT \int_V \left(\cosh\frac{zq\psi}{kT} - 1\right) dv$$

$$= -\frac{nkT}{\kappa^3} \int_V \left[|\nabla_\xi\eta|^2 + 2|\cosh\eta - 1|\right] dv. \tag{3.3.2}$$

From the free energy A, the force acting between the particles can be obtained as follows. Let O and O' be convenient fixed points inside the particles P and P', each of which has the constant surface potential ψ_0. If $\mathbf{s}$ is the vector position of O relative to O', and the particle P is given a uniform translation $\delta\mathbf{s}$, then by expansion*

$$\psi(r, s + ds) = \psi(r, s) + \psi_1(r, s)\delta s + \cdots \tag{3.3.3}$$

*We here revert to scalar notation for simplicity.

As a result of the displacement, the second term in Eq. (3.3.2) changes by

$$\delta A_2 = -2nkT \left[\int_{V^*+\delta V_1} \left\{ \cosh \frac{zq}{kT} (\psi + \psi_1 \delta s) - 1 \right\} dv \right.$$

$$\left. - \int_{V^*+\delta V_2} \left(\cosh \frac{zq\psi}{kT} - 1 \right) dv \right] \quad (3.3.4)$$

where V^* is the volume of the diffuse ionic atmosphere surrounding the two particles which is not occupied by the particles in either configuration, δV_1 is the volume vacated by P when O is displaced to $s + ds$ and which is occupied by the diffuse layer, and δV_2 is the corresponding volume vacated by the diffuse layer and occupied by P due to the displacement. Clearly, $\delta V_1 = \delta V_2$. At any point in δV_1 or δV_2 the potential differs from ψ_0 by an amount of order δs. By expansion of $\cosh (zq/kT)(\psi + \psi_1 \delta s - 1)$ and collection of terms linear in δs, we find

$$\delta A_1 = -2nkT\delta s \int_{V^*} \frac{zq\psi_1}{kT} \sinh \frac{zq\psi}{kT} dv \quad (3.3.5)$$

$$= -\frac{D\delta s}{4\pi} \int_{V^*} \psi_1 \nabla^2 \psi \, dv$$

since ψ satisfies the Poisson–Boltzmann equation. By exactly the same procedure, the change in the first term of Eq. (3.3.2) is

$$\delta A_1 = -\frac{D}{8\pi} \left[2\delta s \int_{V^*} (\nabla \psi \cdot \nabla \psi_1) \, dv - \int_{\delta V_1} |\nabla(\psi + \psi_1 \, ds)|^2 \, dv \right.$$

$$\left. + \int_{\delta V_2} |\nabla \psi|^2 \, dv \right] \quad (3.3.6)$$

thereby giving for the sum of the integrals over V^*,

$$\frac{D\delta s}{4\pi} \int_{V^*} [\psi \nabla^2 \psi_1 - \psi_1 \nabla^2 \psi - \nabla \cdot (\psi \nabla \psi_1)] dv = \frac{D\delta s}{4\pi} \int_{S'_P + S^*_P} \psi_1 \frac{\partial \psi}{\partial \mathbf{n}} dS \quad (3.3.7)$$

where the last step follows from a conversion to a surface integral. As usual $\mathbf{n}$ is the outward normal, S'_P the surface of particle P, and S^*_P the surface which bounds the volume swept out by P when displaced. On S'_P, $\psi_1 = 0$. On S^*_P

$$\psi_1 \, dS = -\frac{\partial \psi}{\partial n} \mathbf{n} \cdot \delta \mathbf{s} \quad (3.3.8)$$

since the potential is ψ_0 when P is at $\mathbf{s}$, and the potential is less than ψ_0 when

P is at $s + \delta s$ because the boundary S_P^* is now in the diffuse layer. The right-hand member of Eq. (3.3.7) then becomes $-2\mathbf{X} \cdot \delta s$ where

$$\mathbf{X} = \frac{D}{8\pi} \int_{S_P} \left(\frac{\partial \psi}{\partial n}\right)^2 \mathbf{n} \, ds = \frac{D}{8\pi} \int_{S_P} \mathscr{E}^2 \mathbf{n} \, ds \qquad (3.3.9)$$

since $S_P^* = S_P$ neglecting terms of order $(\delta s)^2$. To complete the calculation, note that the integrands of the last two terms in Eq. (3.3.6) may be equated to $\mathscr{E}^2$ and $dv = \pm \mathbf{n} \cdot \delta s \cdot ds$. The sum of the last two terms in Eq. (3.3.6) is seen to be $\mathbf{X} \cdot \delta s$ by comparison with Eq. (3.3.9). The total change in free energy is finally

$$\delta A = -\mathbf{X} \cdot \delta s \qquad (3.3.10)$$

and therefore $\mathbf{X}$ is the repulsive force exerted by P' on P.

Equation (3.3.10) is valid when the surface of integration is the same as the surface of the particle. It is, however, more convenient to have an expression for the force acting on a particle when the surface of integration is arbitrary. This is found in the same manner and is

$$\mathbf{X} = \int_{S_c} \left[\frac{D}{4\pi} \{(\mathscr{E} \cdot \mathbf{n})\mathscr{E} - \tfrac{1}{2}\mathscr{E}^2\mathbf{n}\} - (\pi - \pi_0) \right] dS \qquad (3.3.11)$$

where S_c is *any* closed surface surrounding P but excluding P'. On the other hand, the use of the Maxwell stress tensor and the Poisson–Boltzmann equation leads to

$$0 = \int_V [\nabla \cdot \mathbf{\Phi} - \nabla \pi] \, dv = \int_S \left[\frac{D}{4\pi} \{(\mathscr{E} \cdot \mathbf{n})\mathscr{E} - \tfrac{1}{2}\mathscr{E}^2\mathbf{n}\} - (\pi - \pi_0)\mathbf{n} \right] dS \qquad (3.3.12)$$

where

$$\nabla \cdot \mathbf{\Phi} = \rho\mathscr{E}$$
$$= \frac{D}{4\pi} \nabla^2 \psi \nabla \psi \qquad (3.3.13)$$
$$= 2nzq \sinh \eta \nabla \psi$$
$$= \nabla \pi.$$

We now proceed to obtain the force by direct integration. Consider a volume in the shape of a hemisphere with the equatorial plane forming the medial plane bisecting the distance between the particles, but excluding the colloidal particle whose center is at 0 inside the hemisphere. The surface integral over the surface of the hemisphere will vanish in the limit as the radius tends to infinity since ψ tends to zero exponentially as $r \to \infty$. At the surface of the particle the potential η_0 is a constant, whereby

$$\pi - \pi_0 = znkT[\cosh \eta - 1] \tag{3.3.14}$$

is also a constant. Using the fact that $\int \mathbf{n}\, dS = 0$ when integrated over the surface of the particle, S_P, Eq. (3.2.12) becomes

$$\int_{S_m} \left[\frac{D}{4\pi} \{ (\mathscr{E} \cdot \mathscr{E}\mathbf{n} - \tfrac{1}{2}\mathscr{E}^2\mathbf{n}\} - \{\pi - \pi_0)\mathbf{n} \right] dS$$

$$= -\int_{S_P} \left[\frac{D}{4\pi} (\mathscr{E} \cdot \mathbf{n})\mathscr{E} - \tfrac{1}{2}\mathscr{E}^2\mathbf{n} \right] dS \tag{3·3·15}$$

with S_m the surface of the median plane. Due to the axial symmetry, both integrals give vectors pointing along the line of centers, so that the only component of $\mathscr{E}$ which contributes to the integral is the component along the line of centers. Thus, Eq. (3.3.15) becomes

$$\int_{S_m} \left[\frac{D}{8\pi} \mathscr{E}^2 + \pi - \pi_0 \right] dS = \frac{D}{8\pi} \int_{S_P} \mathscr{E}^2 \cos \theta \, dS \tag{3.3.16}$$

where θ is the angle between a radius vector $\mathbf{b}$ and the line of centers. From Eq. (3.3.9) it is seen that the right-hand side of (3.3.16) is just the force the particle experiences in the direction of the line of centers due to the field $\mathscr{E}$. The force of repulsion is, by combination of our previous results,

$$\mathbf{F}(\mathbf{r}) = \frac{D}{4\pi} \int_{S_m} |\nabla\psi|^2 \, dS + 2nkT \int_{S_m} [\cosh \eta - 1] \, dS. \tag{3.3.17}$$

For the purposes of integration, consider the dimensionless force

$$\tilde{\mathbf{f}}(\xi) = \frac{\kappa^2}{nkT} \mathbf{F}(\mathbf{r})$$

$$\tilde{f}(\xi) = \int_{S'_m} [|\nabla\eta|^2 + 2(\cosh \eta - 1)] \, dS'. \tag{3.3.18}$$

At moderate interparticle separations, the potential *at the midplane* is small and $\cosh \eta$ may be expanded as a power series in η thus reducing Eq. (3.3.18) to

$$\tilde{f}(\xi) = \int_{S'_m} [|\nabla\eta|^2 + \eta^2] \, dS. \tag{3.3.19}$$

Let $\overset{\circ}{\xi}_1$ and $\overset{\circ}{\xi}_2$ be the distances from particles one and two to a given point in the median plane. To the same accuracy as the expansion in Eq. (3.3.19),

$$\eta = A \left[\frac{e^{-\overset{\circ}{\xi}_1}}{\overset{\circ}{\xi}_1} + \frac{e^{-\overset{\circ}{\xi}_2}}{\overset{\circ}{\xi}_2} \right] \tag{3.3.20}$$

which represents a superposition of potentials from the two particles. Simple physical arguments as well as Hoskin's numerical integration verify this approximation. The constant A was defined in Eqs. (3.2.3) and (3.2.18). Let the integration of Eq. (3.3.19) be carried out with $\xi_1 = \xi_2$. Then $\eta = 2A\, e^{-\xi}/\xi$ and

$$\tilde{f}(\xi) = 2\pi \int_0^\infty \left[\left(\frac{d\eta}{d\rho} \right)^2 + \eta^2 \right] \rho\, d\rho \qquad (3.3.21)$$

where ρ is the radius vector from the intersection of the line of centers and the median plane to a given point in the median plane. By a suitable change of variable, $\xi^2 = \rho^2 + \tfrac{1}{4}\xi^2$, the integral in Eq. (3.3.21) may be readily evaluated with the result

$$\tilde{f}(\xi) = 8\pi A^2 \frac{e^{-\xi}}{\xi^2}(1 + \xi) \qquad (3.3.22)$$

for the force of repulsion between two rigid impenetrable polyions. We have in this development neglected the changes in the surrounding medium which lead to the term in Eq. (3.1.65).

To compute the free energy of interaction of the two polyions it is sufficient to integrate the force of repulsion from infinity to the distance of separation. Thus, since

$$-\frac{d\tilde{A}}{d\xi} = \tilde{f}(\xi) \qquad (3.3.23)$$

we have

$$\tilde{A}(\xi) = -8\pi A^2 \int_\infty^\xi \frac{e^{-x}}{x^2}(1 + x)\, dx \qquad (3.3.24)$$

whereupon noticing that $(d/dx)(e^{-x}/x) = (e^{-x}/x^2)(1 + x)$, Eq. (3.3.24) may be immediately integrated to give, in conventional units

$$A(r) = \frac{nkT}{\kappa^3}\, \tilde{A}(r)$$

$$= \frac{Dk^2T^2}{z^2q^2\kappa^2}\, A^2 \frac{e^{-\kappa r}}{r}. \qquad (3.3.25)$$

But $(\kappa b)e^{\kappa b}\, c(\kappa b, \eta_0) = A$ whereupon

$$A(r) = Db^2\psi_0^2 c^2(\kappa b, \eta_0)\, \frac{e^{\kappa(2b-r)}}{r}. \qquad (3.3.26)$$

It is now pertinent to examine the relationship between the theory developed and experiment. We have already mentioned that the combination of the electrostatic forces considered herein and van der Waals attractive forces enable Verwey and Overbeek to account for most of the properties of solutions of lyophobic colloids.[1] The details are to be found in their book and we here consider only one qualitative and one quantitative test of the theory.

The qualitative test of the theory arises from a study of the distribution of pairs of particles, carried out by Steiner.[7] Due to the strength of the repulsive forces, there is considerable spatial ordering of the polyions in solution. Consequently, by inversion of the angular distribution of scattered light from suitably chosen colloidal solutions (i.e., those that exhibit diffraction due to interparticle interference) the radial distribution function may be determined experimentally. Steiner has carried out such a program for Ag I sols in which the charge is due to the adsorption of I$^-$. Figure 3.3.1 shows

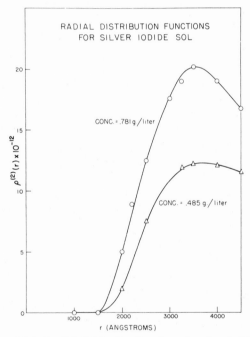

FIG. 3.3.1. Radial distribution functions for silver iodide sol.
Reproduced from Ref. 7.

the radial distribution function deduced. The structure of the solution alters with dilution and ionic strength in the expected manner and it is found that the strong repulsive forces prevent the particles from approaching closer than 1500 A at low ionic strengths. The maximum in the distribution function occurs at about 3000 A, so the long range of the repulsive force is not un-reasonable. Further, the shift in position of the maximum in the distribution function is approximately of the magnitude expected from the concentration change, and the addition of a quantity of bivalent cation produces an effect roughly equivalent to that of one hundred times its concentration of mono-valent cation, in accordance with the Shulz–Hardy rule.[1]

Quantitative examination of the data, however, reveals deviations from the theoretical predictions. From the shape of the radial distribution function it is known that the forces are repulsive. To determine the nature of these forces one may plot log (U/kT) against log r (appropriate to a repulsion proportional to r^{-n}) and log (rU) against r (appropriate to a repulsion of the form $e^{-\kappa r}/r$). Neither of these plots is at all linear, nor does either one appear to be rectifi-able by the addition or subtraction of linear parts. At first glance, then, the repulsive forces calculated by Levine and Verwey and Overbeek do not appear to be of the proper form. Several possibilities must be considered.

1. The colloidal particles were examined in an electron microscope and found to be very polydisperse. Under these conditions the inversion of the light scattering data requires the distribution of particle sizes before g_2 may be calculated.

2. Due to the linearization of the Poisson–Boltzmann equation (Section 3.1) or the superposition assumption (Section 3.3), the form of the potential of mean force is simply an exponential decay superimposed on a Coulomb potential. The particle diameters in the solutions studied by Steiner were about 100 A and therefore $\kappa b \approx 2$–3. With a charge per particle of about 1000 units, the surface potential $\eta_0 > 10^2$ and the approximations used almost certainly are inapplicable.

3. The neglect of third body polyions may be important at the concen-trations investigated, since the experiments indicate that the range of the force is large enough to involve volume elements containing on the average more than two particles.

The experiments of Steiner provide perhaps a too critical test of the theory, since it is the repulsive potential at small distances which makes the dominant contribution to g_2. A suitable quantitative test of part of the theory can be constructed by comparison of experimental and calculated second virial coefficients. It is well known that $B_2(T)$, given by

$$B_2(T) = -2\pi N_0 \int_0^\infty \left[e^{-u(r)/kT} - 1 \right] r^2 \, dr \qquad (3.3.27)$$

is insensitive to the detailed form of the potential. In the particular case considered here, the potential is increasingly positive as $r \to 0$ and the contribu-tion to the integral in (3.3.27) therefore becomes vanishingly small. We may anticipate that the potential expressed in Eq. (3.3.25) will be an adequate approximation at all values of r. It is convenient to define

$$B_2(T) = \frac{2\pi N_0}{\kappa^3} F(N)$$

$$F(N) = \int_0^\infty \left[1 - \exp\left(-\frac{Ne^{-\xi}}{\xi} \right) \right] \xi^2 \, d\xi \qquad (3.3.28)$$

$$N = \frac{DkT}{z^2 q^2 \kappa} A^2$$

and to perform the integration defining $F(N)$ numerically. Some values of $F(N)$ are listed in Table 3.3.6.[6]

TABLE 3.3.1[a]
$F(N)$

N	$F(N)$	$\log F(N)$	$\log N$
1	0.8793	−0.05586	0.00000
10	5.562	0.74523	1.00000
200	34.10	1.53270	2.30103
1000	70.76	1.84976	3.00000
10,000	165.6	2.21916	4.00000
100,000	328.7	2.51675	5.00000
1,000,000	581.2	2.76432	6.00000

[a]Reproduced from ref. 6.

The computation of $B_2(T)$ requires the value of N which involves the radius of the polyion, the electrolyte concentration, and either the surface potential or surface charge density.

Suitable measurements of the second virial coefficient with which the theory can be compared are rather rare. The best data are for the interaction of soap micelles in the concentration region above the critical micelle concentration. In this region both the size of the micelle and the micelle–micelle interactions change with changing external salt concentration, but the effects may be disentangled by a combination of viscometry and light scattering. From the first, the size of the micelle may be estimated, and from the second molecular weight and second virial coefficient determined.

Using the viscosity data of Kushner[8] and the light scattering data of Kushner and Hubbard,[9] Pierce has calculated the number of soap anions in the micelle, n, and the radius of a micelle, b, for the system sodium dodecyl

TABLE 3.3.2[a]

NaCl mole/l	c_0^b mole/l	c mole/l	n	κ	b	A	N	$F(N)$	B_2(calc.) l/mole	B_2(obs.) l/mole
0.0	0.00974	0.00974	39.5	3.246×10^6	2.27×10^{-7}	6.410	176.1	31.6	3490	3190
0.02	0.00558	0.02558	65.9	5.261	2.38	17.60	819.7	64.6	1680	1710
0.03	0.00472	0.03472	75.2	6.116	2.42	25.52	1484	83.2	1380	1400
0.10	0.00146	0.10146	86.0	1.046×10^7	2.43	112.1	16870	195	645	517
0.20	0.00091	0.20091	92.9	1.473	2.43	410.3	158400	376	445	452

[a]Reproduced from ref. 6.
[b]c_0: critical micelle concentration.

sulfate, sodium chloride, and water. These values were then used to compute N via the computation of A from Hoskins machine integration of the Poisson–Boltzmann equation. The results are shown in Table 3.3.2.[6]

As can be seen, the numerical agreement between theory and experiment is excellent. However, it should be noted that it was necessary to use the machine computations to obtain A, in general agreement with the conclusions about the function form of the potential of mean force drawn from Steiner's experiments.

In general we may conclude that the theory is in satisfactory agreement with experiment when the separation between the polyions is large. The agreement between theory and experiment when the separation is small or the surface potential high has not been adequately tested, but preliminary indications are that many-body potentials will have to be considered.

3.4 The Self-Energy of Rigid Polyions

It is observed experimentally that the titration curve of a protein or polyelectrolyte is often considerably flatter than that of the corresponding mixture of simple acids and bases, and that the added electrolyte concentration plays a role in determining the deviation from simple behavior. This is but one of the many possible manifestations of the electrostatic self-energy of the polyion.

It is quite simple to calculate the electrostatic free energy of a polyion which is partially permeable to the medium in the continuous charge approximation.[10] Although the model is unrealistic, the result obtained has value in the extreme limits of the rigid polyion and the completely flexible chain polyelectrolyte as well as in the intermediate semipermeable range. Consider then a spherical polyion of radius b with a net charge $\tilde{Q}$ which may or may not be evenly distributed over its surface. Let there be an equivalent sphere of radius b_0, concentric with the large sphere of radius b. The sphere of volume $\frac{4}{3}\pi b_0^3$ represents the total volume occupied by the mass of the polyion. The space into which the solvent may permeate is then given by the volume $(4\pi/3)(b^3 - b_0^3)$. Let the polyion also have an impenetrable core of radius b_1. It is clear that b_1 cannot exceed b_0, and that the volume $(4\pi/3)(b^3 - b_1^3)$ will contain both solvent and polyion structure. Finally, denote by θ^2 the fraction of the volume occupied by solvent, i.e.,

$$\theta^2 = \frac{b^3 - b_0^3}{b^3 - b_1^3}. \tag{3.4.1}$$

For simplicity, consider only the case when the added electrolyte is 1 : 1 salt. The linearized Poisson–Boltzmann equation must now be solved in three regions, and continuity conditions applied at the boundaries of these regions.

In region I, the volume external to the polyion, the net charge density arises entirely from the unequal distribution of positive and negative counterions and byions. Thus,

$$\psi_1 = C_1 \frac{e^{-\kappa r}}{r}. \tag{3.4.2}$$

In region II, the volume occluded between the surface of the polyion and the surface of the impenetrable core, a net charge density can once again arise from the distribution of small ions. However, account must also be taken of the possible distribution of polyion charges. If the polyion charges are uniformly distributed throughout region II, their density is $3\tilde{Q}/4\pi(b^3 - b_1^3)$ and the linearized Poisson–Boltzmann equation becomes

$$\nabla^2 \psi - \theta^2 \kappa^2 \psi = -\frac{3\tilde{Q}}{D(b^3 - b_1^3)} \tag{3.4.3}$$

since only a fraction θ^2 of the total volume is occupied by the solvent. For the present we take κ^2 to be the same in regions I and II. This corresponds to assuming that the added salt concentration and the dielectric constant are the same in both regions, an approximation valid if θ^2 is large enough. The solution to Eq. (3.4.3) is, of course, the sum of the general solution of the homogeneous equation plus a particular solution. An obvious particular solution is

$$\psi_{IIp} = \frac{3\tilde{Q}}{\theta^2 \kappa^2 D(b^3 - b_1^3)} \tag{3.4.4}$$

so that the general solution of (3.4.3) becomes

$$\psi_{II} = C_4 \frac{e^{-\theta\kappa r}}{r} + C_5 \frac{e^{\theta\kappa r}}{r} + \frac{3\tilde{Q}}{\theta^2 \kappa^2 D(b^3 - b_1^3)}. \tag{3.4.5}$$

Region III consists of the interior of the impenetrable core. Since this region is devoid of charge,

$$\psi_{III} = \text{constant}$$

$$\frac{d\psi_{III}}{dr} = 0. \tag{3.4.6}$$

Of course, if the charge is uniformly distributed on the surface of the polyion, there is no charge in region II and the potential becomes instead

$$\psi_{II,s} = C_2 \frac{e^{-\theta\kappa r}}{r} + C_3 \frac{e^{\theta\kappa r}}{r}. \tag{3.4.7}$$

To evaluate the constants in Eqs. (3.4.2), (3.4.5), and (3.4.7) note that

$$\left(\frac{d\psi_{\text{II}}}{dr}\right)_{r=b_1} = 0 \tag{3.4.8}$$

and

$$\psi_{\text{I}}(b) = \psi_{\text{II}}(b) \tag{3.4.9}$$

$$\int_{b_1}^{b} 4\pi r^2 \rho_{\text{II}}\, dr + \int_{r}^{\infty} 4\pi r^2 \rho_{\text{I}} \cdot dr = -\tilde{Q} \tag{3.4.10}$$

which latter is just the condition of electroneutrality. Since the purpose of the calculation is to evaluate the electrostatic self-energy of the polyion, ψ need only be determined for $r \leqslant b$. Using the boundary conditions indicated, if the polyion charges are uniformly distributed throughout the volume,

$$\psi(r) = \frac{3\tilde{Q}(1+\kappa b)}{\kappa^3 D(b^3 - b_0^3)}\left[\frac{\kappa}{1+\kappa b} - \frac{\dfrac{1+\theta\kappa b_1}{1-\theta\kappa b_1}e^{\theta\kappa(r+b-2b_1)} - e^{\theta\kappa(b-r)}}{r\left\{(1+\theta)\dfrac{1+\theta\kappa b_1}{1-\theta\kappa b_1}e^{2\theta\kappa(b-b_1)} - (1-\theta)\right\}}\right] \tag{3.4.11}$$

whereas if the charge is distributed on the surface,

$$\psi(b) = \frac{\tilde{Q}}{D\kappa b^2}\frac{\dfrac{1+\theta\kappa b_1}{1-\theta\kappa b_1}e^{2\theta\kappa(b-b_1)} - 1}{(1+\theta)\dfrac{1+\theta\kappa b_1}{1-\theta\kappa b_1}e^{2\theta\kappa(b-b_1)} - (1-\theta)} \tag{3.4.12}$$

which can be easily obtained from (3.4.11) by suitable modification of the form of the charge density and evaluation of ψ at $r = b$.

To compute the electrostatic self-energy, the polyion is charged from zero to $\tilde{Q}$, at constant ionic strength. If the charge lies entirely on the surface, then from Eq. (3.4.12)

$$\Delta A_{\text{elec}} = N_0 \int_0^{\tilde{Q}} \psi\, dq$$

$$= \frac{N_0 \tilde{Q}^2}{2D\kappa b^2}\frac{\dfrac{1+\theta\kappa b_1}{1-\theta\kappa b_1}e^{2\theta\kappa(b-b_1)} - 1}{(1+\theta)\dfrac{1+\theta\kappa b_1}{1-\theta\kappa b_1}e^{2\theta\kappa(b-b_1)} - (1-\theta)} \tag{3.4.13}$$

where ΔA_{elec} is expressed in molar units. On the other hand, if the charge is distributed throughout the volume, for each increment dq in charge, there is an increment of $(4\pi r^2 \, dq \, dr)/[(4\pi/3)(b^3 - b_1^3)]$ in a spherical shell of thickness dr. Thus,

$$\Delta A_{\text{elec}} = \frac{N_0}{b^3 - b_1^3} \int_0^{\tilde{Q}} \int_{b_1}^b 3r^2 \psi \, dq \, dr$$

$$= \frac{9N_0\tilde{Q}^2}{2\kappa^2 D(b^3 - b_0^3)} \left[\frac{1}{3} + \frac{1 + \kappa b}{\kappa^3(b^3 - b_0^3)} \times \right.$$

$$\left. \frac{1 - \theta\kappa b \dfrac{1 + \theta\kappa b_1}{1 - \theta\kappa b_1} e^{2\theta\kappa(b - b_1)} - (1 + \theta\kappa b)}{1 + \theta \dfrac{1 + \theta\kappa b_1}{1 - \theta\kappa b_1} e^{2\theta\kappa(b - b_1)} - (1 - \theta)} \right]. \quad (3.4.14)$$

TABLE 3.4.1[a]

VALUES OF $\Delta A_{\text{elec}}/\tilde{Q}^2$ (cal/mole), $b_0 = 25$ A

b	$\Gamma/2$	(34.15)	(34.13) b_1				(34.14) b_1			
			0 A	10 A	20 A	25 A	0 A	10 A	20 A	25 A
35 A	0.001	44.6	43.4	43.4	43.4	43.4	54.9	54.2	51.1	48.8
	0.01	29.3	25.0	25.0	25.0	24.9	33.6	33.2	30.9	29.1
	0.05	19.0	12.9	12.9	12.7	12.6	16.9	16.7	15.7	14.7
	0.15	13.8	7.55	7.51	7.29	7.06	8.43	8.22	9.01	7.61
50 A	0.001	28.1	26.5	26.5	26.5	26.5	43.1	42.5	41.8	40.7
	0.01	16.8	12.7	12.7	12.7	12.6	16.6	16.5	16.1	15.7
	0.05	10.3	5.95	5.94	5.87	5.80	6.37	6.36	6.27	6.17
	0.15	7.26	3.44	3.43	3.38	3.33	2.69	2.69	2.67	2.65

[a]Reproduced from ref. 10.

Equation (3.4.14) may be reduced to simpler forms in two interesting cases. If $b_1 = b = b_0$, the polyion is an impenetrable sphere and

$$\Delta A_{\text{elec}} = \frac{N_0\tilde{Q}^2}{2Db(1 + \kappa b)} \quad (3.4.15)$$

a well-known result. The other extreme possibility is the loosely coiled polymer chain, for which $\theta^2 = 1$, $b_1 = 0$, whereupon

$$\Delta A_{\text{elec}} = \frac{3N_0\tilde{Q}^2}{2Db} \left\{ \frac{1}{\kappa^2 b^2} - \frac{3}{2\kappa^5 b^5} \left[\kappa^2 b^2 - 1 + (1 + \kappa b)^2 \, e^{-2\kappa b} \right] \right\} \quad (3.4.16)$$

which will be seen to be the same as the result of the Hermans–Overbeek theory for flexible polyelectrolytes.[12,13]

A comparison of the calculated energies for the models leading to Eqs. (3.4.13), (3.4.14), and (3.4.15) is made in Table 3.4.1. It is clear that the differences are quite marked, but that the values of ΔA_{elec} depend but little on the choice of b_1. Provided D is the same in regions I and II, it seems unimportant whether the solvent lies in an external layer with θ^2 close to unity, or whether it penetrates into the sphere with a lower resultant value of θ^2.

Before leaving the subject of continuous charge distributions it is necessary to discuss the recent calculations of Wall and Berkowitz,[14] Lifson,[15] and Nagasawa and Kagawa[16] for the case of permeable linear polyelectrolytes. The method used utilizes the near neutrality of the polymeric domain. As usual the starting point is the Poisson–Boltzmann equation,

$$\frac{q}{kT} \nabla^2 \psi - \kappa^2 \left(\sinh \frac{q\psi}{kT} - \frac{cf}{2c_s^0} \right) = 0 \tag{3.4.17}$$

with c_f the local concentration of fixed charges in the polymeric domain and c_s^0 the bulk concentration of simple salt. If the condition of Donnan equilibrium is applied to the macromolecule, assumed to be electrically neutral, we must have

$$\frac{q}{kT} \nabla^2 \psi_0 = 0; \qquad \frac{q\psi_0}{kT} = \sinh^{-1} \frac{cf}{2c_s^0} \tag{3.4.18}$$

which is the approximate solution suggested by Kimball et al.[17] In a real polyion the charge density is not everywhere zero, but it may indeed be small. It is therefore convenient to use a perturbation procedure based on the solution embodied in Eq. (3.4.18). To this end define a correction potential, ψ', by

$$\psi = \psi_0 + \psi'. \tag{3.4.19}$$

The substitution of (3.4.19) into (3.4.17) followed by expansion of the hyperbolic sine leads to

$$\frac{q}{kT} \nabla^2(\psi_0 + \psi') = \frac{q}{kT} \nabla^2 \psi' = \kappa^2 \left(\sinh \frac{q\psi_0}{kT} + \frac{q\psi'}{kT} \cosh \frac{q\psi_0}{kT} + \cdots - \frac{c_f}{2c_s^0} \right)$$

$$= \kappa^2 \frac{q\psi'}{kT} \cosh \frac{q\psi_0}{kT} \tag{3.4.20}$$

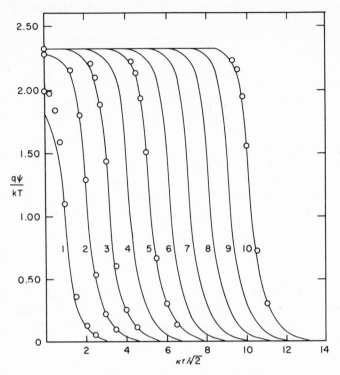

FIG. 3.4.1. The potential as a function of distance as calculated by Lifson [Eqs. (3.4.24), (3.4.25)] and Wall and Berkowitz. The open circles are the calculations of Lifson and the solid lines derive from the machine calculations. The numbers adjacent to each curve refer to the polyion size in reduced units, $\kappa R_s/\sqrt{2}$. All calculations are for the case where $c_f/c_s^0 = 10$. [Reproduced from ref. 15.]

neglecting higher powers of ψ'. Now define

$$\mu^2 = \frac{q\kappa^2}{kT} \cosh \frac{q\psi_0}{kT} = \kappa^2 \left(1 + \frac{c_f^2}{4(c_s^0)^2}\right)^{\frac{1}{2}} \tag{3.4.21}$$

so that

$$\frac{q}{kT} \nabla^2 \psi' = \mu^2 \frac{q\psi'}{kT}. \tag{3.4.22}$$

Lifson[15] assumes that the fixed charge density is constant within the polymeric domain, whereupon

$$\frac{c_f}{2c_s^0} = \text{constant}, \qquad r < R_s$$

$$\frac{c_f}{2c_s^0} = 0, \qquad\qquad r > R_s$$

(3.4.23)

with R_s the average radius of the coil. Using the boundary conditions

$$\psi(0) = \text{finite}, \qquad \psi(\infty) = 0$$

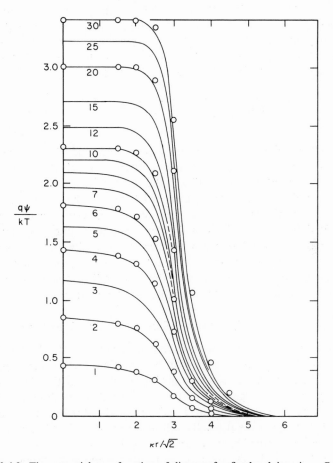

FIG. 3.4.2. The potential as a function of distance for fixed polyion size, $\kappa R_s/\sqrt{2} = 3$. The numbers adjacent to each curve refer to the ratio of concentrations, c_f/c_s^0. The open circles are computed from Eqs. (3.4.24), (3.4.25) and the solid lines derive from the machine computation of Wall and Berkowitz. [Reproduced from ref. 15.]

and the continuity of ψ and $d\psi/dr$ at the boundary $r = R_s$, a solution of Eq. (3.4.22) is seen to be

$$\frac{q\psi}{kT} = \sinh^{-1} \frac{c_f}{2c_s^0} \left[1 - \frac{1 + \kappa r}{1 + (\kappa/\mu)\tanh \mu r} \cdot \frac{1}{\cosh \mu r} \cdot \frac{\sinh \mu r}{\mu r} \right] \qquad r < R_s$$

(3.4.24)

$$\frac{q\psi}{kT} = \frac{q\psi(R_s)}{kT} \frac{R_s}{r} e^{-\kappa(r - R_s)}, \qquad r > R_s \qquad (3.4.25)$$

since μ^2 in (3.4.22) plays the role of κ^2 in the usual linearized Poisson–Boltzmann equation. Note that if κR_s is large enough, the correction term in the right-hand side of Eq. (3.4.24) will vanish over most of the range of r of interest. The region of r which remains is $\sinh \mu r \cong \frac{1}{2}e^{\mu r}$. Under these conditions, Eq. (3.4.24) can be approximated as

$$\frac{q\psi}{kT} = \sinh^{-1} \frac{cf}{2c_s^0} \left[1 - \frac{1 + \kappa R_s}{2(\mu + \kappa)} \cdot \frac{e^{\mu(r - R_s)}}{r} \right] \qquad (r < R_s). \qquad (3.4.26)$$

The result of calculations based on Eqs. (3.4.25) and (3.4.26) are shown in Figs. 3.4.1, and 3.4.2 where they are compared with the computer solutions of Eq. (3.4.17) obtained by Wall and Berkowitz.

In general, as will be seen again in Chapter 7, the polymeric domain is almost electrically neutral and the potential is very much smaller than anticipated from models which do not account for the large internal salt concentration. The relationship between these results and ion binding will be discussed in a later chapter. Suffice it to say for the present that we will find it convenient to distinguish between domain binding (where the electrostatic potential is greater than kT) and a more specific interaction.

3.5   Some Effects of the Discrete Charge Distribution

In the preceding sections we have considered only the case in which the charge distribution on the macroion was continuous. We shall present here some of the mathematical details in the calculation of the electrostatic free energy when the charge distribution is assumed to be discrete. Of course, closed formulas can be obtained only when the distribution of charges is specified, but our general results will also provide valuable physical insight into the nature of the charge–charge interactions.

Consider first the electrostatic work W_{ik} required to bring a macroion i and a small ion k from infinite separation in the pure solvent to a given relative configuration.[18] Since the charge distribution on the polyion need not be isotropic, this configuration is specified by the orientation and separation of the ions. For simplicity let the small ion be a point charge of magnitude $z_k q$. The macroion will be characterized by point charges $z_1 q_1 \cdots z_s q_s$ located in a cavity ω_0, of dielectric constant D_0, in the solvent of dielectric constant D. The potential W_{ik} is related to the work of charging the system in the given configuration, W, and the work of charging the two ions when infinitely far apart, W_0, by the relation

$$W_{ik} = W - W_0. \tag{3.5.1}$$

For the work W we may use the standard formula,

$$W = \tfrac{1}{2}[z_k q \psi_e(\mathbf{r}_k) + \sum_{l=1}^{z} z_l q \psi_i(\mathbf{r}_l)] \tag{3.5.2}$$

with $\psi_i(\mathbf{r}_l)$ the electrostatic potential in the interior of ω_0 at the point $\mathbf{r}_l$ of location of charge $z_l q$, and $\psi_e(\mathbf{r}_k)$ the potential exterior to ω_0 at the point of location $\mathbf{r}_k$ of the ion with charge $z_k q$.

To proceed further it is necessary to determine the electrostatic potentials ψ_e and ψ_i. Both the macroion and the small ion possess net charges and therefore the potentials ψ_e and ψ_i satisfy Poisson's equation in their respective regions of space. We may write

$$\nabla^2 \psi_i = -\frac{4\pi}{D} \rho_i$$

$$\nabla^2 \psi_e = -\frac{4\pi}{D} \rho_e \tag{3.5.3}$$

subject to the boundary conditions

$$\psi_i(b) = \psi_e(b)$$

$$D_0 \mathbf{n} \cdot \nabla \psi_i(b) = D \mathbf{n} \cdot \nabla \psi_e(b) \tag{3.5.4}$$

everywhere on the surface of the cavity ω_0. As usual, solutions to Eqs. (3.5.3) may be constructed by adding a particular solution to the general solution of the corresponding Laplace equation. If the macroion is a sphere of radius b, it may be shown that[18]

$$\psi_i = \sum_{l=1}^{z} \frac{z_l q}{D_0 |\mathbf{r} - \mathbf{r}_l|} + \sum_{n=0}^{\infty} \sum_{m=-n}^{n} B_{nm} r^n P_n^m(\cos \theta) e^{im\phi}$$

$$\psi_e = \frac{z_k q}{D |\mathbf{r} - \mathbf{r}_k|} + \sum_{n=0}^{\infty} \sum_{m=-n}^{n} \frac{A_{nm}}{r^{n+1}} P_n^m(\cos \theta) e^{im\phi} \tag{3.5.5}$$

where the polar coordinates r, θ, ϕ refer to an origin at the center of the macroion. Note that there are singularities at the location of each point charge, a feature of the potential completely lost in the smeared charge model.

As usual, the $P_n^m(\cos \theta)$ are the associated Legendre polynomials of the first kind. From Eqs. (3.5.4)

$$
\left.\begin{array}{l}
\psi_e(b, \theta, \phi) = \psi_i(b, \theta, \phi) \\[2mm]
D\left(\dfrac{\partial \psi_e}{\partial r}\right)_{r=b} = D_0\left(\dfrac{\partial \psi_i}{\partial r}\right)_{r=b}
\end{array}\right\}
\quad
\begin{array}{l}
0 \leqslant \theta \leqslant \pi \\[2mm]
0 \leqslant \phi \leqslant 2\pi
\end{array}
\qquad (3.5.6)
$$

On the surface of the sphere the potential due to the discrete charges may be written in the form

$$
\frac{z_k q}{D|\mathbf{r} - \mathbf{r}_k|} = \sum_{n=0}^{\infty} \sum_{m=-n}^{n} F_{nm} r^n P_n^m(\cos \theta) e^{im\phi}
$$

$$
F_{nm} = \frac{z_k q}{D r_k^{n+1}} \frac{(n - |m|)!}{n(+|m|)!} P_n^m(\cos \theta) e^{-im\phi_k}
$$

$$
\sum_{i=1}^{z} \frac{z_l q}{D_0|\mathbf{r} - \mathbf{r}_l|} = \sum_{n=0}^{\infty} \sum_{m=-n}^{n} \frac{G_{nm}}{r^{n+1}} P_n^m(\cos \theta) e^{im\phi}
$$

$$
G_{nm} = \frac{1}{D_0} \frac{(n - |m|)!}{(n + |m|)!} \sum_{l=1}^{z} z_l q r_l^n P_n^m(\cos \theta_l) e^{-im\phi_l}.
$$

(3.5.7)

By substitution and use of the orthogonality conditions on the Legendre polynomials the relations determining the expansion coefficients A_{nm} and B_{nm} are found to be

$$
A_{nm} + b^{2n+1} F_{nm} = G_{nm} + b^{2n+1} B_{nm}
$$

$$
(n + 1)A_{nm} - nb^{2n+1} F_{nm} = \sigma[(n + 1)G_{nm} - nb^{2n+1} B_{nm}]
$$

$$
\delta = \frac{D_0}{D}
\qquad (3.5.8)
$$

with the solutions

$$
B_{nm} = \frac{2n + 1}{n + 1 + n\delta} F_{nm} + \frac{(n + 1)(\delta - 1)}{n + 1 + n\delta} \frac{G_{nm}}{b^{2n+1}}
$$

$$
A_{nm} = \frac{n(1 - \delta)b^{2n+1}}{n + 1 + n\delta} F_{nm} + \frac{(2n + 1)\delta}{n + 1 + n\delta} G_{nm}.
\qquad (3.5.9)
$$

The substitution of Eqs. (3.5.5) and (3.5.9) into Eq. (3.5.1) gives the required work.

Consider now the simple case when the polyion is electrically neutral on the average. Then[18]

$$W_{ik} = \frac{z_k q}{Dr_k} \sum_{n=1}^{\infty} \sum_{l=1}^{Z} \frac{(2n+1)z_l q}{n+1+n\delta} \left(\frac{r_l}{r_k}\right)^n P_n(\cos\theta_{kl})$$

$$+ (1-\delta)\frac{z_k^2 q^2 b}{2Dr_k^2} \sum_{n=1}^{\infty} \frac{n}{n+1+n\delta} \left(\frac{b}{r_k}\right)^{2n} \quad (3.5.10)$$

with r_k the distance of the ion with charge $z_k q$ from the center of the sphere and θ_{kl} the angle between the vectors $\mathbf{r}_k$ and $\mathbf{r}_l$ from the center of the sphere terminating in the charges $z_k q$ and $z_l q$. When the charge distribution may be regarded as equivalent to a point dipole of magnitude μ_D center of the sphere, Eq. (3.5.10) becomes[18]

$$W_{ik} = \frac{3z_k q\mu_D \cos\theta_k}{(2D+D_0)r_k^2} + \frac{z_k^2 q^2 b^3}{2\pi^4}\left(\frac{D-D_0}{D}\right)\sum_{n=0}^{\infty} \frac{n+1}{(n+2)D+(n+1)D_0}\left(\frac{b}{r}\right)^{2n}.$$

$$(3.5.11)$$

From the general theory presented in Chapter 1,

$$\ln\gamma_i = \sum_{k=1}^{r} \frac{Nc_k^0}{1000}\int_{\omega}^{\infty}(1-e^{-W_{ik}/kT})\,dv \quad (3.5.12)$$

in the low concentration limit. For solvents of high dielectric constant and at high temperatures we make use of the expansion

$$\int_{\omega}^{\infty}(1-e^{-W_{ik}/kT})\,dv = \int_{\omega}^{\infty}\left[\frac{W_{ik}}{kT}-\frac{(W_{ik})^2}{2(kT)^2}+\cdots\right]dv \quad (3.5.12')$$

to yield, after neglect of small terms in D_0/D, and taking note of the fact that the volume of exclusion ω is larger than the polyion volume by a spherical shell of thickness equal to the radius of a simple ion,

$$\ln\gamma_i = -\frac{\pi Nq^2}{1000\,DkT}\left[\frac{3}{2}\frac{\mu_D^2}{DakT}-\frac{b^3}{a}\alpha(\rho)\right]\sum_{k=1}^{r}c_k^0 z_k^2 \quad (3.5.13)$$

with $\rho = (b/a)$, a being the sum of the radii of the polyion sphere and the simple ion, and the function $\alpha(\rho)$ defined by

$$\alpha(\rho) = \frac{\rho^4}{3}[(\rho^3-2)\ln(1+\rho)-(\rho^3+2)\ln(1-\rho)-2\rho^2]. \quad (3.5.14)$$

It is readily seen that there are two contributions in the potential of mean force W_{ik}. The first member is simply the interaction between the charge and the dipole, while the second term represents a repulsion between the real ion and an image distribution in the cavity ω_0 created by the macroion. For the use of Eq. (3.5.13) in the interpretation of the thermodynamic properties of peptides and other dipolar solutes we refer the reader to the well-known text by Cohn and Edsall.[11]

Consider now the more complicated case where the entire charge distribution must be specified. The work of charging the sphere may now be shown to be[19]

$$W = \frac{q^2}{2b} \sum_{k=1}^{Z} \sum_{l=1}^{Z} \eta_k \eta_l (A_{kl} - B_{kl}) - \frac{q^2}{2a} \sum_{k=1}^{Z} \sum_{l=1}^{Z} \eta_k \eta_l C_{kl} \tag{3.5.15}$$

where

$$A_{kl} = \frac{b}{D_0 r_{kl}} \tag{3.5.16}$$

$$B_{kl} = \frac{1}{D_0} \sum_{n=0}^{\infty} \frac{(n+1)(D-D_0)}{(n+1)D + nD_0} \rho_{kl}^n P_n(\cos \theta_{kl}) \tag{3.5.17}$$

$$C_{kl} = \frac{1}{D} \left[\frac{x}{1+x} + \sum_{n=1}^{\infty} \frac{2n+1}{2n-1} \left(\frac{D}{(n+1)D + nD_0} \right)^2 \right.$$

$$\left. \times \frac{x^2 \sigma_{kl}^n P_n(\cos \theta_{kl})}{\frac{K_{n+1}(x)}{K_{n-1}(x)} + \frac{n(D-D_0)}{(n+1)D + nD_0} \left(\frac{b}{a} \right)^{2n+1} \frac{x^2}{4n^2-1}} \right] \tag{3.5.18}$$

$$x = \kappa a$$

$$\rho_{kl} = \frac{r_k r_l}{b^2}$$

$$\sigma_{kl} = \frac{r_k r_l}{a^2}$$

$$K_n(x) = \sum_{s=0}^{n} \frac{2^s n! (2n-s)!}{s! (2n)! (n-s)!} x^s. \tag{3.5.19}$$

As usual we choose the discharged state of the polyion as the zero for energy. The variables η_k are occupation numbers having the values 0, +1, or −1 corresponding to uncharged, positive, and negative groups.

Equation (3.5.15) represents the work of charging as a sum of two contributions: the factor involving the A_{kl} is the work of charging in an unbounded medium of dielectric constant D_0; the factor involving the B_{kl} represents the modification arising from the fact that the polyion is a bounded cavity within a medium of higher dielectric constant D; the factor involving the C_{kl} represents the interaction with the salt ions of the solvent and vanishes at zero salt concentration, when $x = \kappa a = 0$. In the factors containing A_{kl} and

B_{kl} the terms with $k = l$ (i.e., A_{kk} and B_{kk}) are self-energy terms, while those with $k \neq l$ represent pair-wise interaction between the sites. In the ionic strength-dependent factor the terms with $k = l$ represent the excess chemical potential of individual charges due to their interaction with salt ions; terms with $k \neq l$ represent the effect of the salt ions on the pair-wise interactions. A_{kk} and, when $\rho_{kk} = 1$, also B_{kk}, are infinite because of the assumption that the ionizable groups may be represented as point charges.

Tanford and Kirkwood[19] found it convenient to separate the infinite self-energy terms from the remainder and to write

$$W = \frac{q^2}{2b} \sum_{k=1}^{z} \eta_k^2 (A_{kk} - B_{kk}) + W' \tag{3.5.20}$$

$$W' = \frac{q^2}{2b} \sum_{k=1}^{z} \left[\sum_{l \neq k} \eta_k \eta_l (A_{kl} - B_{kl}) - \frac{b}{a} \sum_{l=1}^{m} \eta_k \eta_l C_{kl} \right]. \tag{3.5.21}$$

Using the cosine rule,

$$r_{kl} = b \left[\rho_{kl} \left(\frac{r_k}{r_l} + \frac{r_l}{r_k} - 2 \cos \theta_{kl} \right) \right]^{\frac{1}{2}} \tag{3.5.22}$$

so that A_{kl} may be rewritten as

$$A_{kl} = \frac{1}{D_0} \left[\rho_{kl} \left(\frac{r_k}{r_l} + \frac{r_l}{r_k} - 2 \cos \theta_{kl} \right) \right]^{-\frac{1}{2}}. \tag{3.5.23}$$

Since $D_0 \ll D$, it is possible to expand the expression B_{kl} in increasing powers of $\delta = D_0/D$, retaining only terms up to δ^2. The Legendre polynomials may be generated by expansion of r_{kl}^{-1},

$$\sum_{n=0}^{\infty} \rho_{kl}^n P_n(\cos \theta_{kl}) = (1 - 2\rho_{kl} \cos \theta_{kl} + \rho_{kl}^2)^{-\frac{1}{2}} \tag{3.5.24}$$

whence, by integration with respect to ρ_{kl}

$$\sum_{n=0}^{\infty} \frac{\rho_{kl}^n}{n+1} P_n(\cos \theta_{kl}) = \frac{1}{\rho_{kl}} \ln \left[\frac{(1 - 2\rho_{kl} \cos \theta_{kl} + \rho_{kl}^2)^{\frac{1}{2}} + \rho_{kl} - \cos \theta_{kl}}{1 - \cos \theta_{kl}} \right] \tag{3.5.25}$$

and

$$
\begin{aligned}
B_{kl} = {} & \frac{1 - 2\delta + 2\delta^2}{D_0(1 - 2\rho_{kl} \cos \theta_{kl} + \rho_{kl}^2)^{\frac{1}{2}}} \\
& + \frac{\delta - 3\delta^2}{D_0 \rho_{kl}} \ln \left[\frac{(1 - 2\rho_{kl} \cos \theta_{kl} + \rho_{kl}^2)^{\frac{1}{2}} + \rho_{kl} - \cos \theta_{kl}}{1 - \cos \theta_{kl}} \right] \\
& + \frac{\delta^2}{D_0} \sum_{n=0}^{\infty} \frac{\rho_{kl}^n P_n(\cos \theta_{kl})}{(n+1)^2}.
\end{aligned}
\tag{3.5.26}
$$

The series occurring in Eq. (3.5.26) converges rapidly, so that both A_{kl} and B_{kl} are readily computed.

When a pair of charges is located at the same distance r_k from the center, $A_{kl} - B_{kl}$ depends only on ρ_{kl}, θ_{kl} and δ. Table 3.5.1 shows calculations

TABLE 3.5.1[a]

VALUES OF $A_{kl} - B_{kl}$ IN WATER AT 25°, FOR $r_k = r_l$

		$A_{kl} - B_{kl}$		
r_k/b	$\cos \theta_{kl}$	$\delta = 0.025$ $(D_0 \cong 2)$	$\delta = 0.050$ $(D_0 \cong 4)$	$\delta = 0.125$ $(D_0 \cong 10)$
	−1	0.0040	0.0041	0.0043
	−0.5	0.0050	0.0051	0.0054
1	0	0.0069	0.0070	0.0072
	+0.5	0.0116	0.0116	0.0118
	+0.9	0.0351	0.0349	0.0342
	−1	0.0064	0.0057	0.0054
	−0.5	0.0084	0.0073	0.0068
0.9	0	0.0126	0.0105	0.0094
	+0.5	0.0258	0.0917	0.0161
	+0.9	0.1580	0.0972	0.0598
	−1	0.0135	0.0097	0.0076
	−0.5	0.0189	0.0131	0.0098
0.8	0	0.0307	0.0201	0.0140
	+0.5	0.0710	0.0429	0.0261
	+0.9	0.4519	0.2416	0.1148
	−1	0.1109	0.0600	0.0296
	−0.5	0.1539	0.0821	0.0392
0.5	0	0.2383	0.1252	0.0574
	+0.5	0.4683	0.2415	0.1054
	+0.9	1.6450	0.8314	0.3432

[a]Reproduced from ref. 19.

applicable to this condition. The values apply to aqueous solutions at 25°C and D_0 has been placed equal to 78.5δ. At any other temperature (or for another solvent), with dielectric constant D, the values must be multiplied by $(78.5/D)$. For interpolation it is convenient to plot $1/(A_{kl} - B_{kl})$ versus $\cos \theta_{kl}$, a plot which shows relatively little curvature. When all charges are not located the same distance from the center, the sum $A_{kl} - B_{kl}$ must be corrected by the use of A_{kl} as defined in Eq. (3.5.23). It will be noted from

Table 3.5.1 that $A_{kl} - B_{kl}$ and, hence, the work of charging is markedly sensitive to ρ_{kl}, and when $\rho_{kl} < 1$, markedly sensitive to the choice of δ.

If the charge were uniformily smeared, we would have

$$W = \frac{Z^2 q^2}{2D} \left(\frac{1}{b} - \frac{\kappa}{1 + \kappa a} \right). \tag{3.5.27}$$

Figure 3.5.1 shows a plot of C_{kl} (with $D = 78.5$) versus x, for values of x up

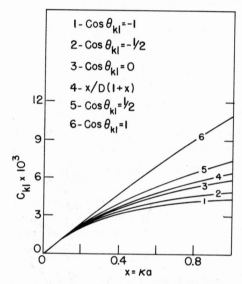

Fig. 3.5.1. Variation of C_{kl} with κa, and the function $x/D(1+x)$. [Reproduced from ref. 19.]

to 1.0. The calculations are based on $\delta = 0.0025$, $b/a = 0.85$, and $\rho_{kl} = 1$. The result is quite insensitive to the values of δ and b/a, and depends only slightly on ρ_{kl}. One may conclude from Fig. 3.5.1 that, where there is a more or less even distribution of the values of $\cos \theta_{kl}$ over the possible range, the substitution of $x/(1 + x)D$ for C_{kl} is a good approximation up to $x \cong 0.5$ with charges near the surface, and to even higher values of x for charges appreciably below the surface. Since $\sum\sum \eta_k \eta_l q^2 = Z^2 q^2$, at very low ionic strength, the concentration dependence is the same as for the smeared charge model. It should be noted further that, as the ionic strength increases to high values, the major contribution to the ionic strength-dependent term will come from terms with $\cos \theta_{kl}$ close to unity. In most instances only the self-energy terms ($\cos \theta_{kl} = 1$) will have $\cos \theta_{kl}$ close to unity; i.e., at high ionic strength the effect of ionic strength becomes largely that caused by the excess chemical potential of the individual charges.

Consider now that the polyion is a rigid polyampholyte. The free energy change resulting from say the addition of protons to basic sites results in a change in chemical free energy as well as a change in electrostatic free energy. Due to the spatial distribution of ionizable sites there are many equivalent arrangements of the charges which are approximately equal in energy. This gives rise to two effects: an entropy of mixing of charged and uncharged sites, and a fluctuation force between molecules. We shall study the latter in Section 36. Here we concentrate on the equilibrium between various charge distributions on the polyion.

Let there be $\Omega_v^{(i)}$ charge configurations of the polyion with v dissociable protons. We have

$$\Omega_v^{(i)} = \prod_j \frac{n_j!}{v_j!(n_j - v_j)!} \tag{3.5.28}$$

where there are v_j protons on a given kind of dissociable site and there are n_j sites of species j. Clearly, not all of these charge configurations have the same energy and therefore there will be different statistical weights corresponding to the different charge arrangements. Consider the partition function

$$Q_v^{(i)} = \sum_{\{v\}} \exp[-W_v^{(i)}/kT] \tag{3.5.29}$$

where the summation extends over all charge configurations. $W_v^{(i)}$ is easily seen to be the electrostatic work, exclusive of self-energies, for charging the species specified. The average contribution to the free energy is then

$$A_v^{(i)} = -kT \ln \sum_{\{v\}} \exp[-W_v^{(i)}/kT]. \tag{3.5.30}$$

If each of the $\Omega_v^{(i)}$ configurations had the same energy, then $W_v^{(i)}$ would simply reduce to $-kT \ln \Omega_v^{(i)} + A_v^{(i)}$. To evaluate $Q_v^{(i)}$ under more general conditions it is convenient to employ a procedure developed by Kirkwood in dealing with cooperative phenomena in alloys.[20] We define an energy $W_v^{(i)*}$ such that

$$-kT \ln \sum_{\{v\}} \exp[-W_v^{(i)}/kT] = -kT \ln \Omega_v^{(i)} + W_v^{(i)*} \tag{3.5.31}$$

and expand the exponentials in Eq. (3.5.31) so that the left-hand side becomes

$$-kT \ln\left[\sum_{\{v\}} \left\{ 1 - \frac{W_v^{(i)}}{kT} + \frac{1}{2}\frac{(W_v^{(i)})^2}{(kT)^2} - \cdots \right\} \right].$$

Since

$$\sum_{\{v\}} 1 = \Omega_v^{(i)}$$

we may add $kT \ln \Omega_v^{(i)}$ to each side, and obtain

$$W_v^{(i)*} = -kT \ln\left[1 - \frac{\langle W \rangle}{kT} + \frac{1}{2}\frac{\langle W^2 \rangle}{(kT)^2} - \cdots \right] \tag{3.5.32}$$

where

$$\langle W \rangle = \sum_{\{v\}} \frac{W_v^{(i)}}{\Omega_v^{(i)}}$$

$$\langle W^2 \rangle = \sum_{\{v\}} \frac{(W_v^{(i)})^2}{\Omega_v^{(i)}}.$$

(3.5.33)

Expanding the logarithm occurring in Eq. (3.5.30) in ascending powers of $(1/kT)$ it is found that

$$W_v^{(i)*} = kT \sum_{n=1}^{\infty} \frac{M_n}{n!} \left(\frac{1}{kT} \right)^n$$

(3.5.34)

where

$$M_1 = \langle W \rangle$$

$$M_2 = \langle W \rangle^2 - \langle W^2 \rangle.$$

(3.5.35)

.

.

.

The calculation of the free energy reduces then to the calculation of the M_n of Eq. (3.5.35) and, hence, the averages of eq. (3.5.33). How difficult this calculation is depends on the number of dissociable sites of the polyion and on the number of terms retained in Eq. (3.5.34). If one retains only the first term, the calculation becomes quite simple, and still represents an approximation considerably superior to the smeared model discussed in the preceding sections. In the calculations made by Tanford two terms in the expansion are retained. It was found that the second term made a relatively small contribution under most conditions, so that neglect of terms higher than the second is probably justified for the low charge density characteristic of proteins. For the case of polyelectrolytes, retention of higher order terms is essential. This is most easily seen by noting that the first term in this moment expansion corresponds to random mixing of charged and uncharged sites. A random mixing approximation is inadequate when the electrostatic forces are strong.[21]

By straightforward considerations it may now be shown that the equilibrium constant $k_v^{(i)}$ for the formation of a protein species $PH^{(i)}$ containing v dissociable protons, distributed in a particular way, such that v_j protons are attached to sites of a particular variety, can be calculated from the relation

$$\log k_v^{(i)} = \sum_j v_j (pK_{\text{int}})_j + \log \Omega_v^{(i)} - \sum_{n=1}^{\infty} \frac{M_n}{2.303 n!} \left(\frac{1}{kT} \right)^n + \frac{W_0}{2.303 kT}.$$

(3.5.36)

In this equation $(K_{int})_j$ is the intrinsic dissociation constant for the jth kind of site, and W_0 is the electrostatic work, exclusive of self-energies, for charging a protein molecule from which all dissociable protons have been removed. By definition,

$$\langle W \rangle = kT \sum_{k=1}^{Z} \sum_{l \neq k} \left[\sum_{\{v\}} \frac{\eta_k \eta_l}{\Omega_v^{(i)}} \right] \phi_{kl} \qquad (3.5.37)$$

$$\langle W^2 \rangle = (kT)^2 \sum_{k=1}^{Z} \sum_{l \neq k} \sum_{k'=1}^{Z} \sum_{l' \neq k'} \left[\sum_{\{v\}} \frac{\eta_k \eta_l \eta_{k'} \eta_{l'}}{\Omega_v^{(i)}} \right] \phi_{kl} \, \phi_{k'l'} \qquad (3.5.38)$$

and

$$\phi_{kl} = \frac{q^2}{2bkT} (A_{kl} - B_{kl}). \qquad (3.5.39)$$

Consider first the calculation of the occupation variables $\sum_{\{v\}} \eta_k \eta_l / \Omega_v^{(i)}$ and $\sum_{\{v\}} \eta_k \eta_l \eta_{k'} \eta_{l'} / \Omega_v^{(i)}$. These parameters are entirely independent of the model chosen, depending only on the number of sites m_j of each kind of group, and on the number μ_j of these which bear a charge.

Tanford has made calculations for models which consist of two kinds of sites, one anionic and one cationic. Let $k = c, c', c''$, etc., represent sites of one kind $(j = 1)$, and $k = n, n', n''$, etc., sites of the second kind $(j = 2)$. In computing $\sum_{\{v\}} \eta_k \eta_l / \Omega_v^{(i)}$ there are pairs of three types only: $k = c, l = c'$; $k = n, l = n'$; $k = c, l = n$. For the first two types $\eta_k \eta_l$ may be $+1$ or 0; for the third type it may be -1 or 0. For each particular pair of sites, then, $\sum_{\{v\}} \eta_k \eta_l$ is just numerically equal to the number of configurations with $\eta_k \eta_l$ not equal to zero, and this is just the number of configurations which have charges at each of the two sites k and l

For instance, for a particular pair c and c', this number is obtained as the number of ways of distributing $\mu_1 - 2$ charges over $m_1 - 2$ sites and μ_2 charges over m_2 sites, $(m_1 - 2)! m_2! / (m_1 - 2 - \mu_1)! (\mu_1 - 2)! (m_2 - \mu_2)! \mu_2!$. Thus,

$$\sum_{\{v\}} \frac{\eta_c \eta_{c'}}{\Omega_v^{(i)}} = \frac{\mu_1(\mu_1 - 1)}{m_1(m_1 - 1)} \qquad (3.5.40)$$

and, in the same way,

$$\sum_{\{v\}} \frac{\eta_n \eta_{n'}}{\Omega_v^{(i)}} = \frac{\mu_2(\mu_2 - 1)}{m_2(m_2 - 1)} \qquad (3.5.41)$$

$$\sum_{\{v\}} \frac{\eta_c \eta_n}{\Omega_v^{(i)}} = - \frac{\mu_1 \mu_2}{m_1 m_2}.$$

To compute $\sum_{\{v\}} \eta_k \eta_l \eta_{k'} \eta_{l'} / \Omega_v^{(i)}$ note that $k \neq l$ and $k' \neq l'$. Each set of possible values of k, l, l', l thus represents two pairs of sites. The two pairs may be identical (k' and l' the same as k and l) in which case the configurations which have $\eta_k \eta_l \eta_{k'} \eta_{l'} = \pm 1$ are the same as those which have $\eta_k \eta_l = \pm 1$, and $\sum \eta_k \eta_l \eta_{k'} \eta_{l'} = \sum \eta_k \eta_l$. If the pairs are different, but share one site, then

$$\sum_{\{v\}} \eta_c \eta_{c'} \eta_c \eta_{c''} / \Omega_v^{(i)} = \frac{\mu_1(\mu_1 - 1)(\mu_1 - 2)}{m_1(m_1 - 1)(m_1 - 2)}$$

$$\sum_{\{v\}} \eta_c \eta_{c'} \eta_c \eta_n / \Omega_v^{(i)} = -\frac{\mu_1(\mu_1 - 1)\mu_2}{m_1(m_1 - 1)m_2}$$

$$\cdot$$
$$\cdot$$
$$\cdot$$

(3.5.42)

or, if all four sites are different,

$$\sum_{\{v\}} \eta_c \eta_{c'} \eta_{c''} \eta_{c'''} / \Omega_v^{(i)} = \frac{\mu_1(\mu_1 - 1)(\mu_1 - 2)(\mu_1 - 3)}{m_1(m_1 - 1)(m_1 - 2)(m_1 - 3)}$$

$$\sum_{\{v\}} \eta_c \eta_{c'} \eta_n \eta_{n'} / \Omega_v^{(i)} = \frac{\mu_1(\mu_1 - 1)\mu_2(\mu_2 - 1)}{m_1(m_1 - 1)m_2(m_2 - 1)}.$$

$$\cdot$$
$$\cdot$$
$$\cdot$$

(3.5.43)

Tanford considers as a simple example the titration of four carboxyl groups on a molecule also containing four amino groups. The latter may be supposed positively charged throughout the titration and there are therefore five different forms of the protein: $v_1 = 0, 1, 2, 3, 4$; the corresponding value of μ_1 being 4, 3, 2, 1, 0. For the form with $\mu_1 = 4$ all occupation variables are equal to unity—there is a single configuration and all sites are charged. For the form with $\mu_1 = 0$ there is also a single configuration, all carboxyl groups being uncharged and all amino groups charged; i.e., $\sum_{\{v\}} \eta_n \eta_{n'} / \Omega$ and all $\sum_{\{v\}} \eta_n \eta_{n'} \eta_{n''} \eta_{n'''} / \Omega$ are equal to unity (regardless of whether n'' or n''' are equal to n or n'), and any occupation variable involving a carboxyl group, i.e., containing a η_c, must be zero. The occupation variables for all possible combinations are shown in Table 3.5.2. The left-hand column of this table shows the type of site pair (or product of pairs) being considered. The small letters c, c', c'', etc., refer to different carboxyl sites, i.e., the symbol $CC \times CC \, cc'$ means that the term considered is a product of two identical pairs of carboxyl groups (k' and l' equal to k and l), the symbol $CC \times CC \, cc' \, cc''$ represents a product of two carboxyl pairs with $k' = k$ or l, etc. Since all $\eta_n = 1$ it is not necessary in this example to distinguish between the $NN \times NN$ products in this way. Occupation variables in which $\eta_k \eta_l$ or $\eta_k \eta_l \eta_{k'} \eta_{l'}$ is negative in sign are so indicated.

It should be emphasized that the occupation variables of Table 3.5.2 are independent of the locations chosen for the sites involved. By contrast, the ϕ_{kl} depend *only* on location, being independent of the values of v or μ.

For a simple model there will be just a limited number of possible values of ϕ_{kl}. For instance, if there are eight sites at the corners of a cube, there will be 56 terms in the sum over k and l [Eq. (3.5.37)], but only three different values of ϕ_{kl} are required.

TABLE 3.5.2[a]

OCCUPATION VARIABLES FOR 4 COOH GROUPS TITRATED
ON A MOLECULE ALSO CONTAINING 4 NH_3^+ GROUPS

			1	2	3
ν_1			1	2	3
μ_1			3	2	1
$\sum_{\{\nu\}} \eta_k \eta_l / \Omega_\nu$					
CC			$\frac{1}{2}$	$\frac{1}{6}$	0
NN			1	1	1
CN			$-\frac{3}{4}$	$-\frac{1}{2}$	$-\frac{1}{4}$
$\sum_{\{\nu\}} \eta_k \eta_l \eta_{k'} \eta_{l'} / \Omega_\nu$					
$CC \times CC$		$cc'\,cc'$	$\frac{1}{2}$	$\frac{1}{6}$	0
		$cc'\,cc''$	$\frac{1}{4}$	0	0
		$cc'\,c''\,c'''$	0	0	0
$NN \times NN$		All terms	1	1	1
$CN \times CN$		$c = c'$	$\frac{3}{4}$	$\frac{1}{2}$	$\frac{1}{4}$
		$c \neq c'$	$\frac{1}{2}$	$\frac{1}{6}$	0
$CC \times NN$			$\frac{1}{2}$	$\frac{1}{6}$	0
$CC \times CN$		$cc'\,cn$	$-\frac{1}{2}$	$-\frac{1}{6}$	0
		$cc'\,c''\,n$	$-\frac{1}{4}$	0	0
$NN \times CN$		All terms	$-\frac{3}{4}$	$-\frac{1}{2}$	$-\frac{1}{4}$

[a]Reproduced from ref. 19.

There will be 24, 24, and 8 terms, respectively, with ϕ_{kl} equal to ϕ_1, ϕ_2, and ϕ_3, where ϕ_1 represents the value of ϕ_{kl} for nearest neighbors, ϕ_2 for sites at the ends of face diagonals, and ϕ_3 for sites at the ends of cube diagonals. For the same model the sum over k, l, k', and l' [Eq. (3.5.38)] will have many more terms, but there will be only six different values of $\phi_{kl}\,\phi_{kl}$. There will be 576 terms in ϕ_1^2, 576 in ϕ_2^2, 64 in ϕ_3^2, 1152 in $\phi_1\phi_2$, 384 in $\phi_1\phi_3$, and 384 in $\phi_2\phi_3$.

Since each term is to be multiplied by the corresponding occupation variable, it is necessary further to subdivide these terms so as to count separately those terms of the sum with different values of the occupation variables. It is at this stage that the positions of the individual sites must be specified. Suppose we continue to consider four carboxyl groups titrated in the presence of four cationic groups which bear a positive charge throughout the pH region in which the carboxyl groups are being titrated. Let the eight sites be placed at the corners of a cube as shown by model A, on which N represents a cationic site and C a carboxyl group. (The points, of course, represent the location of point charges if $\eta_k = \pm 1$.) The subdivision of site pairs and their products, as required by the occupation variables of Table 3.5.1, is then that shown in Table 3.5.3.

It is now a simple matter to compute M_1/kT and $M_2/(kT)^2$ by multiplying each term in Table 3.5.3 by the appropriate occupation variable in Table 3.5.2. The result of the calculation for this particular model, with suitable dimensions, is shown in Table 3.5.4.

It should be noted that the maximum contribution of M_2 to any $\log k_v$ [cf. Eq. (3.5.36)] is about 0.17. This is small compared to the maximum contribution of all of the electrostatic terms, which reaches 3.65. The term in M_2 may *seem* to make a large contribution in the case of $\log k_1$, but this is not so because the total electrostatic effect on $\log k_1$ is very small.

The four equilibrium constants of the present model can now be calculated at once from Eq. (3.5.36) and, hence, the curve for the titration of carboxyl groups.

Tanford has made calculations of the titration curve for several models, as shown in Fig. 3.5.2. It is found that the location of charged sites with

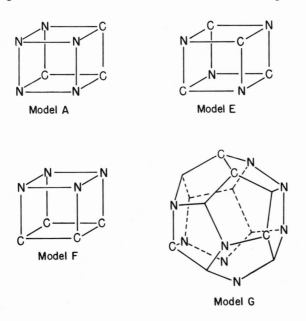

Model A   Model E

Model F

Model G

FIG. 3.5.2. Several models for testing the effect of charge distribution on the titration curve. [Reproduced from ref. 19.]

TABLE 3.5.3[a]

NUMBER OF TERMS OF DIFFERENT ENERGY FOR MODEL A

(a) In evaluation of M_1

	ϕ_1	ϕ_2	ϕ_3
CC	6	4	2
NN	6	4	2
CN	12	16	4

(b) In evaluation of M_2

		ϕ_1^2	ϕ_2^2	ϕ_3^2	$\phi_1\phi_2$	$\phi_1\phi_3$	$\phi_2\phi_3$
$CC \times CC$	$cc'\,cc'$	12	8	4	0	0	0
	$cc'\,cc''$	16	0	0	48	16	16
	$cc'\,c''\,c'''$	8	8	0	0	8	0
$NN \times NN$	All terms	36	16	4	48	24	16
$CN \times CN$	$c = c'$	40	64	8	96	16	32
	$c \neq c'$	104	192	8	288	80	96
$CC \times NN$		72	32	8	96	48	32
$CC \times CN$	$cc'\,cn$	64	64	0	144	64	48
	$cc'\,c''\,n$	80	64	16	144	32	48
$NN \times CN$	All terms	144	128	16	288	96	96

[a]Reproduced from ref. 19.

TABLE 3.5.4[a]

ELECTROSTATIC FREE ENERGY TERMS FOR MODEL A[b, c]

ν_1	μ_1	$\dfrac{\langle W \rangle}{kT}$	$\dfrac{\langle W \rangle^2 - \langle W^2 \rangle}{(kT)^2}$	$\dfrac{\langle W \rangle}{kT} + \dfrac{1}{2}\dfrac{\langle W \rangle^2 - \langle W \rangle^2}{(kT)^2} - \dfrac{W_0}{kT}$
0	4	-2.648[d]	0	0
1	3	-1.987	-0.787	0.268
2	2	-0.362	-0.670	1.951
3	1	$+2.224$	$+0.105$	4.767
4	0	$+5.773$	0	8.421

[a] Reproduced from ref. 19.

[b] For a sphere of radius 8 A, with charges 0.8 A below the surface of the molecule.

[c] By Eq. (3.5.36) the last column of this table is the contribution of all electrostatic energy terms to the various log k_ν. It will be noted that there are no subspecies in this particular model so that the k_ν do not need to be subdivided into various $k_\nu^{(i)}$.

[d] Since we are considering the titration of carboxyl groups only we are in effect considering the cationic protons as not dissociable. Hence this term is W_0/kT for this model.

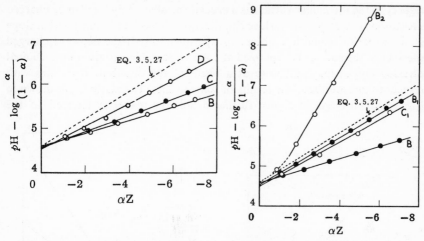

FIG. 3.5.3.a. Logarithmic plots for eight −COOH groups *at the surface* of a sphere of radius 10 A. Sites distributed according to models B, C, and D. [Reproduced from ref. 19.] FIG. 3.5.3.b. Logarithmic plots for eight −COOH groups (models B and C) at various depths within a sphere of radius 10 A. Curve B is for sites at the surface; curves B_1 and C_1 for sites 1 A below the surface; curve B_2 for sites 2 A below the surface. [Reproduced from ref. 19.] Model B: 8 charges at the corners of a cube. Model C: 8 charges randomly placed at the vertices of a dodecahedron. Model D: 8 charges occupying all vertices of two adjacent faces of a dodecahedron.

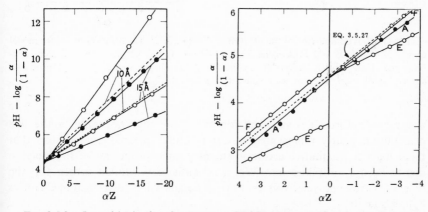

FIG. 3.5.3.c. Logarithmic plots for twenty −COOH groups, uniformly distributed, on spheres of radius 10 and 15 A, respectively. For each value of the radius the upper solid line has the sites 1 A below the surface; the lower solid line has them at the surface; and the dashed line represents a plot according to Eq. (3.5.27). [Reproduced from ref. 19.] FIG. 3.5.3.d. Logarithmic plots for four −COOH groups in the presence of four −NH$_3^+$ groups, at a depth of 0.8 A in a sphere of radius 8 A. The left-hand side shows calculations for models A, E, and F. The right-hand side applies to the same models with the −NH$_3^+$ groups removed. [Reproduced from ref. 19.]

respect to the molecular surface is a crucial variable. Only by placing charged sites about 1 A below the surface can titration curves be obtained with a slope of the order of magnitude observed experimentally. If the depth of charged sites below the surface is kept fixed at 1 A considerable variation in calculated titration curves may still arise as a result of the distribution of sites with respect to one another. The results of some of these calculations are shown in Figs. 3.5.3 through 3.5.5. For further details the reader is referred to the

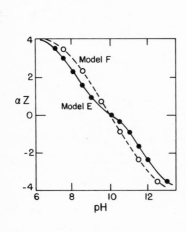

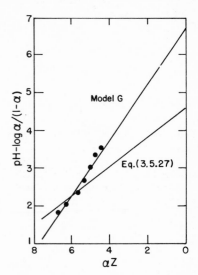

FIG. 3.5.4. A comparison of the titration curves of models E and F (see Fig. 3.5.2). Note that these models differ only in the charge distribution. [Reproduced from ref. 19.]

FIG. 3.5.5. The titration curves for model G (see Fig. 3.5.2) and the equivalent continuous charge distribution. [Reproduced from ref. 19.]

paper by Tanford. It is sufficient for us to note that not only is the site distribution poorly represented as a uniformly smeared charge, but that there may be qualitative errors in the computed properties of the polyion due to such a simplification. We may anticipate similar errors for the case of flexible polyelectrolytes. Further details will be discussed in Chapters 7, 8, 9 and 10.

3.6   Fluctuation Interactions between Rigid Polyampholytes

When consideration is given to the possibility that charges of both signs may exist on the surface of the molecule, a new mechanism of interaction between macroions becomes apparent. If there is a net difference in the numbers of positive and negative charges, at the isoelectric point there will be a number of uncharged groups otherwise identical in chemical character with some of the charged groups. For one of the types of charge, there exists under these conditions a large number of different possible arrangements of the charges on the surface of the molecule. A number of these arrangements are of approximately equal electrostatic energy and one may expect the charges to fluctuate in occupation of first one and then another set of sites. If two such molecules are brought close to each other, the charges on one ion tend to polarize the charges on the other in such a manner as to separate like charges. This is possible only if there exist surface charge distributions of roughly equal energy and the resultant decreased repulsion is, of course, equivalent to a net attraction, since the net charge of each particle is zero.[22]

The forces described in the preceding paragraph are only significant at the isoelectric point. For, when the molecule is far from the isoelectric point, on the average charge of only one sign is present, and the considerations presented in previous sections suffice to describe the interactions. At the isoelectric point, where the average net charge is zero, conditions are rather different.

Consider two polyampholytes with their centers of mass separated by the distance r. The interaction between the two molecules may be written[23]

$$U = \sum_{i=1}^{Z_1} \sum_{j=1}^{Z_2} \frac{\eta_i^{(1)}\eta_j^{(2)}q^2 \exp[-\kappa(r_{ij}^{(12)} - 2b_{12})]}{Dr_{ij}^{(12)}(1 + \kappa b_{12})} \qquad (3.6.1)$$

where there are Z_1 and Z_2 sites on polyions one and two, respectively, q is the magnitude of the charge, $r_{ij}^{(12)}$ is the separation of site i on molecule one from site j on molecule two, and $\eta_i^{(1)}$ and $\eta_j^{(2)}$ are occupation variables which can have the values zero for an uncharged site and plus or minus unity for positive and negative charges on the specified site. Now the potential of mean force is defined by

$$e^{-W/kT} = \langle e^{-U/kT} \rangle \qquad (3.6.2)$$

where the average is taken as unweighted and with molecules one and two fixed. It may be shown readily that the potential of mean force may be expressed in terms of unweighted averages of the potential energy U. We use the truncated form,

$$W(r) = \langle U \rangle - \frac{1}{2kT}(\langle U^2 \rangle - \langle U \rangle^2) \qquad (3.6.3)$$

correct to terms of order $(kT)^{-1}$. At the isoelectric point, the average net charge is zero and the first term vanishes, as do all powers of $\langle U \rangle$. Now define

$$q_i^{(1)} = q\eta_i^{(1)} = \langle q_i^{(1)} \rangle + \Delta q_i^{(1)}$$
$$\Delta q_i^{(1)} = q[\eta_i^{(1)} - \langle \eta_i^{(1)} \rangle] = q\Delta\eta_i^{(1)}. \qquad (3.6.4)$$

The use of Eq. (3.6.1) in Eq. (3.6.3) gives

$$W(r) = -\frac{q^4}{2kTD^2r^2} \sum_{i=1}^{Z_1} \sum_{l=1}^{Z_1} \sum_{k=1}^{Z_2} \sum_{s=1}^{Z_2} \langle \eta_i^{(1)}\eta_l^{(1)}\rangle\langle\eta_k^{(2)}\eta_s^{(2)}\rangle \times$$

$$\frac{\exp[-2\kappa(r-2b_{12})]}{(1+\kappa b_{12})^2} \left\langle\frac{r^2}{r_{ik}^{(12)}r_{ls}^{(12)}}\right\rangle \quad (3.6.5)$$

where the last factor accounts for differences in charge–charge separation due to the distribution of charges on the surface of the molecule. With the substitution of Eq. (3.6.5) into Eq. (3.6.6), the potential of mean force becomes (setting $\langle r^2/r_{ik}^{(12)}r_{ls}^{(12)}\rangle = 1$, valid at large separations)

$$W(r) = -\frac{q^4}{2kTD^2r^2} \sum_{i,l=1}^{Z_1} \sum_{k,s=1}^{Z_2} [\langle\eta_i^{(1)}\eta_l^{(1)}\rangle - \langle\eta_i^{(1)}\rangle\langle\eta_l^{(1)}\rangle] \times$$

$$[\langle\eta_k^{(2)}\eta_s^{(2)}\rangle - \langle\eta_k^{(2)}\rangle\langle\eta_s^{(2)}\rangle] \frac{\exp[-2\kappa(r-2b_{12})]}{(1+\kappa b_{12})^2}. \quad (3.6.6)$$

This term represents the interactions between the fluctuating charges and fluctuating multipoles of both molecules. The charge fluctuations may be obtained easily by differentiation of the Grand Partition Function, as is seen later. To calculate the excess chemical potential, one uses the relation (see Chapter 1)

$$\frac{100M_1}{NkT}\left(\frac{\partial\mu_1^E}{\partial c_1}\right)_{T,p,\kappa_0} = 4\pi\left[\int_0^\infty r^2(g_{10}(r)-1)\,dr - \int_0^\infty r^2(g_{11}(r)-1)\,dr\right]$$

$$(3.6.7)$$

where, in the usual notation,

$$\mu_1 = \mu_1^0(T,p) + kT\ln c_1 + \mu_1^E$$

$$\mu_1^0(T,p) = \lim_{c\to 0}[\mu_1 - kT\ln c_1]$$

$$g_{10}(r) = \exp[-W_{10}/kT]; \quad g_{11}(r) = \exp[-W_{11}/kT] \quad (3.6.8)$$

with W_{10} and W_{11} the potentials of mean force between a macromolecule and a solvent molecule and between two macro molecules, respectively. The dimensions of a solvent molecule are small relative to those of the macromolecule in almost all cases, and the solvent plays the role of a dielectric continuum. One finds, therefore, after substitution and integration, that[23]

$$\frac{\mu_1^E}{kT} = \frac{c_1}{100M_1}\left[\frac{7\pi Nb_{12}^3}{6} - \frac{\pi Nq^4\langle\Delta\eta_i^{(1)2}\rangle}{(DkT)^2\kappa_0(1+\kappa_0 b_{12})^2}\right] \quad (3.6.9)$$

where the macroion contribution to the ionic strength is neglected.

At high salt concentrations, κ_0 is large and the second term is negligible relative to the first term. Under these conditions, the excess chemical potential is positive, representing the net repulsive forces which result in an excluded volume. When the salt concentration is very low, the second term dominates and the excess chemical potential is negative. Careful light-scattering experiments have verified Eq. (3.6.9) quantitatively.[24] The magnitude of the fluctuating charge may be obtained independently from the titration curve of the polyampholyte and is in complete agreement with the values determined from the independent light-scattering experiments. (See Fig. 3.6.1).

A very interesting application of this theory has been presented by Kirkwood[22] in a discussion of enzyme kinetics. Suppose the enzymatic reaction can be described by the Michaelis–Menten relation

$$-\frac{d[S]}{dt} = \frac{k_3[E][S]}{K_m + [S]}$$

(3.6.10)

in which $[S]$ and $[E]$ are the substrate and enzyme concentrations, K_m the Michaelis–Menten constant, and k_3 the intrinsic rate constant for the enzymatic reaction. Now the potential of mean force is the work required to bring a pair of molecules from infinite separation to the distance r. If the standard free energy of formation of a complex is ΔG^0, then this is equal to the potential of mean force at contact; i.e.,

$$\Delta G^0 = W(2b)$$

(3.6.11)

and, therefore,

$$kT \ln K = -W(2b)$$

(3.6.12)

where K is the equilibrium constant for the formation of the complex. Let the complex now undergo reaction through the intermediary of a transition state with free energy of activation $\Delta G^{\ddagger}$. Thus,

$$\Delta G^{\ddagger} = W^{\ddagger}(2b^{\ddagger}) - W(2b).$$

(3.6.13)

Suppose the substrate molecule is specifically bound to the active site of the enzyme by local forces with potential W_0, in which the protein does not participate. In the activated state, the potential is denoted by $W_0^{\ddagger}$, the subscript zero again indicating that protein forces are omitted. Under the influence of the protein, the local forces are different and the corresponding potentials are denoted W' and $W^{\ddagger}$. If K_m and K_m^0 and k_3 and k_3^0 are the Michaelis–Menten constants and intrinsic rates of reaction with and without protein participation, then

$$kT \ln \frac{K_m}{K_m^0} = W' - W_0$$

$$kT \ln \frac{k_3}{k_3^0} = \Delta W_0^{\ddagger} - \Delta W^{\ddagger}$$

(3.6.14)

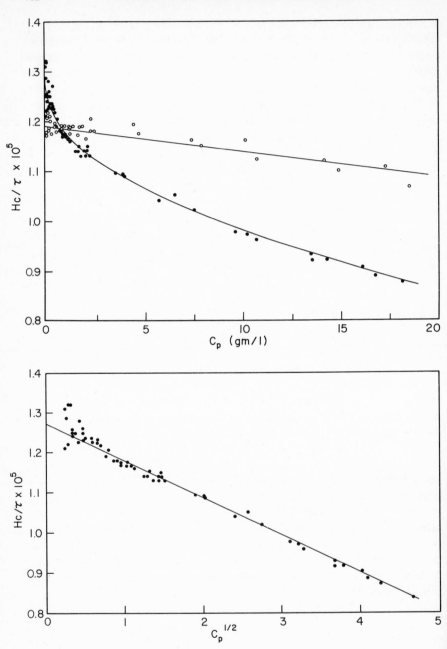

FIG. 3.6.1. The fluctuation contribution to the second virial coefficient. [Reproduced from ref. 24.]

where the intrinsic rate of decomposition of ES into reactants is assumed to be small relative to k_3.

If the protein contains a total of Z groups with acid dissociation constants $\{K_i\}$ and if the substrate molecule has dipole moments μ' and $\mu^{\ddagger}$ in the normal and activated states, then, neglecting all forces except those arising from charge fluctuations leads to

$$W - W_0 = -\frac{1}{2kT} \sum_{i=1}^{Z} \frac{K_i[\mathrm{H}^+]}{([\mathrm{H}^+] + K_i)^2} \frac{q^2\mu^2 \cos^2 \gamma_i}{D_e^2 r_i^4} \tag{3.6.15}$$

where γ_i is the angle between the dipole moment of the substrate molecule and the radius vector $\mathbf{r}_i$ from site i, D_e is the effective dielectric constant, and the first factor arises from the evaluation of $\langle \eta_i^2 \rangle$ as is seen in a subsequent section. Note that electrostatic interactions between charges on the protein again are neglected.

Since the fluctuation potentials fall off as r^{-4}, one may simplify Eq. (3.6.15) by assuming that only nearest neighboring groups are effective. If there are Z_α nearest neighboring groups with dissociation constants K_α,

$$W - W_0 = -\frac{Z_\alpha q^2 \mu^2}{4D_e^2 R_\alpha^4 kT} \cdot \frac{K_\alpha[\mathrm{H}^+]}{([\mathrm{H}^+] + K_\alpha)^2}$$

$$\frac{1}{R_\alpha^4} = \frac{2}{Z_\alpha} \sum \frac{\cos^2 \gamma_i}{r_i^4}. \tag{3.6.16}$$

From Eqs. (3.6.16) and (3.6.14),

$$\log \frac{k_3}{k_3^0} = \frac{Z_\alpha q^2 (\mu^{\ddagger 2} - \mu^2)}{4D_e^2 k^2 T^2 R_\alpha^4} \cdot \frac{[K_\alpha \mathrm{H}^+]}{([\mathrm{H}^+] + K_\alpha)^2}. \tag{3.6.17}$$

Note that, for this simplified model, k_3 has a maximum when the $p\mathrm{H}$ is equal to pK. Further, the $p\mathrm{H}$ dependence predicted by the model is a symmetric bell-shaped curve. If the electrostatic interactions between charges on the surface of the protein are accounted for, then the curve may be skewed since the proton population on the molecular surface, and hence the free energy of activation, are not symmetric functions of the $p\mathrm{H}$. Kirkwood has made a crude comparison of the theory with some experimental data of Bergmann and Fruton[25] on the effect of $p\mathrm{H}$ on the pepsin hydrolysis of carbobenzoxy-l-glutamyl-l-tyrosine. The maximum rate occurs at $p\mathrm{H}$ 4, thereby identifying the participating groups as carboxylates. With an estimated dielectric constant and an estimated dipole moment difference, the fall in rate to one-half of its value in an interval of one $p\mathrm{H}$ unit is accounted for if there are about ten carboxyl groups at an average distance of 5 A from the adsorption site. This picture is qualitatively reasonable. It should be stressed that we have considered only one of many possible contributory forces operative in enzyme reactions and it is not to be implied that the charge fluctuation mechanism is the only or even the most important of the possibilities. The charge fluctuation mechanism, however, does provide a physical picture in qualitative accord with experiment.

REFERENCES

1. E. J. W. VERWEY AND J. T. G. OVERBEEK, " Theory of the Stability of Lyophobic Colloids." Elsevier, New York, 1948.
2. S. LEVINE, *Proc. Roy. Soc.* A170, 165 (1939); *Phil. Mag.* 41, 53 (1950).
3. F. BOOTH, *Proc. Roy. Soc.* A203, 514, 533 (1950).
4. J. G. KIRKWOOD AND J. MAZUR, *J. Polymer Sci.* 9, 519 (1952).
5. N. E. HOSKIN AND S. LEVINE, *Phil. Trans. Roy. Soc.* A248, 449 (1956).
6. P. PIERCE, Ph.D. Dissertation, Department of Chemistry, Yale University, 1958.
7. R. STEINER, Ph.D. Dissertation, Department of Chemistry, Harvard University, 1950.
8. L. M. KUSHNER, B. C. DUNCAN, AND J. I. HOFFMAN, *J. Res. Natl. Bur. Standards* 49, 85 (1952).
9. L. M. KUSHNER AND W. D. HUBBARD, *J. Colloid Sci.* 10, 428 (1955).
10. C. TANFORD, *J. Phys. Chem.* 59, 788 (1955).
11. E. J. COHN AND J. T. EDSALL, " Proteins Amino Acids and Peptides." Reinhold, New York, 1943.
12. J. HERMANS AND J. T. G. OVERBEEK, *Rec. trav. chim.* 67, 761 (1948).
13. F. OOSAWA, N. IMAI, AND I. KAGAWA, *J. Polymer Sci.* 13, 93 (1954).
14. F. T. WALL AND J. BERKOWITZ, *J. Chem. Phys.* 26, 114 (1957).
15. S. LIFSON, *J. Chem. Phys.* 27, 700 (1957).
16. M. NAGASAWA AND I. KAGAWA, *Bull Chem. Soc. Japan* 30, 961 (1957).
17. G. E. KIMBALL, M. CUTLER, AND H. SAMELSON, *J. Phys. Chem.* 56, 57 (1952).
18. J. G. KIRKWOOD, *J. Chem. Phys.* 2, 351 (1934).
19. C. TANFORD AND J. G. KIRKWOOD, *J. Am. Chem. Soc.* 79, 5333 (1957); C. TANFORD, *J. Am. Chem. Soc.* 79, 5340 (1957).
20. J. G. KIRKWOOD, *J. Chem. Phys.* 6, 70 (1938).
21. F. E. HARRIS AND S. A. RICE, *J. Phys. Chem.* 58, 725, 733 (1954).
22. J. G. KIRKWOOD, *Discussions Faraday Soc.* No. 20, 78 (1955).
23. J. G. KIRKWOOD AND J. B. SHUMAKER, *Proc. Natl. Acad. Sci. U.S.* 38, 863 (1952).
24. S. N. TIMASHEFF, H. M. DINTZIS, J. G. KIRKWOOD, AND B. COLEMAN, *J. Am. Chem. Soc.* 79, 782 (1957); S. LOWEY, Ph.D. Dissertation, Department of Chemistry, Yale University, 1957.
25. M. BERGMANN AND J. S. FRUTON, *J. Biol. Chem.* 127, 627 (1939).

Appendix 3A.   A Table of f_3 f_5 and f_7[a, b]

κb	$f_3(\kappa b)$	$f_5(\kappa b)$	$f_7(\kappa b)$
0.00	0.0000000	0.000000	0.00000
0.10	(-3) 1.0038257	(-6) 8.707530	(-8) 9.31754
0.15	(-3) 1.6249817	(-5) 1.681218	(-7) 2.06626
0.20	(-3) 2.2310799	(-5) 2.615957	(-7) 3.58709
0.25	(-3) 2.8102477	(-5) 3.627577	(-7) 5.43936
0.30	(-3) 3.3590318	(-5) 4.685094	(-7) 7.57002
0.35	(-3) 3.9772675	(-5) 5.767469	(-7) 9.93103
0.40	(-3) 4.3661470	(-5) 6.860100	(-6) 1.24803
0.45	(-3) 4.8273970	(-5) 7.952686	(-6) 1.51814
0.50	(-3) 5.2629110	(-5) 9.037926	(-6) 1.80034
0.60	(-3) 6.0641700	(-4) 1.116706	(-6) 2.39082
0.70	(-3) 6.7837330	(-4) 1.322176	(-6) 3.00286
0.80	(-3) 7.4333780	(-4) 1.519009	(-6) 3.62456
0.90	(-3) 8.0229120	(-4) 1.706767	(-6) 4.24750
1.00	(-3) 8.5604520	(-4) 1.885440	(-6) 4.86569
1.10	(-3) 9.0527370	(-4) 2.055260	(-6) 5.47496
1.20	(-3) 9.5053940	(-4) 2.216586	(-6) 6.07243
1.30	(-3) 9.9231470	(-4) 2.369839	(-6) 6.65617
1.40	(-2) 1.0309990	(-4) 2.515470	(-6) 7.22491
1.50	(-2) 1.0669324	(-4) 2.653931	(-6) 7.77793
1.60	(-2) 1.1004057	(-4) 2.785660	(-6) 8.31484
1.80	(-2) 1.1609418	(-4) 3.030592	(-6) 9.34013
2.00	(-2) 1.2142425	(-4) 3.253364	(-5) 1.03020
2.20	(-2) 1.2615586	(-4) 3.456681	(-5) 1.12032
2.40	(-2) 1.3038642	(-4) 3.642876	(-5) 1.20475
2.50	(-2) 1.3233854	(-4) 3.730194	(-5) 1.24495
2.60	(-2) 1.3419299	(-4) 3.813961	(-5) 1.28387
2.80	(-2) 1.3763734	(-4) 3.971657	(-5) 1.35808
3.00	(-2) 1.4076967	(-4) 4.117452	(-5) 1.42776
3.50	(-2) 1.4748887	(-4) 4.437836	(-5) 1.58438
4.00	(-2) 1.5297526	(-4) 4.707154	(-5) 1.71964
4.50	(-2) 1.5754286	(-4) 4.936656	(-5) 1.82741
5.00	(-2) 1.6140668	(-4) 5.134543	(-5) 1.94077
5.50	(-2) 1.6471898	(-4) 5.306916	(-5) 2.03213
6.00	(-2) 1.6759080	(-4) 5.458414	(-5) 2.11343
7.00	(-2) 1.7232518	(-4) 5.712297	(-5) 2.25174
8.00	(-2) 1.7607834	(-4) 5.916680	(-5) 2.36492
9.00	(-2) 1.7910331	(-4) 6.084760	(-5) 2.45921
10.00	(-2) 1.8161428	(-4) 6.225427	(-5) 2.53894
11.00	(-2) 1.8372664	(-4) 6.344988	(-5) 2.60723
12.00	(-2) 1.8552856	(-4) 6.447606	(-5) 2.66638
13.00	(-2) 1.8708394	(-4) 6.536871	(-5) 2.71809
14.00	(-2) 1.8844024	(-4) 6.615167	(-5) 2.76368
15.00	(-2) 1.8963342	(-4) 6.684399	(-5) 2.80419

[a] Reproduced from ref. 6.

[b] The number in parentheses indicates the power of ten by which the result is to be multiplied by. Example: $\kappa b = 9.00$, $f_3(\kappa b) = 1.7910331 \times 10^{-2}$.

4. The Theory of Electrophoresis of Rigid Macroions

4.0 Introduction

Aside from their own intrinsic interest, a study of the transport properties of solutions of rigid macroions is valuable in two respects. From the theoretical point of view, the motion of a particle through a solvent medium is sensitive to details of the molecular configuration and molecular interactions different from those manifested in equilibrium properties. From the practical point of view, many fractionation and/or preparation techniques depend upon differences between the transport coefficients of the various species present in a mixture. In this chapter we shall examine the mathematical theory of electrophoresis of rigid polyions. In view of the excellent reviews of experimental work available,[1, 2] we shall make no attempt to examine data in this chapter, but rather shall confine our attention to the details of the theory.

Before proceeding to a detailed discussion of the theory of electrophoresis, some very elementary remarks appear to be pertinent. Imagine that the polyion is a rigid sphere of radius b and charge Zq. Due to the strong electrostatic forces the polyion is surrounded by an atmosphere of counterions. When an electric field is applied to the system and the particle achieves a steady velocity, the electrical force is just balanced by the viscous forces exerted by the solvent. These viscous forces may be considered to be composed of the frictional force that would be present if the surrounding liquid contained no charges and the forces exerted by that component of the motion of the liquid caused exclusively by the action of the applied field on the charges in the liquid part of the double layer.

Now, the charge density in the double layer must satisfy Poisson's equation. If the particle is large, and therefore the radius of curvature of the surface of the particle large, then the electric field due to the double layer is normal to the surface. Let the charge density in the double layer be denoted ρ and let $\mathbf{n}$ be the outward normal to the surface. The external force exerted on a layer of liquid of thickness dn is $\rho \mathbf{X}\, dn$ with $\mathbf{X}$ the electric field strength. The electric force must be exactly equal to the viscous force on the same shell, so that

$$\frac{d^2\mathbf{v}}{dn^2} = -\frac{\rho \mathbf{X}}{\eta} \tag{4.0.1}$$

187

with v the fluid velocity. Two integrations of Eq. (4.0.1) and use of the boundary conditions $\psi(r \to \infty) = 0$, $(d\psi/dn)_{n \to \infty} = 0$, $(dv/dn)_{n \to \infty} = 0$, $v(n \to \infty) = 0$, gives the simple relation

$$v = -\frac{D\zeta}{4\pi\eta} X \tag{4.0.2}$$

where ζ is the potential at the surface of the rigid macroion. Due to the action of the electric field the macroion has a steady velocity U. If the fluid is assumed incompressible, $\nabla \cdot v = 0$, and since there are no sources of the external field internal to the solution, $\nabla \cdot X = 0$. But v and X are, by Eq. (4.0.2), proportional to one another so that v is everywhere parallel and proportional to the field strength, and the constant of proportionality is $-D\zeta/4\pi\eta$. With respect to a system of coordinates fixed in the particle, the fluid at a great distance from the particle moves with velocity $-U$. Combining the above two results one finds for the electrophoretic velocity

$$U = \frac{D\zeta}{4\pi\eta} X \tag{4.0.3}$$

a result independent of the shape of the polymer if it is sufficiently large. Equation (4.0.3) was first derived by Helmholtz[3] and Smoluchowski.[4]

The preceding arguments depend upon the double layer dimensions being small relative to the particle dimensions and therefore the particle alters the geometry of the applied electric field. In the other extreme case when the double layer dimensions are large compared to the macroion dimensions, the factor 4π must be replaced by 6π, since then the very small deformation of the applied field has almost no influence on the solvent counterflow (the electrophoretic force described in viscous terms above). It will also be noted that, in terms of the theory of conductance of simple electrolytes, forces due to the deformation of the double layer have been neglected. In general, when the double layer dimensions fall into neither of the extreme cases mentioned previously, when relaxation forces and inertial terms must be considered the treatment of the balance of forces and the resultant particle velocity becomes very complex. The theory, especially as developed by Booth,[5-8] is presented in Sections 4.1, 4.2, and 4.3.

It must, of course, be borne in mind that the surface potential is related to the charge on the polyion. In general,

$$\begin{aligned}
Zq &= -\int_b^\infty \rho \, dv \\
&= -\int_b^\infty D \frac{d}{dr}\left(r^2 \frac{d\psi}{dr}\right) dr \\
&= -Db^2\left(\frac{d\psi}{dr}\right)_{r=b}
\end{aligned} \tag{4.0.4}$$

which becomes, in the Debye–Hückel approximation,

$$\psi = \frac{\zeta b}{r} e^{\kappa(b-r)}$$

$$Zq = D\zeta b(1 + \kappa b). \tag{4.0.5}$$

Whether the charge Zq or the potential ζ is to be preferred in describing the macroion is largely a matter of taste. For the case of polyelectrolytes, where there are fixed charge sites, it is the total charge which seems pertinent. In the case of the traditional colloids such as Ag I, the surface charge density varies with the state of the solution and the surface potential seems the more appropriate variable.

We now turn to a consideration of the details of the theory with particular reference to the case of the spherical polyion. We shall use the arguments of Booth in the treatment to follow. We do not mean to neglect the contributions of other workers, especially Overbeek,[1, 2] but believe that the consistency of Booth's theory makes the argument easier to follow.

4.1 Henry's Theory[9]

We first examine the motion under the influence of an electric field of charged spherical particles suspended in a pure solvent or an electrolyte solution. Let the sphere have radius b and conductivity σ_0, while the fluid viscosity, conductance, and dielectric constant are η, σ and D, respectively. The polyion is characterized by a fixed charge Zq and the electrostatic potential ψ outside the sphere is determined in part by the charge density ρ in the fluid around the sphere. The surface potential ζ is determined by the number of charges and the radius of the polyion. Finally, let the polyion be acted upon by an external field of magnitude X, parallel to the x axis. This electric field causes a steady-state velocity U.

When there is no external field, the potential about the polyion is spherically symmetric so that ψ and all its derivatives are functions of r only. Henry assumes that this is also true when X is nonzero, a postulate equivalent to the neglect of the relaxation of the double layer. The removal of this assumption will be considered in Section 4.3. Let ϕ and ϕ_0 be the potentials due to the field X outside and inside the polyion sphere. Because of the assumed spherical symmetry, these potentials are unaffected by the presence of the double layer, so that

$$\nabla^2 \phi = 0$$
$$\nabla^2 \phi_0 = 0 \tag{4.1.1}$$

with the boundary conditions

$$\frac{\partial \phi}{\partial x} = -X; \qquad r \to \infty$$

$$\left. \begin{array}{l} \phi = \phi_0 \\[2mm] \dfrac{\partial \phi}{\partial \theta} = \dfrac{\partial \phi_0}{\partial \theta} \\[3mm] \sigma \dfrac{\partial \phi}{\partial r} = \sigma_0 \dfrac{\partial \phi_0}{\partial r} \end{array} \right\} \qquad r = b \qquad (4.1.2)$$

where we have used polar coordinates r, θ, Ω, centered about the polyion with the polar axis along the x axis. It is readily verified that the solution satisfying Eqs. (4.1.1) and (4.1.2) is

$$\phi = -X\left(r + \frac{\lambda b^3}{r^2}\right) \cos \theta$$

$$\phi_0 = -X(1 + \lambda)r \cos \theta \qquad (4.1.3)$$

$$\lambda = \frac{\sigma - \sigma_0}{2\sigma + \sigma_0}.$$

At any given point, the total electrostatic potential is the sum of the potential arising from the field, and the mean electrostatic potential in the solution, $\phi + \psi$. This potential leads to a mechanical force per unit volume of $-\rho\nabla(\phi + \psi)$. If the inertia terms are neglected, the hydrodynamic equations of motion therefore become

$$\eta\nabla \times \nabla \times \mathbf{v} + \nabla p = -\rho\nabla(\phi + \psi) \qquad (4.1.4)$$

$$\nabla \cdot \mathbf{v} = 0$$

with $\mathbf{v}$ the velocity of flow and p the hydrostatic pressure. The fluid is taken to be incompressible, as is readily ascertained from the form of Eq. (4.1.4). To solve the hydrodynamic equation of motion, the charge density as a function of r must be specified. From the Poisson equation and Eq. (4.1.1),

$$\rho = -\frac{D}{4\pi} \nabla^2(\phi + \psi) = -\frac{D}{4\pi} \nabla^2 \psi. \qquad (4.1.5)$$

It is convenient to define a modified pressure p_1 by the relationship

$$p_1 = p - \int_\infty^r \rho\left(\frac{\partial \psi}{\partial r}\right) dr \qquad (4.1.6)$$

whereupon

$$\nabla p = \nabla p_1 + \nabla \int_\infty^r \rho\left(\frac{\partial \psi}{\partial r}\right) dr. \qquad (4.1.7)$$

Now the second term on the right-hand side of Eq. (4.1.7) has only an r component which is $\rho(\partial\psi/\partial r)$. Under these circumstances the gradient operation may be shown to give the result

$$\nabla p = \nabla p_1 + \rho\nabla\psi \tag{4.1.8}$$

and the equation of motion thereby becomes

$$\eta\nabla \times \nabla \times \mathbf{v} + \nabla p_1 = -\rho\nabla\phi \tag{4.1.9}$$

$$\nabla \cdot \mathbf{v} = 0.$$

To reduce Eq. (4.1.9) we remove the term on the left-hand side by the divergence operation. This yields

$$\begin{aligned}
\nabla^2 p_1 &= -\nabla \cdot (\rho\nabla\phi) \\
&= -\rho\nabla \cdot \nabla\phi - \nabla\phi \cdot \nabla\rho \\
&= -\frac{\partial\phi}{\partial r}\frac{\partial\rho}{\partial r} \\
&= -\frac{DX}{4\pi}\left(1 - 2\lambda\frac{b^3}{r^3}\right)\cos\theta\frac{\partial}{\partial r}\nabla^2\psi
\end{aligned} \tag{4.1.10}$$

using Eqs. (4.1.1), (4.1.3), and (4.1.5). The major remaining problem is the solution of Eq. (4.1.10). As a trial solution take

$$p_1 = -\frac{DX}{4\pi}\cos\theta\left(\frac{\partial\psi}{\partial r} - 2\lambda b^3 P_1\right) \tag{4.1.11}$$

with the function P_1 dependent upon r only. The Laplacian operator applied to (4.1.11) gives

$$\nabla^2 p_1 = -\frac{DX}{4\pi}\cos\theta\left[\frac{\partial}{\partial r}\nabla^2\psi - \frac{2\lambda b^3}{r}\left(\frac{\partial^2(P_1 r)}{\partial r^2} - \frac{2P_1}{r}\right)\right]. \tag{4.1.12}$$

Comparison of Eqs. (4.1.12) and (4.1.10) shows that

$$r^2\frac{\partial^2(P_1 r)}{\partial r^2} - 2P_1 r = \frac{\partial}{\partial r}\nabla^2\psi. \tag{4.1.13}$$

A particular solution of this differential equation is

$$P_1 = r\int^r \frac{dr}{r^4}\nabla^2\psi \tag{4.1.14}$$

which, upon substitution into Eq. (4.1.11) gives

$$p_1 = -\frac{DX}{4\pi}\cos\theta\left[\frac{\partial\psi}{\partial r} - 2\lambda b^3 r\int^r \frac{dr}{r^4}\nabla^2\psi\right]. \tag{4.1.15}$$

Examination of Eq. (4.1.6) shows that p_1 tends to zero as r tends to infinity. Using this limiting condition and setting the lower limit of integration in Eqs. (4.1.14) and (4.1.15) as ∞, we obtain

$$p = \int_\infty^r \rho\left(\frac{\partial\psi}{\partial r}\right) dr - \frac{DX}{4\pi} \cos\theta\left[\frac{\partial\psi}{\partial r} - 2\lambda b^3 r \int_\infty^r \frac{dr}{r^4} \nabla^2\psi\right]. \quad (4.1.16)$$

To complete the solution of the equations of motion, the fluid velocity $\mathbf{v}$ must be determined. Now

$$\nabla \times \nabla p_1 = 0 \quad (4.1.17)$$

so that the application of the curl operation to Eq. (4.1.4) gives

$$\eta\nabla \times \nabla \times \mathbf{W} = -\nabla \times (\rho\nabla\phi)$$
$$\mathbf{W} = \nabla \times \mathbf{v}. \quad (4.1.18)$$

We note that $\mathbf{v}$ has no component along Ω, and the rotation caused by the operation $\nabla \times$ then implies that $\mathbf{W}$ and therefore $\nabla \times \nabla \times \mathbf{W}$ have vanishing r and θ components. The components of $\mathbf{W}$ and $\nabla \times \nabla \times \mathbf{W}$ along Ω are

$$W_\Omega = \frac{1}{r}\left[\frac{\partial}{\partial r}(rv_\theta) - \frac{\partial v_r}{\partial\theta}\right]$$
$$(\nabla \times \nabla \times \mathbf{W})_\Omega = -\frac{1}{r}\left[\frac{\partial^2}{\partial r^2}(rW_\Omega) + \frac{1}{r}\frac{\partial}{\partial\theta}\left(\frac{1}{\sin\theta}\frac{\theta}{\partial\theta}(\sin\theta\, W_\Omega)\right)\right]. \quad (4.1.19)$$

But from the vector identity

$$\nabla \times (\rho\nabla\phi) = \nabla\rho \times \nabla\phi + \rho\nabla \times \nabla\phi \quad (4.1.20)$$

we see that $\nabla \times (\rho\nabla\phi)$ has neither r component nor θ component, and a component of magnitude $(1/r)/(d\rho/\partial r)/(\partial\phi/\partial\theta)$ along Ω. By combination of Eqs. (4.1.18) and (4.1.19) we find

$$(\nabla \times \nabla \times \mathbf{W})_\Omega = -\frac{1}{\eta r}\frac{\partial\rho}{\partial r}\frac{\partial\phi}{\partial\theta} = \frac{DX}{4\pi\eta}\sin\theta\left(1 + \lambda\frac{b^3}{r^3}\right)\frac{\partial}{\partial r}\nabla^2\psi. \quad (4.1.21)$$

To obtain the solution to this differential equation take the trial function

$$W_\Omega = -\frac{DX}{4\pi\eta}\left(\frac{\partial\psi}{\partial r} + \lambda b^3 \hat{W}_1\right)\sin\theta \quad (4.1.22)$$

with $\hat{W}_1$ a function of r only. The substitution of (4.1.22) into (4.1.21) yields

$$(\nabla \times \nabla \times \mathbf{W})_\Omega = \frac{DX}{4\pi\eta}\left[\frac{\partial}{\partial r}\nabla^2\psi + \frac{\lambda b^3}{r}\left(\frac{\partial^2(r\hat{W}_1)}{\partial r^2} - \frac{2\hat{W}_1}{r}\right)\right]. \quad (4.1.23)$$

Again, comparison of Eqs. (4.1.21) and (4.1.23) gives the requirement that the trial function $\hat{W}_1$ satisfy the differential equation, namely

$$r^2 \frac{\partial^2(r\hat{W}_1)}{\partial r^2} - 2r\hat{W}_1 = \frac{\partial}{\partial r} \nabla^2 \psi \tag{4.1.24}$$

from which, as before [compare Eq. (4.1.13)]

$$\hat{W}_1 = r \int^r \frac{dr}{r^4} \nabla^2 \psi \tag{4.1.25}$$

and

$$W_\Omega = -\frac{DX}{4\pi\eta} \sin\theta \left[\frac{\partial\psi}{\partial r} + \lambda b^3 r \int_\infty^r \frac{dr}{r^4} \nabla^2 \psi \right] \tag{4.1.26}$$

where we have again used the condition that the vorticity $\mathbf{W}$ must vanish in the limit as $r \to \infty$.

Consider the new variable ξ defined by

$$\xi = \frac{\partial\psi}{\partial r} + \lambda b^3 r \int_\infty^r \frac{dr}{r^4} \nabla^2 \psi. \tag{4.1.27}$$

In terms of ξ and the first integral of the differential equation, $\hat{W}_1$, Eqs. (4.1.19) become

$$(\nabla \times \mathbf{v})_\Omega \equiv \frac{1}{r}\frac{\partial}{\partial r}(rv_\theta) - \frac{1}{r}\frac{\partial v_r}{\partial\theta} = -\frac{DX\xi}{4\pi\eta}\sin\theta$$

$$\nabla \cdot \mathbf{v} \equiv \frac{1}{r^2}\frac{\partial}{\partial r}(r^2 v_r) + \frac{1}{r\sin\theta}\frac{\partial}{\partial\theta}(\sin\theta\, v_\theta) = 0. \tag{4.1.28}$$

Consider the trial solutions

$$v_r = R_1(r)\cos\theta$$
$$v_\theta = R_2(r)\sin\theta \tag{4.1.29}$$

where, as indicated, R_1 and R_2 are functions of r only. By substitution of Eqs. (4.1.29) into (4.1.28)

$$\frac{1}{r}\frac{\partial}{\partial r}(rR_2) + \frac{1}{r^3}(r^2 R_1) = -\frac{DX\xi}{4\pi\eta}$$

$$\frac{1}{r^2}\frac{\partial}{\partial r}(r^2 R_1) + \frac{2}{r^2}(rR_2) = 0. \tag{4.1.30}$$

This set of equations is closely related in structure to Eq. (4.1.13). By elimination of rR_2 and solution in the same form as Eq. (4.1.14),

$$v_r = \frac{DX}{6\pi\eta}\cos\theta\left[\int^r \xi\, dr - \frac{1}{r^3}\int^r r^3\xi\, dr \right]$$

$$v_\theta = -\frac{DX}{6\pi\eta}\sin\theta\left[\int^r \xi\, dr + \frac{1}{2r^3}\int^r r^3\xi\, dr \right]. \tag{4.1.31}$$

Thus far we have determined only particular solutions. To obtain a complete solution we must now obtain the general function satisfying the homogeneous equations

$$\eta \nabla \times \nabla \times \mathbf{v} + \nabla p = 0$$

$$\nabla \cdot \mathbf{v} = 0. \tag{4.1.32}$$

Because of the symmetry about the polar axis (the x axis) the general solution is of the form

$$p = A_0 + \left(A_1 r + \frac{B_1}{r^2} \right) \cos \theta$$

$$v_r = \left(\frac{A_1 r^2}{10\eta} + \frac{B_1}{\eta r} + A_2 + \frac{B_2}{r^3} \right) \cos \theta \tag{4.1.33}$$

$$v_\theta = \left(-\frac{A_1 r^2}{5\eta} - \frac{B_1}{2\eta r} - A_2 + \frac{B_2}{2r^3} \right) \sin \theta$$

$$v_\Omega = 0.$$

But we require that $\mathbf{v}$ be finite everywhere and this in turn implies that $A_1 = 0$. The complete solution of Eqs. (4.1.9) is then

$$p = \int_\infty^r \rho \left(\frac{\partial \psi}{\partial r} \right) dr + A_0 + \cos \theta \left[\frac{B_1}{r^2} - \frac{DX}{4\pi} \left(3 \frac{\partial \psi}{\partial r} - 2\xi \right) \right]$$

$$v_r = \cos \theta \left[\frac{B_1}{\eta r} + A_2 + \frac{B_2}{r^3} + \frac{DX}{6\pi\eta} \left(\int^r \xi \, dr - \frac{1}{r^3} \int^r r^3 \xi \, dr \right) \right]$$

$$v_\theta = \sin \theta \left[-\frac{B_1}{2\eta r} - A_2 + \frac{B_2}{2r^3} - \frac{DX}{6\pi\eta} \left(\int^r \xi \, dr + \frac{2}{r^3} \int^r r^3 \xi \, dr \right) \right]$$

$$v_\Omega = 0 \tag{4.1.34}$$

subject to the boundary conditions

$$v_r = -U \cos \theta, \qquad v_\theta = U \sin \theta, \qquad \psi = 0; \qquad r \to \infty$$

$$v_r = 0, \qquad v_\theta = 0, \qquad \psi = \zeta; \qquad r = b. \tag{4.1.35}$$

Using these boundary conditions and noting that

$$\lim_{r \to \infty} \frac{1}{r^3} \int^r r^3 \xi \, dr = 0$$

one finds that

$$A_2 = -U - \frac{DX}{6\pi\eta} \int_b^\infty \xi \, dr$$

$$\frac{B_1}{b\eta} = \frac{3}{2} U - \frac{DX}{4\pi\eta} \int_\infty^b \xi \, dr \qquad (4.1.36)$$

$$\frac{B_2}{b^3} = -\frac{1}{2} U + \frac{DX}{12\pi\eta} \int_\infty^b \xi \, dr + \frac{DX}{6\pi\eta b^3} \int_\infty^b r^3 \xi \, dr.$$

Introduction of these constants of integration gives for the complete solution of the equations of motion the lengthy relations

$$p = \int_\infty^r \rho\left(\frac{\partial\psi}{\partial r}\right) dr + \cos\theta \left[\frac{3\eta b}{2r^2} U - \frac{b}{r^2} \frac{DX}{4\pi} \int_\infty^b \xi \, dr - \frac{DX}{4\pi}\left(3\left(\frac{\partial\psi}{\partial r}\right) - 2\xi\right)\right]$$

$$v_r = \cos\theta \left[-\left(1 - \frac{3b}{2r} + \frac{b^3}{2r^3}\right) U - \left(\frac{b}{r} - \frac{b^3}{3r^3}\right)\frac{DX}{4\pi\eta} \int_\infty^b \xi \, dr\right.$$
$$\left. - \frac{DX}{6\pi\eta}\left(\int_\infty^r \xi \, dr - \frac{1}{r^3}\int_b^r r^3\xi \, dr\right)\right]$$

$$v_\theta = \sin\theta \left[\left(1 - \frac{3b}{4r} - \frac{b^3}{4r^3}\right) U + \left(\frac{b}{2r} + \frac{b^3}{6r^3}\right)\frac{DX}{4\pi\eta} \int_\infty^b \xi \, dr\right.$$
$$\left. - \frac{DX}{6\pi\eta}\left(\int_\infty^r \xi \, dr + \frac{1}{2r^3}\int_b^r r^3\xi \, dr\right)\right]$$

$$v_\Omega = 0. \qquad (4.1.37)$$

There remains now only the computation of the force acting on the spherical polyion. Since the external force acts only along the x direction, only the x components of forces need be considered. The x component of the stress normal to a point on the surface of the polyion is

$$\sigma_{rx} = \left(-p + 2\eta \frac{\partial v_r}{\partial r}\right)_{r=b} \cos\theta - \left(\frac{\partial v_\theta}{\partial r} + \frac{1}{r}\frac{\partial v_r}{\partial\theta} - \frac{v_\theta}{r}\right)_{r=b} \sin\theta. \quad (4.1.38)$$

In terms of the solution to the equations of motion, Eqs. (4.1.37), this stress becomes

$$\sigma_{rx} = \cos\theta \int_\infty^b \rho\left(\frac{\partial\psi}{\partial r}\right) dr + \cos^2\theta \left[-\frac{3\eta}{2b} U\right.$$
$$\left. + \frac{DX}{4\pi}\left\{\frac{1}{b}\int_\infty^b \xi \, dr + 3\left(\frac{\partial\psi}{\partial r}\right)_{r=b} - 2\xi_b\right\}\right]$$
$$+ \sin^2\theta \left[-\frac{3\eta}{2b} U + \frac{DX}{4\pi}\left\{\int_\infty^b \xi \, dr + \xi_b\right\}\right]. \quad (4.1.39)$$

The total force is obtained from Eq. (4.1.39) by integration over the entire sphere, and adding the force due to the fixed charge,

$$F_x = 2\pi b^2 \int_0^\pi \sigma_{rx} \sin\theta \, d\theta + ZqX \qquad (4.1.40)$$

or, after quadrature,

$$F_x = -6\pi b\eta U + DXb \int_\infty^b \xi \, dr + DXb^2 \left(\frac{\partial\psi}{\partial r}\right)_{r=b} + ZqX. \qquad (4.1.41)$$

Now, by Gauss' theorem

$$ZqX = -DXb^2 \left(\frac{\partial\psi}{\partial r}\right)_{r=b} \qquad (4.1.42)$$

canceling the second to last term on the right-hand side of Eq. (4.1.41). In the steady state, U is a constant and the force acting on the sphere vanishes. Thus,

$$U = \frac{DX}{6\pi b\eta} \int_\infty^b \xi \, dr$$

$$= \frac{DX}{6\pi\eta} \left[\zeta + \frac{\lambda b^3}{2} \left(b^2 \int_\infty^b \frac{dr}{r^4} \nabla^2\psi \, dr - \int_\infty^b \frac{dr}{r^2} \nabla^2\psi \right) \right]. \qquad (4.1.43)$$

If we set

$$\nabla^2\psi = \frac{1}{r^2} \frac{\partial}{\partial r} \left(r^2 \frac{\partial\psi}{\partial r} \right)$$

and integrate by parts there are obtained two equivalent relationships for the electrophoretic velocity U,

$$U = \frac{DX}{6\pi\eta} \left[\zeta + \lambda b^3 \left(3b^2 \int_\infty^b \frac{dr}{r^5} \left(\frac{\partial\psi}{\partial r}\right) - 2 \int_\infty^b \frac{dr}{r^3} \left(\frac{\partial\psi}{\partial r}\right) \right) \right]$$

$$U = \frac{DX}{6\pi\eta} \left[\zeta(1 + \lambda) + 3\lambda b^3 \left(5b^2 \int_\infty^b \frac{dr}{r^6} \psi - 2 \int_\infty^b \frac{dr}{r^4} \psi \right) \right] \qquad (4.1.44)$$

Henry has tabulated the integral functions of Eq. (4.1.44). In order to analyze the theory of this section we must carefully consider the approximations made. It is this task to which we now turn.

4.2 The Inertial Term[7]

The treatment of the preceding section has involved two principal assumptions. These are the neglect of relaxation effects and the omission of inertial terms in the equations of motion. In the actual computation of the right-hand side of Eq. (4.1.44) Henry also used the linearized Poisson–Boltzmann equation, but that approximation is not necessary to obtain the general result. In this section we examine the effect of the inertia term still neglecting any relaxation effects.

We anticipate that the inertia term will be of importance *only at high frequencies* (of the applied field) or if $Ub\eta$ is not small relative to unity. Even though $Ub\eta \ll 1$ under most experimental conditions, the examination of the formal structure of the theory is of considerable interest. When flexible polyelectrolytes are considered and it is realized that much of the over-all polymer movement is due to segmental motion, the importance of inertial effects becomes more apparent. This study, even though carried out for the case of the rigid sphere will provide considerable physical insight into the general problem.

Choosing the center of the sphere as the origin, Eq. (4.1.4) must be replaced by

$$\eta \nabla \times \nabla \times \mathbf{v} + \mathbf{v} \cdot \nabla \mathbf{v} = -\nabla p - \rho \nabla(\phi + \psi)$$
$$\nabla \cdot \mathbf{v} = 0 \tag{4.2.1}$$

where the fluid is again considered to be incompressible. To solve Eqs. (4.2.1), Booth[7] uses a technique introduced by Oseen, and rewrites (4.2.1) in the form

$$\eta \nabla \times \nabla \times \mathbf{u} + \mathbf{U} \cdot \left(\frac{\partial \mathbf{u}}{\partial x}\right) = -\nabla p - \rho \nabla(\phi + \psi)$$
$$\nabla \cdot \mathbf{u} = 0 \tag{4.2.2}$$
$$\mathbf{v} = \mathbf{U} + \mathbf{u}$$

where, as before, $\mathbf{U}$ is the velocity at infinity which must be imparted to the fluid to maintain the polyion stationary. The solutions of Eq. (4.2.2) in spherical polar coordinates are, when $\rho = 0$, $\phi + \psi = 0$,

$$u_r = -\sum_{n=0}^{\infty} (n+1) \frac{A_n P_n(\mu)}{r^{n+2}} + \frac{1}{2} e^{\mu x} \sum_{n=0}^{\infty} B_n [\chi'_n(x) - \mu\chi_n(x)] P_n(\mu)$$

$$u_\theta = -\sin\theta \sum_{n=1}^{\infty} \frac{A_n P'_n(\mu)}{r^{n+2}} - \frac{1}{2} e^{\mu x} \sin\theta \sum_{n=0}^{\infty} B_n \chi_n(x) \left[\frac{P'_n(\mu)}{x} - P_n(\mu)\right]$$

$$u_\Omega = 0$$

$$p = -U \frac{\partial}{\partial x} \sum_{n=0}^{\infty} \frac{A_n P_n(\mu)}{r^{n+1}} \tag{4.2.3}$$

where

$$\mu = \cos \theta$$
$$x = kr \tag{4.2.4}$$
$$k = U/2\eta.$$

As usual the $P_n(\mu)$ are Legendre polynomials while the A_n and B_n are integration constants and the functions χ_n are defined by [10]

$$\chi_n = \frac{(2n + 1)}{(\pi/2x)^{1/2}} K_{n+1/2}(x) \tag{4.2.5}$$

with $K_{n + 1/2}(x)$ the modified Bessel function of the second kind. The prime on χ'_n indicates a differentiation with respect to the variable x.

In the Debye–Hückel approximation

$$\nabla^2(\phi + \psi) = -\frac{4\pi}{D} \rho$$

$$\psi = \frac{b\zeta}{r} e^{\kappa(b - r)} \tag{4.2.6}$$

which incidentally defines the ζ potential in terms of the surface charges and the size of the polyion. Once again Eqs. (4.1.6) and (4.1.1) are used to obtain a particular solution, here of the form

$$p_1 = \frac{DX\zeta}{4\pi r} \cos \theta \, C_0(\kappa r) \tag{4.2.7}$$

with

$$C_0(\kappa r) = \kappa b \, e^{\kappa(b - r)}$$

$$\times \left[1 + \frac{1}{\kappa r} + \lambda(\kappa b)^3 \left(-\frac{1}{2(\kappa r)^2} + \frac{1}{6\kappa r} - \frac{1}{12} + \frac{\kappa r}{12} - \frac{(\kappa r)^2}{12} e^{\kappa r} Ei(\kappa r) \right) \right]. \tag{4.2.8}$$

As in the preceding section we must also investigate particular solutions for u_r and u_θ. Define a variable $\mathbf{W}_2$ by

$$\mathbf{W}_2 = e^{Ux/2\eta} \nabla \times \mathbf{u} \tag{4.2.9}$$

so that from Eq. (4.2.2) one finds

$$\nabla \times \nabla \times \mathbf{W}_2 + k^2 \mathbf{W}_2 = -\frac{e^{-kx}}{\eta} \nabla \times (\rho \nabla(\phi + \psi))$$

$$k = \frac{U}{2\eta}. \tag{4.2.10}$$

Since the polar axis is along x and the motion is therefore symmetrical about θ, W_{2r}, and $W_{2\theta}$ are zero. For $W_{2\Omega}$ Booth assumes

$$W_{2\Omega} = \sin \theta \sum_{n=1}^{\infty} C_n(r) P'_n(\mu) \tag{4.2.11}$$

$$C_n(r) = \frac{(-)^n DX}{2\pi^2 \eta (2n+1)^2} \int_r^{\infty} r\psi_n(kz) \times$$

$$[\psi_n(kz)\chi_n(kr) - \psi_n(kr)\chi_n(kz)]\left[1 + \frac{\lambda b^3}{z^3}\right] \frac{d\psi}{dz} dz \tag{4.2.12}$$

with (4.2.12) obtained by substitution of (4.2.11) into (4.2.10) and equating of like powers of $\sin \theta \, P'_n(\mu)$. Similarly

$$e^{-kx} = \sum_{n=0}^{\infty} (-)^{n+1} \frac{G_n(kr)}{kr} P'_n(\mu) \tag{4.2.13}$$

with the G_n defined by[10]

$$G_n(x) = (2n+1)\left(\frac{\pi}{2x}\right)^{\frac{1}{2}} J_{n+\frac{1}{2}}(x) \tag{4.2.14}$$

with $J_{n+\frac{1}{2}}(x)$ a Bessel function of the first kind and of the indicated order. For questions related to the convergence of the integral of Eq. (4.2.12) we refer the reader to Booth's paper.[7]

The development thus far contains all of the physical information necessary for the numerical solution of the problem. To actually make a calculation of U Booth expands the C_n as a series in ascending powers of k or kr. The substitution of the expansion, explicit integration and use of the boundary conditions leads after much tedious algebra to the result

$$U = \frac{DX(1+\lambda)\zeta}{6\pi\eta}\left[1 + \frac{67}{11,520\pi^2}\left(\frac{bDX\zeta(1+\lambda)}{\eta^2}\right)^2 + \cdots\right] \tag{4.2.15}$$

and the correction term is small under almost all conditions. It is most important for nonconducting particles ($\lambda = \frac{1}{2}$), and increases in value as b and ζ increase in magnitude. This calculation confirms in detail the validity of the neglect of the inertial term.

4.3 The Relaxation Effect[5]

In the theory of conductance of solutions of simple electrolytes it is found that the contributions to the retarding force due to the electrophoretic and relaxation effects are comparable. Having established that inertial effects are negligible, we now turn to a study of the effect of double layer relaxation on the cataphoretic velocity. As in the preceding section we shall follow in detail the analysis of Booth[5] which, incidentally, also removes the approximation that the potential is given by the linearized Poisson–Boltzmann equation.

As before we shall make use of the assumptions that the inertial term in the equation of motion is negligible, that the dielectric constant D, viscosity η, and ionic mobilities ω_i, are uniform throughout the fluid phase, that the velocity of the polyion is proportional to the electric field strength, and that the electrolyte is incompressible. For simplicity it is further assumed that the electrolyte contains equal numbers of positive and negative ions and that the polyion is nonconducting. To fully utilize the boundary conditions at the surface of the polyion it is necessary to know some of the properties of the surface of the polyion. Booth assumes that the thickness of the surface phase is small relative to the polyion diameter, that the surface charge is immobile, and that the surface charge densities in the presence and absence of a field, v_X and v_0 are related by

$$v_X(\mathbf{s}) = -\frac{3XD_0}{8\pi}\cos\theta + v_0(\mathbf{s}) \qquad (4.3.1)$$

with D_0 the dielectric constant of the solid and θ the angle between the radius vector to $\mathbf{s}$ and the field direction. The first term on the right-hand side of Eq. (4.3.1) represents a small charge due to the difference between the conductivities of solid and liquid phases. Let the coordinate system be fixed with respect to the polyion and the motion of the fluid be steady so that the equation of motion is given by Eq. (4.1.4). If ω_i is the mobility of an ion, the mean velocity of ions of type i is

$$\langle \mathbf{u}_i \rangle = \mathbf{v} - \omega_i\left(z_i q\nabla\psi + \frac{kT}{c_i}\nabla c_i\right) \qquad (4.3.2)$$

where c_i is the concentration of the ions. The equation of continuity applied to the ith ionic species gives

$$\nabla\cdot(c_i\langle\mathbf{u}_i\rangle) = 0 \qquad (4.3.3)$$

and Eqs. (4.1.4), (4.3.2), and (4.3.3) provide, together with the boundary conditions, a soluble set of equations. Booth adopted the following iterative procedure for the solution of this set of equations. Equations (4.3.2) and (4.3.3) are solved with $\mathbf{v} = 0$ and $\mathbf{X} = 0$. This solution for ψ, denoted ψ_1, is the potential near the surface when the particle is at rest and no field is acting upon it. The potential ψ_1 is then substituted into (4.1.4) giving a solution $\mathbf{v}_1$ for the velocity and by adjusting the velocity at infinity such that the total force on the sphere vanishes, a first estimate $\mathbf{U}_1$ of the macroion velocity is obtained. The process is repeated by substitution of $\mathbf{v}_1$ for $\mathbf{v}$ in (4.3.2) and (4.3.3.), and so forth.

In the first approximation we have

$$c_{1,i} = c_i^0 \exp(-qz_i\psi_1/kT) \tag{4.3.4}$$

with c_i^0 the bulk concentration of ion i, and $c_{1,i}$ denotes the first approximation to the ionic concentration c_i. Poisson's equation reads

$$\nabla^2\psi_1 = -\frac{4\pi q}{D}\sum_1^r c_i z_i \exp(-qZ_i\psi_1/kT) \tag{4.3.5}$$

with boundary conditions

$$\psi_1 = 0; \qquad r \to \infty$$
$$\frac{d\psi_1}{dr} = -\frac{Zq}{Db^2}; \qquad r = b \tag{4.3.6}$$

where r is the distance from the center of the sphere whose radius is b, and Z is the number of charges on the sphere. Expansion of the exponential on the right-hand side of Eq. (4.3.5) and solution of the equation for the potential in power series form gives

$$\psi_1(\kappa r) = \sum_{v=1}^{\infty} \frac{q^{2v-1}}{D^v b^v}(-kT)^{1-v}Z^v\lambda_v(\kappa r) \tag{4.3.7}$$

and where the functions λ_v are relatively easy to calculate. The first four are:

$$\lambda_1(\kappa r) = \frac{\kappa b}{1+\kappa b}\frac{e^{\kappa(b-r)}}{\kappa r}$$

$$\lambda_2(\kappa r) = 0$$

$$\lambda_3(\kappa r) = -\frac{\hat{q}_3}{6}\left(\frac{\kappa b}{1+\kappa b}\right)^3\frac{e^{3\kappa b}}{\kappa r}[\hat{\alpha}(\kappa b)e^{-\kappa r}$$
$$-2\,e^{\kappa r}Ei(4\kappa r) + e^{-\kappa r}Ei(2\kappa r)] \tag{4.3.8}$$

with the functions $\hat{q}_m$ and $\hat{\alpha}$ defined by

$$\hat{q}_m = \frac{\sum_1^r c_i z_i^{m+1}}{\sum_1^r c_i z_i^2}$$

$$\hat{\alpha}(\kappa b) = \frac{e^{-\kappa b}}{1+\kappa b} - Ei(2\kappa b) + 2\left(\frac{1-\kappa b}{1+\kappa b}\right)e^{2\kappa b}Ei(4\kappa b). \tag{4.3.9}$$

In this first approximation the potential due to the surface charge and the surrounding electrolyte has spherical symmetry. Under these circumstances Henry's results can be used to calculate U_1, the first approximation to the cataphoretic velocity. From the second of Eqs. (4.1.44) specialized to the case when the polyion is nonconducting

$$\lambda = \frac{\sigma - \sigma_0}{2\sigma + \sigma_0} = \frac{1}{2}$$

$$U_1 = \frac{DX}{6\pi\eta}\left[\frac{3}{2}\,\psi_1(\kappa b) + 3(\kappa b)^3\int_{\kappa b}^{\infty}\frac{\psi_1(\kappa r)}{(\kappa r)^4}\,d(\kappa r)\right.$$

$$\left. - \frac{15}{2}(\kappa b)^5\int_{\kappa b}^{\infty}\frac{\psi_1(\kappa r)}{(\kappa r)^6}\,d(\kappa r)\right]. \qquad (4.3.10)$$

If we now write

$$U_\mu = \sum_{\nu=1}^{\infty} m_{\mu,\nu}Z^\nu = \sum_{\nu=1}^{\infty} d_{\mu,\nu}\zeta^\nu \qquad (4.3.11)$$

then the substitution of Eqs. (4.3.7) and (4.3.8) into (4.3.10) and (4.3.11) leads to the following results for the expansion coefficients:

$$m_{1,1} = \frac{Xq}{6\pi\eta b}\,\hat{X}_1(\kappa b) \qquad (4.3.12)$$

$$\hat{X}_1(\kappa b) = \frac{1}{(1+\kappa b)}\left[1 + \frac{(\kappa b)^2}{16} - \frac{(5b\kappa)^3}{48} - \frac{(\kappa b)^4}{96} - \frac{(\kappa b)^5}{96}\right.$$

$$\left. + \frac{(\kappa b)^4}{8}\,e^{\kappa b}Ei(\kappa b)\left(1 - \frac{(\kappa b)^2}{12}\right)\right] \qquad (4.3.13)$$

$$m_{1,2} = 0 \qquad (4.3.14)$$

$$m_{1,3} = \frac{Xq^5}{6\pi\eta b^3(DkT)^2}\,\hat{q}_3\hat{X}_3(\kappa b) \qquad (4.3.15)$$

$$\hat{X}_3(\kappa b) = -\frac{1}{6}\left(\frac{\kappa b}{1+\kappa b}\right)^3\left[\frac{\hat{\alpha}(\kappa b)\hat{X}_1(\kappa b)(1+\kappa b)}{\kappa b}\right.$$

$$+ e^{4\kappa b}Ei(4\kappa b)\left(-\frac{2}{\kappa b} - \frac{\kappa b}{8} - \frac{5(\kappa b)^2}{24} + \frac{(\kappa b)^3}{48} + \frac{(\kappa b)^4}{48}\right)$$

$$- \frac{45}{32}(\kappa b)^3\,e^{3\kappa b}Ei(3\kappa b)\left(1 - \frac{79}{300}(\kappa b)^2\right)$$

$$+ e^{2\kappa b}Ei(2\kappa b)\left(\frac{1}{\kappa b} + \frac{\kappa b}{16} - \frac{5(\kappa b)^2}{48} - \frac{(\kappa b)^3}{96} + \frac{(\kappa b)^4}{96}\right)$$

$$+ \frac{1}{8}(\kappa b)^3\left(1 - \frac{(\kappa b)^2}{12}\right)\left(\int_{\kappa b}^{\infty}\frac{e^{-\kappa r}Ei(2\kappa r)}{\kappa r}\,d(\kappa r)\right.$$

$$\left. - 2\int_{\kappa b}^{\infty}\frac{e^{\kappa r}Ei(4\kappa r)}{\kappa r}\,d(\kappa r)\right)$$

$$- \frac{1}{48(\kappa b)} + \frac{3}{80} - \frac{9\kappa b}{320} + \frac{427}{960}(\kappa b)^2$$

$$+ \frac{59}{1920}(\kappa b)^3 - \frac{79}{640}(\kappa b)^4\right] \qquad (4.3.16)$$

$$m_{1,4} = 0. \tag{4.3.17}$$

Following the general method suggested, the first approximation to the fluid velocity is used in Eqs. (4.3.2) and (4.3.3) to obtain the second approximation ψ_2 to the potential. The second approximation to the potential is then used to calculate v_2 and U_2. The calculations are long and tedious but reasonably straightforward. The same types of considerations that arose in Section 4.1 must also be used in this more complicated case. Booth has carried the calculations through to the second approximation corresponding to the dependence of the cataphoretic velocity on the fourth power of ζ. The reader is referred to the original paper for details and we here merely quote the result

$$U_2 = U_1 - \frac{Dz}{18\pi\eta} \int_b^\infty \int_r^\infty \left(\frac{d\psi_1(\kappa z)}{dz} \Delta_1^{(z)} \Phi(z) \right.$$

$$\left. - \Phi(z) \frac{d(\Delta_0^{(z)} \psi_1(\kappa z))}{dz} \right) \left(\frac{z^2}{r^2} - \frac{r}{z} \right) dr \, dz \tag{4.3.18}$$

$$\Phi = \frac{q^4 Z^2}{b(DkT)^2} \lambda_{2,1} + \frac{q^6 Z^3}{b^2(DkT)^3} \lambda_{3,1} \tag{4.3.19}$$

$$\Delta_n^{(z)} = \frac{d^2}{dz^2} + \frac{2}{z} \frac{d}{dz} - \frac{n(n+1)}{z^2} \tag{4.3.20}$$

and $\Delta_1^{(x)}\lambda_{2,1}$, and $\Delta_1^{(x)}\lambda_{3,1}$, are determined by

$$(\Delta_1^{(x)} - 1)\lambda_{2,1}(x) = -\hat{q}_3 \left[F_2(\kappa b, x) + \frac{(\kappa b)^3}{2x^2} \left(F_2'(\kappa b, \kappa b) \right. \right.$$

$$\left. \left. - \frac{1}{2} F_1'(\kappa b, \kappa b) - \frac{1}{\kappa b} F_1(\kappa b, \kappa b) \right) \right]$$

$$- \hat{q}_3^* \left[F_3(\kappa b, x) + \frac{(\kappa b)^3}{2x^2} F_3'(\kappa b, \kappa b) \right] \tag{4.3.21}$$

$$(\Delta_1^{(x)} - 1)\lambda_{3,1} = -\hat{q}_4^* \left[F_6(\kappa b, x) + \frac{(\kappa b)^3}{2x^2} \left(F_6'(\kappa b, \kappa b) \right. \right.$$

$$\left. \left. + \left(\frac{d\lambda_1}{dx} \right)_{\kappa b = x} \left\{ F_3(\kappa b, \kappa b) + \frac{\kappa b}{2} F_3'(\kappa b, \kappa b) \right\} \right) \right] \tag{4.3.22}$$

$$\hat{q}_m^* = DkT \frac{\displaystyle\sum_{i=1}^{s} c_i \omega_i^{-1} z_i^{m-1}}{q^2 \pi\eta \displaystyle\sum_{i=1}^{s} c_i z_i^2} \tag{4.3.23}$$

where the F_i are complicated functions of r and b. The function F_1 is quoted directly in Booth's paper, but the $F_i(i > 1)$ must be worked out from the analysis sketched. As usual F' is the derivative of F with respect to x and is to be evaluated at the point $x = \kappa b$.

We are now ready to analyze the results of this complex analysis. The general formula for the cataphoretic velocity may be written

$$U = \frac{XZq}{6\pi\eta b}\hat{X}_1(\kappa b) + \frac{X}{6\pi\eta}\sum_{\gamma=2}^{\infty}\frac{Z^\nu q^{2\nu-1}}{b^\nu(DkT)^{\nu-1}}[\hat{X}_\nu^\dagger + \hat{Y}_\nu^\dagger + \hat{Z}_\nu^\dagger] \quad (4.3.24)$$

or alternatively as a power series in ζ. The first term on the right-hand side of Eq. (4.3.24) represents the polyion velocity when the two approximations are made that the linearized Poisson–Boltzmann equation is adequate and moreover that the potential is spherically symmetric. The terms $\hat{X}_\nu^\dagger$, $\hat{Y}_\nu^\dagger$, and $\hat{Z}_\nu^\dagger$ represent corrections due to relaxation of the double layer and to deviations from the simple screened Coulomb potential. In particular, $\hat{X}_\nu^\dagger$ represents a contribution to the electrophoretic velocity due to the symmetrical part of the field around the polyion arising from the higher order terms in the expanded Poisson–Boltzmann equation which are neglected in the linearized approximation. Both $\hat{Y}_\nu^\dagger$ and $\hat{Z}_\nu^\dagger$ arise from the distortion of the double layer about the polyion, with $\hat{Y}_\nu^\dagger$ independent of the ionic mobilities. The functions $\hat{Z}_\nu^\dagger$ depend upon the reciprocals of the ionic mobilities and are therefore greater the smaller the ionic mobilities. In general the importance of all the correction terms relative to the leading terms increases as the ionic valencies increase.

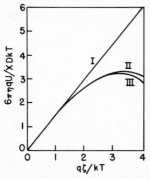

FIG. 4.3.1. The electrophoretic mobility as a function of poly-ion charge. I: Henry's theory. II: Booth's theory neglecting terms of order ζ^4. III: Booth's theory neglecting terms of order ζ^5. [Reproduced from ref. 5.]

The magnitude of the correction to Henry's formula depends upon many factors, the surface potential, ionic mobilities and concentrations, etc. If Z is taken as a constant independent of κ, the correction is a maximum when $\kappa b \approx 1$, whereas if ζ is assumed independent of κ, the maximum occurs when $\kappa b \approx 6$. As can be seen in Fig. 4.3.1, the correction is quite large when $q\zeta/kT > 2$.

By substitution of numerical values it is found that the corrections are appreciable in water at 25°C when ζ is 100 millivolts. Since zeta potentials often exceed this figure the correction terms are of great importance in the analysis of experimental data of systems containing rigid particles.

REFERENCES

1. J. T. G. OVERBEEK, *in* "Advances in Colloid Science," Vol. III, p. 97. Interscience, New York, 1950.
2. M. BIER, " Electrophoresis." Academic Press, New York, (1959.
3. H. HELMHOLTZ, *Ann. Physik* **7,** 337 (1879).
4. M. SMOLUCHOWSKI, *Bull. acad. sci. Cracovie,* p. 182 (1903).
5. F. BOOTH, *Proc. Roy. Soc.* **A203,** 514 (1950).
6. F. BOOTH, *Proc. Roy. Soc.* **A203,** 533 (1950).
7. F. BOOTH, *J. Chem. Phys.* **18,** 1361 (1950).
8. F. BOOTH, *J. Chem. Phys.* **22,** 1956 (1954).
9. D. C. HENRY, *Proc. Roy. Soc.* **A133,** 106 (1931).
10. S. GOLDSTEIN, *Proc. Roy. Soc.* **A123,** 216 (1929).

5. Theoretical Aspects of Chain Configuration and Counterion Distribution in Solutions of Flexible Chain Polyelectrolytes*

HERBERT MORAWETZ †

5.0 Introduction

The rapid growth of polymer chemistry over the past three decades has been due in large part to a deepening awareness of the relationship between the physical characteristics of polymers and the structure of chain molecules. It became apparent at an early stage of the development that such molecular properties as chain length, chain stiffness, degree of branching, the number of polar groups, and the regularity in their spacing are more significant than the detailed chemical make-up of the polymers. The clarification of the relation between such various features of " molecular architecture " and the physical characteristics of polymers led to the synthesis of polymers which not only duplicated, but often improved upon the useful properties of such natural materials as *Hevea* rubber, cellulosic fibers, and silk fibroins.

The distinction, in classical chemistry, between uncharged molecules and electrolytes composed of ions has its counterpart in the field of high polymers. In fact, the attachment of large numbers of ionized groups to the backbone of a molecular chain produces such striking modifications in the polymer characteristics that the contributions of other properties of the chain to the properties of polyelectrolytes have often been neglected. Many natural products or their derivatives such as gelatin, alginic acid, and various de-esterified pectins are polyelectrolytes. They find wide industrial application and a detailed understanding of polyelectrolyte behavior will, therefore, be essential to the design of synthetic polyelectrolytes with similar useful characteristics.

On the other hand, the question has been raised whether the analogy between natural polymers of very high complexity and their synthetic analogs,

* Financial assistance received from the National Science Foundation Grant 2271–NSF–7503 during the writing of this chapter is gratefully acknowledged.
† Dept. of Chemistry, Polytechnic Institute of Brooklyn, Brooklyn N.Y.

whose behavior might lend itself to an easier analysis, could not serve to clarify problems of biological interest. Staudinger pointed out[1] that synthetic polymers carrying ionizable groups could be considered models of proteins or nucleic acids and this suggestion gave the first impetus to the study of synthetic polyelectrolytes. The flexibility of such chain molecules was not appreciated at that time and it was therefore not realized that the analogy with globular proteins (and also with nucleic acids which were shown much later to consist of very " stiff " particles) is extremely tenuous.

The detailed study of the physical chemistry of dilute polymer solutions soon led to the realization that chain molecules are more or less coiled, their extension in the solvent medium being dependent on the energetic interaction of the polymer segments with the solvent molecules. Whenever an uncharged polymer could be converted—by titration or by chemical modification—to a chain molecule carrying large numbers of ionized groups attached to the polymer " backbone," chain extensions could be arrived at, which were far beyond the range attainable by changes of solvent. But the ionic charges fixed to the polymer chain do not only affect the configuration of the macro-molecule. They also create a high local charge density, which must strongly affect the properties of the simple ions present in the solution. Thus, the study of flexible polyelectrolytes is concerned with two types of problems. On the one hand we wish to know how the density of ionized groups along the polymer chain (and other chain properties) affects the extension of the polymeric coil and any physical properties dependent on chain extension. On the other hand we are concerned with the electrochemistry of polyelectrolyte solutions, with the effect of the polyion on ionic activity coefficients, ion pair formation, electrophoretic phenomena, etc. The first problem is new to the flexible polyelectrolyte; the second involves modifications of treatments in the preceding chapters for rigid ions. It is clear that both types of problems cannot be strictly separated. The interaction of the polyion with simple ions and the resulting ionic distribution will modify the repulsion between the fixed polymer charges and thus alter the extension of the macromolecule. Conversely, not only the charge but also the shape of the polyion will deter-mine its effect on the properties of the simple electrolyte. It is this inter-dependence which is responsible for the complexities—and the fascination—of this field.

In the following discussions, it will be necessary to consider the behavior of uncharged polymer chains before the modifications due to the ionic charges of polyelectrolytes can be taken up. It seemed inappropriate to repeat here a detailed treatment of the physical chemistry of polymer solutions and we are, therefore, giving only its broad outline, referring the reader to the excellent monographs by Mark and Tobolsky,[2] Flory,[3] Stuart,[4] and Tompa[5] for more information.

5.I   The Molecular Configuration of Flexible Polyelectrolyte Chains

5.1.1 CONFIGURATION OF UNCHARGED POLYMER CHAINS

When we represent n-butane by the formula

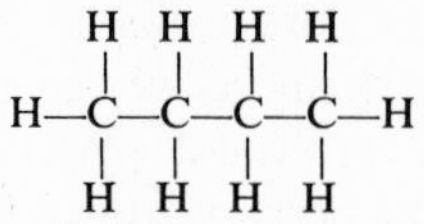

$$
\begin{array}{ccccc}
& H & H & H & H \\
& | & | & | & | \\
H- & C- & C- & C- & C- H \\
& | & | & | & | \\
& H & H & H & H
\end{array}
$$

we are quite aware that this schematic picture is not intended to represent the actual distribution of the atoms in the butane molecule. We actually possess fairly detailed information about the structure of such molecules: data obtained from X-ray diffraction fix the bond distances at about 1.54 A for the C—C and 1.09 A for the C—H bonds. In addition, the four carbon valences are known to be distributed symmetrically in space, so that any two of them form a tetrahedral angle of about 109.5°. The frequencies of absorption lines in the infrared may be interpreted in terms of the forces resisting the elongation of bonds or the distortion of bond angles. Information on the latter point may also be obtained by a comparison of the heats of combustion of normal paraffins with those of cycloparaffins in which the bond angles are different from their normal values. When we consider the position in space of the substituents of two carbon atoms joined to each other, we meet with restrictions in addition to the bond angle requirements. These arise due to the fact that even in such a simple molecule as ethane rotation of the methyl groups is not entirely free. This " hindered rotation " effect was first deduced from the " anomalous " heat capacity of ethane vapor[6] and was originally thought to be a consequence of the mutual repulsion of the hydrogen atoms. The situation is visualized most conveniently by the schematic representation introduced by Newman[7] and shown in Fig. 5.1.1, in which the two carbons

(a) Staggered Conformation   (b) Eclipsed Conformation

FIG. 5.1.1. The conformations of ethane.

of ethane are symbolized by circles and the bond between them lies in the line of sight. The extremes in the relative positions of the hydrogens are then given by the " conformation " a, in which they occupy staggered positions and are at a maximum distance from each other and b, where the hydrogens lie behind each other ("eclipsed conformation") resulting in the greatest mutual interference. The energy barrier for rotation of the C—C bond is about 3000 cal/mole[6] corresponding at ordinary temperatures to approximately 10^{11} jumps per second from one energy minimum to the next one. On the other hand, the energy barrier is sufficiently high to render the staggered conformation a hundred times more probable than the eclipsed one. Calculations seem to indicate that, in ethane as well as in other molecules, only about half of the energy barrier is due to van der Waals repulsion of the nonbonded atoms,[8] the remainder resulting perhaps from a bond orientation effect caused by quadrupole and higher multipole moments of the bonding electrons.[9]

In the case of the butane molecule, two stable conformations may be distinguished as illustrated in Fig. 5.1.2, since the two skew structures are

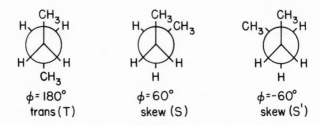

FIG. 5.1.2. Conformations of n–butane.

equivalent. It has been estimated that the skew conformation has a potential energy 800 cal/mole above that of the trans conformation,[10, 11] which would make, at ordinary temperatures, two-thirds of the butane molecules trans.

When we pass on to consider the conformations of pentane, we must characterize not only the relative spatial distribution of the substituents of carbon 2 and 3, but also of carbons 3 and 4. We have then, in addition to the conformation TT, TS, and ST, two different conformations consisting of a sequence of two skew arrangements, namely, SS and SS'. For a long hydrocarbon chain, the number of stable conformations—which we shall henceforth call configurations in agreement with general usage in this field—becomes extremely large and can be treated only statistically. It is obvious that it is not practicable to specify the location of each of the carbon atoms in the chain and it becomes necessary, therefore, to describe the chain configuration by some parameter which has an easily visualized physical significance.

The most frequently adopted choice is the end-to-end distance of the chain h, which is characterized by a probability distribution function $W(h)$ defined so that $\int_0^\infty W(h)\, dh = 1$. For many purposes the significant quantity is the mean square displacement $\langle h^2 \rangle$:

$$\langle h^2 \rangle = \int_0^\infty h^2 W(h)\, dh. \tag{5.1.1.}$$

Alternatively, the chain extension is characterized by the mean square radius of gyration $\langle R_G^2 \rangle$ defined by:

$$\langle R_G^2 \rangle = \frac{1}{n} \sum_1^n \langle r_i^2 \rangle \tag{5.1.2}$$

where n is the number of chain elements and $\langle r_i^2 \rangle$ is the mean square distance of the ith element from the center of gravity of the chain.

In theoretical treatments of chain configurations it is convenient to start with an idealized model in which the chain consists of links represented by mathematical lines of zero volume and all angles in space between successive links are equally probable. We shall denote quantities relating to the root mean square dimensions of such chains by the subscript 0. If the length of the link is b, the probability distribution of h has been shown by Kuhn and Grün[12] to be given by:

$$W_0(h)\, dh = \left(\frac{3}{2\pi nb^2}\right)^{3/2} \left[\exp\left(-\frac{1}{b}\right) \int_0^h L^*\, dh\right] 4\pi h^2\, dh \tag{5.1.3}$$

where $h_{\max} = nb$ is the maximum chain extension and L^*, the inverse Langevin function of the argument $(h/h_{\max})$, is defined by

$$L^* \left[\coth (h/h_{\max}) - h_{\max}/h\right] = h/h_{\max}. \tag{5.1.4}$$

If the number of links is large, the overwhelming majority of possible configurations will have end-to-end separations h much smaller than $h_{\max}$. For $h \ll h_{\max}$, relation (5.1.4) reduces to

$$W_0(h)\, dh = \left(\frac{3}{2\pi nb^2}\right)^{3/2} \exp\left(\frac{-3h^2}{2nb^2}\right) 4\pi h^2\, dh. \tag{5.1.5}$$

This gives for the mean square end-to-end displacement $\langle h_0^2 \rangle$ and for the mean square radius of gyration $\langle R_{0G}^2 \rangle$ of random chains

$$\langle h_0^2 \rangle = nb^2 \tag{5.1.6}$$

$$\langle R_{0G}^2 \rangle = \tfrac{1}{6} nb^2. \tag{5.1.7}$$

The effect of fixed bond angles θ on the extension of polymer chains was studied by Wall[13] who concluded that for large numbers of links and bond angles not too close to 180° (specifically, $n(1 + \cos \theta) \gg 1$) the mean square chain end-to-end separation may be approximated by[5]

$$\langle h_0^2 \rangle = nb^2 \frac{(1 - \cos\theta)}{(1 + \cos\theta)}. \tag{5.1.8}$$

An additional complication arises due to hindered rotation around the links of the chain. The potential energy of any sequence of four carbon atoms, such as those represented in Fig. 5.1.2, will be a function of the angle ϕ and since the extended trans conformation represents, in general, a lower energy state, the deviation from randomness in the distribution of ϕ values will lead to an increase in $\langle h_0^2 \rangle$. Benoit has shown[14] that for cases where the trans conformation is only slightly favored

$$\langle h_0^2 \rangle = nb^2 \frac{(1 - \cos \theta)}{(1 + \cos \theta)} \times \frac{(1 - \langle \cos \phi \rangle)}{(1 + \langle \cos \phi \rangle)} \qquad (5.1.9)$$

and Benoit and Sadron[15] have analyzed critically the errors introduced by the approximations used in deriving this relation.

However, in the fully extended chain shown in Fig. 5.1.3, the length per link is $b \sin (\theta/2)$ and we can thus see that the extensions of chains with fixed bond angles cannot be represented by the distribution function of Eq. (5.1.3). For instance, the requirement that there be a fixed tetrahedral bond angle between successive links will increase $\langle h_0^2 \rangle$ by a factor whose exact value depends on $\langle \cos \phi \rangle$, but will reduce h_{max} by a factor of $(\frac{2}{3})^{1/2}$.

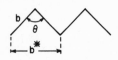

FIG. 5.1.3. Section of carbon chain in all-trans conformation.

There are additional restrictions on the chain configuration due to the volume occupied by groups attached to the carbon skeleton. With bulky chain substituents, the spatial interference of substituents attached to neighboring monomer units may become so pronounced that it will allow only a very small fraction of the configurations which would be possible if all of the three conformations illustrated in Fig. 5.1.2 were accessible to every link of the polymer backbone.

This subject has received a great deal of attention in recent years, particularly due to the pioneering investigations of Volkenshtein, Ptitsyn and their collaborators, whose results have recently been summarized by Flory.[15a] Other important contributions to this problem are due to Lifson,[15b, c] Nagai,[15d, e] and Fordham.[15f] It is clear that the spatial interference of substituents of a chain of type $(-CH_2-CHR-)_n$ restricts the number of accessible conformations which are no longer independent of one another. Neighboring conformations have been found to be governed by the following rules:[15a]

1. A sequence of two skew conformations with an opposite direction of rotation of the bond angle (SS') results in severe spatial interference even in polyethylene and it is, therefore, also very improbable in polymers of vinyl derivatives.

2. An isotactic chain (i.e., one in which all the asymmetric carbons have the same configuration) favors the alternation of trans and skew conformations. A segment of such a chain with two R substituents is represented by

and it may be shown that in a regularly alternating TS sequence the chain backbone generates a helix with three monomer units per turn. In solutions of isotactic polymers, " kinks " are introduced into the chain by an occasional presence of a TT or SS pair, so that the polymer conformation becomes something like —$TSTSTSTSTSSTSTSTTSTSTS$—.

3. With syndiotactic polymers (in which asymmetric carbons have regularly alternating configurations), the low energy binary conformational sequences are TT and SS represented by

syndiotactic, TT syndiotactic, SS

However, a sequence of more than two S conformations is excluded, so that a typical conformation of the polymer chain backbone becomes .. $TTTTTTSSTTTT$ ...

It can be shown that in polyethylene even more distant than nearest neighbor conformations must be governed by restrictive correlations[15g] and such must, therefore, also be the case in vinylic chains. With polymers of the $(—CH_2—CR_1R_2—)_n$ type the steric interference of chain substituents is so severe that the bond angles become severely distorted[15h] and the potential energy minima corresponding to $\phi = \pm 60°$ and $\phi = 180°$ may be entirely obliterated. For example, in polyisobutene the bond angle is expanded to 114° and the ϕ angle corresponding to the lowest energy of the chain seems to be 98°.[15a, h]

At present we have little knowledge of the relative energy associated with various sequences of conformations in real polymer chains. But even if such knowledge were available, a rigorous treatment would be prohibitively difficult. The problem is treated most conveniently by the concept of an " equivalent chain " introduced by Kuhn.[16] In this model, a group of s successive monomers stretched to their maximum extension are welded into a " statistical chain element " of length $A = sb^*$, where b^* is the component of the length of the monomer unit in the direction of maximum chain extension, and N such elements are freely jointed without bond angle restrictions.

The values of A and N are chosen so that the real and the equivalent chain have the same maximum extension and the same mean end-to-end displacements, i.e.,

$$AN = h_{max} \tag{5.1.10}$$

$$A^2N = \langle h_0^2 \rangle. \tag{5.1.11}$$

For $h \ll h_{max}$, the distribution function of the end-to-end displacement of the equivalent chain is given by analogy with (5.1.5) as:

$$W_0(h) \, dh = \left(\frac{3}{2\pi NA^2}\right)^{3/2} \exp\left(\frac{-3h^2}{2NA^2}\right) 4\pi h^2 \, dh \tag{5.1.12}$$

and this treatment has been shown to account properly for the effect of variations in the chain stiffness on $W_0(h)$ in the range of small chain extensions.[17] It has also been assumed[18] that for highly extended chains $W_0(h)$ is given in analogy with (5.1.3) by

$$W_0(h) \, dh = \left(\frac{3}{2\pi NA^2}\right)^{3/2} \exp\left[-\left(\frac{1}{A}\right)\int_0^h L^* \, dh\right] 4\pi h^2 \, dh \tag{5.1.13}$$

but the applicability of the equivalent chain concept in this range appears to be much less certain.

From (5.1.10) and (5.1.11)

$$N = h_{max}^2 / \langle h_0^2 \rangle \tag{5.1.14}$$

and since $h_{max} = b^*Z$ and $N = Z/s$ where $Z = n/2$ is the degree of polymerization,

$$s = Z\langle h_0^2 \rangle / h_{max}^2. \tag{5.1.15}$$

The value of $\langle h_0^2 \rangle$ may be obtained from light scattering data and it is usually assumed that h_{max} has the value:

$$h_{max} = 2b \sin (\theta/2) Z \tag{5.1.16}$$

corresponding to the staggered configuration represented by Fig. 5.1.3. Combining (5.1.9), (5.1.15), and (5.1.16) and substituting $\sin^2(\theta/2) = (1 - \cos \theta)/2$,

$$s = \frac{1}{1 + \cos \theta} \frac{1 - \langle \cos \phi \rangle}{1 + \langle \cos \phi \rangle}. \tag{5.1.17}$$

For the tetrahedral bond angle $\cos \theta = -\frac{1}{3}$ and $\langle \cos \phi \rangle$ may be calculated if the dependence of potential energy on the angle ϕ is known. Taylor [19] has estimated for a polymethylene chain the values $\langle \cos \phi \rangle = -\frac{1}{2}$ at 0°C and $-\frac{3}{7}$ at 50°C, leading to $s = 4.5$ and $s = 3.75$ at these two temperatures. These should represent minimum values of s, which would be increased by all chain substituents. It has, however, been pointed out above that in consequence of steric interference of chain substituents the all-trans conformation of the chain backbone as represented in Fig. 5.1.3 is possible only for syndiotactic vinylic chains. For a perfectly isotactic chain with moderately bulky substituents, the maximum extension corresponds to the helix generated by a regular alternation of trans and skew conformations, which shortens b^* to 2.06 A from the value of 2.53 A characteristic of the all-trans chains. The h_{max} values are altered, of course, in the same proportion. With more bulky chain substituents, the bond angles may be distorted to lead to a further chain contraction. Assuming that the configurations of polymer chains in a crystal lattice correspond to maximum chain extension, we obtain a value as low as $b^* = 1.71$ A for isotactic poly(3-methyl butene-1).[20] As for polymers of the $(-CH_2-CR_1R_2)$ type, we may cite as typical examples $b^* = 2.33$ A[15h] for polyisobutene and $b^* = 2.10$ A for isotactic poly(methyl methacrylate).[20a] For atactic polymers (in which the sequence of d and l configurations of the asymmetric carbons in the chain backbone is random) the b^* and h_{max} values should be intermediate between those expected for isotactic and syndiotactic chains.

If the chain extension is distorted from its most probable value to a new arbitrary value h, the retractive force f will be given by

$$f = \partial A/\partial h \qquad (5.1.18)$$

where A, the free energy of the chain is given by $A = -TS$, if chain extension does not alter the volume and the internal energy is independent of chain configuration.† Since the configurational contribution to the entropy of the chain is

$$S_{conf} = k \ln W(h) \qquad (5.1.19)$$

we have, by using the probability distribution for low chain extensions as given by Eqs. (5.1.5) and (5.1.6),

$$f = \frac{kT}{\langle h_0^2 \rangle^{1/2}} \left(\frac{3h}{\langle h_0^2 \rangle^{1/2}} - \frac{2\langle h_0^2 \rangle^{1/2}}{h} \right). \qquad (5.1.20)$$

For a given ratio of chain extension over the rms extension of the relaxed random chain, the retracting force is here inversely proportional to $\langle h_0^2 \rangle^{1/2}$ (or

†Strictly speaking, any chain extension will change bond conformations and will, therefore, affect the internal energy if the potential energy of the trans and the skew conformations are significantly different. This effect was first treated both theoretically and experimentally by Volkenstein and Ptitsyn[21] and more recently it has been shown by Flory, Hoeve, and Ciferri[22] that the temperature dependence of the modulus of elasticity of polyethylene behaves as would be expected from the lower potential energy of trans conformations.

the square root of the number of statistical chain elements). For high degrees
of extension, for which the distribution function (5.1.13) is applicable,

$$f = kT \left(\frac{1}{A} L^* - \frac{2}{h} \right) = kT \left(\frac{3h\lambda}{\langle h_0^2 \rangle} - \frac{2}{h} \right) \tag{5.1.21}$$

where the correction factor $\lambda = h_{max} L^*/3h$ may be closely approximated[23] by

$$\lambda = 1 + \frac{3h^2}{s(h_{max}^2 - h^2)} \tag{5.1.22}$$

so that $\lambda \to 1$ for $h \ll h_{max}$ and $\lambda \to \infty$ as $h \to h_{max}$. For extensions approaching the maximum value, the retractive force is found by (5.1.20) to be inversely
proportional to the length of the statistical chain element (characterizing
chain stiffness) and proportional to a universal function of h/h_{max}. It is
instructive to evaluate f for a typical case: with $A = 10^{-7}$ cm, $N = 100$, the
retractive force is 4×10^{-7} dyne when the chain extension is half its maximum value. It may be noted that such a force would stretch a carbon–carbon
bond by only 0.01 % and that the extension due to bond angle distortion would
be only slightly higher. Thus, we are fully justified in neglecting changes in bond
lengths and valence angles in computing the retractive force in stretched
polymer chains.

It will be noted that our discussion of chain configurations started with a
model in which the volume of the chain elements was neglected and that we
considered later the effects of the volumes of chain substituents only insofar
as they led to restrictions in the configuration of neighboring segments and an
increase in the stiffness of a chain. However, the equivalent chain is still
treated as a mathematical line which may return to within an arbitrarily small
distance of a point occupied by a previous section. If we consider this line to
be the axis of a real chain characterized by a finite thickness, then many of the
configurations considered previously will correspond to an overlap of volumes
assigned to different chain segments and they will, therefore, have to be
excluded. It is obvious, that the proportion of such excluded configurations
will be highest if the chain is relatively tightly coiled and that it will decrease
with increasing chain extension. As a consequence, this long-range spatial
interference, referred to commonly as the " excluded volume effect," will
tend to shift the probability distribution function towards higher values of h.

Let us now consider the effect produced by the nature of the solvent
medium on the configuration of a dissolved polymer chain. If heat is evolved
in the dissolution of the polymer, configurations in which polymer segments
are in contact with each other will represent states of higher energy. If the
increase in energy is ΔE_c for the formation of a contact between two solvated
polymer segments, then the probabilities of configurations with c such contacts are reduced by $\exp(-c \, \Delta E_c/kT)$. The net effect is equivalent to an

increase in the excluded volume which, however, becomes temperature-dependent. For very high values of ΔE_c, no contacts between polymer segments are allowed, and the chain behaves as if a sheath with half the thickness of a monomolecular layer of solvent were an integral part of it. The extension of such a chain will have the highest value which can be attained with uncharged polymers in which energetic interactions do not extend beyond nearest neighbors.

If a polymer dissolves in a solvent with the absorption of heat, configurations in which chain segments are in contact with each other represent states of lower energy and the increased probabilities of such configurations counteract, to some extent, the excluded volume effect. The situation is analogous to that of a van der Waals gas, where the expansion due to the finite volume of the molecules is counteracted by the attractive forces between them. The equation of state of such a gas is (written to terms of the order of the second virial coefficient)

$$pv = RT + p\left(b - \frac{a}{RT}\right) + \cdots \qquad (5.1.23)$$

where b is the excluded volume and a the parameter characterizing intermolecular forces. At the Boyle temperature $b - (a/RT) = 0$, and the ideal gas laws are obeyed over a considerable pressure range. Similarly, conditions may be realized with solutions of polymer chains when the excluded volume effect is canceled by the excess of polymer-polymer over polymer-solvent interaction. Under such circumstances each polymer chain acts independently of the presence of other chains (i.e., the ideal solution laws are obeyed) and the chain configurations are unaffected by long-range interference (i.e., the equivalent random chain treatment is applicable). The close relation between the departure from ideal solution behavior and perturbations of the random chain configuration was first pointed out by Flory and Krigbaum[24] who introduced the terms "Θ solvent" or "Θ temperature" to describe solutions in which the second osmotic virial coefficient vanishes and the mean square chain end displacement is proportional to the square root of the chain length as required by (5.1.11). This correlation has since been checked with excellent results on a large number of polymer-solvent systems.

For solvents which interact with the polymer segments more strongly than the Θ solvent the chain will be more highly extended. Using the notation

$$\langle h^2 \rangle^{1/2} = \alpha_E \langle h_0^2 \rangle^{1/2} \qquad (5.1.24)$$

where α_E is the factor by which all linear chain dimensions are expanded over their corresponding values in a Θ solvent, it is found[25] that

$$\alpha_E^3 - 1 = \text{const} \times B_2 \times N^{1/2} \qquad (5.1.25)$$

where B_2 is the osmotic second virial coefficient. Thus, the chain end displacement increases in all but Θ solvents more rapidly than the square root of the chain length and $\langle h^2 \rangle/N$ increases without limit.

5.1.2 THEORIES OF CONFIGURATION OF FLEXIBLE POLYIONS

We have seen that the configurations of chain molecules in solution may be affected to a considerable extent by the variation in the enthalpy of the system with the unfolding of the polymer coil. When the polymer is uncharged, such changes in energy may be considered to be entirely due to nearest neighbor interactions. We are then only concerned with the question of how the number of contacts between polymer segments depends on the extension of the molecule.

When ionic charges are attached to the macromolecules, the forces between the charges are very much larger than the dispersion forces or the dipole-dipole interactions acting between uncharged groups. Moreover, these forces act over relatively large distances so that estimates of the potential energy E of the chain no longer merely involve information about the number of contacts between polymer segments, but require an evaluation of the Coulombic interaction between all pairs of charges attached to the polymer.

From the theoretical point of view, the simplest case is that of a freely jointed chain (or, the Kuhn equivalent chain discussed in the previous section) carrying $v = \alpha Z$ charges of the magnitude q at a constant spacing along its length, while all counterions are far removed from the region occupied by the polyion. Kuhn, Künzle, and Katchalsky[18] have investigated this model, believing that it is a reasonable approximation to conditions existing in dilute polyelectrolyte solutions. It will be shown later that, for any experimentally accessible system, the polyion holds a considerable proportion of the counterions at distances which are small compared to the chain extension and it may, therefore, be questioned whether the calculations of Kuhn et al. should be compared with experimental results. Nevertheless, this model is of great theoretical interest, and can probably be dealt with more precisely than models in which the counterion distribution has to be taken into account.

Qualitatively, we can see that an expansion of the polymer chain will tend to increase the distance between the fixed charges and thus decrease the electrostatic potential energy E_{el}. If it is assumed that E_{el} is a function of h, then the probability distribution of the chain end displacements is obtained by multiplying the a priori probability given by (5.1.12) or (5.1.13) by the appropriate Boltzmann exponential

$$W(h)\, dh = W_0(h) \exp(-E_{el}/kT)\, dh. \qquad (5.1.26)$$

The new distribution of h may then be characterized by its most probable value h^* obtained from $d\,W(h)/dh = 0$, or

$$kT\left(\frac{d\ln W_0}{dh}\right)_{h=h^*} = \left(\frac{dE_{el}}{dh}\right)_{h=h^*}. \tag{5.1.27}$$

Alternatively, the mean square chain end displacement of the charged chain may be evaluated from

$$\langle h^2 \rangle = \int h^2 W(h)\,dh/\int W(h)\,dh. \tag{5.1.28}$$

For the unperturbed random chain $(h^*)^2 = (\frac{2}{3})\langle h^2 \rangle$, but as the chain is extended, the probability distribution of h becomes increasingly sharper and $(h^*)^2/\langle h^2 \rangle$ approaches unity.

Kuhn, Künzle, and Katchalsky[18] use a model in which the ionic charges of the chain are distributed in equal amounts at the junction points of consecutive chain elements and the ends of the chain. With s monomer units in a statistical element and a fraction α of the monomers carrying the electronic charge q, the charge at each junction becomes $s\alpha qN/(N+1) = \xi$. The effective dielectric constant governing charge interactions is assumed to be adequately approximated by the dielectric constant of water D_w. The electrostatic potential energy is then given by

$$E_{el} = \sum_i \sum_j \frac{\xi_i \xi_j}{D_w r_{ij}} \tag{5.1.29}$$

where r_{ij} is the distance between the ith and the jth charge. Since there are $N^2/2$ interacting pairs of charges, $\xi_i \xi_j/D_w$ is constant, $N\xi = vq$ and for long chains unity is negligible in comparison with N,

$$E_{el} = \frac{v^2 q^2}{2D_w}\left\langle\frac{1}{r}\right\rangle_h \tag{5.1.30}$$

where $\langle 1/r \rangle_h$ is the mean value of the reciprocal distance of two charges selected at random from all chain configurations with the end-to-end distance h held constant. Using the distribution function (5.1.12) for the relation between the distance of two junction points and the number of chain segments separating them, Katchalsky, Künzle, and Kuhn obtain[23]

$$E_{el} = -\frac{v^2 q^2}{D_w h}\int_0^\infty \frac{j}{(j+1)^2}\cdot I\sqrt{\frac{3h^2}{2j\langle h_0^2 \rangle}}\,dj$$

$$I(x) = \frac{2}{\pi}\int_0^x e^{-n^2}\,dn \tag{5.1.31}$$

which reduces for large values of $h/\langle h_0^2 \rangle^{1/2}$ to

$$E_{el} = (v^2 q^2/D_w h)\ln(3h^2/2\langle h_0^2 \rangle). \tag{5.1.32}$$

The introduction of these values of E_{el} into (5.1.26) or (5.1.27) leads to extremely high extensions of the polyion. For instance, for a polymerization degree of 400, with four monomer units forming a statistical segment of 10 A length, and at 60% ionization, the polyion stripped of counter-ions would be stretched to 95% of its maximum extension.

We shall see later that even in dilute polyelectrolyte solutions containing no added salt a considerable fraction of counter-ions are held in the proximity of the polyion. The presence of these counter-ions partially neutralizes the effect of the fixed charges and thus reduces the expansion of the chain. The forces between the ionized groups of the polymer are further sharply reduced if salts are added to the solutions, raising still higher the concentration of counter-ions in the vicinity of the polyion. All existing theories of polyelectrolyte configuration try to treat the problem of the screened Coulomb potential by assuming that the electrostatic potential in the neighborhood of a charge carried by the polyion varies with distance in a manner similar to that prescribed by the Debye-Hückel theory for potentials in the neighborhood of the ions of simple salts, i.e., as $\exp(-\kappa r)/r$. In the theory of Katchalsky and Lifson[26] which treats polyelectrolyte solutions in which all small ions are univalent, κ is defined by

$$\kappa^2 = \frac{4\pi q^2 \sum c_i^0}{DkT} \tag{5.1.33}$$

where $\sum c_i^0$ is the concentration of all the free ions per ml of solution. Each fixed charge is treated as surrounded by a counter-ion atmosphere which remains unchanged during chain expansion, so that the only electrical work involved in changes of the chain configuration is due to the interactions of the fixed charges with each other. The electrostatic energy due to interaction of the ith and jth charge is then given by

$$E_{ij} = q^2 \exp(-\kappa r_{ij})/D_w r_{ij} \tag{5.1.34}$$

and summing over all pairs of charges, the total electrostatic energy due to interactions between the fixed charges of the chain is

$$E_{el} = \left(\frac{v^2 q^2}{2D_w}\right)\left\langle\frac{\exp(-\kappa r)}{r}\right\rangle. \tag{5.1.35}$$

The evaluation of the average magnitude of $\exp(-\kappa r)/r$ for all pairs of charges and all configurations consistent with a fixed end-to-end distance was performed by Katchalsky and Lifson using the probability distribution of random chain configurations. The result is

$$E_{el} = \left(\frac{v^2 q^2}{D_w h}\right) \ln\left[1 + \left(\frac{6h}{\kappa\langle h_0^2\rangle}\right)\right]. \tag{5.1.36}$$

The most probable chain end displacement was then obtained by equating the electrostatic expanding force dE_{el}/dh to the elastic retractive force from relation (5.1.21), giving

$$(h^*)^2/\langle h_0^2 \rangle - \frac{2}{3} = \frac{v^2 q^2}{3D_w kTh^*} \left[\ln \left(1 + \frac{6h^*}{\kappa \langle h_0^2 \rangle} \right) - \frac{6h^*/\kappa \langle h_0^2 \rangle}{1 + 6h^*/\kappa \langle h_0^2 \rangle} \right].$$

$$(5.1.37)$$

Katchalsky and Lifson are careful to point out the complications inherent in the definition of $\langle h_0^2 \rangle$. The length of the statistical chain element describing the behavior of the partially ionized polymer is not necessarily equal to that of the corresponding polymer in the uncharged state, since solvation of both uncharged and charged groups may alter appreciably the effective chain stiffness. It is assumed that the length of a statistical chain segment is a linear function of the degree of ionization so that $\langle h_0^2 \rangle$ would also be linear in α. This procedure introduces two adjustable parameters which cannot be checked by any independent measurement and the theory has been criticized on that account.[27] However, there seems to be little doubt that the intrinsic stiffness of polymer chains must be altered when the constituent monomer units are ionized, not only because of solvation effects, but also because of the effect the electrical charges must have on the relative probabilities of the trans and the skew conformations.[35]† Neglecting this effect, which we shall discuss later in more detail, may lead to a theory in which the adjustable parameters are eliminated, but this will be accomplished only by ignoring one of the physical realities.

The derivation of (5.1.36) and (5.1.37) assumes that $h\kappa > 1$, i.e., that the chain extension is larger than the mean thickness of the counter-ion atmosphere. This requires that for aqueous solutions of univalent ions at ordinary temperatures the free ion concentrations exceed 0.002 and 0.001 equivalents per liter for chain extensions of 100 A and 400 A, respectively. Expansion of the bracketed term in (5.1.37) leads to

$$\frac{(h^*)^2}{\langle h_0^2 \rangle} - \frac{2}{3} = \frac{v^2 q^2}{3D_w kTh^*} \left[\frac{1}{2} \left(\frac{6h^*}{\kappa \langle h_0^2 \rangle} \right)^2 - \frac{2}{3} \left(\frac{6h^*}{\kappa \langle h_0^2 \rangle} \right)^3 + \cdots \right] \qquad (5.1.38)$$

and for dilute polyelectrolyte solutions containing uni-univalent electrolyte of concentration $c_s^0 = \frac{1}{2} \sum c_i^0$ we obtain for a chain of sufficient length so that $6h^*/\kappa \langle h_0^2 \rangle \ll 1$ the simple relation

$$(h^*/\langle h_0^2 \rangle^{1/2}) - (2\langle h_0^2 \rangle^{1/2}/3h^*) = 3v^2/4\pi \langle h_0^2 \rangle^{3/2} c_s^0. \qquad (5.1.39)$$

†Actually, the variation of the polyion solvation with the degree of neutralization seems to depend in a complicated manner on the concentration of simple electrolyte, since Pinner found[28] poly(methacrylic acid) to be salted out of solution most easily at pH 4, corresponding to about 10% ionization.

Since $\langle h_0^2 \rangle$ is proportional to N, this result implies that, for $h/\langle h_0^2 \rangle^{1/2}$ sufficiently large to render the second term on the left negligible, at constant charge density (v/N) along the chain and at constant salt concentration, the chain end displacement becomes proportional to N, i.e., the actual chain end displacement would be a constant fraction of the maximum chain extension. Such a conclusion is certainly not plausible for solutions containing sufficient salt to reduce the range of charge interactions to distances much smaller than the extension of the polymer molecule.

The theory of Kuhn, Künzle, and Katchalsky[18] contains the assumption that the probability distribution of chain configurations for any given chain end displacement is independent of the degree of ionization. In the theory of Katchalsky and Lifson[26] this position is modified considerably by the gradual lengthening of the statistical chain element with increasing ionization from 9 to 24 monomer units for un-ionized and ionized poly(methacrylic acid), respectively, but Harris and Rice[29] have pointed out that configurations of different energy are still given equal statistical weight in the computation of $\langle \exp(-\kappa r)/r \rangle$ in Eq. (5.1.35). Since configurations of lower energy will be occupied preferentially, the actual energy of the system will be less than calculated. The point may be illustrated by considering n_t/n_s, the ratio in the number of doubly ionized succinate ions which exist in the trans and the skew conformations. This will be

$$(n_t/n_s) = (n_t^0/n_s^0) \exp[- q^2(1/r_t - 1/r_s)/DkT] \qquad (5.1.40)$$

where (n_t^0/n_s^0) would be the ratio of the two conformations if all factors other than Coulombic charge interaction were taken into account and r_t, r_s are the distances between the charges corresponding to the two conformations. Thus, the mean distance between the ionized carboxyls becomes larger than it would be in the absence of the charges and the electrostatic energy $E_{el} = (q^2/D)\langle 1/r \rangle$ is smaller than would be calculated using the relative probability of the two conformations to be expected in an uncharged molecule.

The error resulting from neglecting the effect of charge interactions on the distribution of molecular configurations at any given chain end displacement concerns, however, not only the energy, but also the configurational entropy. As the interacting charges reduce the accessibility of more and more configurations, the entropy of the chain at any fixed end-to-end displacement will be progressively reduced. We see, therefore, that in calculating the free energy of the chain, part of the error in the estimate of its energy will be canceled by the error in the estimated entropy. (Hildebrand and Scott[30] have previously drawn attention to the general occurrence of such cancellations of errors in the estimate of free energies from approximate models.) We are, however, not concerned with the absolute value of the free energy but only with its variation as the chain is being extended at constant charge density. Obviously, the fraction of configurations for a fixed chain end displacement, in which different chain segments come to lie close to each other and which represent states of high energy if the chain is charged, decreases with increasing chain extension. We must conclude, therefore, that the electrostatic energy is being overestimated more strongly for small chain extensions and that $\partial E_{el}/\partial h$ is in fact less negative than calculated by Katchalsky and Lifson.[26] The error in $\partial A/\partial h$ would be expected to be smaller but in the same direction, so that the equilibrium chain expansion would be overestimated by the theory.

The elimination of this error in the treatment of the model of Katchalsky and Lifson would meet with formidable mathematical difficulties. This led Rice and Harris[31] to a different approach to the problem of polyion configuration. They consider an equivalent chain with the ionic charges carried by the monomer units concentrated at the midpoints of each statistical chain element. The extension of the chain may then be represented as a function of the average angle γ formed between adjoining chain elements. In the absence of charges, $\langle \cos \gamma \rangle = 0$, but on ionization the average angle between the chain elements will grow beyond $90°$ and $\langle \cos \gamma \rangle$ will assume increasingly negative values. The factor by which the mean square chain end displacement of the polyion is expanded over the value characterizing the uncharged chain is then given in analogy with (5.1.8) by

$$\langle h^2 \rangle / \langle h_0^2 \rangle = \frac{1 - \langle \cos \gamma \rangle}{1 + \langle \cos \gamma \rangle}. \tag{5.1.41}$$

As a first approximation, $\langle \cos \gamma \rangle$ is calculated considering only $E_{i,i+1}$, the potential energy due to the repulsion of charges carried by adjoining chain elements. For any assumed dependence of $E_{i,i+1}$ on charge separation, $\langle \cos \gamma \rangle$ may then be rigorously obtained from

$$\langle \cos \gamma \rangle = \frac{\int \cos \gamma \, \exp[-E_{i,i+j}(\gamma)/kT] \, d\Omega}{\int \exp[-E_{i,i+1}(\gamma)/kT] \, d\Omega} \tag{5.1.42}$$

where $d\Omega = 2\pi \sin \gamma \, d\gamma$ is the element of solid angle. For $E_{i,i+1}$ given by the screened Coulomb potential and with a distance $A \sin (\gamma/2)$ between the midpoints of adjoining chain elements,

$$E_{i,i+1} = Q^2 \, \exp[-\kappa A \sin (\gamma/2)]/D_E A \sin (\gamma/2) \tag{5.1.43}$$

where Q is the charge carried by the statistical chain element. The definition of Q differs significantly from $s\alpha q$ used by Katchalsky and Lifson,[26] since the possibility of ion-pair formation between the fixed charges and their counterions is being considered, but a discussion of this factor will be deferred to a later section. Harris and Rice[29] point out that the proper value to be used for the effective dielectric constant D_E when dealing with the electrostatic interactions of charges spaced relatively closely along the polymer chain should be much lower the dielectric constant of water since a large proportion of the lines of force will run through the nonpolar regions of the polymer chain.

In this they follow the reasoning of the theory formulated by Kirkwood and Westheimer[32] to account for the ratio of the dissociation constants of dibasic acids. The value $D_E = 5.5$ used by Harris and Rice was taken from Hasted *et al.*[33] who used it to characterize water molecules not free to orient in an electric field (i.e., in an alternating field of infinite frequency). This estimate was adopted for the interpretation of polyelectrolyte behavior, since a single water molecule was believed to find space between adjoining charges of, e.g., highly ionized poly-(methacrylic acid) and such water molecules would not be free to orient coopera-tively in the manner characteristic of the behavior of water in bulk. However, estimates of D_E obtained from the ratio of the ionization constants of a dibasic acid, which would seem to be much more closely related to polyelectrolyte be-havior, lead to much higher values. Westheimer and Shookhoff[34] found for glu-taric acid $D_E = 72$, only slightly below the dielectric constant of water. As would be expected, D_E dropped off sharply on alkyl substitution of glutaric acid, but even for β, β, diethylglutaric acid (which may represent a reasonable model of a short section of, e.g., poly(acrylic acid)) D_E retained a value of 15.

Actually, the results obtained by the Rice and Harris theory for $\langle \cos \gamma \rangle$ and $\langle h^2 \rangle / \langle h_0^2 \rangle$ are rather insensitive to the assumed value of the effective dielectric constant, since in the relation (5.1.43) both numerator and de-nominator increase with D_E, and the electrostatic potential energy $E_{i,i+1}$ goes through a flat maximum when the value for the dielectric constant is chosen so as to make $1/\kappa$ equal to half the charge separation. For example, if statistical chain segments of 10 A length carry one ionic charge and the polyion is immersed in $0.04M$ uni-univalent electrolyte, the calculated $\langle h^2 \rangle / \langle h_0^2 \rangle$ will increase only from 2.76 to 3.42 as D_E is lowered from 15 to 5.5.*

The choice of a proper value for the effective dielectric constant involves, nevertheless, some serious difficulties. Low values of D_E imply that the charges are rather deeply buried within a relatively nonpolar region[32] and under these circumstances the potential would be affected by the concen-tration of added electrolyte much less than assumed in (5.1.43). Moreover, the model used by Rice and Harris artificially fixes the distance over which neighboring charges interact to a value close to the length of a statistical chain element. In reality, the spacing of interacting charges depends on the degree of ionization of the polymer, and with increasing charge separation D_E should rapidly approach the macroscopic dielectric constant of the medium.

It will also be noted that according to (5.1.41) the expansion factor $\langle h^2 \rangle / \langle h_0^2 \rangle$ due to the charging of the polyion to any given charge density (v/N) is independent of chain length, i.e., the charged polymer is still assumed to behave as a random chain with an end-to-end separation increasing as $N^{1/2}$. This conclusion is certainly erroneous, since the charged polyion finds itself in a medium which is very far from Flory's "Θ solvent." In fact, it is clear that for chains carrying ionic charges the excluded volume effect must be particu-

---

* Harris and Rice used the value $D_E = 80$ to compute $\langle \cos \gamma \rangle$ (see p. 330). This discussion refers to possible other choices for D_E.

larly large, since energetic interactions do not involve only polymer segments in direct contact with each other, as may be assumed for solutions of uncharged polymers, but extend over distances large compared with the size of solvent molecules. Thus we see that the theories of Katchalsky and Lifson[26] on the one hand and of Rice and Harris[31] on the other err in opposite directions, the first one overestimating and the second underestimating the dependence of the polyion extension on the chain length.

In contrast to Katchalsky and Lifson who assume that the total electrostatic energy is available for expansion, Rice and Harris assume that the interaction of charges placed on neighboring monomer units can have no significant effect on chain expansion. Both conclusions are now known to be incorrect, as can be demonstrated by considering the effect of the electrostatic repulsion of neighboring charged groups on the relative frequency of conformational sequences in a material such as poly(acrylic acid). A section of this polymer, isotactic or syndiotactic, is shown in Fig. 5.1.4 in the trans-trans (TT), trans-skew (TS) and skew-skew (SS) conformations. (As stated previously, the SS' sequence corresponds to such serious spatial interference that it need not be considered). It is obvious on inspection of Fig. 5.1.4 that the most extended TT conformation forces the ionized groups into closest proximity to each other and that the mutual electrostatic repulsions of neighboring ionized groups will tend to favor conformational sequences corresponding to a more contracted shape of the chain. Lifson[35] has carried out a general analysis of neighboring group interactions as they affect the correlation of successive conformations in a polyelectrolyte chain and he concluded that the contraction due to neighboring group effects leads to a reduction of $\langle h_0^2 \rangle$ with increasing ionization, the effect being more pronounced with syndiotactic polymers. Long range electrostatic interactions leading to chain expansion are then superimposed on this effect.

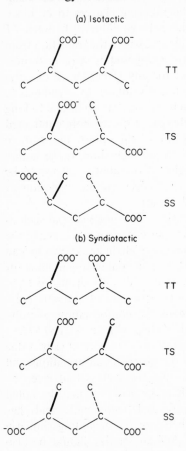

FIG. 5.1.4. Conformations of a segment of ionized polyacrylic acid.

While Lifson treated the problem in a completely general way, it may be noted that the results may be modified by considering simultaneously the steric interference and the electrostatic interaction of the chain substituents. In particular, it is known that polymers of isotactic vinyl derivatives exist mainly in TS sequences, since both TT and SS lead to serious spatial interference of chain substituents.[15a,b,c] The effect of neighboring charges on the chain conformation should, therefore, be relatively small. Conversely, syndiotactic polymers favor both TT and SS, i.e., the most expanded and the most contracted form of the chain segment. Thus we should expect the chain contracting effect due to the repulsion of neighboring charges to be much more pronounced in syndiotactic chains, where the equilibrium between TT and SS is easily perturbed, than in isotactic chains, where the transition from the predominant TS to the contracted SS form is opposed by steric restraints. However, it has been shown previously that for any given degree of polymerization of a vinylic chain, h_{max} is significantly longer for syndiotactic than for isotactic chains. Thus, the syndiotactic chain, may pass, due to long range electrostatic repulsion of the fixed charges, from a more contracted to a more expanded form than the isotactic chain. Since electrostatic effects due to nearest neighbor interaction are of such short range nature that they can probably not be shielded by electrolyte addition, it would be expected that salt addition will have a larger effect on the extension of syndiotactic than that of isotactic poly(acrylic acid).

An analysis of the nearest neighbor interaction effect in a material such as poly(methacrylic acid) is much more complex, since steric interference of the two substituents on every second carbon atom of the chain are known to lead to considerable bond angle distortions[15h] and the potential energy minima for rotation of the C—C—C bond angle may be considerably displaced from their normal values. Assuming that the crystal structure of isotactic poly-(methyl methacrylate)[20a] gives a valid clue to the conformation of poly-(methacrylic acid) in its fully extended state (with a $..TSTSTS..$ conformation), it would appear that ϕ is shifted from its normal skew value of $60°$ to $36°$. Without more knowledge of the possible conformations and their correlations in chains of this type an analysis such as that given by Lifson cannot be carried out for poly(methacrylic acid). It may be noted that predominantly isotactic and syndiotactic poly(methacrylic acid) has recently been shown to differ in titration behavior,[35a] suggesting similarly significant differences in chain expansion. Similar studies should be also carried out to compare the behavior of isotactic poly(acrylic acid)[35b] with that of the atactic polymer.

At the same time that polyelectrolyte theories were being formulated in which the chainlike character of the polyion was taken explicitly into account, the problem was also investigated with the use of models in which the polyion was represented by a spherically symmetrical charge cloud. If the polymer coil is immersed in a large volume of uni-univalent salt, the local charge density at any point within the region occupied by the polymer will be

$$\rho = q[c_f + c_s^0 e^{-q\psi/kT} - c_s^0 e^{+q\psi/kT}]$$

where the three terms on the right represent concentrations of charges fixed to the polymer backbone, byions and counterions, respectively. The Poisson–Boltzmann equation then has the form

$$\nabla^2\psi = \frac{4\pi q}{D}[-c_f + 2c_s^0 \sinh(q\psi/kT)]. \tag{5.1.44}$$

Relation (5.1.44) was first utilized by Hermans and Overbeek[36] to interpret the expansion of polymer chains during ionization. They considered the case when $q\psi/kT \ll 1$, so that $\sinh(q\psi/kT) \approx q\psi/kT$. The fixed charges were treated as a continuous distribution within the volume of a sphere of radius R_s, which is hydrodynamically equivalent to the polymer coil. The results obtained were found to be quite insensitive to the nature of the charge distribution, being almost identical when it was assumed constant (i.e., $c_f = 3v/4\pi R_s^3$) or when it was given by a Gaussian distribution. Equation (5.1.44) could then be solved for the potential distribution and the electrical energy of the coil was approximated by

$$E_{el} = \frac{3}{5}\frac{v^2 q^2}{DR_s[1 + 0.6\kappa R_s + 0.4\kappa^2 R_s^2]}. \tag{5.1.45}$$

Hermans and Overbeek used for R_s the value obtained by Debye[37] for a sphere which is hydrodynamically equivalent to a free draining coil. In that case, R_s is related to the mean square radius of gyration by $R_s^2 = \frac{5}{3}\langle R_G^2\rangle$. Using random chain statistics, the dependence of R_s on the end-to-end extension of the chain was shown to be given by

$$R_s^2 = (5/36)(\langle h_0^2\rangle + \langle h^2\rangle) \tag{5.1.46}$$

and substitution of (5.1.45) and (5.1.46) into (5.1.26) and (5.1.28) led to

$$y^3\frac{y^2 - 2}{y^2 - 1} = \frac{6v^2 q^2}{5\sqrt{5}\langle h_0^2\rangle^{1/2}DkT}\frac{1 + 1.2\kappa R_s + 1.2\kappa^2 R_s^2}{(1 + 0.6\kappa R_s + 0.4\kappa^2 R_s^2)^2} \tag{5.1.47}$$

$$y^2 = (\langle h^2\rangle/\langle h_0^2\rangle) + 1.$$

The significance of (5.1.47) is not easy to visualize. However, in recent years a great deal of evidence has indicated that the hydrodynamic behavior of polymer coils may be represented as equivalent to that of spheres whose radii are proportional to the chain end displacement. Thus, we may set in (5.1.46) $R_s = ah^*$ and substituting this relation into (5.1.27) we obtain

$$(h^*)^2/\langle h_0^2 \rangle - \tfrac{2}{3} = \frac{v^2 q^2}{DkTah^*} \frac{1 + 1.2\kappa a h^* + 1.2\kappa^2 a^2 (h^*)^2}{5(1 + 0.6\kappa a h^* + 0.4\kappa^2 a^2 (h^*)^2)^2}. \quad (5.1.48)$$

For the usual case, where $\kappa a h^* \gg 1$, (5.1.48) simplifies to

$$(h^*)^2/\langle h_0^2 \rangle - \tfrac{2}{3} = v^2/4V_s c_s^0 = (v/4)(c_f/c_s^0) \qquad (5.1.49)$$

where V_s is the volume of the equivalent sphere.

The theory of Hermans and Overbeek is limited in its application by the requirement that potentials within the region occupied by the polyion be sufficiently low to justify the use of the Debye–Hückel approximation. Kimball, Cutler, and Samelson have suggested that smaller errors will, in general, be introduced if the Donnan assumption of electroneutrality within the polymer coil is used.[38] For $\rho = 0$, Eq. (5.1.44) reduces to

$$\sinh(q\psi/kT) = c_f/2c_s^0 \qquad (5.1.50)$$

which may be solved to yield the distribution of the electrostatic potential and the electrostatic energy of the coil, provided we know the distribution of the fixed charge density.

Kimball et al. also point out that the distribution of chain segment concentrations in unperturbed random chains is of the form

$$c(r) = \text{const} \exp(-3r^2/2h^2) \qquad (5.1.51)$$

and that this distribution is equivalent to the Boltzmann distribution in a field in which the potential is given by $U(r) = 3kTr^2/2h^2$. In their procedure, the electrostatic potential energy $q\psi(r)$ of a fixed charge is added to the " configurational potential " $U(r)$ to obtain a second relation between c_f and ψ. The theory predicts that the ionization of groups attached to a polymer chain will expand the end-to-end distance by no more than a factor of $\sqrt{2}$—a result which is certainly much lower than the experimentally observed effects. The cause of the error was traced by Flory[39] and by Lifson[40] to the fact that the distribution of each polymer segment was treated as if it were independent of the location of other segments. This results in a vast overestimate of the entropy of polymer expansion reducing the relative importance of the electrostatic effects.

Although the theory of Kimball, Cutler, and Samelson led to faulty conclusions with respect to the expansion of charged chain molecules, their suggestion to treat the conditions within the swollen polyion in terms of the Donnan equilibrium has been most fruitful. It is instructive to see to what extent the result of Hermans and Overbeek for a uniform fixed charge density within an equivalent sphere will be altered, if the Donnan assumption is used instead of the Debye–Hückel approximation. The electrostatic energy of a coil carrying v fiixed charges may then be obtained from (5.1.50) as[40]

$$E_{el} = vkT \int_0^1 \sinh^{-1}(c_f\lambda/2c_s^0) \, d\lambda$$

$$= vkT[\sinh^{-1}(c_f/2c_s^0) - \sqrt{1 + (2c_s^0/c_f)^2} + (2c_s^0/c_f)] \quad (5.1.52)$$

and after substituting $c_f = 3v/4\pi(ah^*)^3$ we may, as previously, equate the electrostatic expanding force dE_{el}/dh to the elastic contracting force $kT \, d \ln W_0/dh$. This leads to

$$(h^*)^2/\langle h_0^2 \rangle - \tfrac{2}{3} = 2c_s^0 V_s(\sqrt{1 + (v/2c_s^0 V_s)^2} - 1). \quad (5.1.53)$$

As long as $v/2c_s^0 V_s \ll 1$, i.e., when the number of fixed ions on the polymer chain is small compared to the salt ions which would be contained in the volume of the equivalent sphere in the absence of the polymer, (5.1.53) reduces to (5.1.49). Thus we see that the Donnan and the Debye–Hückel approximations lead to identical results in that limiting case. As the concentration of fixed charges within the polymer coil becomes comparable to the ion concentration in the surrounding medium, the expansion predicted by (5.1.53) becomes considerably smaller than that calculated from (5.1.49). At high ratios of $v/2c_s^0 V_s$ (5.1.53) would imply that the chain expansion is independent of salt concentration, but the Donnan assumption is obviously inapplicable in that case.

Flory[39] has pointed out that, whenever the Donnan assumption is applicable, the swelling of a polyelectrolyte coil may be predicted without introducing the concept of electrostatic potential. This is physically obvious when we consider instead of the change of shape of an isolated molecule, the swelling of a macroscopic gel for which the conditions of electroneutrality must hold much more precisely. The driving force for the swelling process may then be assigned with equal justification, to osmotic pressure or to the electrostatic repulsion of the charges fixed to the gel structure. Either point of view, if properly applied, must lead to the same prediction of the swelling equilibrium.

The calculation of the expansion of polyelectrolyte molecules from osmotic considerations may be illustrated by a simple example. We shall represent the polyelectrolyte again by a sphere with a constant charge density, immersed in a large volume of uni-univalent electrolyte of concentration c_s^0. If we neglect the volume occupied by the polymer, the sphere will contain V_s/v_1 molecules of solvent. Neglecting deviations from ideal solution behavior, the difference in the free energy of the solvent molecules outside and inside the sphere is given by

$$A_0 - A_i = (V_s/v_1)kT\{\ln (1 - 2v_1c_s^0) - \ln [1 - v_1(2c_s^0 + \Delta c)]\} \quad (5.1.54)$$

where Δc is the excess of total ion concentration inside the sphere. It can easily be shown that the Donnan condition requires for systems of polyions and univalent ions

$$\Delta c = 2c_s^0[\sqrt{(c_f/2c_s^0)^2 + 1} - 1] \tag{5.1.55a}$$

and that for $c_f \ll 2c_s^0$ (5.1.54) and (5.1.55a) yield $A_0 - A_i = kTvc_f/4c_s^0$, which is identical with the value obtained for E_{el} from (5.1.52) when the fixed charge concentration within the sphere is small compared with the concentration of the mobile ions.

Flory's treatment of the problem[39] is considerably more elaborate in that it takes into account the Gaussian distribution of the fixed charges within the polymer coil. The charge density is treated as uniform over the volumes of concentric spherical shells with the mobile ions distributed so as to fulfill the Donnan condition in each shell. Mobile ions of all valence types are taken into account, which requires (5.1.55a) to be replaced by

$$\Delta c = (c_f^2/\sum c_i^0 z_i^2)[1/2 - (v_+ - v_-)c_f/3v_- z_- c_s^0(v_+ + v_-) + \cdots] \tag{5.1.55b}$$

where v_i and z_i are the numbers and valences of the ions constituting the simple electrolyte, the subscripts $+$ and $-$ referring to byions and counterions of the polyion. Assuming that all linear dimensions are expanded by the same factor and equating the osmotic expanding force due both to polymer segments and mobile ions to the elastic retractive force of the chain, Flory obtains the result

$$\langle h^2 \rangle / \langle h_0^2 \rangle = \langle h_1^2 \rangle / \langle h_0^2 \rangle + v \left\{ \frac{v}{1.16 \langle h^2 \rangle^{3/2} \sum c_i^0 z_i^2} + (z_+ - z_-) \times \right.$$
$$\left. \left(\frac{v}{0.81 \langle h^2 \rangle^{3/2} \sum c_i^0 z_i^2} \right)^2 + \cdots \right\} \tag{5.1.56}$$

where $\langle h_0^2 \rangle$ and $\langle h_1^2 \rangle$ correspond to the mean square chain end separations of the uncharged polymer in a Θ solvent and the given solvent, respectively, while $\langle h^2 \rangle$ refers to the polymer carrying v ionic charges.

If the medium is a " Θ solvent" for the uncharged chain, the first term on the right vanishes and we obtain, for low charge densities, a result equivalent to (5.1.49) or (5.1.53), the effective volume of the coil being replaced by $0.44 \langle h^2 \rangle^{3/2}$. However, Flory's result gives us an interesting insight into the manner in which swelling due to the excluded volume effect of the uncharged polymer is superimposed on the Donnan effect. By rearranging (5.1.56) we obtain

$$(\langle h^2 \rangle / \langle h_1^2 \rangle) - 1 = (\langle h_0^2 \rangle / \langle h_1^2 \rangle) v \left\{ \frac{v}{1.16 \langle h^2 \rangle^{3/2} \sum c_i^0 z_i^2} + \cdots \right\} \tag{5.1.57}$$

showing that the relative expansion of the polymer due to the Donnan effect, or the dependence of polyelectrolyte dimensions on the concentration of added salts is inversely proportional to the expansion of the uncharged polymer dimensions in a given medium over those in a Θ solvent. This factor has usually not been taken into account in interpreting experimental results; yet the high polarity of ionizable polymers would lead one to suspect that their aqueous solutions are generally very far from the Θ point, even when the chain carries no charge.

Another interesting aspect of the Flory theory is the light it sheds on the effect of the valences of the mobile ions on polyelectrolyte behavior to be expected if the assumptions underlying the treatment are valid. Inspection of (5.1.55) shows that for low values of c_f/c_s^0 the excess of ion concentration in the region occupied by the polyion is a function of ionic strength, e.g., for di-univalent electrolyte of a given concentration, the same result should be obtained whether the divalent ion bears a charge of the same or the opposite sign as the polymer. Only at higher charge densities, where the second term on the right becomes significant, would the Donnan treatment predict a difference of polyelectrolyte response to divalent counterions and divalent byions. This conclusion of the theory is certainly at variance with experimental indications that the properties of polyelectrolyte solutions are much more sensitive to the valence of counterions than of byions.

We may now compare the conclusions to be reached from theories in which the charges carried by a polyelectrolyte are smeared out to continuous charge densities with conclusions resulting from models which retain specifically the chain character of the polyion in estimating the electrostatic expanding force.

In principle, theories considering the detailed chain configuration of the backbone to which the charges of the polyion are attached should have the advantage of allowing a treatment of the chain expansion from its initial highly coiled state up to its full extension, while theories dealing with continuous charge distributions are restricted to the expansion range in which the distribution of chain segments remains spherically symmetrical. However, this advantage may be more apparent than real since, as was pointed out earlier, we have no certain knowledge of the length of the fully extended polymer chain. Also, the probability distribution of configurations approaching full extension is liable to be in error due to the fact that changes of chain extension in this range certainly involve, even for uncharged chains, an appreciable change of potential energy.[21]

When models with continuous charge distributions are used, the square of the chain expansion ratio is found to be linear in the reciprocal electrolyte concentration, while the model of Katchalsky and Lifson[26] leads, for very long chains and at fairly high electrolyte content, to a linear dependence of $h*/\langle h_0^2 \rangle^{1/2}$ on $1/c_s^0$. The theory of Hermans and Overbeek[36] as modified in Eq. (5.1.48) predicts that, for a given charge density of the polyion and at low

c_f/c_s^0 ratios, the chain end displacement is proportional to the 3/5 power of the chain length.† In Flory's treatment,[39] this conclusion corresponds only to the case where the medium is a Θ solvent for the polymer in the discharged state—for better solvent media, a more rapid increase of $\langle h^2 \rangle^{1/2}$ with N is predicted. These predictions seem physically much more plausible than those of Katchalsky and Lifson[26] or Harris and Rice,[27] which, as was pointed out earlier, predict a first power and half-power dependence of $\langle h^2 \rangle^{1/2}$ on N.

On the other hand, the treatment of a polyelectrolyte coil by a continuous charge distribution has some obvious limitations. If we use Flory's model[39] and consider a very long polymer chain, we find that the average density of polymer segments decreases rapidly to extremely low values at large distances from the center of the coil. By smearing out the polymer charges over the volume of a spherical shell, the Donnan effect providing the driving force for further expansion assumes very low values for these outlying portions of the coil. Yet, since the charges are constrained to positions of close proximity along the polymer chain, it is clear that there will be a lower limit to the expansion force no matter how thinly the chain may be distributed over a given volume. It is also essential to consider the narrow spacing of ionic charges along the polymer backbone if we want to arrive at a proper appreciation of the high values of the electrostatic potential in the immediate neighborhood of the chain. The resulting high attraction for counterions may be described in terms of reduced activity coefficients or as ion-localization—in either case the driving force towards coil expansion will be substantially reduced. But all of the theoretical estimates of polyelectrolyte expansion which we have discussed are either independent or only slightly dependent on the dielectric constant of the medium, reflecting our present inability to account properly for ionic activity coefficients or ion association.

5.2   The Counterion Distribution

5.2.1 LONG-RANGE ELECTROSTATIC INTERACTIONS

The preceding discussion showed that the extent to which a flexible chain molecule will expand due to the mutual repulsion of ionic charges attached to the molecular backbone cannot be assessed unless the location of the mobile counterions is known. These counterions tend to be concentrated in the regions of high electrostatic potential surrounding the polyion and they provide partial shielding for the fixed charges, so that the expansion of the chain molecule is much less than it would be if the counterions could be entirely removed. If we have only polyions and their counterions, but no additional electrolyte in the system, and if we gradually expand the total volume of the system, the probability that the counterions will escape from

† Provided that $(h^*)^2/\langle h_0^2 \rangle \gg \frac{2}{3}$.

the field of the polyions will be correspondingly increased. However, this process is severely limited for two reasons. First, it will be shown that at high charge densities of the polymer molecule the counterion concentration in the adjoining region falls off only very slowly with the expansion of the system. Secondly, the electrostatic potentials which we encounter are such that even the hydrogen or hydroxyl ions, due to the self-ionization of water, provide a sufficient reservoir of counterions to ensure that the charges of polyelectrolytes with high densities of ionizable groups will be partially shielded even at extreme dilutions.

Just as the distribution of the counterions determines the extent to which flexible polyions will expand, so the expansion of the polyion modifies the electrostatic potential and with it the counterion distribution. This mutual interdependence should, strictly speaking, be dealt with in any model of polyelectrolyte solutions. On the other hand, the mathematical complexities of a treatment of such models are at present insurmountable and it would appear that much can be gained by obtaining precise results for the counterion distribution even in models containing rather artificial assumptions about the arrangement of the fixed polymer charges.

The first such attempt was made by Alfrey et al.[41] and independently by Fuoss et al.[42] utilizing a model in which the polyion was represented by a long rod with a uniform surface charge. The solution was supposed to contain only such polyions and their counterions, but no simple electrolyte. Such a system will have a rather sharp potential energy minimum when the rods are parallel to each other and distributed in a hexagonal array as shown in Fig. 5.2.1 and if the rods are sufficiently long and carry a high charge density we may neglect all fluctuations from this lowest energy configuration. The counterions of any one rod will then be contained in a channel with hexagonal cross section as represented in Fig. 5.2.1 by cross-hatching.

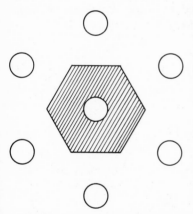

FIG. 5.2.1. Close-packed hexagonal arrangement of rod-shaped polyions (cross-hatching indicates the volume occupied by the counter-ions of the central rod).

It is clear that a problem involving such hexagonal symmetry would not lend itself to mathematical treatment, but it may be assumed (probably with small error) that the potential distribution around a given rod is only slightly altered if we introduce cylindrical symmetry. This may also be justified as representing a time-averaged distribution of rods.

We may now describe our model as follows: a polyanion is represented by a rod of radius a with one ionizable group for every section of length b^* of the chain. The degree of ionization is α so that the rod has a surface charge density of $-\alpha q/2\pi ab*$. Univalent counterions are distributed in a cylinder of radius R

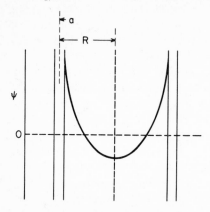

concentric with the rod. If c_P is the bulk average number of ionizable polymer groups per unit volume, $R=(1/\pi c_P b*)^{1/2}$. The electrostatic potential falls from high values at the surface of the rod to a minimum at the surface of the cylinder enclosing the counterions and increases again at larger distances due to the approach to neighboring polyanions. A schematic representation is given in Fig. 5.2.2.

FIG. 5.2.2. Distribution of the electrostatic potential in a system containing rod-shaped polyions.

It is most convenient to define the potential ψ so that $\psi = 0$ in a region in which the local counterion concentration assumes its bulk-average value. The variation of the electrical potential with the distance r from the axis of the charged rod may then be obtained from the Poisson–Boltzmann equation in the form appropriate for cylindrical symmetry,

$$\frac{1}{r}\frac{d}{dr}\left(r\frac{d\psi}{dr}\right) = \frac{4\pi c_P \alpha q}{D} e^{q\psi/kT} \tag{5.2.1}$$

which may be integrated to

$$\psi(r) = \frac{kT}{q} \ln\left\{\frac{\delta^2 DkT}{2\pi c_p \alpha q^2}r^2 \cos^2[\delta(\ln r + \beta)]\right\} \tag{5.2.2}$$

where δ and β are constants to be determined from the boundary conditions. The total counterion charge is equal in magnitude and opposite in sign to that of the polyion, so that

$$\frac{-\alpha q}{b*} = \int_{r=a}^{r=R} 2\pi r \rho(r)\, dr. \tag{5.2.3}$$

Also, the electrical potential passes through a minimum at the surface of the cylinder enclosing the counterions,

$$\left(\frac{d\psi}{dr}\right)_{r=R} = 0. \tag{5.2.4}$$

These boundary conditions lead to

$$\delta = \cot[\delta(\ln R + \beta)]$$

$$\delta(\ln R + \beta) = \tan^{-1}\left[\frac{1 - (\alpha q^2/b^*DkT)}{\delta}\right] + \delta \ln\left(\frac{R}{a}\right) \qquad (5.2.5)$$

from which the dependence of δ and β on the various parameters describing the model can be computed.

It should be emphasized that the model described above has only a rather tenuous relation to the properties of a real polyion. Real polyions do not seem to be fully stretched, so that their charge would be more concentrated than in the model. The fact that the ionic charge is localized rather than smeared out uniformly over the surface of a rod may lead, as we shall see later, to ion pair and complex ion formation. The dielectric constant D which is effective in the neighborhood of the polyion certainly has lower values than the macroscopic value. But we may note that in each case our model tends to underestimate the interaction of polyion and counterions, so that the computed results may be considered as highly significant in that they represent a lower bound to the interactions to be expected in real polyelectrolyte solutions.

It is in that sense only that we should consider the results of calculations based on the rod model. If we represent the backbone of poly(acrylic acid) by a carbon chain with all-trans conformation, the parameter b^* may be set at 2.53×10^{-8} cm. Fortunately, the results of the calculations are quite insensitive to the value chosen for a. Using $a = 6 \times 10^{-8}$ cm, $\alpha = 0.5$, $D = 80$, and $T = 290$, the distribution of electrostatic potential is given in Fig. 5.2.3. It

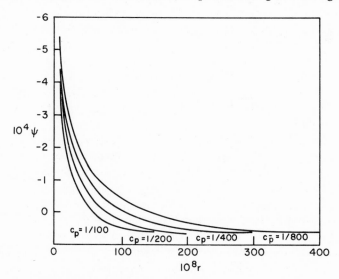

FIG. 5.2.3. Electrostatic potential in solutions of half-neutralized polyacrylic acid ($a = 6 \times 10^{-8}$). [Reproduced from reference 41.]

may be seen that ψ rises to values of about 4×10^{-4}, so that at ordinary temperatures $q\psi/kT \approx 5$ showing clearly the inapplicability of treatments based on the assumption that $q\psi$ is much smaller than kT. The integral counterion distributions plotted in Fig. 5.2.4 reveal another most character-

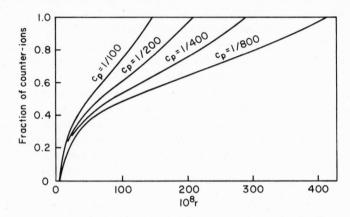

FIG. 5.2.4. Integral counterion distribution of half-neutralized polyacrylic acid at various solution concentrations ($a = 6 \times 10^{-8}$). [Reproduced from reference 41.]

istic feature of the model: an eightfold dilution of the polyelectrolyte concentration (from a base molarity of 1/100 to 1/800) results in relatively little change in the concentrations of counterions held in close proximity to the polyion. For instance, the fraction of counterions held at $r \leqslant 15 \times 10^{-8}$ drops only from 0.29 to 0.25 for this eightfold dilution of the system. It also is significant that for the 1/800 base molar solution with $R = 425$ A, one-third of the counterions are located at distances less than 30 A from the polyion, making it abundantly clear that, even in the most dilute solutions lending themselves to experimental study, the polyion is far from being separated from its counterions.

Before concluding our remarks on the charged rod model, attention should be drawn to some of its inherent limitations. It cannot be applied to the representation of solutions which contain simple electrolytes in addition to polyelectrolytes, since Eq. (5.2.1) can be integrated only if the right-hand side contains no term due to the distribution of the charge of byions. This excludes, of course, also the case of a single polyion and its counterions in an infinite volume of water, since the hydrogen and hydroxyl ions formed by self-ionization of the solvent medium would have to be taken into account. The model also loses logical self-consistency if the charge density of the rods is low, since it is then no longer reasonable to assume that the rods will be parallel to each other and in positions corresponding to an electrostatic potential energy minimum. It may then fairly be asked, what charge density is required to make the model plausible. Unfortunately, we do not know, but the answer will certainly depend on the dilution of the system.

The discussion in the preceding section, of the configuration of poly-electrolyte chains demonstrated the advantages to be gained from the use of a model based on the Donnan approximation. If we consider a macroscopic portion of a system, e.g., a polyelectrolyte solution contained in a dialysis bag in equilibrium with a solution of simple electrolytes, then it is obvious that the Donnan condition must apply, i.e., the net charge of the solution in the bag can only be an infinitesimal fraction of the total fixed charges on the polymer. At the other extreme, if we consider a single molecule carrying only a few ionized groups, then we cannot be certain that an overwhelming fraction of the counterions is held close to the molecule. Polyions of fairly high mole-cular weight and charge density represent an intermediate case and it will be most instructive to inquire how accurately the Donnan condition may be expected to hold under various conditions.

This problem was first posed by Oosawa et al.[43] In their model the polyion is represented by spheres of radius R_s with a uniform density of fixed negative charges c_f immersed in a medium containing univalent salt of bulk average concentration c_s^0. Although the authors give the general treatment for this case, we shall concern ourselves for the present only with the solution applic-able for dilute polyelectrolytes. A function $\lambda(r)$ representing the ratio of the local net charge density to the charge density due to fixed charges is defined by

$$c_f\lambda(r) = c_f + c_s^0 \exp(+q\psi/kT) - c_s^0 \exp(-q\psi/kT) \qquad (5.2.6)$$

and it is assumed that $\lambda \ll 1$. Solving the Poisson–Boltzmann equation with $c_f\lambda q$ as the local charge density, setting $d\psi/dr = 0$ at $r = \infty$, and making both ψ and $d\psi/dr$ continuous at $r = R_s$, an explicit expression is obtained for $\lambda(r)$:

$$\lambda(r) = \sqrt{1 + (1/f^2)}\{\ln[f + \sqrt{1 + f^2}] - \ln B\}\frac{\sinh \kappa'r}{\kappa'r \cosh \kappa'R_s} \qquad (5.2.7)$$

where $f = c_f/2c_s^0$, B is an integration constant close to unity and κ' defined by

$$\kappa' = \left[\frac{4\pi q^2 c_f}{DkT}\sqrt{1 + (1/f)^2}\right]^{\frac{1}{2}} \qquad (5.2.8)$$

is a characteristic reciprocal shielding length for the interior of the polyion analogous to κ of the Debye–Hückel theory. The fraction of the fixed polyion charge which is not compensated by counterions is given by the value of λ averaged over the volume of the sphere,

$$\langle \lambda \rangle = \frac{3DkT}{4\pi q^2 R_s^2 c_f} \ln(f + \sqrt{1+f^2}). \tag{5.2.9}$$

Let us now consider some characteristics of these results. For a given R_s and c_f, $\langle \lambda \rangle$ will be approximately a linear decreasing function of c_s^0. In a typical case, with $R_s = 10^{-6}$ cm, $c_f = 10^{20}$ cm^{-3} and $c_s^0 = 10^{19}$ cm^{-3} (corresponding to 0.16 molar fixed charge concentration within the sphere representing the polyion and $0.016M$ concentration of simple electrolyte) the value of $\langle \lambda \rangle$ is only 0.08, with $\lambda(r)$ increasing from 0.00005 at the center of the sphere to 0.25 at its surface. According to (5.2.7), the value of λ at the surface of the sphere will decrease linearly with R_s^2, while $\langle \lambda \rangle$ is according to (5.2.9) proportional to $1/R_s^2$, so that we may conclude that the Donnan approximation corresponds quite closely to the physical situation for polyelectrolyte coils with diameters of several hundred Å.

An indication of the manner in which $\langle \lambda \rangle$ would be expected to change with variations in the concentration of polyelectrolyte was presented by Oosawa.[43a] In this approximation the solution containing polyelectrolyte and simple electrolyte is divided into spherical regions containing a single spherical macroion with a uniform net charge density in the center. The counter-charge is also assumed to be uniformly distributed in the space surrounding the macroion. It is then found that a decrease in the volume fraction ϕ occupied by the macroion produces an increase in $\langle \lambda \rangle$ given by

$$\frac{1 - \langle \lambda \rangle}{\langle \lambda \rangle + (c_s^0/c_f)} = \frac{\phi}{1 - \phi} \exp\left[\langle \lambda \rangle \frac{vq^2}{DkTR_s} (1 - \phi)^{\frac{1}{3}} \right] \tag{5.2.9'}$$

Using again the example given above, $R_s = 10^{-6}$ cm, $c_f = 10^{20}$ cm^{-3}, $c_s^0 = 10^{19}$ cm^{-3}, we find that the variation in the polymer charge with polymer concentration as predicted above is very slight. In fact, $\langle \lambda \rangle$ changes from 0.23 to 0.27 and 0.33 as ϕ is reduced from 0.1 to 0.01 and 0.001. It may also be noted that Eq. (5.2.9') predicts that the net charge of the polyion should approach the total of the fixed charges with infinite dilution (i.e., $\langle \lambda \rangle \to 1$ as $\phi \to 0$). This is certainly incorrect for solutions containing simple electrolyte as demonstrated by the theory of Oosawa et al.[43] and treatments to be discussed below.

A model similar to that used by Oosawa, Imai, and Kagawa was also investigated by Wall and Berkowitz,[44] who used an electronic computer to solve Eq. (5.2.8). In this manner a general solution could be obtained, unrestricted by the requirement that the net charge density be small compared to the fixed charge density within the polyion sphere for a system containing a low concentration of the polyions in a large volume of electrolyte solution. Figure 5.2.5 gives their results for $\langle \lambda \rangle$ as a function of f at various values of

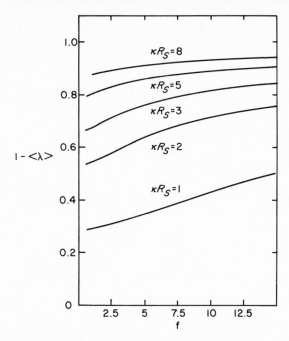

Fig. 5.2.5. Percent ion association for first model as function of initial charge density for polyions of different radii.   [Reproduced from reference 44.]

κR_s. Lifson[45] succeeded in obtaining an analytic solution to the problem of the distribution of electrostatic potentials in systems containing spherical regions of high fixed charge density immersed in solutions of simple electrolytes. It will be recalled from the discussion presented in Chapter 3 that he took the Donnan potential as a first approximation and assumed that a correction potential was small so as to satisfy the Debye–Hückel approximation. Requiring the correction potential and its derivative to be continuous at the surface of the polyion, he obtained for systems in which the radius R_s of the polyion was large compared to $1/\kappa$, potentials within the polyion sphere given by

$$\psi = \psi_{\text{Donnan}}\left[1 - \frac{1 + \kappa R_s}{2\kappa r[1 + (1 + f^2)^{\frac{1}{2}}]}\right] \exp[\kappa(1 + f^2)^{\frac{1}{2}}(r - R_s)] \qquad (5.2.10a)$$

$$r < R_s.$$

The potentials beyond the polyion are in analogy with classical electrolyte theory

$$\psi(r) = \psi(R_s) \cdot \frac{\exp(-\kappa r)/r}{\exp(-\kappa R_s)/R_s}, \qquad r > R_s. \qquad (5.2.10b)$$

The results obtained by Lifson's procedure are in remarkably close agreement with the results of the numerical computations of Wall and Berkowitz. Both show very convincingly that the deviation of ψ from the Donnan potential is, for cases of interest in polyelectrolyte studies, appreciable only in the immediate neighborhood of the surface of the polyion.

5.2.2 SITE BINDING OF COUNTERIONS BY ION PAIR AND BY COMPLEX ION FORMATION

In the preceding treatments of the spatial distribution of simple ions in polyelectrolyte solutions the ionic charges were either smeared out to a uniform surface charge of a rodlike polyion with cylindrically symmetrical counter-charges, or else, the polyion was represented by a uniform fixed charge density over a spherical volume. Such treatments will necessarily obliterate all effects which are due to the site localization of the individual ionic charges. Also, the close proximity of neighboring charges on the polyion bears little relation to the mean distances of charged groups within the region occupied by the chain molecule. A specific example will illustrate the problems encountered: Schneider and Doty[46] found that the end-to-end distance of the sodium salt of carboxymethylcellulose of a polymerization degree of 3300 in $0.05M$ sodium chloride was 2400 A as against a contour length of 17,000 A for the fully stretched chain. For a uniform distribution of of the fixed charges within an equivalent sphere (taken to have a radius equal to the end-to-end distance of the coiled chain) we should obtain a distance of 270 A between neighboring fixed charges. Since, in fact, the charges are attached to the polymer chain, the distance between neighboring charges is only of the order of 5 A. The characteristic shielding length $1/\kappa'$ given by (5.2.8) is about 14 A. Thus, the smearing out of the fixed charges over the volume occupied by the polyion would lead us to conclude that the electrostatic forces between the charges of the polymer are largely eliminated by the shielding action of the added salt, while in fact the mobile ions will have little influence on the interactions of neighboring charges on the polymer chain.

Two important aspects of the shortcoming of all theories based on continuous charge distributions should be particularly emphasized. (1) Since the volume occupied by swollen polymer chains in any given solvent medium increases more rapidly than the chain length, a smeared-out density of fixed charges will tend to decline with increasing chain length of the polyion. (2) Using the more sophisticated procedure of Flory,[39] the fixed charge density is smeared out over concentric spherical shells, rather than considered uniform over the entire equivalent sphere. This causes the fixed charge density to fall off as a Gaussian function of the distance from the center of the chain. Yet it is obvious that the distance of neighboring ionized groups attached to the polyion chain depends neither on the chain length nor on the position of the chain segment within the polymer coil.

The localization and close proximity of the fixed charges may lead to interactions with counterions which may be broadly classified as ion-pair formation or the formation of complex ions. This type of counterion association is quite distinct from the consequences of long-range electrostatic field effects operative between the polyion and the small mobile ions present in the system. On the other hand, the distinction between ion pairs and complex ions may not always lend itself to a nonambiguous experimental demonstration.

The concept of ion pair formation was first defined by Bjerrum[47] who pointed out that for ions of opposite charge the number for ions of charge z_2 to be found in successive concentric shells around a given reference ion of charge z_1 passes through a minimum at a characteristic distance $r_{min} = -q^2 z_1 z_2 / 2DkT$. Bjerrum proposed that all ions at a distance less than r_{min} be considered as associated in ion pairs. With singly charged ions, such association is usually important only in media of low dielectric constant, although a very slight extent of ion pair formation (with a dissociation constant of 2 for the ion pair of sodium bromate) has been inferred by Fuoss and Kraus[48] from conductance data even in an aqueous medium. As would be expected, ion association is greatly magnified in solutions of " bolaform electrolytes," in which one of the ions carries two charges constrained by covalent bonds to remain close to one another. Typical examples are the bis-quaternary ammonium salts represented below, which were studied by Fuoss and Chu.[49]

$$Cl^- H_3C - \overset{\overset{CH_3}{|+}}{N} - (CH_2)_n - \overset{\overset{CH_3}{|+}}{N} - CH_3 \; Cl^-$$
$$\underset{CH_3}{|} \qquad \underset{CH_3}{|}$$

It was found that one of the chloride ions was held fairly strongly by this bis-quaternary ammonium ion, the dissociation constants for the ion pairs being estimated from conductance data as 0.125, 0.13, and 0.135 for $n = 3, 4$, and 5. It is particularly interesting to note the surprisingly slight dependence of the dissociation constant on the number of methylene groups separating the cationic charges, indicating how poorly the spatial relationships in the molecule are suggested by the chemical formulae of the ions.

It is apparent that the same causes which lead to ionic association in the case of bolaform electrolytes will be even more effective in the case of polyelectrolytes carrying closely spaced charges. The analogy has been discussed at length by Rice[50] and we shall revert to this subject in discussions of the experimental methods by which counterion association may be quantitatively evaluated (see Chapter 9). Suffice it here to point out that the concept of ion-pair formation considers the solvated ion as an integral unit, so that the extent of ion-pair formation by the fixed charges of polyelectrolytes would be expected to increase with a decreasing radius of the *hydrated* counterion.

The formation of complex ions differs from ion-pair formation in that the association of the species is now governed by interactions which lead to chemical bond formation. This has several important consequences. (1) The geometry of complex ions is determined by the spatial disposition of the hybrid bond orbitals of the cation, e.g., the tetrahedral sp^3 bonds of Zn^{++}, the square planar dsp^2 bonds of Cu^{++}, and the octahedral d^2sp^3 bonds of Co^{+++}. (2) There are large, characteristic differences in the stability of complex ions, which depend on the nature of the cation and the ligand but may also exhibit pronounced specificities in cation-ligand affinities. Since the complexation involves the displacement of a water molecule from the coordination shell by the ligand group, the stability of the complex will, other things being equal, tend to increase with a decreasing radius of the *unsolvated ion*. (3) The interaction between the cation and the ligand may perturb the spectra of both constituents of the complex. Such spectral shifts, when observable, provide a clear distinction between an ion pair and a complex ion.

The theoretical treatment of complex ions has proceeded along two lines: Pauling[51] classified them as ionic or covalent and emphasized the magnetic properties of the complex ion as criteria of the extent of covalent bond formation between the cation and the ligands. In recent years, great effort has been concentrated on interpreting the behavior of complex ions by the crystal field theory, which deals with the splitting of the energy levels of the *d*-electron orbitals of transition metal and rare earth cations by the electrical field of the surrounding ligands. This theory, which has been reviewed by Moffitt and Ballhausen,[52] has had particularly spectacular success in interpreting the spectra of complex ions.

It is instructive to compare the equilibria involved in the dissociation of a carboxylic acid and the dissociation of a carboxylate complex with a metal cation.

$$
R-\overset{\overset{\displaystyle O}{\parallel}}{C}-OH \rightleftharpoons R-\overset{\overset{\displaystyle O}{\parallel}}{C}-O \cdots H^+ \rightleftharpoons R-\overset{\overset{\displaystyle O}{\parallel}}{C}-O^- + H^+ \qquad (5.2.11a)
$$

$$
[R-\overset{\overset{\displaystyle O}{\parallel}}{C}-O \rightarrow M]^{+\nu-1} \rightleftharpoons R-\overset{\overset{\displaystyle O}{\parallel}}{C}-O^- \cdots M^{+\nu} \rightleftharpoons R-\overset{\overset{\displaystyle O}{\parallel}}{C}-O^- + M^{+\nu} \quad (5.2.11b)
$$

In the dissociation of the carboxylic acid, it is quite safe to consider the concentration of the ion pair negligible compared with that of the covalently bonded undissociated acid. However, the same assumption is by no means safe in the case of the complex ion. Thus, Nancollas[53] derived dissociation constants of acetate complexes of alkaline earth ions of the order of magnitude 0.1. This is quite similar to the ion-pair dissociation constants found by Fuoss and Chu[49] for bolaform electrolytes and since the charge is more highly concentrated in the doubly charged alkaline earth ions, Nancolas' results may reflect no more than the expected tendency towards ion pairing. It is safer to interpret the behavior of silver acetate with a dissociation constant of 0.2[54] as due to complex formation, since the cation bears only a single charge. In the case of the cupric ion, the nature of its association with carboxylates is revealed by a variety of evidence to be due to complex formation: The monoacetato complex has a dissociation constant of only 0.02,[55] the negatively charged triacetato complex has a considerable affinity for a fourth carboxylate in spite of the electrostatic repulsion which has to be overcome,[55] and the complexes exhibit characteristic spectral shifts both in the visible region[56] (explicable by the crystal field theory) and in the ultraviolet[57] (which is due to a shift of the carboxyl absorption band, normally found at $180\text{--}220m\mu$ to longer wavelengths as a consequence of electronic transitions between the ligand and the cation[58]).

When a molecule carries several ligand groups which may coordinate simultaneously with a cation, the resulting complex is called a chelate. Since polyelectrolytes will carry, in general, large numbers of potential ligand groups, a careful consideration of the conditions of chelate formation is in order. As a first approximation it is reasonable to assume that the coordination of a cation with several ligand groups will be characterized by an enthalpy change, which does not depend on whether the ligand groups form part of a single molecule. This assumption has been found to be generally— but not always—borne out by experimental data in a study of Spike and Parry,[59] who compared the coordination to various transition metal cations of two molecules of ammonia and a molecule of ethylene diamine, respectively. However, the free energies of reactions such as

$$Cd(CH_3NH_2)_2{}^{++} + H_2NCH_2CH_2NH_2 \rightleftharpoons \left[Cd \begin{smallmatrix} H_2 \\ N-CH_2 \\ | \\ | \\ N-CH_2 \\ H_2 \end{smallmatrix} \right]^{++} + 2CH_3NH_2 \qquad (5.2.12)$$

are invariably negative due to a gain of entropy. Writing (5.2.12) as a two step process

$$Cd(CH_3NH_2)_2{}^{++} + H_2NCH_2CH_2NH_2 \rightleftharpoons [CH_3NH_2CdNH_2CH_2CH_2NH_2]^{++} + CH_3NH_2 \qquad (5.2.13a)$$

$$[CH_3NH_2CdNH_2CH_2CH_2NH_2]^{++} \rightleftharpoons \left[Cd \begin{smallmatrix} H_2 \\ N-CH_2 \\ | \\ | \\ N-CH_2 \\ H_2 \end{smallmatrix} \right]^{++} + CH_3NH_2 \qquad (5.2.13b)$$

it is reasonable to assume that (5.2.13a) involves no entropy change and the results may then be stated to mean that the release of a methylamine molecule in (5.2.13b) more than compensates for the entropy loss in ring closure.

In assessing the complex-forming tendency of polymeric acids, it is essential to bear in mind that many functional groups, which by themselves have no demonstrable ability to coordinate with cations, may participate in chelate formation when they are part of a molecule containing strong ligand groups. Examples of such participation in chelate formation by hydroxyl, ether, and peptide groups are illustrations of this phenomenon. It has been found, for instance, that citrate binds calcium more than twenty times as strongly as tricarballylate,[60] which differs from citrate in lacking a hydroxyl group. A particularly striking case was reported by Schwarzenbach et al.[61] who compared the Ca^{++} chelation with the two compounds shown below.

$$HOOCCH_2 \diagdown CH_2COOH$$
$$NCH_2CH_2CH_2CH_2CH_2N$$
$$HOOCCH_2 \diagup CH_2COOH$$

$$HOOCCH_2 \diagdown CH_2COOH$$
$$NCH_2CH_2OCH_2CH_2N$$
$$HOOCCH_2 \diagup CH_2COOH$$

It was found that the substitution of the central methylene group by an ether linkage increased the chelate stability by a factor of 2.5×10^5. It has also been shown[62] that Cu^{++} coordinated to the terminal amine group of glycylglycine may form a chelate by displacing the peptide hydrogen.

The implications of our knowledge of complex ion and chelate formation with low molecular weight species to the behavior of polymers in the presence of complex forming cations has been discussed in detail by Morawetz.[63] To simplify the analysis of experimental data, it is desirable to carry out complexation studies involving polyelectrolytes in solutions containing a high concentration of simple electrolytes, so as to eliminate, as far as possible, effects due to the long-range electrostatic field of the polyion and isolate effects due to the specific affinities of complex forming cations and the ligand groups of the polymer.

If the polymer carries groups of ligands, sufficient to satisfy the complexation capabilities of a given cation and if these groups of ligands are well separated from one another along the polymer chain, they will behave independently of one another in a manner analogous to a low molecular weight chelating agent. An example of this case is the binding of alkaline earth cations by hydrolyzed copolymers of maleic anhydride studied by Morawetz et al.[64] Such copolymers may be made so as to contain the monomer units in a regularly alternating sequence

$$(-CH-CH-CH_2-CH-)_n$$
$$\underset{COOH}{|}\;\underset{COOH}{|}\qquad\underset{X}{|}$$

and since alkaline earth ions coordinate commonly with two ligand groups only, the carboxylate pairs along the polymer chain behave, as long as the chelate units are rather sparsely spaced, in a manner very similar to that of the succinate ion.

A much more complicated situation arises when the ligand groups are spread evenly along the polymer chain and particularly when the complex forming cation may coordinate with more than two such groups. A typical case of this type is the complexation of Cu^{++} by polyacrylic acid.[65-67] It is instructive to compare the behavior of the polycarboxylic acid with that of a monofunctional analog, e.g., acetate ion. In that case, we have to consider the equilibria[55]

$$Cu^{++} + OAc^- \rightleftharpoons CuOAc^+$$
$$CuOAc^+ + OAc^- \rightleftharpoons Cu(OAc)_2$$
$$Cu(OAc)_2 + OAc^- \rightleftharpoons Cu(OAc)_3^- \qquad (5.2.14)$$
$$Cu(OAc)_3^- + OAc^- \rightleftharpoons Cu(OAc)_4^=.$$

However, in the case of dilute solutions of the polymeric acid, it is obvious that only the first association step will be governed by the over-all concentration of carboxylate groups. Once the cupric ion becomes attached to a polymer chain, any further group coordination equilibria will be governed merely by conditions existing within the isolated polymer coils, being independent of their number in solution. This is due to the fact that association with two or more ligand groups will favor overwhelmingly ligands forming part of the same molecule. We may describe the situation somewhat loosely by a model containing small highly concentrated droplets of carboxylate solution suspended in a medium containing no ligand groups.

This situation has three important consequences. First, we may note that, in contrast to conditions in acetate solution, the *kind* of complex formed with a polycarboxylic acid will be independent of the carboxylate concentration. This has been confirmed by the observations that the absorption maxima of both the visible and the ultraviolet spectra of Cu^{++} in partially ionized poly-carboxylic acid solutions are independent of the degree of ionization of the polymer or its stoichiometric concentration.[65, 66, 57] Secondly, the very high local concentration of carboxylate residues within the swollen coil of, e.g., poly(methacrylic acid) tends to favor the participation of many ligand groups in chelate formation. Comparisons of spectra of Cu^{++} in acetate and partially ionized poly(methacrylic acid) indicate that the cation is coordinated with four carboxylate groups even if the polymer is as dilute as $0.01N$, although the tetraacetato complex forms only partially even in $3N$ acetate solution.[68]

Finally, the equilibrium between free and bound copper may be treated as an adsorption isotherm. As a result, we should expect a linear relation between the polymer concentration (i.e., the number of adsorbent particles) and the ratio of bound to free copper, as long as the polymers bind few cations so that there is no energetic interference between them. This result, which has been confirmed experimentally,[66] is in sharp contrast to conditions in acetate solution, where the formation of the higher complexes involves a higher power of ligand concentration.

In the chemistry of low molecular weight materials, the formation of ion chelates is confined almost exclusively to cases where a five- or six-membered ring may be formed. This is due to steric hindrance effects in rings of seven to twelve members and the relatively low probability of closure for larger ring sizes. With polymers, there is no such limitation on the possible size of chelate rings. The reason for this difference is easy to see: since a large number of ligand groups are confined to the region of the swollen polymer coil, there will be a multiplicity of ligand pairs which may cooperate in the formation of chelate rings. Thus, although the formation of any one of these rings may be extremely improbable, the aggregate probability of forming *any* of the many possible rings may be very high indeed. It has been shown that extensive chelate ring formation took place in the complexation of a cation with a polymer, where the smallest possible ring size involved 20 atoms.[68] It might be more descriptive to think of such cases in terms of a model, where the polymer is represented by a concentrated solution of the ligands, rather than by classical chelating agents with their closely defined spatial relationships of the constituent ligand groups.

This discussion has so far been confined entirely to complexation involving cations. Yet, there are indications that anions may also form complexes with organic functional groups. Such phenomena in proteins are suggested by the specific activation of the enzyme amylase by chloride ions[69] and the strong affinity for anions of serum albumin. This protein binds not only anions of long chain fatty acids which might be accounted for by the general tendency of nonpolar materials to cluster in aqueous media, but also anions such as chloride and thiocyanate.[70] That similar effects can be encountered with synthetic polymers is indicated by the observation of Strauss, that quaternized polyvinylpyridine may bind an excess of bromide counterions, so that the polymer in concentrated KBr solutions carries a net negative charge.[71] This phenomenon is probably due to charge transfer complex formation of the bromide ion with the pyridine nucleus, similar to that observed spectroscopically for 1-methyl pyridinium iodide.[72] The un-ionized polymer polyvinyl pyrrolidone has also been shown to bind strongly a variety of anions,[73, 74] and it is tempting to speculate whether the complex is here similar to that characterizing anion binding by serum albumin.

REFERENCES

1. H. STAUDINGER, " Die hochmolekularen organischen Verbindungen," p. 39. Springer, Berlin, 1932.
2. H. MARK AND A. V. TOBOLSKY, " Physical Chemistry of High Polymeric Systems," (2nd ed.). Interscience, New York, 1950.
3. P. J. FLORY, "Principles of Polymer Chemistry." Cornell Univ. Press, Ithaca, New York, 1953.
4. H. A. STUART, " Die Physik der Hochpolymeren, II. Das Makromolekül in Lösungen." Springer-Verlag, Berlin, 1953.
5. H. TOMPA, " Polymer Solutions," Academic Press, New York, 1956.
6. J. D. KEMP AND K. S. PITZER, J. Am. Chem. Soc. 59, 276 (1937).
7. M. S. NEWMAN, J. Chem. Educ. 32, 344 (1955).
8. E. A. MASON AND M. M. KREEVOY, J. Am. Chem. Soc. 77, 5808 (1955).
9. L. J. OOSTERHOFF, Discussions Faraday Soc. 10, 79, 87 (1951).
10. J. D. KEMP, Chem. Revs. 27, 39 (1940).
11. K. ITO, J. Am. Chem. Soc. 75, 2430 (1953).
12. W. KUHN AND F. GRÜN, Kolloid-Z. 101, 248 (1942).
13. F. T. WALL, J. Chem. Phys. 11, 67 (1943).
14. H. BENOIT, J. Polymer Sci. 3, 376 (1948).
15. H. BENOIT AND C. SADRON, J. Polymer Sci. 4, 473 (1949).
15a. P. J. FLORY, J. Polymer Sci., 49, 105 (1961).
15b. S. LIFSON, J. Chem. Phys. 29, 80 (1958).
15c. S. LIFSON AND I. OPPENHEIM, J. Chem. Phys. 33, 109 (1960).
15d. K. NAGAI, J. Chem. Phys. 30, 660 (1959).
15e. K. NAGAI, J. Chem. Phys. 31, 1169 (1959).
15f. J. W. L. FORDHAM, J. Polymer Sci. 39, 321 (1959).
15g. S. LIFSON, J. Chem. Phys. 30, 964 (1959).
15h. A. M. LIQUORI, Acta Cryst. 8, 345 (1955).
16. W. KUHN, Kolloid-Z. 76, 258 (1936); Ibid. 87, 3 (1939).
17. B. H. ZIMM, W. H. STOCKMAYER, AND M. FIXMAN, J. Chem. Phys. 21, 1716 (1953).
18. W. KUHN, O. KÜNZLE, AND A. KATCHALSKY, Helv. Chim. Acta 31, 1994 (1948).
19. W. J. TAYLOR, J. Am. Chem. Soc. 16, 257 (1948).
20. G. NATTA, Modern Plastics 34, No. 4, 169 (1956).
20a. J. D. STROUPE AND R. E. HUGHES, J. Am. Chem. Soc. 80, 2341 (1958).
21. M. V. VOLKENSTEIN, J. Polymer Sci. 29, 441 (1958).
22. P. J. FLORY, C. A. J. HOEVE, AND A. CIFERRI, J. Polymer Sci. 34, 337 (1959).
23. A. KATACHALSKY, O. KÜNZLE, AND W. KUHN, J. Polymer Sci. 5, 283 (1950).
24. P. J. FLORY AND W. R. KRIGBAUM, J. Chem. Phys. 18, 1086 (1950).
25. W. R. KRIGBAUM, J. Polymer Sci. 18, 315 (1955).
26. A. KATCHALSKY AND S. LIFSON, J. Polymer Sci. 11, 409 (1956).
27. F. E. HARRIS AND S. A. RICE, J. Polymer Sci. 15, 151 (1955).
28. S. H. PINNER AND T. ALFREY, JR., J. Polymer Sci. 9, 478 (1952).
29. F. E. HARRIS AND S. A. RICE, J. Phys. Chem. 58, 725 (1954).
30. J. H. HILDEBRAND AND R. L. SCOTT, " The Solubility of Nonelectrolytes," 3rd ed., pp. 135–136. Reinhold, New York, 1950.
31. S. A. RICE AND F. E. HARRIS, J. Phys. Chem. 58, 733 (1954).
32. J. G. KIRKWOOD AND F. H. WESTHEIMER, J. Chem. Phys. 6, 506, 513 (1938).
33. J. B. HASTED, D. M. RITSON, AND C. H. COLLIE, J. Chem. Phys. 16, 1 (1948).
34. F. H. WESTHEIMER AND M. W. SHOOKHOFF, J. Am. Chem. Soc. 61, 555 (1939).
35. S. LIFSON, J. Chem. Phys. 29, 89 (1958).

35a. E. M. LOEBL AND J. J. O'NEILL, *J. Polymer Sci.*, **45**, 538 (1960).

35b. M. L. MILLER, M. C. BOTTY, AND C. E. RAUHUT, *J. Colloid Sci.*, **15**, 83 (1960).

36. J. J. HERMANS AND J. T. G. OVERBEEK, *Rec. trav. chim.* **67**, 761 (1948).

37. P. DEBYE, *J. Chem. Phys.* **14**, 636 (1946).

38. G. E. KIMBALL, M. CUTLER, AND H. SAMELSON, *J. Phys. Chem.* **56**, 57 (1952).

39. P. J. FLORY, *J. Chem. Phys.* **21**, 162 (1953).

40. S. LIFSON, *J. Polymer Sci.* **23**, 431 (1957).

41. T. ALFREY, JR., P. W. BERG, AND H. MORAWETZ, *J. Polymer Sci.* **7**, 543 (1951).

42. R. M. FUOSS, A. KATACHALSKY, AND S. LIFSON, *Proc. Natl. Acad. Sci. U.S.* **37**, 579 (1951).

43. F. OOSAWA, N. IMAI, AND I. KAGAWA, *J. Polymer Sci.* **13**, 93 (1954).

43a. F. OOSAWA, *J. Polymer Sci.* **23**, 421 (1957).

44. F. T. WALL AND J. BERKOWITZ, *J. Chem. Phys.* **26**, 114 (1957).

45. S. LIFSON, *J. Chem. Phys.* **27**, 700 (1957).

46. N. S. SCHNEIDER AND P. DOTY, *J. Phys. Chem.* **58**, 762 (1954).

47. N. BJERRUM, *Kgl. Danske Vidensk. Selskab.* **7**, No. 9 (1926).

48. R. M. FUOSS AND C. A. KRAUS, *J. Am. Chem. Soc.* **79**, 3304 (1957).

49. R. M. FUOSS AND V. H. CHU, *J. Am. Chem. Soc.* **73**, 949 (1951).

50. S. A. RICE, *J. Am. Chem. Soc.* **78**, 5247 (1956).

51. L. PAULING, " The Nature of the Chemical Bond," pp. 92–123. Cornell Univ. Press, Ithaca, New York, 1947.

52. W. MOFFITT AND C. J. BALLHAUSEN, *Ann. Rev. Phys. Chem.* **7**, 107 (1956).

53. G. H. NANCOLLAS, *J. Chem. Soc.*, 744 (1956).

54. F. H. MACDOUGALL AND S. PETERSON, *J. Phys. Chem.* **51**, 1346 (1947).

55. S. FRONAEUS, *Acta Chem. Scand.* **5**, 859 (1951).

56. I. M. KLOTZ, I. L. FULLER, AND J. M. URQUHART, *J. Phys. & Colloid Chem.* **54**, 18 (1950).

57. H. MORAWETZ, *J. Polymer Sci.* **17**, 442 (1955).

58. M. LINHART AND M. WEIGEL, *Z. anorg. Chem.* **264**, 321 (1951); *Z. physik. Chem.*, N.F. **5**, 20 (1955).

59. C. G. SPIKE AND R. W. PARRY, *J. Am. Chem. Soc.* **75**, 2726, 3770 (1953).

60. J. SCHUBERT AND A. LINDENBAUM, *J. Am. Chem. Soc.* **74**, 3529 (1952).

61. G. SCHWARZENBACH, H. SENN, AND G. ANDEREGG, *Helv. Chim. Acta* **40**, 1886 (1957).

62. H. DOBBIE, W. O. KERMACK, AND H. LEES, *Biochem. J.* **59**, 246, 257 (1955); S. P. DATTA AND B. R. RABIN, *Biochim. et Biophys. Acta* **19**, 574 (1956); *Trans. Faraday Soc.* **52**, 1123 (1956); A. R. MANYCK, C. B. MURPHY, AND A. E. MARTELL, *Arch. Biochem. Biophys.* **59**, 373 (1955).

63. H. MORAWETZ, *Fortschr. Hochpolym. Forsch.* **1**, 1 (1958).

64. H.MORAWETZ, A. M. KOTLIAR, AND H. MARK, *J. Phys. Chem.* **58**, 619 (1954).

65. F. T. WALL AND S. J. GILL, *J. Phys. Chem.* **58**, 1128 (1954).

66. A. M. KOTLIAR AND H. MORAWETZ, *J. Am. Chem. Soc.* **77**, 3692 (1955).

67. H. P. GREGOR, L. B. LUTTINGER, AND E. M. LOEBL, *J. Phys. Chem.* **59**, 34 (1955).

68. H. MORAWETZ AND E. SAMMAK, *J. Phys. Chem.* **61**, 1357 (1957).

69. K. MYRBÄCK, *Z. Physiol. Chem.* **159**, 1 (1926).

70. G. SCATCHARD, I. SCHEINBERG, AND S. ARMSTRONG, *J. Am. Chem. Soc.* **72**, 535, 540 (1950).

71. U. P. STRAUSS, N. L. GERSHFELD, AND H. SPIERA, *J. Am. Chem. Soc.* **76**, 5909 (1954).

72. E, M. KOSOWER, AND J. C. BURBACH, *J. Am. Chem. Soc.* **78**, 5838 (1956).

73. S. BARKIN, H. P. FRANK, AND F. R. EIRICH, *Ricerca Sci.* **25A**, 844 (1955).

74. H. P. FRANK, S. BARKIN, AND F. R. EIRICH, *J. Phys. Chem.* **61**, 1375 (1957).

6. Bolaform Electrolytes

6.0 Introduction

It is well known that the primary (K_1) and secondary (K_2) dissociation constants of a dibasic acid usually differ by more than a factor of four. If the ionization processes at the two acid groups were independent, K_1/K_2 would be exactly four, since both the un-ionized and double ionized species have symmetry numbers of two and the single ionized intermediate has a symmetry number of unity. A theoretical explanation of the deviation of K_1/K_2 from the value 4 was given by Bjerrum[1] in terms of the elestrostatic forces operative within the molecule. Subsequent refinement of this model by Kirkwood and Westheimer[2] has shown that the concepts originally introduced are adequate to account quantitatively for the observations.

In recent years, interest in this problem has been reawakened from two diverse points of view. Fuoss and co-workers[3-6] have extensively studied the properties of a class of compounds known as bolaform electrolytes. (It will be recalled that a bolaform electrolyte is one in which the charges are separated from one another by a chain of atoms.) It is interesting to note that these compounds exhibit ion-pair formation between the bolaform ion (bolion) and its counterions, even in solvents such as water.[3-6] A few selected results are entered in Table 6.0.1. Further interest derives from the observation that the counterions to a polyelectrolyte in solution are intimately associated with

TABLE 6.0.1

SECONDARY DISSOCIATION CONSTANTS OF SOME BOLAFORM ELECTROLYTES[a]

Compound	$K_2 \times 10^3$
1. Br [Py(CH$_2$)$_2$Py] Br	0.98
2. Br [Py(CH$_2$)$_4$Py] Br	1.16
3. Br [Py(CH$_2$)$_{10}$Py] Br	3.30
4. Cl [Py(CH$_2$)$_2$O(CH$_2$)$_2$Py] Cl	1.95
5. I [Me$_3$N(CH$_2$)$_2$O(CH$_2$)$_2$NMe$_3$] I	0.99
6. I [Me$_3$N(CH$_2$)$_2$O(CH$_2$)$_2$N Et$_3$] I	0.99
7. I [Et$_2$MeN(CH$_2$)$_2$O(CH$_2$)$_2$NMe Et$_2$] I	1.00
8. MeSO$_4$ [Me$_3$N(CH$_2$)$_2$O(CH$_2$)$_2$NMe$_3$] MeSO$_4$	1.02

[a]Reproduced from R. M. FUOSS, *J. Am. Chem. Soc.* **77**, 198 (1955).

it. It has been suggested that this intimate association is due to the formation of ion pairs at specific groups along the polymer chain.[7, 8] Thus, the two types of ion-pair formation in bolaform electrolytes and in polyelectrolytes are postulated to be closely related. We shall later see that the analysis of ion association to polyions requires the rejection of ion-pair formation at specific sites. For the present we shall see whether or not this assumption permits the construction of a consistent theory of bolaform electrolytes.

The original calculations of Bjerrum and of Kirkwood and Westheimer are not well adapted to the discussion of the mechanism of ion-pair formation and its extension to more complex systems. We shall show that the Bjerrum relation (or its subsequent modification) may be derived easily in a manner which emphasizes the physical processes leading to ion-pair formation and the subsequent deviation of K_1/K_2 from the value four.

6.1 Some General Considerations

We consider a volume V containing N bolions, with the total number of dissociable groups $2N$. Let the solution also contain, in addition to the counterions to the bolions, additional electrolyte consisting of L_j ions of charge q_j where $\sum_j L_j q_j = 0$. We shall deal mostly with 1 : 1 electrolytes, though no restriction will be made limiting the development to this case. Each of the $2N$ dissociable groups may be occupied by a charged group or a bolion-counterion pair. We shall assume, as implied, that the sites at which ion-pair formation occurs are the charged groups.

It is convenient to divide the free energy of the solution into three parts. The first portion, A_1 is the electrostatic free energy arising from the interactions between charged groups on the same bolion. The remainder of the electrical free energy, A_2, results from the interaction between all charged species in solution, regarding each as a structureless particle. Finally, the last portion of the free energy, A_3, is the chemical free energy associated with the state of ion-pair formation and is computed from a reference state in which the bolion is in its completely ion-paired form. This choice of reference state emphasizes the calculation of the dissociation constant rather than the association constant. Obviously the choice of the completely dissociated state would have been equally suitable. With the convention adopted above, and letting α be the degree of dissociation of the bolion, we note that the total free energy of the solution may be written

$$A = A_1(\alpha) + A_2(\{a_i\}) + A_3(\alpha) \tag{6.1.1}$$

with a_i the activity of species i. Of the contributions to A_2, we may distinguish between A_{21} which is the electrical free energy of the free counterions which may bind to a bolion and form ion pairs and A_{22} which contains the remainder of the free energy of interaction between all ions in the solution treated as structureless particles. To compute A_1 we shall regard ion pairs as uncharged.

By minimizing the free energy of the solution with respect to the degree of dissociation at constant external salt activity, the equilibrium counterion activity may be obtained. Performing the indicated differentiation, one obtains

$$\left(\frac{\partial A_1}{\partial \alpha}\right)_{\{a_j\}} + 2N\mu_{C-} + 2N(\mu_{B+}^0 - \mu_{B+C-}^0) = 0 \tag{6.1.2}$$

where we have used the fact that the various contributions to the free energy may be written as

$$A_1 = A_1(\alpha)$$
$$A_2 = A_{21} + A_{22} = 2N\alpha\mu_{C-} + N\mu_{BB} + \sum_j L_j\mu_j$$
$$A_3 = 2N\alpha(\mu_{B+}^0 - \mu_{B+C-}^0)$$
$$\sum_i N_i \, d\mu_i = 0 \tag{6.1.3}$$

and where μ_i is the chemical potential of species i. We use the notation μ_{BB} to refer to the chemical potential of the bolion and μ_{B+} to refer to the chemical potential of the group at which ion pairs form. Use of the relations

$$\mu_i = \mu_i^0 + kT \ln a_i$$
$$-kT \ln K_s^0 = \mu_{B+}^0 + \mu_{C-}^0 - \mu_{B+C-}^0 \tag{6.1.4}$$

enables us to write, after substitution of (6.1.4) into (6.1.2),

$$\left(\frac{\partial A_1}{\partial \alpha}\right)_{\{a_j\}} + 2NkT \ln a_{C-} - 2NkT \ln K_s^0 = 0. \tag{6.1.5}$$

6.2 The Electrical Free Energy

In this section we consider the calculation of the electrostatic free energy of the electrolyte solution under discussion. If a model is used which represents the bolion as having two discrete charges and occupying a spherically symmetric region of space from which other bolions are partially excluded, but into which the small ions and counterions may penetrate, it is possible to solve the Poisson–Boltzmann equation with only the Debye–Hückel approximations.[9] That is, the equation is linearized, and the potential of the bolion represented as spherically symmetric. Though this model makes the approximation of representing the mass distribution as continuous and spherically symmetric, this is far less serious than would be the approximation of replacing the discrete charges by a homogeneous charge distribution. In dilute solution, where the average distance between ions is very large relative to the separation of the charges within one bolion, the approximation of spherical symmetry ought to be quite good. In fact, as the solution approaches the state of infinite dilution all effects due to the separation of the charges within the bolion must disappear and the properties of the solution become independent of both the sizes of the ions and their charge distributions. The model described above also accounts for the removal of counterions from the solution by the formation of ion pairs with bolions. As might have been anticipated, it is found that the screened Coulomb potential is a valid first approximation to the potential of mean force between two charges in the solution. In general, due to the partial exclusion of other bolions from the region between the charges of any one bolion, the screening constant for the interaction of the fixed groups on the bolion differs from that in the bulk of the solution. This effect is very large for polyelectrolyte solutions where the region occupied by any one polyion contains a very large number of charges constrained by the polymer backbone to remain close together. However, owing to the small number of charges on a bolion, and to the fact that the region between charges is, in reality, readily accessible to all the small ions and to most of the charge due to others of its kind, the difference in screening constants will be small. To a good approximation, we may therefore neglect the difference and treat all charge-charge interactions with one screening constant. These considerations enable us to write for the electrical free energy of the solution, the expression

$$A_{elec} = \frac{N}{2} \sum_{i \neq j} \frac{q^2 e^{-\kappa|r_{ij}|}}{D_0 |r_{ij}|} - \frac{\kappa}{3D} \sum_j L_j q_j^2 - \frac{2N\alpha q^2 \kappa}{3D} - \frac{4N\alpha^2 q^2 \kappa}{3D}$$

$$\kappa^2 = \frac{4\pi}{DVkT} \left(\sum L_j q_j^2 + 2N\alpha q^2 + 4N\alpha^2 q^2 \right) \qquad (6.2.1)$$

where D_0 (which may differ from D) is the effective dielectric constant for the interaction of two groups on the same ion when separated by a distance $|r_{ij}|$. Note that, aside from the first term representing the self-energy of the bolions, the expression for the electrostatic free energy is exactly that which might have been deduced from the Debye–Hückel theory of ordinary electrolytes. Thus the second term in Eq. (6.2.1) is the electrostatic free energy of the added electrolyte, the third is that of the counterions, and the last is due to the bolions interacting as structureless particles. The first term corresponds to A_1 and the last three to A_2. Of the three terms comprising A_2, the electrostatic free energy of the counterions may be also visualized as the electrical free energy change on building ion atmospheres about each of the charged groups on a bolion. We now turn to a further specification of the term A_1.

The sum in Eq. (6.2.1) is to be carried out over all charged groups within the same molecule. In this section we shall develop a practical method of evaluating the sum and obtaining the degree of dissociation. We shall use the method of the Grand Partition Function since it is especially simple and elegant in the case under consideration. Moreover, the use of a Grand Partition Function [10] enables us to calculate the degree of dissociation, the extent of charge fluctuation, etc., by simple differentiation.

It will be seen that, due to the manner in which the free energy of the solution has been subdivided, the quantity to be calculated in this section is the electrical free energy due to the self-interactions of N *independent* particles. Now consider each of the $2N$ ionizable groups in the solution, each of which may be in one of two states, ionized or ion paired. With each group let us associate a state variable η_i, $i = 1, 1', 2, 2', \cdots, N, N'$, where we temporarily regard the ions as distinguishable. The η_i may have two values, 0 and 1, $\eta_i = 0$ referring to an ion paired group and $\eta_i = 1$ to an ionized group. The total electrostatic energy may therefore be written

$$E = \sum_{i=1}^{N} \frac{\eta_i \eta_i' q^2 e^{-\kappa|r_{ij}|}}{D_0 |r_{ij}|} = \sum_{i=1}^{N} \chi_i \eta_i \eta_i' \tag{6.2.2}$$

which states that the total energy is the sum of the energy of all the bolions. Equation (6.2.2) is equivalent to the statement of the independence of the ions made above. We next introduce a semigrand partition function $\Xi(\lambda, T)$ defined by the relation

$$\Xi(\lambda, T) = \sum_{\{\eta_i\}} e^{-E/kT} \lambda^{\Sigma \eta_i + \eta_i'} \tag{6.2.3}$$

where λ plays the role of an absolute activity whose value is determined by the total charge due to the bolions and where the summation is to be carried out over all possible values of the η_i. Introducing (6.2.2) into (6.2.3) we immediately obtain

$$\Xi(\lambda, T) = \left(\sum_{\{\eta_i \eta_i'\}} e^{-\chi \eta_i \eta_i'/kT} \lambda^{\eta_i + \eta_i'} \right)^N \tag{6.2.4}$$

which may be evaluated by inspection, yielding the relation

$$\ln \Xi(\lambda, T) = N \ln (1 + 2\lambda + \lambda^2 e^{-\chi/kT}). \tag{6.2.5}$$

In Eq. (6.2.4) and (6.2.5) it has been assumed that the charge separation is the same in all bolions. When the number of atoms in the chain that separates the charges is small, this is a very good approximation. As the chain length between charges increases, the distribution of charge separations broadens. In that instance, it is easiest to assume that, for the purpose of calculating the electrostatic energy, all bolions have their charges separated by the average charge separation. This should be a quite adequate approximation for our purposes (see Appendix 6A).

Returning to the definition of $\Xi(\lambda, T)$, it is possible to see that

$$\left(\frac{\partial \ln \Xi}{\partial \ln \lambda}\right)_T = 2\alpha N \tag{6.2.6}$$

using the fact that the most probable value of $\sum \eta_i + \eta'_i$ may be identified with $2\alpha N$, the total charge in solution residing on the bolions. The free energy A_1 may now be expressed as

$$A_1 = -kT \ln \Xi(\lambda, T) + 2\alpha N k T \ln \lambda. \tag{6.2.7}$$

Standard procedures then enable us to obtain the relations

$$\alpha = \frac{\lambda + \lambda^2 e^{-\chi/kT}}{1 + 2\lambda + \lambda^2 e^{-\chi/kT}} \tag{6.2.8}$$

$$\frac{S_1(\alpha)}{2Nk} = \tfrac{1}{2} \ln \left(1 + 2\lambda + \lambda^2 e^{-\chi/kT}\right) - \alpha \ln \lambda + \frac{\chi}{2kT} \frac{\lambda^2 e^{-\chi/kT}}{1 + 2\lambda + \lambda^2 e^{-\chi/kT}} \tag{6.2.9}$$

$$\frac{1}{2Nk}\left(\frac{\partial S_1(\alpha)}{\partial \alpha}\right) = -\ln \lambda + \frac{\chi}{kT} \frac{\lambda(1 + \lambda) e^{-\chi/kT}}{1 + 2\lambda e^{-\chi/kT} + \lambda^2 e^{-\chi/kT}} \tag{6.2.10}$$

where $S_1(\alpha) = -(\partial A_1(\alpha)/\partial T)$ is the entropy arising from the distribution and interaction of the charges. From Eq. (6.2.8) it can be seen that when the interaction between charges on the same molecule tends to zero, then $\exp(-\chi/kT)$ tends to unity, and the degree of neutralization is given by

$$\alpha = \frac{\lambda}{1 + \lambda}. \tag{6.2.11}$$

By substitution of (6.2.5) and (6.2.7) in (6.1.5) we thus find for the equilibrium condition (6.1.5) the relation

$$\ln a_{c^-} + \ln \lambda - \ln K_s^0 = 0. \tag{6.2.12}$$

We can now calculate the equilibrium counterion activity or, if this is known, the equilibrium degree of dissociation, presuming only that we know the intrinsic dissociation constant K_s^0. In dilute solution, to a sufficient approximation, the activity coefficient term appearing in Eq. (6.2.12) may be neglected relative to the contributions from the other terms. At equilibrium then (in very dilute solutions)

$$\lambda = \frac{K_s^0}{c_{c^-}}. \tag{6.2.13}$$

The equilibrium constant for the secondary dissociation reaction may be written

$$K_2 = c_{c^-} \frac{K_s^0(2\alpha - 1) + \alpha c_{c^-}}{2K_s^0(1 - \alpha)} \qquad (6.2.14)$$

where we have put $K_1 = 2K_s^0$. Note that Eq. (6.2.14) is defined for the hypothetical ideal solution, or for a solution sufficiently dilute that activity coefficients may be approximated by unity. Since K_2 must obviously be independent of concentration, it will have the same value in all real solutions, and the activity coefficients are just those factors which correct K_2 to this constant value. We may now find by substitution of (6.2.13) into (6.2.8)

$$K_2 = \frac{K_s^0}{2} \exp\left(-\frac{\chi}{kT}\right) \qquad (6.2.15)$$

which is the Bjerrum relation. The value of the dielectric constant for inter-charge interaction is given by the Kirkwood–Westheimer theory, but the functional form (6.2.15) remains unaltered.

6.3   The Secondary Dissociation Constant as a Function of Chain Length

We consider first the ion pair dissociation constants and the intercharge separations required to fit the experimental data. The dissociation equilibrium discussed may be characterized by the mass action expression (6.2.14) and (6.2.8) and (6.2.12). These equations suffice for the calculation of the counter-ion activity or, if this is known, of the degree of dissociation presuming only that the intrinsic dissociation constant K_s^0 and the interaction energy χ are known. In sufficiently dilute solution the activity of the counterions may be approximated by the counterion concentration with the result expressed in (6.2.13) under the same conditions wherein Eq. (6.2.14) is accurate.

There are two matters which must be considered before the calculations can be made. These are the assignment of the dielectric constant for the interaction of the two charges on the same bolion and the magnitude of the intrinsic dissociation constant. The work of Hasted et al.[11] indicates that the dielectric constant in the immediate vicinity of a charge is very much lower than the bulk value. From studies of the dielectric properties of electrolytes in water these investigators found that the dielectric constant was as low as 5 at a distance of about 1 A and that it rose to the bulk value only when the distance from the ion had increased to about 4–5 A. In the particular case of bolaform electrolytes most of the data in the literature[3-6] refer to solutions in alcohols and other organic solvents of moderate to low dielectric constant. The dissociation data for the compounds chosen for comparison in this chapter refer to methanol solutions. Alcohols in general have less " quasi-crystalline " structure than water and though one would expect a similar decrease in dielectric constant as the ion is approached, this effect should not persist to as great distances from the ion as is the case in water. We shall use the bulk dielectric constant for the interaction of all charged groups separated by more than 4 A in methanol. Effective dielectric constants computed from the theory of Kirkwood and Westheimer[12] are in agreement with this choice. The assignment of the intrinsic dissociation constant presents problems of a different nature. Of course, we are guided by the knowledge that for a strong electrolyte the intrinsic dissociation constant must be large. There is, however, no reliable method of estimation available inasmuch as the dissociation constant may depend upon other factors than the bulk dielectric properties of the medium.[13] Therefore, the experimental value of the dissociation constant for one compound together with an assumed charge separation has been used to compare a value of K_s^0 and then this value of K_s^0 employed to calculate the dissociation constants for other homologous compounds. The values of K_s^0 and the assumed distance are not arbitrary since they must satisfy a self-consistency condition which will become apparent later.

Using 31.5 for the dielectric constant of methanol, and an assumed charge separation of 8.8 A, the observed dissociation constant (4.4×10^{-3}) of $Br(Me_3N)CH_2)_2CO_2(CH_2)_2CO_2(CH_2)_2NMe_3)Br$ leads to a computed intrinsic dissociation constant of 6.6×10^{-2}. This is comparable with experimental data for similar type dissociations of other quaternary ammonium salts. In Table 6.3.1 are listed the charge separations and the corresponding calculated values of K_2 for two similar compounds.

TABLE 6.3.1[a]

Compound	K_2(expt.)[3-6]	K_2(calcd.)	Charge separation A
$BrMe_3N(CH_2)_3NMe_3Br$	1.04×10^{-3}	1.05×10^{-3}	5.1
$BrMe_3N(CH_2)_5NMe_3Br$	1.58×10^{-3}	1.52×10^{-3}	5.7

[a] Reproduced from S. A. RICE, *J. Am. Chem. Soc.* **78**. 5247 (1956).

The implication of the computed distances will be discussed in the next section. Here we merely note that the increment of 0.3 A/CH_2 group indicated is far less than that expected for a completely extended chain.

Before leaving this section it is pertinent to make some comments about the nature of the equilibrium constant K_2. Fuoss[6] has proposed that K_2 is not a true equilibrium constant determined by the usual criterion of thermodynamic equilibrium but rather depends upon some geometric factors which determine the kinetics of the association reaction at the charge site.

It is important to note that on general grounds the use of a steric factor cannot be correct since: (a) the use of the steric factor suggested by Fuoss would lead to a difference between primary and secondary dissociation constants in a molecule with noninteracting groups other than that due to the symmetry properties of the molecule, and (b) effective volume (kinetic) blocked by the molecule should be effectively independent of charge separation. Deduction (a) is in direct conflict with general and well-established statistical principles and the comment (b) leads to conclusions at variance with experiment. Further, if it is to be claimed that the steric factor appears in both the primary and secondary dissociations, then it will cancel when their ratio is taken. Since the Bjerrum theory discusses only the ratio of dissociation constants, it cannot be used to deduce evidence for such a steric effect. For these reasons Fuoss' proposal must be rejected.

6.4 Configurational Properties of the Molecule

We now turn to an examination of the change in chain configuration with a change in charge state or ionic strength. As previously noted, Eq. (6.2.12) is valid only if there is no change in the dimensions of the molecule when charged. If such a change does occur, then the free energy arising from the interaction between charged groups on the same bolion consists of both the electrostatic self-energy of the ion and the free energy change on expansion. Under these conditions Eq. (6.2.12) is modified to read

$$\ln a_{c-} + \ln \lambda - \ln K_s^0 + \left(\frac{\partial A_{\text{expansion}}}{\partial \alpha}\right) = 0. \qquad (6.4.1)$$

Under many conditions the expansion free energy is small relative to the other contributions[7] so that the dominant forces at equilibrium are just those already discussed. If the molecules are sufficiently long the distribution of chain lengths will be Gaussian. For a Gaussian distribution of end-to-end separations it may be shown readily that

$$\frac{h_1^*}{h_0^*} = \left[1 + \frac{v^2 q^2 e^{-\kappa h_1^*}}{3DkTh_1^*}(1 + \kappa h_1^*)\right]^{\frac{1}{2}} \qquad (6.4.2)$$

with v the number of charges per chain end, h_1^* the most probable separation of the chain ends in the given charge state, and h_0^* the unperturbed (uncharged molecule) most probable separation of the chain ends. The difference between Eq. (6.4.2) and the similar relation describing the configurational properties of polyelectrolytes[14] arises from a difference in the distribution of charges along the molecular skeleton.

The applicability of Eq. (6.4.2) to the compounds under discussion is questionable since it is well known that chains containing small numbers of

links do not have a Gaussian distribution of end to end distances. In connection with a different problem, machine computations have been made by
other investigators of the probability of a given end-to-end distance for
chains of 5, 6, 7, and 8 links, with a fixed bond angle of 110° and where one
link in each case is 2.0 A long and all the others are 1.5 A in length.[15] Some
of the results are plotted in Figs. 6.4.1, 6.4.2. In Fig. 6.4.1 there also

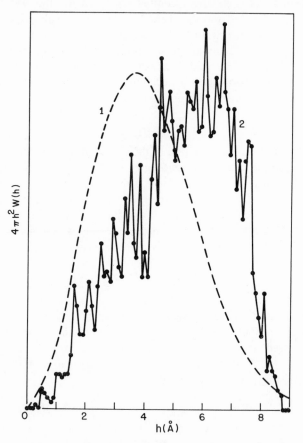

FIG. 6.4.1. 8 links, $\langle h^2 \rangle^{\frac{1}{2}}_{\text{Gaussian}} = 4.44$ A; contour length $= 10.3$ A; curve 1—
Gaussian distribution with free rotation; curve 2—machine calculations.

appears the probability of a given end-to-end distance as computed from
random statistics with an average bond length dictated by the chain composition mentioned above. As expected, in all cases the machine calculations
indicate that the chain is more extended than the equivalent random chain.
Of greater importance is the manner in which the distribution of chain lengths

changes with distance. From Fig. 6.4.1 it is seen that the probability of a given end-to-end distance tends to zero extremely rapidly once the most probable end-to-end distance has been exceeded. This decay is much faster than the similar decline for the equivalent randomly coiled chain. Thus, it will require a much greater force to extend the real chain (machine calculations) than the randomly coiled chain. The expansion of the equivalent random chain may be estimated from Eq. (6.4.2). When $\kappa h_1^* \approx 0$ so that the electrostatic interactions are at maximum strength, a random chain of seven links without fixed

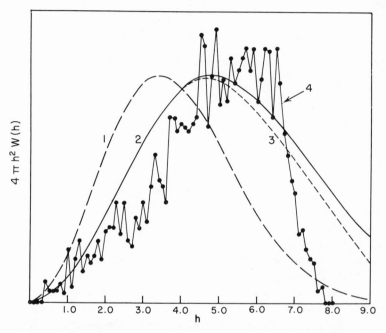

FIG. 6.4.2. 7 links, $\langle h^2 \rangle_{\text{Gaussian}}^{1/2} = 4.18$ A; contour length $= 9.06$ A. Curve 1—Gaussian distribution with free rotation; curve 2—Gaussian distribution with fixed bond angles; curve 3—inverse Langerin function; curve 4—machine calculation. [Reproduced from S. A. RICE, *J. Am. Chem. Soc.* **80**, 3207 (1958).]

bond angles expands from 4.18 to 5.89 A, a factor of 1.41, whereas a random chain of seven links with fixed bond angles of 110° but free rotation about each bond would expand from 5.90 to 7.7 A, a factor of 1.31. Both of these computations greatly overestimate the expansion due to the fact that a Gaussian distribution of end-to-end separations is only valid when $h_1^*/h_{\max} \leqslant 0.3$. Since the contour length of the molecule $h_{\max}$ is only 9.06 A, this condition is clearly violated. The physical reason for the overexpansions is that the Gaussian distribution overweights configurations in which the chain ends are far apart by failing to take cognizance of the finite contour length of the molecule. An estimate of the error incurred is obtained easily. The Gaussian distribution of end-to-end separations is only an approximation to the correct distribution function defined by

$$W(h_1)\,dh_1 = \text{const} \exp\left[-\frac{h_{max}}{Zb^2}\int_0^{h_1} L^*(x)\,dx\right]4\pi h_1^2\,dh_1 \qquad (6.4.3)$$

where b is the length of the monomer and Z the degree of polymerization. With the approximation[16]

$$L^*(x) = 3x\left[1 + \frac{0\cdot6x^2}{1-x^2}\right] \qquad (6.4.4)$$

where $L^*(x)$ is the inverse Langevin function, the free energy change on stretching the molecule from h_0^* to h_1^* becomes

$$A_{expansion} = \frac{3kT}{2}\left[0\cdot4\left(\frac{h_1^*}{h_0^*}\right)^2 - 1 - 0\cdot6\left(\frac{h_{max}}{h_0^*}\right)^2\ln\left\{1 - \left(\frac{h_1^*}{h_{max}}\right)^2\right\}\right] \qquad (6.4.5)$$

which leads to the relation

$$\frac{h_1^*}{h_0^*} = \left[\frac{1 + v^2q^2e^{-\kappa h_1^*}(1 + \kappa h_1^*/3DkTh_1^*)}{0\cdot4 + 0\cdot6/[1-(h_1^*/h_{max})^2]}\right]^{1/2}. \qquad (6.4.6)$$

Using a charge separation of 5.90 A for the unperturbed state, Eq. (6.4.6) predicts that the chain will only expand by a factor of 1.13 to 6.63 A. Moreover, when κh_1^* is not small, these expansions will be markedly reduced. The physical basis for the trend in expansions indicated above is qualitatively clear from an examination of Fig. 6.4.1. The probability of large end-to-end separations is smaller for the distribution function (6.4.3) than for a Gaussian distribution function. For a real chain (machine calculations) the probability of large chain extensions is still smaller than for the chain described by the distribution function (6.4.3). It is clear that, when the weighting of large chain extensions is decreased, the root mean square separation of the chain ends must also decrease. The obvious conclusion to be drawn is that to a first approximation the equilibrium end to end separation of short chains remains effectively constant. Care must be exercised in the use of this deduction which may be an adequate approximation for equilibrium properties but which also must certainly be more inaccurate for the description of transport phenomena.

The internal consistency of the proposed bolion model requires that the mean end-to-end separation of the chain ends be the same as computed either from the secondary dissociation constant K_2 or from the distribution functions plotted in Figs. 6.4.1 and 6.4.2. From Fig. 6.4.1 it is noted that the equilibrium separation of the chain ends of the uncharged chain of seven links is between 5.5 and 6.5 A. This corresponds very well to the computed charge separation

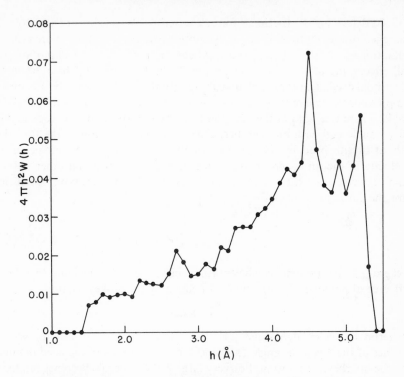

FIG. 6.4.3. 5 links, $\langle h^2 \rangle^{\frac{1}{2}}_{\text{Gaussian}} = 3.73$ A; contour length $= 6.85$ A. [Reproduced from S. A. RICE, *J. Am. Chem. Soc.* **80**, 3207 (1958).]

of 5.7 A. Figure 6.4.3 is a plot of the probability of a given end-to-end distance for a chain of five links. It is seen readily that the probability that the chain be maximally extended is very large. Therefore, for the compound BrMe₃N(CH₂)₃NMe₃Br, consisting of four links, it may reasonably be assumed that the chain is maximally extended. This would correspond to a charge separation of 4.93A in good agreement with the computed value of 5.1 A. This is the first of our internal checks. The use of a different value of K_s^0 would not result in such good agreement so that both the arbitrary choice of charge separation and K_s^0 is severely restricted.

6.5   Transport Properties at Infinite Dilution

In this section we shall discuss briefly the conductance of solutions of bolaform electrolytes with a view to ultimate application of similar ideas to polyelectrolyte solutions. Our attention will be focused on testing the relation between conductance and frictional coefficient for consistency with the charge separations already computed. We reserve for the following section our comments about the effect to be expected at finite concentrations.

To determine the frictional forces exerted by the solvent upon a molecule dragged through it the following assumptions are made: (a) the hydrodynamic situation can be described by the usual equations of viscous fluid motion with all effects due to the molecular nature of the flow process absorbed into a frictional coefficient for the chain subunit. (b) the molecule may be adequately approximated by a model in which a series of beads are connected by massless rods. There is a bead at the junction of every two rods, and the rods make a fixed angle $\pi-\theta$ with one another. Due to the perturbation of the flow field about the jth bead by the presence of the other parts of the molecule, the velocity of the jth bead contains components arising from all other parts of the molecule. For this model it can be shown that the frictional coefficient for the entire molecule is[17]

$$\zeta = \frac{Z\zeta_0}{1 + (a_0/Z) \sum_p [(Z - p)/r_p]} \tag{6.5.1}$$

where ζ_0 is the frictional coefficient for a single bead, Z is the number of beads (degree of polymerization), and a_0 is defined by the Stokes relation

$$\zeta_0 = 6\pi a_0 \eta_0. \tag{6.5.2}$$

Peterlin has calculated values of $Z\zeta_0/\zeta$ for $Z = 1$ to $Z = 100$ for various values of the " valence angle "$\pi-\theta$ and for several ratios of the bead diameter to the interbead separation. For very large Z the molecule becomes randomly coiled and the frictional coefficient becomes

$$\zeta = \frac{Z\zeta_0}{1 + 8/3\sqrt{6/\pi}(Za_0/\langle h_1^2 \rangle^{1/2})} \tag{6.5.3}$$

whereas if the molecules were rodlike the frictional coefficient would be[18]

$$\zeta = \frac{6\pi\eta_0 b_0 Z}{\ln Z - [1 - (6\pi\eta_0 b_0/2\zeta_0)]} \tag{6.5.4}$$

where b_0 is the interbead separation. Note the difference between this model and the more familiar rigid ellipsoid for which [19]

$$\zeta = \frac{6\pi\eta_0(b'^2 - a'^2)^{1/2}}{\ln[(b'/a') + \sqrt{(b'/a')^2 - 1}]} \xrightarrow{b' \gg a'} \frac{6\pi\eta_0 b'}{\ln(2b'/a')} \tag{6.5.5}$$

where b' and a' are the semimajor and semiminor axes of the ellipsoid and η_0 is the viscosity of the solvent.

It usually has been assumed tacitly that molecules as small as those under consideration can be adequately approximated as rigid ellipsoids. Accordingly, if the parameter a' or ζ_0 is fixed by one experimental point, then the lengths or the conductances of a homologous series can be calculated from Eq. (6.5.4) or Eq. (6.5.5). For the conductance of the polymethylene salts in methanol, fixing a' in Eq. (6.5.5) with the experimental conductance of $BrMe_3N(CH_2)_5NMe_3Br$ and the length 9.8 A (5.7 A for the charge separation plus twice the covalent radius of a methyl group), an excellent correlation is found between the computed lengths and the number of atoms in the chain. The conductance data require that there be an increase of 0.89 A per CH_2 group added. The increment to be expected if the chain were fully extended is 1.32 A. This is inconsistent with the increment of 0.3 A per CH_2 obtained from the charge separations required to fit the secondary dissociation constant. The discrepancy may be attributed to the fact that even molecules as small as these are far from rigid ellipsoids.

A better model to use for these compounds is a short stiff chain as described previously. Arbitrarily choosing $a_0/b_0 = 0.5$ (other choices have little effect) and using the same compound as above to fix the frictional coefficient per bead, ζ_0, the conductances for all the members of the series that have been measured are shown in Table 6.5.1. Two different choices of the " valence

TABLE 6.5.1[a]

n	Λ_∞^{++}(expt.)	Λ_∞^{++}(rigid ellipsoid)[b] [Eq. (6.5.5)]	Λ_∞^{++}(short chain) [Eq. (6.1.2)] $\cos \theta = 0.5$	$\cos \theta = 0.3$	$2b'$ (A, rigid ellipsoid)[c]
3	(85)	80.7	90.7	89.7	(8.1)
4	81.1	80.0	85.5	85.0	9.0
6	75.4	78.2	75.3	76.3	11.0
10	66.8	74.5	65.0	66.3	14.3

[a] Reproduced from S. A. RICE, *J. Am. Chem. Soc.* **80**, 3207 (1958).
[b] Computed using 0.3 A. CH_2 group obtained from the charge separations for $n = 3.5$.
[c] Length required to obtain exact agreement between Eq. (6.5.5) and experiment.

angle " have been made, $\cos \theta = 0.5$ corresponding to a moderate barrier to rotation when the true valence angle is $110°$, and $\cos \theta = 0.3$ corresponding to no barrier to rotation for the same valence angle.

The rigid ellipsoid model predicts a rate of change of the limiting conductance with concentration which is much too small to fit the data. On the other hand, despite the discrepancies between the observed and calculated conductances for the shortest chains, the chain model is in much better overall agreement with experiment. What is of even greater moment is that the use of the chain model is internally consistent with the charge separations computed from the secondary dissociation constant, whereas a glance at the last column of Table 6.5.1 will demonstrate strikingly the extent of the inconsistency between the rigid ellipsoid model and the considerations previously cited. For instance, for $n = 3$ using the charge separation from Table 6.5.1 plus twice the covalent radius of a methyl group leads to a prediction of $2b' = 9.2 \, A$, whereas experiment requires $2b' = 8.1 \, A$ on a rod model.

6.6   The Diffusion Coefficient in Dilute Solution

At any finite concentration the speed with which a bolion moves under the influence of an applied force depends upon the well-known ion atmosphere relaxation and electrophoretic effects. Only at extreme dilution where $1/\kappa \gg h_1^*$ is it a valid approximation to assume that the bolion is a divalent ion and that the ion atmosphere is spherically symmetric. When $1/\kappa \leqslant h_1^*/2$ it is a more reasonable approximation to assume that the atmospheres about each of the charges are almost independent. If it be further assumed that these atmospheres are nearly spherically symmetric, then the relaxation and electrophoretic effects for a bolion are just twice those for an ordinary ion. In the intermediate range the ion atmosphere is not spherically symmetric. Whereas the limiting behavior at infinite dilution is the same for bolions and ordinary divalent ions, this similarity will not persist at higher concentrations. To the effects already mentioned must be added the change in frictional coefficient due to changes in the intercharge separation. Since the intercharge separation is concentration-dependent (even though only slightly so for short chains), there will be a continuously increasing difference between a bolion and a divalent ion as the concentration is increased from that corresponding to infinite dilution. These same considerations arise in the discussion of the conductance of polyelectrolytes. Thus, if the number of charges is small and $1/\kappa < R/2$, where R is the charge separation, the atmospheres may be thought of as independent. However, when $1/\kappa \sim R$ the ion atmospheres overlap and form a sort of sausage casing around the polyion. Concerning the other limiting case, it is doubtful if solutions can be obtained which are sufficiently dilute that the total atmosphere of a polyion is spherically symmetric. Moreover, the effects of changing polyion dimension with changing concentration are very large and probably dominate the other two terms in the frictional force.

Consider a completely dissociated bolion. The effect of association will be discussed later. To calculate the frictional coefficient or the diffusion coefficient of a bolion, it is necessary to consider the mutual perturbation of the flow fields about each ion due to the presence of other ions. This perturbation is of the same origin as the perturbation to the flow field about one bead arising from another bead on the same molecule. Let the fluid be moving with a uniform velocity $\mathbf{C}$ in the x direction. If a force $\mathbf{F}$ with components F_x, F_y,

F_z acts at the origin, then the increment in the velocity of the fluid at the point r is, in component form,

$$\delta u = \frac{1}{8\pi\eta}\left(F_x\nabla^2\phi - F_x\frac{\partial^2\phi}{\partial x^2} - F_y\frac{\partial^2\phi}{\partial x\,\partial y} - F_z\frac{\partial^2\phi}{\partial x\,\partial z}\right)$$

$$\delta v = \frac{1}{8\pi\eta}\left(F_y\nabla^2\phi - F_x\frac{\partial^2\phi}{\partial x\,\partial y} - F_y\frac{\partial^2\phi}{\partial y^2} - F_z\frac{\partial^2\phi}{\partial y\,\partial z}\right) \qquad (6.6.1)$$

$$\delta w = \frac{1}{8\pi\eta}\left(F_z\nabla^2\phi - F_x\frac{\partial^2\phi}{\partial x\,\partial z} - F_y\frac{\partial^2\phi}{\partial y\,\partial z} - F_z\frac{\partial^2\phi}{\partial z^2}\right)$$

where, in general, ϕ is a function of r, the fluid density ρ, the fluid velocity $\mathbf{C}$, and the viscosity η given by

$$\phi = \frac{2\eta}{\rho C}\int_0^{\rho C(r-x)/2\eta}\frac{1-e^{-t}}{t}\,dt. \qquad (6.6.2)$$

When $\rho(\mathbf{C}(r-x)/2\eta)$ is small, ϕ may be expanded in a Taylor's series and only the first two terms retained. The result of this expansion when substituted into Eq. (6.6.1) is the well-known equation of motion[20]

$$\delta u = \frac{1}{8\pi\eta}\left(\frac{F_x}{r} + \frac{\mathbf{F}\cdot\mathbf{r}}{r^3}x\right)$$

$$\delta v = \frac{1}{8\pi\eta}\left(\frac{F_y}{r} + \frac{\mathbf{F}\cdot\mathbf{r}}{r^3}y\right) \qquad (6.6.3)$$

$$\delta w = \frac{1}{8\pi\eta}\left(\frac{F_z}{r} + \frac{\mathbf{F}\cdot\mathbf{r}}{r^3}z\right).$$

Consider now some one bolion surrounded by its counterions. Let the origin be located at the center of mass of the bolion. The total velocity imparted at the origin by the forces acting on the surrounding counterions is simply the integral of Eq. (6.6.3) over all space, i.e.,

$$u = \frac{1}{8\pi\eta}\int\left(\frac{F_x}{r} + \frac{\mathbf{F}\cdot\mathbf{r}}{r^3}\cdot x\right)d\tau$$

$$v = \frac{1}{8\pi\eta}\int\left(\frac{F_y}{r} + \frac{\mathbf{F}\cdot\mathbf{r}}{r^3}y\right)d\tau \qquad (6.6.4)$$

$$w = \frac{1}{8\pi\eta}\int\left(\frac{F_z}{r} + \frac{\mathbf{F}\cdot\mathbf{r}}{r^3}z\right)d\tau.$$

In the case of diffusion the bulk velocity of the fluid is zero and the driving force is just the gradient of the chemical potential. We thus have

$$\mathbf{F} \, d\tau = \sum_{i=1}^{r} (c_i \nabla \mu_i + c_i q_i \nabla \Psi) \, d\tau \qquad (6.6.5)$$

where the conventional but arbitrary separation of the chemical potential into " chemical " and " electrical " parts has been made, c_i is the concentration of species i (moles per unit volume of solvent), q_i the charge on the ion, and Ψ the diffusion potential.

To determine the force acting the electrostatic potential must be known. It can be shown readily that, to terms of order κ, the potential at a point P, a distance r_1 and r_2 from the two charges of the bolion, is given by[9, 21]

$$\psi = \frac{q_1 e^{-\kappa r_1}}{D r_1} + \frac{q_1 e^{-\kappa r_2}}{D r_2}$$

$$\kappa^2 = \frac{4\pi}{DkT} \sum_{i=1}^{r} c_i^0 q_i^2 \qquad (6.6.6)$$

where c_i^0 is the bulk concentration of species i and D is the dielectric constant of the system. Throughout we are treating the charge sites as point ions. To the same approximation that Eq. (6.6.6) represents the potential, the chemical potential of the bolion is[21]

$$\mu_1 = \mu_1^0(T, p) + kT \ln c_1 - \frac{q_1^2}{D} \left[\kappa - \frac{1}{h_1^*} (e^{-\kappa h_1^*} - 1) \right] \qquad (6.6.7)$$

where, as before, h_1^* is the most probable separation of the charges of a bolion. Equation (6.6.6) may be used in conjunction with the Boltzmann relation in the usual manner to determine the mean concentration of species i about the bolion. Thus

$$c_i = c_i^0 e^{-q_i \psi / kT}$$

$$= c_i^0 \left[1 - \frac{q_1 q_i e^{-\kappa r_1}}{D r_1 kT} - \frac{q_1 q_i e^{-\kappa r_2}}{D r_2 kT} \right] \qquad (6.6.8)$$

$$+ \cdots$$

and substitution of Eq. (6.6.8) into Eq. (6.6.5) gives

$$\mathbf{F} = \sum_{i=1}^{r} \left\{ c_i^0 \left[\frac{q_1 q_i e^{-\kappa r_1}}{D r_1 kT} + \frac{q_1 q_i e^{-\kappa r_2}}{D r_2 kT} \right] \nabla \mu + \left[\frac{q_1 q_i e^{-\kappa r_1}}{D r_1 kT} + \frac{q_1 q_i e^{-\kappa r_2}}{D r_2 kT} \right] c_i^0 q \, \nabla \Psi_i \right\}$$

$$(6.6.9)$$

where the Gibbs–Duhem relation and the condition of electro-neutrality have been employed. After carrying out the indicated differentiations and using Eq. (6.6.7) some algebraic manipulation leads to the result

$$\mathbf{F} = \frac{q_1}{DkT} \left[\frac{e^{-\kappa r_1}}{r_1} + \frac{e^{-\kappa r_2}}{r_2} \right] \left[(q_1^2 c_1^0 + q_2^2 c_2^0) \nabla \Psi \right] \tag{6.6.10}$$

where a term in $\nabla \kappa$ has been neglected as being very much smaller than the term retained.

It is convenient at this point to make an approximation in the hydro-dynamic treatment and replace the potential used in Eq. (6.6.6) by an average potential at a distance r from the center of mass of the bolion. If the electro-static potential of Eq. (6.6.6) is averaged over the surface of a sphere of radius r, the result is

$$\langle \psi \rangle = \frac{2q_1}{Dh_1^* r \kappa} e^{-\kappa r} [e^{\kappa h_1 */2} - e^{-\kappa h_1 */2}]. \tag{6.6.11}$$

Note that Eq. (6.6.11) reduces correctly to

$$\langle \psi \rangle_{h_1 *=0} = \frac{2q_1}{Dr} e^{-\kappa r} \tag{6.6.12}$$

in the limit as h_1^* tends to zero and the bolion becomes a divalent ion. The force $\mathbf{F}$ may now be written in the form

$$\mathbf{F} = \frac{4q_1}{DkT} \frac{e^{-\kappa r}}{h_1^* \kappa r} \sinh \frac{\kappa h_1^*}{2} (q_1^2 c_1^0 + q_2^2 c_2^0) \nabla \Psi. \tag{6.6.13}$$

Let the x axis be along $\nabla \Psi$. From Eq. (6.6.4) it is readily found that

$$\mathbf{u} = \frac{1}{2\pi\eta} \frac{q_1 \sinh(\kappa h_1^*/2)}{DkT h_1^* \kappa} (q_1^2 c_1^0 + q_2^2 c_2^0) \nabla \Psi \int \frac{e^{-\kappa r}}{r} \left(\frac{1}{r} + \frac{x^2}{r^3} \right) d\tau \tag{6.6.14}$$

and this integral is valuated easily in polar coordinates with the result that

$$\mathbf{u} = \frac{2q_1 \sinh(\kappa h_1^*/2)}{3\pi\eta h_1^*} \nabla \Psi. \tag{6.6.15}$$

This velocity represents the amount by which the motion of the central ion is decreased due to the fact that the flow field about it is perturbed by its atmo-sphere. The equations of motion for the bolion and a small ion now assume the form

$$\zeta_1 \left[\langle \mathbf{C}' \rangle - \frac{2q_1 \sinh(\kappa h_1^*/2)}{3\pi\eta c_1^0} \nabla \Psi \right] = -\nabla \mu_1 - 2q_1 \nabla \Psi$$

$$\zeta_2 \left[\langle \mathbf{C}' \rangle - \frac{q_2 \kappa}{6\pi\eta} \nabla \Psi \right] = -\nabla \mu_2 - q_2 \nabla \Psi \tag{6.6.16}$$

with ζ_1 and ζ_2 the frictional coefficients of the bolion and counterion, respectively, and $\langle C' \rangle$ the mean velocity (the same for bolion and counterions) with which the ions move in the absence of any perturbation. Equations (6.6.16) may be solved for $n\langle C' \rangle$ where

$$nq^2 = c_1^0 q_1^2 + c_2^0 q_2^2. \tag{6.6.17}$$

After some tedious algebra it is finally found that

$$n\langle C' \rangle = -kT\nabla n \, \frac{q_2 - 2q_1}{\zeta_1 q_2 - 2\zeta_2 q_1} \left[1 + \frac{2q_2 q_1 \kappa}{4DkT} + \frac{2q_2 q_1 \kappa^2 h_1^*}{8DkT} \right] \tag{6.6.18}$$

where a term in κ has been neglected as being negligible relative to the terms retained. In the limit as the separation of the charges tends to zero, Eq. (6.6.18) reduces to the well-known result for small electrolytes. The last term on the right-hand side contains a factor κ^2 but since h_1^* is of order κ^{-1} especially for the larger bolions, the total magnitude of this term is of order κ. The diffusion coefficient finally becomes

$$\mathscr{D} = \mathscr{D}_0 \left(1 + \frac{q_2 q_1 \kappa}{2DkT} + \frac{q_2 q_1 \kappa^2 h_1^*}{4DkT} \right) \tag{6.6.19}$$

where $\mathscr{D}_0$ the diffusion coefficient of the ideal solution, is itself concentration dependent through the dependence of the frictional coefficient ζ_1 on the intercharge separation. The asymmetry of the ion atmosphere is seen to lead to an additional retarding force on the bolion which ultimately reduces the flux of matter by about $\mathscr{D}_0 |\kappa^2 h_1^* q_2 q_1 / 4DkT|$. This term is of the same order of magnitude as the term in κ arising for ordinary electrolytes. The ratio of the third term to the second term is, in fact, $\kappa h_1^*/2$, which has the value 0.2 for a $10^{-3}N$ solution of the $n = 10$ compound (see Table 6.5.1) in methanol at 25°. It should further be noted that since the frictional coefficient of the bolion increases as the charge separation increases, the diffusion coefficient of a bolaform electrolyte will always be smaller than the diffusion coefficient of a corresponding ordinary 2 : 1 electrolyte.

When ion association is taken into account, it is necessary to note that much of the transport of solute occurs via singly charged and uncharged species. The electrophoretic effect is reduced to that at the appropriate ionic concentration, but a similar hydrodynamic interaction, due to the correlation in the relative motion of any two bodies, also exists for the uncharged species. The concentration dependence of $\mathscr{D}$ becomes still more complicated by the inclusion of this effect. If only the singly and doubly charged ions are present in the solution, then the electrophoretic effect may be calculated for each as in the preceding. The total diffusion coefficient may then be approximated as

$$\mathscr{D} = \alpha \mathscr{D}_{B++} + \tfrac{3}{2}(1 - \alpha)\mathscr{D}_{BC+} \tag{6.6.20}$$

and each term contains both the electrophoretic correction and the appropriate interaction correction, both tending to retard the motion of the diffusing entity.

6.7 Discussion

The purpose of this chapter has been to interpret the properties of solutions of bolaform electrolytes in terms of an internally consistent model with emphasis on the special role played by the intercharge separation and its variation with concentration. The extent to which self-consistency is attainable with reference to an available model is best judged by the reader in terms of the numerical examples cited.

A most important feature of solutions of bolaform electrolytes is the presence of ion pairs whose dissociation is described by the constant K_2. The problem of the molecular nature of the ion paired state is an old and still unsettled one. Nevertheless, recent considerations by Gilkerson[22] suggest that the most appropriate model describing an ion pair is one in which the ions are much more intimately associated than in the simple spatial classification scheme proposed by Bjerrum.[23] Whether or not this association involves the contact of solvated ions is immaterial. The important feature of the free volume model is that the ion pairs are much closer together than ion atmosphere dimensions and exclude a negligible portion of the configurations usually considered in the Debye–Hückel theory. Gilkerson's detailed molecular considerations supplement thermodynamic arguments previously presented for an intimate association of the ions in pair formation.

There is, of course, considerable interplay between the intercharge separation and the extent of ion pairing. Because the molecules under consideration are not rigid, the total free energy change on ionization must include contributions arising from the change in entropy when the molecular configuration is changed. Energy changes, which have not been considered in this chapter, may or may not accompany configurational changes depending upon whether or not there is considerable overlap of hydrogen atoms or other groups or whether internal hydrogen bonding exists, and so forth. Since the total free energy must be minimized simultaneously with respect to h_1 and α, it is not strictly correct to use the simple Bjerrum relation which assumes that there is no change in molecular configuration on ionization. The error involved may not be large for small molecules, but it certainly becomes larger as the limiting case of the polyion is approached.

In this chapter the theory of transport in solutions of bolaform electrolytes has been approached with emphasis on the asymmetric nature of the ion atmosphere. It has been shown that there exists a retarding force due to the hydrodynamic interaction of the charge cloud on one end of the bolion with the charge on the other end. This retarding force is in addition to the retarding forces that would exist for a spherically symmetric divalent ion. Moreover, both the charge-charge separation and the mean charge on a bolion are concentration-dependent (the latter due to incomplete dissociation). Since the steady state established in diffusion or conductance is the result of a complicated interplay between competing forces, the resultant concentration dependence of these properties can be very different from the pattern typical of small strong electrolytes. To treat the conductance it would be necessary to compute the retarding force due to ion atmosphere lag and relaxation, an effect which does not alter the diffusion coefficient because all ions move with the same mean velocity. It is anticipated that the asymmetry of the ion atmosphere plays a relatively larger role in relaxation phenomena, which themselves result from atmosphere asymmetry, than in electrophoretic phenomena which result from the bulk motion of the entire atmosphere. The corresponding problems for polyelectrolytes are still more complex.

Consider again the equilibrium properties of the solution, especially the question of ion pairing. The obvious cause of ion-pair formation is the net increase in energy when a singly ionized bolion becomes doubly ionized. If this increase in energy is larger than the gain in translational free energy for the freed counterion, then ion pairing is relatively favored. The extent of ion-pair formation depends on the concentration through the entropic part of the free energy since the gain in translational energy per counterion is larger the more dilute the solution.

More insight into the forces responsible for ion pair formation can be obtained from an examination of the entropy changes arising from the charge-charge interactions within one bolion. A schematic plot of Eq. (6.2.9) (Fig. 6.7.1) shows two peaks, corresponding to the placement of charge on one site and then on the other. Note that this entropy is lower than that for a bolion with no interaction between the charges. The physical changes corresponding to the entropy changes shown in Fig. 6.7.1 may be visualized as follows. At small degrees of ionization there are predominantly uncharged and single charged species in solution. Since these species differ, we may regard them as forming a mixture and there is a corresponding entropy of mixing. Alternatively, the entropy change may be regarded as arising from the number of ways in which $2\alpha N$ charges may be placed on a total of $2N$ possible sites. In the case of a bolion with noninteracting charges, the entropy of mixing passes through a single maximum as α increases, just as for an ideal solution. This is due to the fact that it requires the same amount of energy to remove a charge for all values of α. In the presence of charge-charge interactions, this situation is modified. As α increases, more and more bolions become singly ionized. In contrast to the previous case, to form a doubly ionized bolion requires that the energy of the solution be markedly increased. The concentration of doubly ionized ions remains low, therefore, and the

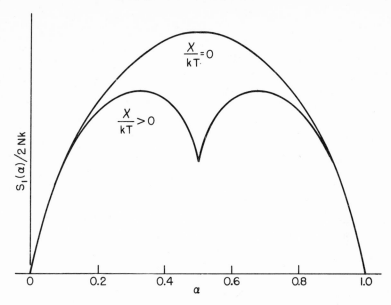

FIG. 6.7.1. A schematic representation of the entropy arising from the charge distribution in a bolaform electrolyte, in the presence and absence of interactions. [Reproduced from S. A. RICE, *J. Am. Chem. Soc.* **78**, 5247 (1956).]

increase in charge as α increases is due predominantly to an increasing concentration of the singly ionized species. Now, however, the solution contains a large number of singly ionized molecules and only a small number of doubly ionized or uncharged molecules. There is, therefore, a sharp decrease in the entropy of mixing as α approaches 0·5 due to the increasing chemical homogeneity of the solution. When α exceeds 0·5, there are once again a large number of energetically equivalent sites at which a charge can be placed. Doubly ionized species make their appearance and mix with the rest of the solution and the entropy goes through a second maximum as α tends to unity.

The difference between the entropy as a function of the degree of neutralization when there are charge-charge interactions and when there are no interactions indicates that a bolion with interactions will resist ionization more than one without interactions. The equilibrium constant for the secondary dissociation is accordingly lowered. If the bolaform electrolyte happens to be a dibasic acid, then as chemical neutralization is increased (by the addition of base, say) the bolion may bind the cation of the base to minimize its electrostatic free energy. In the case of the neutralization of a polyacid (or diprotic acid) in water, counterions may possibly be bound rather than protons because the proton concentration is dictated by the supplementary equilibrium governing the ion product of water. The proton concentration is therefore low. Only counterions are available to be bound to the bolion (or to the polymer if indeed there is any ion-pair formation with the polyion) and thereby lower its free energy. For a strong bolaform electrolyte, this binding appears as a smaller dissociation constant, K_2.

Thus, there appears to be a direct relationship between the mechanism of ion-pair formation and the lowering of K_2 for a bolaform electrolyte. Throughout the preceding analysis we have implicitly assumed that the mechanisms of ion-pair formation and acid radical-proton addition are the same. When we consider polyelectrolytes we shall see that, despite the success of this assumption for bolaform electrolytes, the evidence is against ion-pair formation in the sense implied in this chapter. Indeed, we shall find it necessary to consider in detail the distribution of counterions about a polyion and to interpret very carefully the physical basis for counterion localization.

Appendix 6A. The Effect of Chain Flexibility and the Calculation of Charge Fluctuations

Throughout the text we have assumed that the bolion is a rigid molecule, i.e., there is no change of configuration due to changes in charge state or ionic strength. If such a change in configuration does occur, the free energy A_1 consists of both the self-energy of the ion, as previously calculated, and the free energy change on expansion. Under these conditions, Eq. (6.2.12) is modified to read

$$\ln a_{c-} + \ln \lambda - \ln K_s^0 + \left(\frac{\partial A_{\exp}}{\partial \alpha}\right) = 0. \tag{6A.1}$$

As an example, we shall calculate $(\partial A_{\exp}/\partial \alpha)$ for a randomly coiled chain with allowance for the finite contour length of the molecule. The probability density that an uncharged chain will assume and end-to-end distance h_1 is given by the relation

$$W(h_1)\, dh_1 = \text{const} \exp\left\{-\frac{h_{\max}}{\langle h_0^2\rangle}\int_0^{h_1} L^*\left(\frac{h}{h_{\max}}\right) dh\right\} 4\pi h_1^2\, dh_1 \tag{6A.2}$$

where $h_{\max}$ is the contour length of the molecule, $\langle h_0^2\rangle$ its unperturbed mean square length, and $L^*(x)$ is the inverse Langevin function.[24] We shall approximate $L^*(x)$ by an expression employed by Kuhn and Grun, namely

$$L^*(x) = 3x\left(1 + \frac{0.6x^2}{1 - x^2}\right). \tag{6A.3}$$

If the bolion chain conatins v charges on each end, the free energy of expansion will be (after replacement of root mean square end-to-end distances by the equivalent most probable end-to-end distances)

$$A_{\text{exp}} = \frac{3kT}{2}\left[0.4\left(\frac{h_1^*}{h_0^*}\right)^2 - 1 - 0.6\left(\frac{h_{\max}}{h_0^*}\right)^2 \ln\left(1 - \left(\frac{h_1^*}{h_{\max}}\right)^2\right)\right]$$

$$- 3kT\ln\left(\frac{h_1^*}{h_0^*}\right) + \frac{\alpha^2 v^2 q^2}{D}\left[\frac{e^{-\kappa h_1^*}}{h_1^*} - \frac{e^{-\kappa h_0^*}}{h_0^*}\right] \quad (6A.4)$$

leading to the result

$$\left(\frac{\partial A_{\text{exp}}}{\partial \alpha}\right) = 3kT\left[0.4\,\frac{h_1^*}{h_0^{*2}} + 0.6\left(\frac{h_{\max}}{h_0^*}\right)^2 \frac{(h_1^*/h_{\max}^2)}{1 - (h_1^*/h_{\max})^2} - \frac{1}{h_1^*}\right]\frac{\partial h_1^*}{\partial \alpha}$$

$$+ \frac{2\alpha v^2 q^2}{D}\left[\frac{e^{-\kappa h_1^*}}{h_1^*} - \frac{e^{-\kappa h_0^*}}{h_0^*}\right] \quad (6A.5)$$

$$- \frac{\alpha^2 v^2 q^2}{D}\,e^{-\kappa h_1^*}\left(\frac{1 + \kappa h_1^*}{h_1^{*2}}\right)\frac{\partial h_1^*}{\partial \alpha}.$$

If the free energy of expansion, Eq. (6A.4), is minimized with respect to h_1^*, the equilibrium end-to-end separation becomes

$$\left(\frac{h_1^*}{h_0^*}\right)^2 = \left[1 + \frac{v^2\alpha^2 q^2 e^{-\kappa h_1^*}(1 + \kappa h_1^*)}{3DkTh_1^*}\right]\left[0.4 + \frac{0.6}{1 - (h_1^*/h_{\max})^2}\right]^{-1} \quad (6A.6)$$

from which $(\partial h_1^*/\partial \alpha)$ may be evaluated readily. Numerical evaluation reveals that the free energy of expansion is small relative to the other contributions to the free energy (see reference 7 for details of a similar calculation).

As a final note, it is of interest to calculate the charge fluctuations. Differentiation of $\Xi(\lambda, T)$ leads to the relations

$$\frac{1}{\Xi(\lambda, T)}\,\frac{\partial \Xi(\lambda, T)}{\partial \ln \lambda} = \langle Q \rangle$$

$$\frac{1}{\Xi(\lambda, T)}\,\frac{\partial^2 \Xi(\lambda, T)}{\partial (\ln \lambda)^2} = \langle Q^2 \rangle \quad (6A.7)$$

from which one readily obtains for the fluctuations in charge

$$\Delta^2 = \frac{\langle Q^2 \rangle - \langle Q \rangle^2}{N} = 4\alpha(1 - \alpha) - \frac{\lambda}{1 + 2\lambda + \lambda^2 e^{-\chi/kT}} \quad (6A.8)$$

where $\langle Q \rangle$ and $\langle Q^2 \rangle$ are the average and average square total charges. Note that this effect also exists in a bolion in which there are no interactions between the charges. Δ^2 is zero only when $\alpha = 0$, or $\alpha = 1$, when fluctuations are obviously physically impossible.

Appendix 6B.   Some Remarks about Conductance

A treatment of the conductance of solutions of bolaform electrolytes can be based on the same sphericalization approximation embodied in Eq. (6.6.11). The net effect of this approximation is the replacement of the true charge by an effective charge

$$q_{eff} = \frac{q^*}{h_1^* \kappa} \left[e^{\kappa h_1^*/2} - e^{-\kappa h_1^*/2} \right] \tag{6B.1}$$

which is concentration-dependent. To the first order of approximation, this effective charge can be used in the ordinary theory of conductance. The results

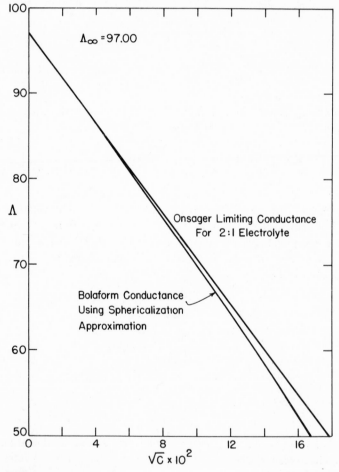

FIG. 6.B.1. A comparison of the theoretical conductances of the effective charge model and an ordinary 2:1 electrolyte.

of such a computation are shown in Fig. 6.B.1 for the case that h_1^* is 13 A, and the limiting conductance Λ_∞ is 97.00. Also shown in Fig. 6.B.1 is the conductance of the bolaform electrolyte treated as a 2 : 1 salt. There is a marked difference between the two curves, but the approximation is probably not valid at the higher concentrations depicted in the figure. We are indebted to Dr. V. S. Vaidhyanathan for the numerical computations quoted.

Appendix 6C. A Simple Model of Doubly Charged Ions

Beginning with a paper on zwitterions in 1932,[21] Kirkwood has published extensively on the extension of the Debye–Hückel theory to complex electrolytes.[25] Most of the models dealt with have been of spherical symmetry and those for which more general geometries were considered have been restricted to the case of zero net charge. In the following we present a simple model of the doubly charged ion, idealized as an ellipsoid of revolution. For the substances considered in this chapter, such a model is superior to those considered by other authors. The primary reason for this superiority is the omission of the assumption of spherical symmetry, either directly or in an approximation such as Eqs. (6.6.11) and (6B.1).

Consider then an ellipsoid of revolution with one charge at each focus. Inside the ellipsoidal surface and excluding the singularities due to the charges at the foci, the potential satisfies the Laplace equation,

$$\nabla^2 \psi = 0 \qquad (6C.1)$$

whereas external to the ellipsoidal surface the potential satisfied the linearized Poisson–Boltzmann equation,

$$\nabla^2 \psi = \kappa^2 \psi \qquad (6C.2)$$

$$\kappa^2 = \frac{4\pi}{DkT} \sum_i c_i^0 (z_i q)^2 \qquad (6C.3)$$

with, as usual, D the dielectric constant of the medium, q the unit charge, z_i the valence and c_i^0 the bulk concentration of the ionic species labeled i. Let the interfocal axis of the ellipsoid be $\mathfrak{m}$. We seek solutions of Eqs. (6C.1) and (6C.2) in prolate spheroidal coordinates defined by

$$\xi = \frac{r_1 + r_2}{\mathfrak{m}}; \quad \mu = \cos \vartheta = \frac{r_1 - r_2}{\mathfrak{m}} \qquad (6C.4)$$

with r_1 and r_2 the distances of a given point from the charges (foci). In this coordinate system the Laplacian operator ∇^2 becomes

$$\nabla^2 = \frac{4}{\mathfrak{m}^2 (\xi^2 - \mu^2)} \left[\frac{\partial}{\partial \xi} (\xi^2 - 1) \frac{\partial}{\partial \xi} + \frac{\partial}{\partial \mu} (1 - \mu^2) \frac{\partial}{\partial \mu} \right]. \qquad (6C.5)$$

The use of Eq. (6C.5) in Eq. (6C.2) and the assumption that

$$\psi(\xi, \mu) = R(\xi)S(\mu) \tag{6C.6}$$

leads to the equations

$$\frac{d}{d\xi}\left[(\xi^2 - 1)\frac{dR}{d\xi}\right] - [B - \mathfrak{p}^2\xi^2]R = 0 \tag{6C.7}$$

$$\frac{d}{d\mu}\left[(1 - \mu^2)\frac{dS}{d\mu}\right] + [B - \mathfrak{p}^2\mu^2]S = 0 \tag{6C.8}$$

where B is a separation constant and

$$2\mathfrak{p} = im\kappa \tag{6C.9}$$

with "i" the imaginary unit. When $\kappa = 0$, we obtain the formulation of the Laplace equation in this coordinate system. In this case we take the separation constant B to be $n(n + 1)$ whereupon Eqs. (6C.7) and (6C.8) reduce to the Legendre equation with solutions $P_n^{(0)}(\xi)$ and $P_n^{(0)}(\mu)$ respectively. The general solution inside the ellipsoid is then

$$\psi'' = \sum_{n=0}^{\infty} A_n P_n^{(0)}(\xi)P_n^{(0)}(\mu). \tag{6C.10}$$

For our purposes we require the potential ψ' defined by

$$\psi' = \sum_{n=0}^{\infty} A_n P_n^{(0)}(\xi)P_n^{(0)}(\mu) + \psi_s \tag{6C.11}$$

where ψ_s is the potential due to the point charges at the foci. Outside the ion, where $\kappa \neq 0$, the angular solution is[26]

$$S_{0l} = \sum_{\nu=0}^{\infty} d_{2\nu}(\mathfrak{p}/l)T_{2\nu}^{(0)}(\mu); \quad l \text{ even}$$

$$= \sum_{\nu=0}^{\infty} d_{2\nu+1}(\mathfrak{p}/l)T_{2\nu+1}^{(0)}(\mu); \quad l \text{ odd} \tag{6C.12}$$

where the d's are coefficients of a series expansion explicitly displayed in Morse and Feshbach,[26] and the functions $T^{(0)}$ are tesseral harmonics. For the radial solutions we take a linear combination of spherical Bessel functions, j_n, and Neumann functions, n_n, i.e.

$$R_{0l} = he_{0l} = \sum_{n}' (i)^{n-l} d_n(\mathfrak{p}/l)[j_n + in_n]$$

$$= je_{0l}(\mathfrak{p}\xi) + ine_{0l}(\mathfrak{p}\xi). \tag{6C.13}$$

The solution indicated converges to $(\kappa m \xi/2)^{-1} \exp(-\kappa m \xi/2)$ in the limit as $\xi \to \infty$, which is the desired behavior. In Eq. (6C.13) the prime on the summation indicates that n runs over even integers if l is even and over odd integers if l is odd. The combination in brackets on the right-hand side of Eq. (6C.12) is the Hänkel function of order n with generating function

$$h_n(y) = -iy^n(-)^n \left(\frac{d}{y\,dy}\right)^n \left(\frac{e^{iy}}{y}\right)$$

$$h_n = j_n + in_n. \tag{6C.14}$$

In our case, $y = (i\kappa m \xi/2)$ whereupon the first few Hänkel functions are

$$h_0 = -\frac{2}{\kappa m \xi} e^{-\kappa m \xi/2}$$

$$h_1 = \frac{2\kappa m \xi + 4}{(\kappa m \xi)^2} e^{-\kappa m \xi/2}$$

$$h_2 = -\frac{24 + 12\kappa m \xi + 2(\kappa m \xi)^2}{(\kappa m \xi)^3} e^{-\kappa m \xi/2}. \tag{6C.15}$$

For small values of $\mathfrak{p}$, Morse and Feshbach[26] give approximate formulas for the d_n from which we find

$$d_0 \cong 1 - \frac{(\kappa m)^2}{36} + \frac{7}{16,200} (\kappa m)^4$$

$$d_1 \cong 1 - \frac{(\kappa m)^2}{100} + \frac{144}{16 \times 55,125} (\kappa m)^4$$

$$d_2 \cong \frac{(\kappa m)^2}{36} + \frac{5}{16 \times 567} (\kappa m)^4. \tag{6C.16}$$

The complete solution outside the ellipsoid is then

$$\psi = \sum_{l=0}^{\infty} C_l S_{0l}(\mu) R_{0l}(\xi). \tag{6C.17}$$

We shall assume that it is a sufficiently accurate approximation to truncate the potential defined by Eq. (6C.17) after the $l = 1$ term,

$$\psi = C_0 S_{00} R_{00} + C_1 S_{01} R_{01}. \tag{6C.18}$$

With the neglect of powers of κm higher than the second,

$$S_{00} = \left(1 - \frac{(\kappa m)^2}{36}\right)P_0^{(0)}(\mu) + \frac{(\kappa m)^2}{36}P_2^{(0)}(\mu)$$

$$S_{01} = \left(1 - \frac{(\kappa m)^2}{100}\right)P_1^{(0)} + \frac{(\kappa m)^2}{100}P_3^{(0)}(\mu)$$

$$R_{00} = -2\left(1 - \frac{(\kappa m)^2}{36}\right)\frac{e^{-\kappa m \xi/2}}{\kappa m \xi} - \frac{(\kappa m)^2}{36}\left[24 + 12\kappa m \xi + 2(\kappa m \xi)^2\right]$$

$$\times \frac{e^{-\kappa m \xi/2}}{(\kappa m \xi)^3}$$

$$R_{01} = 4\left(1 - \frac{(\kappa m)^2}{100}\right)\left(1 + \frac{\kappa m \xi}{2}\right)\frac{e^{-\kappa m \xi/2}}{(\kappa m \xi)^2} \tag{6C.19}$$

which may be substituted into Eq. (6C.18) to obtain the potential ψ. The corresponding expansion of the potential inside the ellipsoid yields

$$\psi' = \frac{4zq}{Dm}\frac{\xi}{\xi^2 - \mu^2} + A_0 + A_1\xi\mu. \tag{6C.20}$$

To evaluate the coefficients (A's and C's) we neglect powers of μ greater than the first and use the two boundary conditions,

$$\psi'(\xi_0 = 2) = \psi(\xi_0 = 2)$$

$$\nabla\psi'(\xi_0 = 2) = \nabla\psi(\xi_0 = 2) \tag{6C.21}$$

at the surface of the ellipsoid. This gives the relations,

$$A_0 \cong \frac{4zq}{Dm}\left[\frac{39 + 3\kappa m}{90 + 90\kappa m + 10(\kappa m)^2} - \frac{1}{2}\right]$$

$$C_0 \cong -\frac{4zq}{Dm}\left[\frac{72\kappa m e^{-\kappa m}}{180 + 180\kappa m + 15(\kappa m)^2}\right]. \tag{6C.22}$$

In the limit as $\kappa \to 0$, the coefficient A_0 should vanish. In fact, due to the truncation error in the several expansions used, A_0 tends to the small value 0.06 as $\kappa \to 0$. This is a measure of the limited accuracy of our two term expansion for the potential.

To utilize the preceding results we require a method for determining the chemical potentials of the ionic species in the solution. For the case under consideration the usual Debye charging process is extremely difficult to apply since the potential is a complex function of κ. It is possible to use an alternative argument which is valid within the same range as the approximations already made in solving the Poisson–Boltzmann equation.[25] We note that the definition of the chemical potential of the bolion, μ_{BB},

$$\frac{\partial A}{\partial N_{BB}} = \mu_{BB} \qquad (6C.23)$$

corresponds to the integral relation

$$\int_{A_0}^{A} dA = \int_{N_{BB}=0}^{N_{BB}} \mu_{BB} \, dN_{BB} \qquad (6C.24)$$

where A_0 is the Helmholtz free energy in the absence of the bolion. If μ_{BB} varies only slowly with N_{BB} we can make the following approximation,

$$A = A_0 + N_{BB}\mu_{BB} \qquad (6C.25)$$

which should be very good when there are several ionic species present. It is now easy to find a potential function for the counterion to the bolion, for, by definition,

$$\psi_i = \frac{\partial A}{\partial q_i} = \psi_i^0 + N_{BB}\frac{\partial \mu_{BB}}{\partial q_i}. \qquad (6C.26)$$

To complete the calculation we require the chemical potential of the bolion. It may be shown that[21]

$$kT \ln \gamma_p^{(2)} = 2zqA_0. \qquad (6C.27)$$

From this we find

$$kT \ln \gamma_{Cl^-}^{(2)} = -\frac{z^2 q^2 \kappa}{12D}\left[\frac{30 + \frac{13}{2}\kappa m + \frac{1}{6}(\kappa m)^2}{(\frac{5}{2} + \frac{5}{2}\kappa m + \frac{1}{6}(\kappa m)^2)^2}\right] - \frac{z^2 q^2}{2D}\frac{\kappa}{1 + \kappa a_{Cl^-}} \qquad (6C.28)$$

for the activity coefficient of the counterion to the bolion.

Lapanje et al.[27] have measured the activity coefficient of chloride ion as the counterion to neutralized ethylene diamine (ED), diethylene triamine (DT), triethylene tetramine (TT), and tetra ethylene pentamine (TP). The results for the bolaform case are shown in Table 6.C.1. The numerical values will be further discussed in a later chapter. For the present we wish to test Eq. (6C.28).

Eq. (6C.28) has been used to compute numerical values for $\ln \gamma_{Cl^-}$ as a function of concentration and of separation between the two charges of the bolion. Consider first the variation in γ_{Cl^-} due to changing the charge separation in the bolion. The experimental data are given in Table 6.C.1 and curve B of Fig. 6.C.1. Curve A of Fig. 6.C.1 is the computed variation of $\ln \gamma_{Cl^-}^{(2)}$. We observe that the theoretical slope is in quantitative agreement

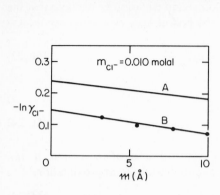

FIG. 6.C.1. Dependence of $\ln\gamma_{Cl^-}$ on the length $\mathfrak{m}$ of the interfocal axis of the two-charge ellipsoidal model. $\kappa = 0.050 \times 10^8$ cm^{-1}.

FIG. 6.C.2. Concentration dependence of $\ln\gamma_{Cl^-}^{(2)}$; curve T — theoretical, and curve E — experimental, for ED (HCl)$_2$.

TABLE 6.C.1

MOLAL ACTIVITY COEFFICIENTS OF CHLORIDE IONS
IN SOLUTIONS OF DIHYDROCHLORIDES

	Molality	ED(HCl)$_2$	DT(HCl)$_2$	TT(HCl)$_2$	TP(HCl)$_2$
1	1.0	0.437	0.428	—	—
2	0.8	0.453	0.451	—	—
3	0.6	0.479	0.482	—	—
4	0.5	0.497	0.501	0.484	—
5	0.4	0.525	0.525	—	—
6	0.3	0.565	0.557	—	—
7	0.2	0.610	0.602	0.601	0.591
8	0.1	0.683	0.686	0.680	0.685
9	0.08	0.703	0.709	0.702	0.712
10	0.06	0.736	0.741	0.735	0.749
11	0.05	0.754	0.757	0.756	0.767
12	0.04	0.774	0.778	0.770	0.792
13	0.03	0.796	0.800	0.801	0.823
14	0.02	0.820	0.828	0.836	0.856
15	0.01	0.873	0.879	0.890	0.900
16	0.009	0.875	0.886	0.896	0.907
17	0.008	0.878	0.896	0.901	0.918
18	0.007	0.882	0.908	0.914	0.929
19	0.006	0.891	0.921	0.925	0.943
20	0.005	0.904	0.936	0.942	0.960

with experiment, but that $\gamma_{Cl^-}^{(2)}$ (theory) is about a factor of 1.10 too small at the concentration for which the data are plotted. This error of 10% arises from the truncation error in the potential and the resultant inaccuracy in the coefficients determined from the boundary conditions. It is our opinion that this agreement is satisfactory. As can be seen in Fig. 6.C.2, the predicted concentration dependence of $\gamma_{Cl^-}^{(2)}$ is almost parallel to but displaced from the experimental values. The discrepancy is seen to be approximately linear in $\sqrt{m_{Cl^-}}$, again suggesting that the dominant source of error is the truncated potential.

REFERENCES

1. N. BJERRUM, *Z. physik. Chem.* **106**, 219 (1923).
2. J. G. KIRKWOOD AND F. H. WESTHEIMER, *J. Chem. Phys.* **6**, 506 (1958).
3. R. M. FUOSS AND D. EDELSON, *J. Am. Chem. Soc.* **73**, 269 (1951).
4. R. M. FUOSS AND V. H. CHU, *J. Am. Chem. Soc.* **73**, 949 (1951).
5. H. EISENBERG AND R. M. FUOSS, *J. Am. Chem. Soc.* **75**, 2914 (1953).
6. O. V. BRODY AND R. M. FUOSS, *J. Phys. Chem.* **60**, 156 (1956).
7. F. E. HARRIS AND S. A. RICE, *J. Phys. Chem.* **58**, 725, 733 (1954).
8. S. A. RICE AND F. E. HARRIS, *J. Chem. Phys.* **24**, 326, 336 (1956).
9. F. E. HARRIS AND S. A. RICE, *J. Chem. Phys.* **25**, 955 (1956).
10. G. S. RUSHBROOKE, "Introduction to Statistical Mechanics." Oxford Univ. Press New York, 1949.
11. J. B. HASTED, D. M. RITSON, AND C. H. COLLIE, *J. Chem. Phys.* **16**, 1 (1948).
12. J. G. KIRKWOOD AND F. H. WESTHEIMER, *J. Chem. Phys.* **6**, 513 (1938).
13. C. A. KRAUS, *Ann. N. Y. Acad. Sci.* **51**, 789 (1949).
14. F. E. HARRIS and S. A. RICE, *J. Polymer Sci.* **15**, 151 (1955).
15. F. A. COTTON AND F. E. HARRIS, *J. Phys. Chem.* **60**, 1451 (1956).
16. W. KUHN AND F. GRÜN, *Kolloid-Z.* **101**, 248 (1942).
17. A. PETERLIN, *J. chim. phys.* **47**, 7–8, 669 (1950); *Ibid.* **48**, 1–2, 13 (1951).
18. J. RISEMAN AND J. G. KIRKWOOD, *J. Chem. Phys.* **18**, 512 (1950).
19. F. PERRIN, *J. phys. radium* **7**, 1 (1936).
20. J. J. HERMANS, *J. Polymer Sci.* **18**, 257 (1956).
21. G. SCATCHARD AND J. G. KIRKWOOD, *Physik. Z.* **33**, 297 (1932).
22. W. R. GILKERSON, *J. Chem. Phys.* **26**, 1199 (1957).
23. N. BJERRUM, *Kgl. Danske Videnskab. Selskab. Mat.-fys. Medd.* **7**, No. 9 (1926).
24. P. J. FLORY, "Principles of Polymer Chemistry." Cornell Univ. Press, Ithaca, New York, 1953.
25. J. G. KIRKWOOD, *J. Chem. Phys.* **2**, 351 (1934); "Proteins, Amino Acids and Peptides," (E. J. Cohn and J. T. Edsall, eds.), Chap. 12. Reinhold, New York, 1943.
26. P. M. MORSE AND H. FESHBACH, "Methods of Theoretical Physics." McGraw–Hill, New York, 1953.
27. S. LAPANJE, J. HAEBIG, H. T. DAVIS, AND S. A. RICE, *J. Am. Chem. Soc.* **83**, 1590 (1961).

7. The Electrostatic Free Energy of Flexible Polyelectrolytes

7.0 Introduction

We have already indicated the dominant role played by the electrostatic free energy of a polyion in determining the properties of polyelectrolyte solutions by giving in Chapter 5 a qualitative discussion of several models and in Chapter 6 detailed calculations for the especially simple case of bolaform electrolytes. In this chapter we consider in detail the calculation of the electrostatic free energy using the methods introduced by Harris and Rice.[1-3] The details of this chapter together with the qualitative survey of Chapter 5 will give the reader an over-all picture of the status of the theory. We do not think it necessary to reproduce all of the calculations that have been published, rather trusting to the general understanding of common features which is generated by the detailed study of any one model.

Our procedure shall be to follow very closely the arguments originally set forth by Harris and Rice. In this exposition we shall make extensive use of the concept of ion localization, referred to loosely as site binding, and defer all critical comment to Chapter 9. Similarly we shall make only crude calculations of the polyion expansion, again deferring critical comment to Chapter 10. In this way it will be possible to present an over-all picture of the theory in a reasonably connected fashion. Once such a picture is constructed its various elements can be analyzed in relation to one another.

A simple theory of the electrostatic contributions to the free energy of polyion solutions must differ significantly from the corresponding theory for solutions totally composed of small ions and molecules.[4] Although the same basic assumptions may be made in either case, the polymer solutions will differ from ordinary solutions because of the nonnegligible volume occupied by the polymer, and because numerous charges will be constrained to definite relative positions by virtue of their attachment to the polymer chains. These differences create and magnify effects whose consideration is unnecessary in ordinary electrolytic solutions. In this chapter is developed a first-order theory for polyion solutions which is analogous to the Debye–Hückel theory for ordinary solutions. One of the major differences is that the theory predicts, in the absence of added salt, that many of the counterions will be found inside the polyion at the concentrations of polymer ordinarily employed.

Some of the ideas discussed by Harris and Rice were first recognized by others as of importance in formulating a complete theory.[5-7] Marcus[8] reviewed some aspects of the problem, but his discussion was restricted to polyelectrolyte models which we feel are intrinsically inadequate to describe properly the free energy changes accompanying alteration of the configuration or charge of the polymeric ions. The treatment of Harris and Rice differs from some of the earlier work in that it considers the electrostatic free energy of polymers containing discrete charges, rather than continuous charge distributions. It is just this difference which permits a more realistic discussion[1] of the free energy changes accompanying alteration of the configuration or charge of the polymer. Katchalsky and co-workers[9] did consider chain polymers with discrete charges attached. The work to be discussed represents an improvement upon their approach in that the localization of counterions can be explicitly considered, and in that one of the primary objectives will be to justify the use of the screened Coulomb potential in calculating the interaction between charges at various locations on the polymer chains.

Another major aim of the theory is to set forth, explicitly, the physical nature of the electrostatic contributions to the free energy included in the formulas obtained here. It will then be possible to define unambiguously the other contributions to the free energy of any sufficiently completely specified polyion solution. The role of osmotic pressure, particularly in ion exchange resin systems, where there are fairly well-defined regions " inside " and " outside " the resin, may therefore be determined unequivocally.

7.1 General Considerations

The initial step in the formulation of the theory will be the determination of the electrostatic potential in the polyion solution as a function of position relative to both polymeric ions and mobile ions. From the potential distribution it will be possible to calculate the distribution of small ions relative to the polyions and to obtain, by a charging process, the electrostatic free energy of the system. It is not possible to proceed with complete generality, but the particular model to be adopted has been chosen so as to restrict the range of usefulness of the theory as little as possible.

The polyions will be regarded as occupying a spherically symmetric region, each volume element of which is partially available for occupancy by small ions. However, we make no assumptions about the manner in which the elements of the polymer are interconnected. The fixed charges of the polyions will be placed at definite positions within the polymer sphere. Changes in the configuration of the polyions will not be considered here, as the electrostatic free energy is to be computed for specified configuration of the polymer. The polyion spheres will be assumed not to be interpenetrable, and will therefore represent regions of exclusion to other polyions. The main weakness of this model is the replacement of a markedly inhomogeneous polymer at a given configuration with a homogeneous sphere. This replacement, however, is far less serious than would be replacement of the charge fixed on the polyion by a homogeneous distribution. For, replacement of the discrete charges by a homogeneous distribution not only renders impossible a consistent discussion of ion localization, but also precludes consideration of the electric saturation in the hydrated regions about the charges. Since as much as 75% of the charge of some polymers is neutralized by intimate association with counterions, and since water of hydration has a dielectric constant of approximately 5, a factor of 16 less than free water, it is reasonable to assert that the properties of a system of discrete charges will behave quite differently than will the same gross charge density distributed homogeneously. On the other hand, the most significant effect of homogenizing the mass distribution of a polymer can only be to remove sources of inhomogeneity in the ion atmospheres about the polymer charges. The distances between near neighboring polymer charges are not large relative to the ion atmosphere dimensions, so that the maximum possible inaccuracy from the mass homogenization cannot be very large. In addition to the approximation just discussed, the model here introduced also suffers from the disadvantage that the size of the equivalent polymer spheres is somewhat ill-defined and correspondingly difficult to characterize. We shall restrict the discussion to systems in which the polymeric ions occupy a small fraction of the total volume, so that it will turn out that, except at very low ionic strengths, the thermodynamic properties of the solution will not depend upon the value of this parameter. The methods of obtaining a numerical value for the radius of the equivalent sphere depend upon the nature of the polymeric species and will be discussed when necessary.

The potential will be calculated using the Poisson–Boltzmann equation, with the Debye–Hückel approximations.[4] However, the polyion theory will require a more critical examination of the approximation to the potential of average force, which governs the distribution of ions relative to a fixed charge distribution. The concentration of ions of a type j in a solution of small ions containing an ion i fixed in position may be expressed in the form

$$c_j = c_j^0 e^{-W_{ij}/kT} \tag{7.1.1}$$

where W_{ij} is the potential of average force between ions i and j, and c_j^0 is the average concentration of ions of type j. In the Debye–Hückel theory, the approximation is made that

$$W_{ij} = q_j \psi_i \tag{7.1.2}$$

where q_j is the charge of an ion of type j and ψ_i is the potential at the position of ion i calculated assuming that only the position of i is specified. This form for W_{ij} has been subjected to considerable analysis, particularly by Onsager and Kirkwood,[10] and is presumably valid at sufficiently low concentrations. It is implicitly assumed when using Eq. (7.1.2) that the free energy necessary to create an ion atmosphere about an ion of species j is independent of its position relative to the fixed ion i. Application of an equation similar to Eq. (7.1.1), to describe the distribution of mobile ions of type j relative to a polymeric ion, involves consideration of the potential of average force between an ion of type j and all the fixed charges of a polyion. However, in a polymeric system, the free energy of creating an atmosphere about an ion depends upon whether it is inside, or outside, the equivalent polymer sphere, as the local concentration of small ions will be affected by the volume excluded to small ions by the polymer. Therefore, the potential of average force must be modified so as to include atmosphere contributions which depend upon position.

Another matter which must be considered in using the Poisson–Boltzmann equation is the range of its validity, especially since deviations occur at increasingly small concentrations as the valency of the ions increases. Experience with small ion solutions indicates that use of the Poisson–Boltzmann equation in the ordinary manner for polyions will be expected to lead to significant error, since the large potentials created by the highly charged polymeric ions will seriously upset the statistics governing the distribution of the small ions. The Poisson–Boltzmann equation will be used therefore in the following way when calculating the potential relative to a polyion fixed in position. Each other polyion will be regarded as associated with the average number of small ions it will contain within its equivalent sphere and will be treated as a single charge of a magnitude equal to the sum of the charges of the polyion and associated ions. This effective charge in general will be much smaller than the fixed charge, and therefore will make the use of the Poisson–Boltzmann equation valid up to considerably higher concentrations than would otherwise be the case.

Next, consider the linearization approximation of the Debye–Hückel theory. Because the polymeric ions are relatively large and can be permeated by counterions, the potential does not reach the high values which otherwise would be associated with ions of such large charge. The most important regions of seriously high potential are just those in the neighborhood of individual sites. Since we wish to consider solutions where the bulk mobile ion concentrations are low enough that the Debye–Hückel approximations apply, we shall assume that these high potential regions near each fixed ion will not introduce serious error, except with regard to the interactions between the fixed charge groups of the polymer. The main effect of these interactions will be to place an abnormally large number of counterions in position where they will quite effectively neutralize the fixed charges of the polymer. To allow for this, we have added to the theory provision for the formation of ion pairs, each consisting of a counterion and a polymer charge in such close proximity that the pair exerts a negligible electric field at the positions of the neighboring polymer charges. A definite free energy change is associated with the formation of each ion pair, and polymer charges thus paired may be regarded as uncharged groups for the purposes of the subsequent development of this chapter. The ion pairing we have just introduced is different in nature and in purpose from that introduced by Bjerrum[11] to extend the Debye–Hückel theory to higher concentrations. The members of our ion pairs are much closer together than ion atmosphere dimensions, and exclude a negligible proportion of the configurations ordinarily considered in the Debye–Hückel theory. For this reason we need make no further explicit reference to the ion pairing in the development of this chapter. Introduction of this ion pairing need not be as arbitrary as it may seem, for it describes, to first order, the free energy contributions of the paired counterions and polymer charges. Application of the Poisson–Boltzmann equation to the modified system then enables the remainder of the electrostatic free energy to be caluated more accurately.

The above paragraph indicates that the electrostatic free energy should be computed on the assumption that any polymer charge involved in a site-bound ion pair should be regarded as uncharged for the purposes of this chapter, and that the estimation of the free energies of formation of the ion pairs may be deferred for consideration elsewhere. We wish to emphasize that the theory to be developed depends only upon the net charge being less than the stoichiometric charge and not on the existence or nonexistence of " real ion pairs." In fact we shall later see that the hypothesis that ion pairs are formed at charge sites must be rejected. Because of its pictorial simplicity, we shall continue to use the terminology of ion binding.

7.1.1 ELECTROSTATIC POTENTIAL

Consider a system of volume V containing N polymeric ions, each possessing n charges of magnitude $-q$ at points $\mathbf{r}_i (i = 1, \cdots n)$ relative to its center of mass. Let the system also contain Nn counterions, each of charge q, and additional electrolyte consisting of L_α ions of charge q_α, where $\sum_\alpha L_\alpha q_\alpha = 0$. We restrict the development to 1–1 electrolytes, so that $q_\alpha = +q$ or $-q$. Within a radius R_s from the center of mass, the polyion will contain a volume fraction ϕ available for counterions. In polymers composed of flexible units and therefore of variable total volume, ϕ will depend upon the total volume and will be the volume fraction not occupied by the polymer units themselves. The volume V will be sufficiently large that the net volume occupied by the polyions will be an insignificant fraction of the total volume. Let m be the average number of counterions to be found within a polyion, and m_α the average number of other ions of charge q_α. The effective charge of the polyion will therefore be $Q = -nq + mq + \sum_\alpha m_\alpha q_\alpha$.

Finally, let the dielectric constant be assumed to have constant values D_0 and D inside and outside the polymer, respectively. The potential of average force between a mobile ion α at the point $\mathbf{r}$ and a fixed polyion will be written in the form

$$W_\alpha(\mathbf{r}) = q_\alpha \psi(\mathbf{r}) + \chi(\mathbf{r}) \qquad (7.1.3)$$

where $\psi(\mathbf{r})$ is the average electrostatic potential at the point $\mathbf{r}$ in the presence of the fixed polyion. Since $\psi(\mathbf{r})$ is to be computed from a Poisson–Boltzmann equation, as in the conventional formulation of the Debye–Hückel theory, it refers to the potential when the point $\mathbf{r}$ is not specified as being occupied by a mobile ion. If the point $\mathbf{r}$ is occupied by an ion α, the potential will differ from ψ by the amount contributed by the atmosphere of that ion. If the atmosphere contribution is a function of position, it will contribute to the force required to move the ion α. In the ordinary Debye–Hückel theory, the atmosphere contribution is position-independent and need not be explicitly considered. However, in the present problem we have two regions of space of which one is partially unavailable for ion occupancy, and therefore we will have to consider the atmosphere term, which in Eq. (7.1.3) is represented by χ.

If we take the zero of W_α, and hence of χ, at large $\mathbf{r}$ in the external solution, $\chi(\mathbf{r})$ will be the work required to destroy an ion atmosphere in the external solution and to build one at the point $\mathbf{r}$. The computation, which will be described in more detail later, depends only upon the bulk concentration and dielectric constant of the region around the point $\mathbf{r}$, so that $\chi(\mathbf{r})$ will have a constant value χ inside the polymer, and will have the constant value zero outside. This property of $\chi(\mathbf{r})$ is what permits its neglect when the entire space under consideration is of the same gross properties, as in the ordinary application of the Debye–Hückel theory.

From the plan and definitions outlined in the preceding paragraphs, the Poisson–Boltzmann equation for the average electrostatic potential at points $\mathbf{r}$ relative to the center of mass of a fixed polyion becomes

$$\nabla^2 \psi(\mathbf{r}) = -\frac{4\pi}{DV}\left(\frac{D\phi}{D_0}\,e^{-\chi/kT}\right)\left[N(n-m)q\,e^{-q\psi/kT} + \sum_\alpha (L_\alpha - Nm_\alpha)q_\alpha e^{-q_\alpha\psi/kT}\right]$$
$$(r < R_s) \qquad (7.1.4a)$$

$$\nabla^2 \psi(\mathbf{r}) = -\frac{4\pi}{DV}\left[N(n-m)q\,e^{-q\psi/kT} + \sum_\alpha (L_\alpha - Nm_\alpha)q_\alpha e^{-q_\alpha\psi/kT} + NQ\,e^{-Q\psi/kT}\right]$$
$$(r > R_s) \qquad (7.1.4b)$$

The right side of Eq. (7.1.4a) describes the average distribution of small ions within a polyion. Note the factor $(D/D_0)\phi\,\exp(-\chi/kT)$, which arises because the available volume is less than the geometric volume of the polymer and because the dielectric constant and ion atmospheres differ inside and outside the polyion. The coefficients of the various terms on the right side of Eq. (7.1.4a) reflect the fact that some of the counterions and other electrolyte are distributed inside other polyions. Equation (7.1.4b), which describes the distribution outside the polymer, contains a contribution from other polyions as well as from the small ions. In both of Eqs. (7.1.4) N has been assumed to be large relative to unity and terms not of order N have been neglected. The reader may notice that Eqs. (7.1.4) are inconsistent in the sense that the collisions radius for polymer-polymer encounters is taken to be equal to the radius R_s of each polymer, rather than $2R_s$. We shall find that the free energy is mainly determined by the interior solution (7.1.4a) under nearly all practical conditions, and that the position and nature of the boundary affect the electrostatic free energy but little. Since, in addition, there is no uniquely defined polymer radius, the mathematical complications of a more rigorous approach seem unwarranted. It should be noted that the present discussion refers only to the deviations from ideality caused by electrostatic interactions, as in conventional applications of the Debye–Hückel theory. In any real solution there will of course be additional deviations from ideality due to short-range interactions, such as those which are describable in terms of volume exclusions. For these latter forces the position and nature of the boundary is clearly of great importance.

If equations analogous to Eqs. (7.1.4) are to be applied to a system containing a single polyion (as will be done in a subsequent chapter of this book dealing with ion exchange resins) Eqs. (7.1.4) must be replaced by

$$\nabla^2 \psi(\mathbf{r}) = -\frac{4\pi}{DV}\left(\frac{D\phi}{D_0}e^{-\chi/kT}\right)\left[nq\,e^{-q\psi/kT} + \sum_\alpha L_\alpha q_\alpha e^{-q_\alpha\psi/kT}\right] \quad (r < R_s) \tag{7.1.4a'}$$

$$\nabla^2 \psi(\mathbf{r}) = -\frac{4\pi}{DV}\left[nq\,e^{-q\psi/kT} + \sum_\alpha L_\alpha q_\alpha e^{-q_\alpha\psi/kT}\right]. \quad (r > R_s) \tag{7.1.4b'}$$

Equations (7.1.4) or (7.1.4') are solved with standard electrostatic boundary conditions at $r = R_s$, and singularities of the form $-q/D_0|\mathbf{r} - \mathbf{r}_i|$ at the positions $\mathbf{r}_i$ of the fixed charges of magnitude $-q$.

The Poisson–Boltzmann Eqs. (7.1.4) and (7.1.4') can be linearized, just as in the conventional development of the Debye–Hückel theory. The linearized form of Eqs. (7.1.4) is

$$\nabla^2 \psi = (\kappa')^2 \psi + C' \quad (r < R_s) \tag{7.1.5a}$$

$$\nabla^2 \psi = \kappa^2 \psi \quad (r > R_s) \tag{7.1.5b}$$

where

$$\kappa^2 = \frac{4\pi}{DVkT} \left[N(n-m)q^2 + \sum_{\alpha} (L_\alpha - Nm_\alpha)q_\alpha^2 + NQ^2 \right] \quad (7.1.6a)$$

$$(\kappa')^2 = \frac{4\pi}{DVkT} \left(\frac{D\phi}{D_0} e^{-\chi/kT} \right) \left[N(n-m)q^2 + \sum_{\alpha} (L_\alpha - Nm_\alpha)q_\alpha^2 \right] \quad (7.1.6b)$$

$$C' = -\frac{4\pi NQ}{DV} \left(\frac{D\phi}{D_0} e^{-\chi/kT} \right). \quad (7.1.6c)$$

Equations (7.1.4′) yield expressions of the form (7.1.5), but with the definitions (7.1.6) of the form

$$\kappa^2 = \frac{4\pi}{DVkT} \left[nq^2 + \sum_{\alpha} L_\alpha q_\alpha^2 \right] \quad (7.1.6a')$$

$$(\kappa')^2 = \frac{4\pi}{DVkT} \left(\frac{D\phi}{D_0} e^{-\chi/kT} \right) \left[nq^2 + \sum_{\alpha} L_\alpha q_\alpha \right] \quad (7.1.6b')$$

$$C' = -\frac{4\pi nq}{DV} \left(\frac{D\phi}{D_0} e^{-\chi/kT} \right). \quad (7.1.6c')$$

It may be seen that the expression (7.1.5a) differs from the corresponding equation of the Debye–Hückel theory by virtue of the presence of the constant term C'. In the Debye–Hückel theory, the corresponding constant term is negligible in the limit of systems of large volume, as it must be in order that the potential vanish at large distances from a single fixed ion. Quantitative considerations indicate that the contribution of this constant term is of order $1/N$ relative to the remainder of the potential. Similar considerations show that the constant term in Eq. (7.1.5b) may be neglected, as in the ordinary Debye–Hückel theory. However, in Eq. (7.1.5a) the constant term is not of negligible magnitude relative to the remaining terms influencing the potential, and must be retained for further consideration.

Solution of Eqs. (7.1.4) for the potential requires expressions for m, the m_α, and χ. Using the potential of average force for positive ions, which we indicate by W_+, one may write

$$m = \phi \frac{N(n-m)}{V} \int_0^{R_s} e^{-W_+/kT} 4\pi r^2 \, dr$$

$$= \phi e^{-\chi/kT} \frac{N(n-m)}{V} \int_0^{R_s} 4\pi r^2 \left(1 - \frac{q\psi}{kT} \right) dr \quad (7.1.7)$$

where the final expression results from the substitution of Eq. (7.1.3), followed by linearization of the exponential. Expressions analogous to Eq. (7.1.7) apply to the m_α.

To calculate the atmosphere free energies χ, consider a process of the following kind. The ion under consideration is first discharged in the solution at large, then moved across the boundary into the polyion (to a position where ψ is the same as its value at the point of discharge), and finally recharged. This process is carried out keeping the remainder of the solution completely charged, since the quantity to be computed is the partial molal free energy change for this charge transfer. The problem is very complex if the atmosphere free energy is assumed to depend significantly upon the position of an ion relative to the boundary of the polymer. Therefore, we shall assume that the boundary does not distort the ion atmospheres at all. Actually, when the polymers are sufficiently large that $\kappa' R_s \gg 1$, the boundary will contribute negligibly to the atmosphere free energy of most of the ions. As will be seen subsequently, this condition is frequently met under the usual experimental conditions. With this approximation, the free energy change on discharge of the ion and its subsequent recharging may be computed by the method used by Guntelberg for obtaining activity coefficients in the Debye–Hückel theory;[12] the result is

$$\chi = \frac{q^2}{2} \left(\frac{\kappa}{D} - \frac{\kappa'}{D_0} \right). \tag{7.1.8}$$

Since the q_α are $\pm q$, Eq. (7.1.8) applies for all the mobile ions in the solution.

7.1.2 SOLUTION OF THE EQUATIONS

The inhomogeneous Eq. (7.1.5a) has solutions which may be written as the sum of the general solution to the corresponding homogeneous equation and a particular solution to the inhomogeneous equation itself. Since a particular solution to Eq. (7.1.5a) is $-C'/(\kappa')^2$, it is only necessary to consider methods of solution of the homogeneous equation. Expression of the solution in this form also indicates that the criterion for neglect of C' depends upon its magnitude with respect to $(\kappa')^2$.

The homogeneous equations obtained from Eq. (7.1.5) are of a well-known form.[13] The most general solution to Eq. (7.1.5a) regular for $r < R_s$, except for the specified singularities at the $\mathbf{r}_i$, is

$$\psi = \sum_i \frac{1}{D_0 |\mathbf{r}_i - \mathbf{r}'_i|} (A_i \, e^{-\kappa'|\mathbf{r}_i - \mathbf{r}'_i|} - (q + A_i) \, e^{\kappa'|\mathbf{r}_i - \mathbf{r}'_i|}) - \frac{C'}{(\kappa')^2}.$$
$$(r < R_s) \tag{7.1.9a}$$

The A_i are constants to be determined from the boundary conditions at $r = R_s$. The most general solution to Eq. (7.1.5b) regular at infinity may be written as a series of terms involving tesseral harmonics and spherical Bessel functions. However, it will be assumed that the distribution of charge of the polyion is sufficiently symmetric that only the spherically symmetric solution of Eq. (7.1.5b) is physically important. This approximation will be best for large values of R_s and large concentrations of small ions. Accordingly, the solution to Eq. (7.1.5b) will be assumed to have the form

$$\psi = \frac{Q\,e^{-\kappa(r-R_s)}}{(1+\kappa R_s)Dr}. \qquad (r > R_s) \qquad (7.1.9b)$$

Since an approximate solution (7.1.9b) is to be used for $r > R_s$, it will not be possible to determine the A_i and Q without a corresponding approximate treatment of the solution (7.1.9a) for the purpose of estimating the boundary conditions. Accordingly, we temporarily assume that the fixed charges are disposed with spherical symmetry about the origin. It may be shown that the A_i then enter the expression for the potential at any point within the polymer sphere only in one weighted sum over all charges i. For this reason, the boundary condition may be satisfied with all the A_i at the same value A.

To prove this, first calculate the sum of Eq. (7.1.9a) for a spherical surface of charge of radius a and thickness da at distance r from the origin. The spherical symmetry allows us to conclude that the A_i all have the same value $A(a)$ for charges on this surface. Converting the sum to a surface integral, and designating the surface charge density as $\rho(a)$, one may perform the integration in polar coordinates, getting

$$\psi(r, a)\,da = \frac{2\pi A(a)}{\kappa' D_0}\,\rho a^2\,da\left(\frac{e^{\kappa' r} - e^{-\kappa' r}}{r}\right)\left(\frac{e^{-\kappa' a} - e^{\kappa' a}}{a}\right) + F(a, r)$$

where

$$F(a, r) = -\frac{2\pi \rho a^2\,da}{\kappa' D_0}\left(\frac{e^{\kappa' a}}{a}\right)\left(\frac{e^{\kappa' r} - e^{-\kappa' r}}{r}\right) \qquad (r < a)$$

$$F(a, r) = -\frac{2\pi \rho a^2\,da}{\kappa' D_0}\left(\frac{e^{\kappa' r}}{r}\right)\left(\frac{e^{\kappa' a} - e^{-\kappa' a}}{a}\right). \qquad (r > a)$$

The potential of the entire charge distribution at any distance r from the center is then given by $\int_0^{R_s} \psi(r, a)\,da$. The first term of this integral is the only place the A_i occur and it is clear that only one of the A_i independently affects ψ. Thus a single value A may be chosen such that

$$A = \frac{1}{H}\int_0^{R_s} A(a)\rho a^2\,da\left(\frac{e^{-\kappa' a} - e^{-\kappa a}}{a}\right)$$

where H is the quantity defined in Eq. (7.1.9c).

The potential and its derivative on the boundary are then given by the expressions

$$\psi = -\frac{H}{D}\left[q\frac{e^{\kappa'R_s}}{R_s} + A\left(\frac{e^{\kappa'R_s} - e^{-\kappa'R_s}}{R_s}\right)\right] - \frac{C'}{(\kappa')^2} \qquad (7.1.9a')$$

$$\frac{\partial\psi}{\partial r} = -\frac{H}{D}\left[q\kappa'\frac{e^{\kappa'R_s}}{R_s} + A\kappa'\left(\frac{e^{\kappa'R_s} - e^{-\kappa'R_s}}{R_s}\right)\right]$$
$$-\frac{qe^{\kappa'R_s}}{R_s^2} - A\left(\frac{e^{\kappa'R_s} - e^{-\kappa'R_s}}{R_s^2}\right) \qquad (7.1.9a'')$$

with

$$H = \frac{2\pi}{\kappa'}\int_0^{R_s} r\rho_f(r)(e^{\kappa'r} - e^{-\kappa'r})\,dr \qquad (7.1.9c)$$

where $\rho_f(r)$ is the number density of fixed charges. Note that $\rho_f(r)$ may be a discontinuous function subject only to the restriction that it be integrable. The appropriate boundary conditions at $r = R_s$ are now: (a) $\psi(r = R_{s-}) = \psi(r = R_{s+})$; (b) $D_0(\partial\psi/\partial r)_{r=R_{s-}} = D(\partial\psi/\partial r)_{r=R_{s+}}$. Combining Eqs. (7.1.9a′), (7.1.9a″), (7.1.9b), and these boundary conditions, the following set of equations for A and Q may be obtained:

$$\frac{D_0 Q}{DH(1 + \kappa R_s)} = A\,e^{-\kappa'R_s} - (q + A)\,e^{\kappa'R_s} - \frac{D_0 C'R_s}{H(\kappa')^2} \qquad (7.1.10a)$$

$$\frac{Q}{H} = A(1 + \kappa'R_s)\,e^{-\kappa'R_s} - (q + A)(1 - \kappa'R_s)\,e^{\kappa'R_s}. \qquad (7.1.10b)$$

Note that the solution depends upon the sphericalized charge distribution only for the purpose of estimating the effect of the boundary. Since most of the charges are far nearer to neighboring charges than to the boundary, and since it will be found that, even in unfavorable cases, the free energy is rather insensitive to the location of the boundary, the spherical approximation will not lead to serious error. Thus the approximate treatment of the boundary does not adversely affect the discrete charge formulation we present. In fact, the present approximation is as unrestrictive as possible if departures from external spherical symmetry are to be assumed of negligible effect. The solution formally obtained for the potential by the method just outlined is rather complicated, and in general it will be necessary to use successive approximation methods to obtain an expression for ψ consistent with the subsidiary conditions (7.1.7) and (7.1.8). Rather than discuss the nature of the solutions in their most general form, we shall consider two cases of particular physical interest.

7.1.3 LARGE EXCESS OF ADDED SALT

We here consider the situation when the number of ions due to added salt is far greater than the number of counterions of the polyions, and the concentration of mobile ions is such that $\kappa' R_s \gg 1$. The analysis of this section applies both to solutions containing many polyions [Eqs. (7.1.5) with Eqs. (7.1.6)] and solutions of a single polyion [Eqs. (7.1.5) with Eqs. (7.1.6')]. In the situation under consideration, $C'/(\kappa')^2 \to 0$, and $(\kappa')^2$ and κ^2 may each be written as a sum of two terms, the first and much larger of which represents the ionic strength of added salt. While the remainder of $(\kappa')^2$ and κ^2 does not approach zero as the quantity of added salt increases, κ' and κ do approach the values calculated neglecting the counterions and polymer binding,

$$\kappa' = \left[\frac{D\phi}{D_0} e^{-\chi/kT}\right]^{1/2} \kappa \qquad (7.1.11a)$$

$$\kappa = \left[\frac{4\pi}{DVkT} \sum_\alpha L_\alpha q_\alpha^2\right]^{1/2}. \qquad (7.1.11b)$$

Equation (7.1.11b) shows that κ is given by the expression describing the usual screening due to the added salt. Equations (7.1.11) and (7.1.8) may be solved simultaneously by an iterative procedure to yield values of κ, κ', and χ.

Solving Eqs. (7.1.10) under the conditions of this section, and substituting the results into Eq. (7.1.9),

$$\psi = -\frac{q}{D} \sum_i \frac{e^{-\kappa'|\mathbf{r}-\mathbf{r}_i|}}{|\mathbf{r}-\mathbf{r}_i|} \qquad (r < R_s) \quad (7.1.12a)$$

$$\psi = -\frac{Hq}{D} e^{-\kappa' R_s}\left[\frac{2\kappa'}{\kappa + \kappa'(D_0/D)}\right]\frac{e^{-\kappa(r-R_s)}}{r}. \qquad (r > R_s) \quad (7.1.12b)$$

We now consider an approximate value for H. It turns out that Eq. (7.1.12b) is of little effect on the thermodynamic functions, so that the approximation to H is not critical. From Eq. (7.1.9b), putting $\rho_f = \text{constant} = 3n/4\pi R_s^3$, one obtains, for $\kappa' R_s \gg 1$

$$H \approx \frac{3n\, e^{\kappa' R_s}}{2(\kappa')^2 R_s^2}. \qquad (7.1.12c)$$

Again, note that an approximate charge distribution is introduced for calculating effects outside the charged region. Nowhere has it been necessary to use an approximate distribution to calculate quantities dependent upon the potential in the neighborhood of the fixed charges. Equation (7.1.12c), together with Eq. (7.1.12b), indicates that the entire region $r < R_s$ has a net charge far less than that of the fixed charges alone when $\kappa' R_s \gg 1$. This result is also consistent with what one might expect for the limiting cases where ϕ approaches zero or unity. For $\phi = 0$, corresponding to a completely impenetrable polymer sphere, $\kappa' = 0$, and solution for the potential (necessarily relaxing the condition $\kappa'R_s \gg 1$) leads to the Debye–Hückel result for ions of radius R_s,

$$\psi = -\frac{nq}{D}\left(\frac{e^{\kappa R_s}}{1 + \kappa R_s}\right)\frac{e^{-\kappa r}}{r}. \qquad (7.1.13)$$

7.1.4 No Added Salt

Even in polyion solutions without added electrolyte, the ordinary experimental conditions are frequently such that $e^{-\kappa'R_s} \ll 1$. While the relation between κ' and the stoichiometric number of counterions depends upon m for solutions containing many polyions, it is frequently possible to verify that $e^{-\kappa'R_s}$ remains small relative to unity. This is not immediately obvious, since a large value of κ' corresponds to most of the atmospheres of most of the fixed ions falling within the polymer. This would in turn decrease the number of ions available to maintain the ionic strength. It must be recognized that some relatively dense polyelectrolytes, such as globular proteins, possess such low values of D_0 or ϕ that the conditions of this paragraph are not satisfied. The simplified expressions to be discussed below should not be applied to such systems (see Chapter 3).

Let us begin, for solutions containing many polyions, by computing m, the average number of counterions in each polymer. The simplest way to obtain m is by solving Eqs. (7.1.10) for $Q = -(n - m)q$. This is more direct than first obtaining ψ and then using Eq. (7.1.7). Introducing the approximation (7.1.12c) for H, and performing the necessary algebraic operations upon Eq. (7.1.10), one may obtain, subject to the condition $e^{-\kappa'R_s} \ll e^{\kappa'R_s}$ which was discussed above,

$$\frac{n-m}{n} \approx \left[\frac{(D_0 R_s kT/q^2 n)(1 - \kappa'R_s)(1 + \kappa R_s) + \{3(1 + \kappa'R_s)/\kappa'R_s\}}{1 + \kappa R_s - (D_0/D)(1 - \kappa'R_s)}\right]. \qquad (7.1.14)$$

The first term in the numerator of Eq. (7.1.14) is frequently small, and we may in such cases approximate

$$\frac{n-m}{n} \approx \frac{3(1 + \kappa'R_s)}{\kappa'R_s[1 + \kappa R_s - (D_0/D)(1 - \kappa'R_s)]}. \qquad (7.1.15)$$

From Eqs. (7.1.6), we have, neglecting $(n - m)$ against its square,

$$\kappa^2 = \left(\frac{4\pi Nnq^2}{DVkT}\right)n\left(\frac{n-m}{n}\right)^2 = \kappa_0^2 n\left(\frac{n-m}{n}\right)^2 \qquad (7.1.16)$$

$$(\kappa')^2 = \kappa_0^2\left(\frac{D\phi}{D_0}e^{-\chi/kT}\right)\left(\frac{n-m}{n}\right). \qquad (7.1.17)$$

Equation (7.1.16) defines κ_0 as the screening parameter descriptive of the stoichiometric number of counterions. Equations (7.1.15) to (7.1.17) can be solved simultaneously by successive approximations to give the screening parameters and counterion distribution.

We illustrate the methods of this section by an example: Consider an open chain polymer for which $D_0 = D$ and $\phi = 1$. Then also $\chi = 0$. The difference between κ and κ' in this example results from the exclusion of other polyions within $r = R_s$. Let $R_s = 1000$ A, $(\kappa_0)^{-1} = 30$ A, and $n = 625$. This corresponds to a 5×10^{-3} base molar solution of charges in, for example, a polymer of D. P.* 10,000 with an effective bond length of 20 A, which is 6% charged. With but few iterations we obtain $(\kappa)^{-1} = 118$ A, $(\kappa')^{-1} = 18.8$ A, with $(n - m)/n = 0.064$. In this case we see that most of the counterions remain within the polymer sphere so that the region occupied by this polymer is nearly neutral, even with no added salt. The results obtained may be seen to be in accord with the numerical approximations used. For more highly charged polymers the approximation is even better (see Chapter 3 for a different computation of this effect).

The electrostatic potential in the salt-free systems can be computed from the values of κ and κ' determined by the methods described in the foregoing. We shall not make further use of the expressions, which are rather complicated, and therefore we will not exhibit them here.

7.1.5 FREE ENERGY COMPUTATION

The electrostatic potential computed in the preceding sections can be used to calculate the difference in free energy (excluding self-energy terms) between the actual polymer solution and a similar solution in which none of the ions are charged. The uncharged solution will contain the required numbers of ions of each kind, distributed in the manner appropriate to the uncharged state. The charging free energy, then, will include contributions associated with the rearrangement of the local concentrations of the various ions which must accompany the process of charging the ions to their final charge states. This definition of the electrostatic free energy is the one ordinarily used and permits a well-defined accounting of all the contributions to the free energies of polyion systems.

When a system containing charges q_α is altered by increasing the charges by amounts dq_α, the free energy will alter by an amount

$$dA = \sum_\alpha \bar{\psi}_\alpha \, dq_\alpha \tag{7.1.18}$$

where $\bar{\psi}_\alpha$ is the potential at the position of q_α (except for the infinite self-potential of q_α itself). Since $\bar{\psi}_\alpha$ is to be computed from the distribution of

* Degree of polymerization.

charges with magnitudes q_α, Eq. (7.1.18), when integrated along any path from zero charge to the final charge values, will give the electrostatic free energy as defined in the preceding paragraph. The use of Eq. (7.1.18) with all the q_α increasing simultaneously by equal fractions of their final values is the charging process used by Debye in his treatment of the free energy of ordinary electrolyte solutions. The choice of charging process to be used here will differ according to the nature of the system.

It must be pointed out that the value of $\bar{\psi}_\alpha$ to be used in Eq. (7.1.18) is to be the average value of ψ at the position of ion α, as distinguished from the average value of ψ at the same point without specifying the occupation of that point by ion α. This implies that the values of ψ as computed in preceding sections are appropriate for computing the potential at the charges of the polyions, but that the potential at the locations of counterions and added electrolyte ions must include, in addition to ψ as computed in the foregoing, the potential associated with their ion atmospheres.

We now consider the free energy of the systems for which the potential was fully discussed.

7.1.6 Large Excess of Added Salt

The simplest method of computing the free energy is first to charge all the ions due to added salt, keeping the polymer and counterions neutral. Since all but an insignificant number of these ions are in the region external to the polymer, application of the Debye–Hückel theory leads straightforwardly to the result

$$\Delta A_{\text{salt}} = -\frac{\kappa}{3D} \sum_\alpha L_\alpha q_\alpha^2 \tag{7.1.19}$$

where κ is given by Eq. (7.1.11b). Now, the polyions and counterions can be simultaneously charged, at constant κ and κ'. The contribution of the counterions may be expressed in terms of the partial molal free energy in a system characterized by κ, since the counterions are so few relative to the added salt. It may thus be seen that

$$\Delta A_{\text{c.i.}} = -\frac{\kappa n q^2}{2D}. \tag{7.1.20}$$

One might at first think that Eqs. (7.1.19) and (7.1.20) do not allow for the migrations of ions across the boundary of the polymer which must necessarily accompany charging of the fixed charges. However, in the presence of a large excess of added salt the net number of charges within the polymer remains essentially constant, as at any point the number of ions of one sign in excess of the average concentration is, to within the Debye–Hückel approximation, equal to the deficiency in the number of ions of the opposite sign. To compute the free energy for charging the fixed ions of the polymer, we require the limiting value of the potential, less the self-potential, at the position of each fixed charge. This is, using Eq. (7.1.12a),

$$\bar{\psi}_j = -\sum_{i \neq j} \frac{q \exp(-\kappa'|\mathbf{r}_i - \mathbf{r}_j|)}{D_0|\mathbf{r}_i - \mathbf{r}_j|} + \frac{\kappa' q}{D_0}. \tag{7.1.21}$$

To perform the integration required after insertion of Eq. (7.1.21) into Eq. (7.1.18), let the instantaneous value of q_i be λq_i, so that the charging process corresponds to increasing λ from zero to unity. Then $dq_i = q_i \, d\lambda$, The integration leads to the result

$$\Delta A_{\text{polymer}} = N \left[\frac{1}{2} \sum_{i \neq j} \frac{q^2 \exp(-\kappa'|\mathbf{r}_i - \mathbf{r}_j|)}{D_0|\mathbf{r}_i - \mathbf{r}_j|} - \frac{\kappa' n q^2}{2D_0} \right]. \tag{7.1.22}$$

Summing Eqs. (7.1.19), (7.1.20), and (7.1.22), the total electrostatic free energy of the polymer solution assumes the form

$$\Delta A_{\text{elec}} = -\frac{\kappa'}{3D} \sum_{\alpha} L_{\alpha} q_{\alpha}^2 + \frac{N}{2} \sum_{i \neq j} \frac{q^2 \exp(-\kappa'|\mathbf{r}_i - \mathbf{r}_j|)}{D_0|\mathbf{r}_i - \mathbf{r}_j|} - \frac{\kappa n N q^2}{2D} - \frac{\kappa' n N q^2}{2D_0}. \tag{7.1.23}$$

Note that the coefficient in the term corresponding to the counterions is $\frac{1}{2}$, rather than the customary $\frac{1}{3}$. The usual factor $\frac{1}{3}$ arises from the fact that κ changes during the charging process. It can be shown that the omission of the counterions from κ in Eq. (7.1.23) is compensated for by the change in coefficient.

7.1.7 No Added Salt

When there is no added electrolyte the calculation of the free energy is very much more difficult because many of the mobile ions are within the polymer. It is not difficult to use Eqs. (7.1.16) or (7.1.17), whichever is appropriate, to compute $\bar{\psi}_i$ and therefrom obtain $\Delta A_{\text{polymer}}$. However, computations of $\Delta A_{\text{c.i.}}$ will depend upon a detailed consideration of the small ion distribution within the polymer. Such a computation, though in principle obvious, is rather tedious and has not been attempted inasmuch as few reliable data for such systems are known.

The model adopted here has several features which distinguish it from other attempts to characterize the electrostatic effects in systems containing polymer ions. Most of these other discussions[5-7] treat the polyion as a continuous charge distribution, rather than specifically describing the inhomogeneous nature of the charge state of the polyion. As can be seen from

Eq. (7.1.23), we have now obtained an expression for the electrostatic free energy of a polymer in excess salt, without replacing the polymer charges by a continuous distribution. The free energy Eq. (7.1.23) corresponds only to the interaction at constant polymer configuration, and must be combined with all other relevant factors to yield a complete thermodynamic description of a polyelectrolyte. Such treatment corresponds to using screened Coulomb potentials for describing the forces between elements of the polymer, and making computations not including any terms descriptive of concentration rearrangements directly caused by the charges. Such a chain model implicitly includes all osmotic or other contributions associated with the small ions.

A second feature of the Harris–Rice development is the inclusion in the theory of effects resulting from the occupancy of space by the polymer. This improvement should enable meaningful computations to be made for highly cross-linked polymers in which the nonaqueous material is at relatively high local concentration.

7.2   Mathematical Interlude†

To complete our discussion of the electrostatic free energy and actually make use of Eq. (7.1.23) we must evaluate the sum over the discrete charge distribution of the polymer. We shall evaluate this sum by a matrix method which has been used to study a variety of statistical problems. However, because the studies referred to are spread rather thinly through the literature, there is no single source which is at the same time adapted to an introductory study of the technique and its applications. Further, since the references are few and far between, a brief discussion of Markoff chains and the statistical mechanics of linear assemblies will be considered here.

Let there be a chain of events each of which might lead to one of v possible results, and which are correlated such that the probability of successive events leading to a chain of results

$$\alpha_1, \alpha_2, \cdots \alpha_n \tag{7.2.1}$$

is proportional to

$$P_n(\alpha_1, \cdots \alpha_n). \tag{7.2.2}$$

The probability of a given function $F(\alpha_1, \cdots \alpha_n)$ having a value corresponding to a sequence of $\{\alpha_i\}$ is then proportional to

$$F(\alpha_1, \cdots \alpha_n)P_n(\alpha_1, \cdots \alpha_n) \tag{7.2.3}$$

and its average value taken over all the configurations of the chain of events

† The details of this section are drawn largely from E. Montroll, *J. Chem. Phys.* **9,** 706 (1941); **10,** 61 (1942); and *Ann. Math. Statist.* **18,** 18 (1947).

corresponding to the sequence of $\{\alpha_i\}$ is

$$\langle F \rangle = \frac{F_1}{F_0} = \frac{\sum_{\{\alpha_i\}} F(\{\alpha_i\})P_n(\{\alpha_i\})}{\sum_{\{\alpha_i\}} P_n(\{\alpha_i\})} \tag{7.2.4}$$

with F_m defined by

$$F_m = \sum_{\{\alpha_i\}} [F(\{\alpha_i\})]^m P_n(\{\alpha_i\}). \tag{7.2.5}$$

We may now ask for the probability of a result α_1 of the first event leading to a result α_n of the nth event. Since we do not care what the intervening events are, this probability may be expressed as

$$P_n(\alpha_1, \alpha_n) = \frac{1}{F_0} \sum_{\alpha_2, \cdots \alpha_{n-1}} P_n(\alpha_1, \cdots \alpha_n). \tag{7.2.6}$$

Now, the characteristic function of a distribution function, $F(x)$, is defined as

$$\phi(t) = \int_{-\infty}^{\infty} e^{itx} \, dF(x). \tag{7.2.7}$$

It may be shown that, if the kth moment of the distribution function exists, the characteristic function may be expanded in a MacLaurin's series for small values of t,

$$\phi(t) = 1 + \sum_1^k \frac{a_m}{m!} (it)^m + O(t^k). \tag{7.2.8}$$

Consider the expansion

$$\ln(1+z) = z - \frac{z^2}{2} + \cdots \pm \frac{z^k}{k} + O(z^k). \tag{7.2.9}$$

If we replace $1 + z$ by $\phi(t)$ and rearrange, the result is

$$\ln \phi(t) = \sum_1^k \frac{\Lambda_m}{m!} (it)^m + O(t^k). \tag{7.2.10}$$

The coefficients Λ_m are known as Thiele semi-invariants. To elucidate the relationship between the moments of the distribution $\{a_i\}$ and the Thiele semi-invariants, put

$$\ln \phi(t) = \ln\left(1 + \sum_1^\infty \frac{a_m}{m!} (it)^m\right) = \sum_1^\infty \frac{\Lambda_m}{m!} (it)^m \tag{7.2.11}$$

so that

$$\phi(t) = 1 + \sum_1^\infty \frac{a_m}{m!} (it)^m = \exp\left(\sum_1^\infty \frac{\Lambda_m}{m!} (it)^m\right). \tag{7.2.12}$$

Defining a function

$$Z_n(x) = \sum_{\{\alpha_i\}} P_n(\alpha_1 \cdots \alpha_n) \, e^{xF(\alpha_1 \cdots \alpha_n)} \tag{7.2.13}$$

it is easy to see that

$$F_m(\alpha) = \lim_{x \to 0} \frac{\partial^m Z_n(x)}{\partial x^m}. \tag{7.2.14}$$

On the other hand, the characteristic function of the distribution F is by definition

$$\phi(t) = Z_n(it)/Z_n(0) \tag{7.2.15}$$

so that the mth Thiele semi-invariant is

$$\Lambda_m = \lim_{x \to 0} \frac{\partial^m Z_n(x)}{\partial x^m}. \tag{7.2.16}$$

Let $G(z)$ be defined such that $G(\xi + h) - G(\xi)$ is the probability that the function $F(\alpha, \cdots, \alpha_n)$ has a value between

$$\xi \leqslant F(\alpha_1 \cdots \alpha_n) \leqslant \xi + h \tag{7.2.17}$$

and if $G(z)$ is continuous at $\xi = x$ and $x = \xi + h$ then it can be shown that

$$G(\xi + h) - G(\xi) = \frac{1}{2\pi i} \lim_{T \to \infty} \int_{-T}^{T} \frac{(1 - e^{-it\xi}) e^{-it\xi}}{t} \, \phi(t) \, dt \tag{7.2.18}$$

where, as before,

$$\ln \phi(t) = \sum_{1}^{k} \frac{\Lambda_m}{m!} (it)^m + O(t^k). \tag{7.2.19}$$

Furthermore,

$$\phi(\xi) \, d\xi = \left(\frac{\partial G}{\partial \xi} \right) d\xi = \frac{d\xi}{2\pi} \lim_{T \to \infty} \int_{-T}^{T} \exp \left[\sum_{1}^{\infty} \frac{\Lambda_m}{m!} (it)^m \right] e^{-it\xi} \, dt \tag{7.2.20}$$

is the probability that $F(\alpha_1, \cdots, \alpha_n)$ has a value between ξ, $\xi + d\xi$. From the definition of F_m [Eq. (7.2.5)] it is seen that

$$\sum_{1}^{\infty} \frac{\Lambda_m (it)^m}{m!} = -\ln Z_n(0) + \lim_{x \to 0} e^{-it(\partial/\partial x)} \ln Z_n(x). \tag{7.2.21}$$

Now, for a constant independent of x,

$$e^{c(\partial/\partial x)} f(x) = f(x + c) \tag{7.2.22}$$

and thereby

$$\sum_1^\infty \frac{\Lambda_m}{m!}(it)^m = \ln \frac{Z_n(it)}{Z_n(0)} \qquad (7.2.23)$$

and

$$G(\xi + h) - G(\xi) = \frac{1}{2\pi i} \lim_{T \to \infty} \int_{-T}^T \frac{e^{-it\xi}(1 - e^{-ith})Z_n(it)}{tZ_n(0)} \, dt. \qquad (7.2.24)$$

Thus, from a knowledge of $Z_n(x)$ one can obtain the desired information concerning a chain of correlated events.

7.2.1 SIMPLE CHAINS

By a simple chain is meant a sequence of events, each of which leads to ν possible results and which occur in such a manner that if the result of the kth event is α_k, the probability of the $(k + 1)$st event yielding a result α_{k+1} is proportional to $p(\alpha_k, \alpha_{k+1})$. Symbolically stated, the probability of an occurrence of the sequence of results

$$\alpha_1, \cdots, \alpha_n$$

is

$$\frac{\prod_{i=1}^{n-1} p(\alpha_i, \alpha_{i+1})}{\sum_{\{\alpha_i\}} \prod_{i=1}^{n-1} p(\alpha_i, \alpha_{i+1})} \qquad (7.2.25)$$

where

$$P_n(\alpha_1 \cdots \alpha_n) = \prod_{i=1}^{n-1} p(\alpha_i, \alpha_{i+1}). \qquad (7.2.26)$$

Furthermore, the probability of a first result α_1 leading to an nth result α_n is

$$P(\alpha_1, \alpha_n) = \frac{\sum_{\alpha_2, \cdots \alpha_{n-1}} \prod_{i=1}^{n-1} p(\alpha_i, \alpha_{i+1})}{\sum_{\alpha_1 \cdots \alpha_n} \prod_{i=1}^{n-1} p(\alpha_i, \alpha_{i+1})}. \qquad (7.2.27)$$

From Eq. (7.2.4), the average value of a function $F(\alpha_1, \cdots, \alpha_n)$ can be determined from

$$\langle F \rangle = \frac{F_1}{F_0} = \frac{\sum_{\alpha_1 \cdots \alpha_n} F(\alpha_1 \cdots \alpha_n) \prod_{i=0}^{n-1} p(\alpha_i, \alpha_{i+1})}{\sum_{\alpha_1 \cdots \alpha_n} \prod_{i=0}^{n-1} p(\alpha_i, \alpha_{i+1})}. \qquad (7.2.28)$$

In many instances, the functions $F(\alpha_1, \cdots, \alpha_n)$ are either additive or multiplicative and of one of the forms $F(\alpha_1, \cdots, \alpha_n) = h(\alpha_1, \alpha_2) + h(\alpha_2, \alpha_3) + \cdots + h(\alpha_{n-1}, \alpha_n)$ or $F(\alpha_1, \cdots, \alpha_n) = g(\alpha_1, \alpha_2)g(\alpha_2, \alpha_3) \cdots g(\alpha_{n-1}, \alpha_n)$. For the second case, consider the transformation

$$g(\alpha_i, \alpha_{i+1}) = \exp[xh(\alpha_i, \alpha_{i+1})] \tag{7.2.29}$$

and for both cases it is convenient to consider a function of the form

$$Z_n(x) = \sum_{\{\alpha_i\}} \prod_{i=1}^{n-1} p(\alpha_i, \alpha_{i+1}) \exp[xh(\alpha_i, \alpha_{i+1})]. \tag{7.2.30}$$

We may now observe that $Z_n(x)$ is the sum of the elements of the nth power of the matrix

$$\mathbf{P}_x = \begin{Vmatrix} p_x(1, 1) \cdots p_x(1, v) \\ \cdot \\ \cdot \\ \cdot \\ p_x(v, 1) \cdots p_x(v, v) \end{Vmatrix} \tag{7.2.31}$$

where the elements $p_x(\alpha, \beta)$ are defined as

$$p_x(\alpha, \beta) = p(\alpha, \beta) \, e^{xh(\alpha, \beta)} \tag{7.2.32}$$

and α and β range over the same set of values as one of the " result " parameters α_1. We may write the eigenvectors of $\mathbf{P}_x$ as

$$\mathbf{\Phi}_{i,x} = \{\phi_{i,x}(1) \cdots \phi_{i,x}(v)\}$$

$$\mathbf{\Psi}_{i,x} = \begin{pmatrix} \psi_{i,x}(1) \\ \cdot \\ \cdot \\ \cdot \\ \psi_{i,x}(v) \end{pmatrix} \tag{7.2.33}$$

so that

$$\mathbf{\Phi}_{i,x} \cdot \mathbf{P}_x = \lambda_{i,x}\mathbf{\Phi}_{i,x}$$
$$\mathbf{P}_x \cdot \mathbf{\Psi}_{i,x} = \lambda_{i,x}\mathbf{\Psi}_{i,x} \tag{7.2.34}$$

where $\lambda_{i,x}$ is the ith eigenvalue of $\mathbf{P}_x$. If the eigenvectors are orthonormal

$$\mathbf{\Phi}_{i,x} \cdot \mathbf{\Psi}_{j,x} = \sum_{\alpha=1}^{v} \phi_{i,x}(\alpha)\psi_{j,x}(\alpha) = \delta_{ij}. \tag{7.2.35}$$

Now, it is well known that

$$p_x(\alpha, \beta) = \sum_{i=1}^{v} \lambda_{i,x}\phi_{i,x}(\beta)\psi_{i,x}(\alpha) \tag{7.2.36}$$

and from Eq. (7.2.34) we easily find

$$\lambda_{i,x} = \mathbf{\Phi}_{i,x} \cdot \mathbf{P}_x \cdot \mathbf{\Psi}_{i,x} = (\lambda_{i,xi}\mathbf{\Phi}_{i,x}) \cdot \mathbf{\Psi}_{i,x}. \tag{7.2.37}$$

Substituting into the expression for $Z_n(x)$, it is easy to show that

$$Z_n(x) = \sum_{i=1}^{v} \lambda_{i,x}^{n-1} \left(\sum_{\beta=1}^{v} \phi_{i,x}(\beta) \right) \left(\sum_{\alpha=1}^{v} \psi_{i,x}(\alpha) \right)$$

$$= \sum_{i=1}^{v} \lambda_{i,x}^{n-1} (\mathbf{\Phi}_{i,x} \cdot \mathbf{1})(\mathbf{1} \cdot \mathbf{\Psi}_{i,x}). \tag{7.2.38}$$

If there is one eigenvalue $\lambda_{L,x}$ greater than all others, then

$$\lim_{n \to \infty} \left[\frac{Z_n(x)}{\lambda_{L,x}^{n-1}(\mathbf{\Phi}_{i,x} \cdot \mathbf{1})(\mathbf{1} \cdot \mathbf{\Psi}_{i,x})} - 1 \right] = 0. \tag{7.2.39}$$

We shall prove relation (7.2.39) for a symmetric matrix for simplicity. The exact analog for a nonsymmetric matrix can be derived but is somewhat more tedious. For a symmetric matrix, $\phi_{i,x}(\alpha) = \psi_{i,x}(\beta)$ and all the eigenvalues are real. Then

$$Z_n(x) = \lambda_{L,x}^{n-1}(\mathbf{\Phi}_{L,x} \cdot \mathbf{1})^2 + \sum_{i \neq L} \lambda_{i,x}^{n-1}(\mathbf{\Phi}_{i,x} \cdot \mathbf{1})^2 \tag{7.2.40}$$

and

$$|\mathbf{\Phi}_{i,x} \cdot \mathbf{1}|^2 = \left| \sum_{\alpha=1}^{v} \phi_{i,x}(\alpha) \right|^2 \leqslant \left[\sum_{\alpha=1}^{v} \phi_{i,x}^2(\alpha) \right] \left[\sum_{\alpha=1}^{v} 1 \right] = v \tag{7.2.41}$$

whereupon

$$\left| \sum_{i \neq L} \lambda_{i,x}^{n-1}(\phi_{i,x}^2) \right| \leqslant v \left| \sum_{i \neq L} \lambda_{i,x}^{n-1} \right| \leqslant v(v-1)|\lambda_{s,x}^{n-1}| \tag{7.2.42}$$

where $\lambda_{s,x}$ is the second largest eigenvalue. Combining,

$$\left| \frac{Z_n(x)}{\lambda_{L,x}^{n-1}(\mathbf{\Phi}_{L,x} \cdot \mathbf{1})^2} - 1 \right| \leqslant \frac{v(v-1)}{(\mathbf{\Phi}_{i,x} \cdot \mathbf{1})^2} \left| \left(\frac{\lambda_{s,x}}{\lambda_{L,x}} \right)^{n-1} \right|. \tag{7.2.43}$$

But since $|(\lambda_{s,x}/\lambda_{L,x})| < 1$, it is easy to see that

$$\lim_{n \to \infty} (\lambda_{s,x}/\lambda_{L,x})^{n-1} = 0 \tag{7.2.44}$$

and finally

$$Z_n(x) \approx \lambda_{L,x}^{n-1}(\mathbf{\Phi}_{L,x} \cdot \mathbf{1})(\mathbf{1} \cdot \mathbf{\Psi}_{L,x}). \tag{7.2.45}$$

When the chain is very long, as might be expected, the probability of $F(\{\alpha_i\})$ having a value between ξ and $\xi + h$ asymptotically goes to a Gaussian distribution.

At this point, weighed down by mathematical formalism, the reader may not see the forest for the trees. In the following we shall give an example that we hope will be illuminating. Barring this, it will at least be useful since it provides the genesis of the concepts to be developed in the succeeding sections of this chapter.

Let us consider a linear lattice of particles. Associated with each particle is an intrinsic energy $v(\alpha_i)$. Moreover, we assume that interactions between nearest neighbors on the lattice are the only important ones. This interaction contributes terms of the form $v(\alpha_i, \alpha_{i+1})$ to the energy. The partition function may now be written

$$Q = \sum_{\alpha_1 \cdots \alpha_N} \left\{ \exp - \frac{1}{kT} [v(\alpha_1) + \cdots + v(\alpha_N) + v(\alpha_1, \alpha_2) + \cdots + v(\alpha_{N-1}, \alpha_N)] \right\}. \tag{7.2.46}$$

Consider the substitution

$$V(\alpha_i, \alpha_j) = \frac{v(\alpha_i)}{2} + v(\alpha_i, \alpha_j) + \frac{v(\alpha_j)}{2} \tag{7.2.47}$$

so that Eq. (7.2.46) becomes

$$Q = \sum_{\alpha_1 \cdots \alpha_N} \prod_{i=1}^{N-1} \exp \left[-\frac{V(\alpha_i, \alpha_{i+1})}{kT} \right] \exp \left[-\frac{v(\alpha_1) + v(\alpha_N)}{2kT} \right]. \tag{7.2.48}$$

Since $(v(\alpha_1) + v(\alpha_N))/2$ is just half the intrinsic energy of the terminal particles, it may be neglected if the chain is long. Equation (7.2.48) then reduces to

$$Q = \sum_{\alpha_1 \cdots \alpha_N} \prod_{i=1}^{N-1} \exp \left[-\frac{V(\alpha_i, \alpha_{i+1})}{kT} \right]. \tag{7.2.49}$$

Consider now the matrix

$$\exp\left(-\frac{\mathbf{V}}{kT}\right) = \left\| \begin{array}{ccc} \exp[-V(\alpha_1, \alpha_1)/kT] & \cdots & \exp[-V(\alpha_N, \alpha_1)/kT] \\ & \cdot & \\ & \cdot & \\ & \cdot & \\ \exp[-V(\alpha_N, \alpha_1)/kT] & \cdots & \exp[-V(\alpha_N, \alpha_N)/kT] \end{array} \right\| \tag{7.2.50}$$

from which we easily see that in the case of the one-dimensional lattice (where the nearest neighbor interactions are of magnitude ε)

$$\exp\left(-\frac{\mathbf{V}}{kT}\right) = \left\| \begin{array}{cc} e^{\varepsilon/kT} & e^{-\varepsilon/kT} \\ e^{-\varepsilon/kT} & e^{\varepsilon/kT} \end{array} \right\| \tag{7.2.51}$$

where

$$V(\alpha, \alpha') = \varepsilon \qquad \alpha \neq \alpha'$$
$$V(\alpha, \alpha') = -\varepsilon \qquad \alpha = \alpha'.$$

These results are obtained simply by considering the following scheme

$$
\begin{array}{c|cc}
\diagdown \, \alpha_i & & \\
\alpha_{i+1} \diagdown & 0 & 1 \\
\hline
0 & \alpha = \alpha' & \alpha \neq \alpha' \\
1 & \alpha \neq \alpha' & \alpha = \alpha'
\end{array}
\Rightarrow
\left\| \begin{array}{cc}
e^{\varepsilon/kT} & e^{-\varepsilon/kT} \\
e^{-\varepsilon/kT} & e^{\varepsilon/kT}
\end{array} \right\|.
$$

To find the maximum eigenvalue of the matrix (7.2.51) we must solve the secular equation

$$
\left| \begin{array}{cc}
e^{\varepsilon/kT} - \phi & e^{-\varepsilon/kT} \\
e^{-\varepsilon/kT} & e^{\varepsilon/kT} - \phi
\end{array} \right| = 0 \tag{7.2.52}
$$

$$
\phi = \frac{2e^{\varepsilon/kT} \pm [4e^{2\varepsilon/kT} - 4(e^{2\varepsilon/kT} - e^{-2\varepsilon/kT})]^{\frac{1}{2}}}{2} \tag{7.2.53}
$$

$$
\phi_{\max} = e^{\varepsilon/kT} + e^{-\varepsilon/kT} = 2\cosh(\varepsilon/kT)
$$

$$
Q = \phi_{\max}^{N-1} = 2^{N-1} \cosh^{N-1}(\varepsilon/kT).
$$

This completes our illustrative example. We may merely note that chains of more complex structure can be broken down into a smaller number of simple " events." More precisely, in a chain of N events, in which the result of each event depends on n of its predecessors ($N \gg n$) we divide N events into N/n sets of " grand events." If each simple event could lead to v results, one grand event can lead to v^n results and a complicated chain becomes a simple chain of grand events with the result of each grand event dependent only on the preceding grand event. The calculations are thus exactly the same as those discussed above.

7.3   Evaluation of the Electrostatic Energy

In this section we shall use the results of Section 7.2 to calculate the electrostatic energy with a discrete charge distribution. It is convenient to proceed via a lengthy reinterpretation of the various terms in Eq. (7.1.23). While not logical, this procedure does provide more insight into the nature of the several contributions.

We shall discuss for definiteness the properties of an aqueous solution of a polyacid. The extension to polymeric bases is straightforward. Let the polymer chain be composed of Z dissociable groups, or, as they will be subsequently referred to, sites. Each of these sites is then assumed to be occupied either by an un-ionized group, a charged group or a bound ion counterion pair. It is important to notice that in the foregoing assumption is included the notion that the sites for bound pairs are the same sites as those for charged and for un-ionized groups.

What do we mean when, in the preceding paragraph, we talk of ion pairs at discrete sites. In general we shall see that the counterion distribution is very highly peaked adjacent to the polymer skeleton. Just as in the case of simple electrolytes, and using a definition similar to that introduced by Bjerrum and discussed in Chapters 2 and 5, we may classify the counterions as bound and free. Now the bound counterions defined in this manner may be visualized as forming a " mobile monolayer " adjacent to the polymer skeleton. That is, because of the large electrostatic field, the counterion is relatively free to move parallel to the polymer skeleton, but relatively restricted in any motion perpendicular to the polymer skeleton. The properties of such a " mobile monolayer " are difficult to compute rigorously. Since we wish to compute the energy, a very good approximation replaces the mobile counterions by a one-dimensional cell model in which the counterions are localized. The logical choices of cell size and cell center are one monomer length and the charged site, respectively. In this approximation each counterion and charged site may be regarded as an ion pair. From the extensive calculations (see Chapter 1) already performed it is found that the dominant terms in the energy arise from the static potential wherein each particle is at the center of its cell on its lattice site. Motions away from the lattice site affect the motional entropy, but to first order leave the energy unchanged. The distributional entropy, i.e., the number of ways of arranging occupied and unoccupied cells is, of course, also unaffected by localizing the particles at cell centers.

In the following, then, we interpret the words " site binding " to mean a cell model of the localized counterion distribution, and interpret the ion pair dissociation constant as the unknown cell partition function.

Consider the free energy A of the polyelectrolyte solution to be divided into three parts, one of which, A_1, describes the free energy of the free hydrogen ions and counterions eligible to take part in binding and neutralization phenomena. The remainder of the free energy will then be the contribution of the polyion itself. The first portion A_2 of this free energy will be the electrical free energy of the net charge of the polyion, regarding the ion pairs as uncharged sites completely equivalent to un-ionized groups. Thus A_2 will include the interaction energy between portions of the net charge of the polymer, and the entropy associated with mixing charged and uncharged groups on the polymer chain. The remainder A_3 of the polyion free energy therefore includes the chemical free energy associated with the state of ionization and binding of the polymer, computed from a reference state in which the polyacid is in its completely undissociated form. Also included in A_3 will be the free energy of mixing of the bound ion pairs and unionized groups among the uncharged sites of the chain.

Let us define α to be the degree of neutralization of the polymeric acid, and f as the fraction of the dissociated sites occupied by bound ion pairs. It is then clear that the net fractional charge of the polyion, which we shall refer to as α', will be $\alpha(1 - f)$, and the fraction of bound sites, referred to the total number of sites of the polymer, will be αf. From the way in which the various contributions to the free energy have been defined, we see that A_2 will be a function of the net charge of the polymer only, but that A_3 will depend upon the numbers of un-ionized sites and ion pair sites separately. A_1 is of course expressible in terms of the activities a_{H^+}, a_{C^+} of the free hydrogen ions and counterions in the solution. We shall for simplicity assume the counterions to be all univalent and of the same species. One may accordingly write

$$A = A_1(a_{H^+}, a_{C^+}) + A_2(\alpha') + A_3(\alpha, \alpha f). \tag{7.3.1}$$

The equilibrium hydrogen ion activity may now be determined by minimizing the free energy of the polymer solution with respect to the degree of neutralization of the polyacid. If the derivative is taken regarding α and αf as the two independent variables,

$$\left(\frac{\partial A_1}{\partial \alpha}\right)_{\alpha f} + A_2' + \left(\frac{\partial A_3}{\partial \alpha}\right)_{\alpha f} = 0 \tag{7.3.2}$$

where the notation A_2' refers to the derivative A_2' with respect to its argument. Equation (7.3.2) may be simplified by introduction of the chemical potential of the hydrogen ions, μ_{H^+}. One may see that $(\partial A_1/\partial \alpha)_{\alpha f} = Z\mu_{H^+}$ since a change in α implies a change in the number of hydrogen ions in the solution, keeping the number of counterions constant. The identification of $(\partial A_1/\partial \alpha)_{\alpha f}$ with $Z\mu_{H^+}$ is valid if changes in the charge of the polymer do not directly affect the activities of the small ions. Such a conclusion is reasonable under conditions such that the Debye–Hückel limiting law applies and that the charge attached to the polymer is not mobile and cannot contribute to the effective ionic strength of the solution.

In an entirely analogous manner the equilibrium degree of binding f can be determined from the requirement that the free energy be a minimum with respect to αf, at constant degree of neutralization α. In this case we recognize $-(\partial A_1/\partial \alpha f)_\alpha = Z\mu_{C^+}$ where the remarks of the preceding paragraph again apply. The negative sign arises from the definition of f, since f increases when the number of free counterions diminishes. The equation corresponding to (7.3.2) is

$$\left(\frac{\partial A_1}{\partial \alpha f}\right)_\alpha - A_2' + \left(\frac{\partial A_3}{\partial \alpha f}\right)_\alpha = 0 \tag{7.3.3}$$

and the simplifications discussed above enable (7.3.2) and (7.3.3) to be written as

$$Z\mu_{H^+} + A_2' + \left(\frac{\partial A_3}{\partial \alpha}\right)_{\alpha f} = 0$$

$$-Z\mu_{C^+} - A_2' + \left(\frac{\partial A_3}{\partial \alpha f}\right)_{\alpha} = 0. \tag{7.3.4}$$

To proceed further it is necessary to introduce the specific nature of the free energy A_3. We shall write A_3 in the form

$$A_3 = Z\alpha(\mu^0_{COO^-} - \mu^0_{COOH}) + Z\alpha f(\mu_{COO-C^+} - \mu_{COO^-}) - TS_m \tag{7.3.5}$$

where S_m denotes the entropy of mixing of the ion pairs and unionized groups. The standard chemical potentials μ^0 refer to states whose free energy does not include any interaction between the ions of the polymer and the atmospheres of small ions around them. These electrical interactions are, therefore, to be included in A_2. We may simplify (7.3.5) by introducing equilibrium constants for the acid dissociation and ion pair formation processes. (The reader is reminded that the ion-pair dissociation constant is to be interpreted as a cell partition function.) The acid dissociation constant can be determined by extrapolation of pH measurements to low values of α, for which it will be shown the theory can be greatly simplified. If we let K_a^0 be the intrinsic dissociation constant for the acid and K_s^0 for the ion pair, we may reduce A_3 to

$$A_3 = Z\alpha(-kT \ln K_a^0 - \mu^0_{H^+}) + Z\alpha f(kT \ln K_s^0 + \mu^0_{C^+}) - TS_m. \tag{7.3.6}$$

We next use the definitions $\mu_{H^+} = \mu^0_{H^+} + kT \ln a_{H^+}$, $\mu_{C^+} = \mu^0_{C^+} + kT \ln a_{C^+}$ and differentiate Eq. (7.3.6) with respect to α and αf. Combining the result with Eq. (7.3.4)

$$A_2' - ZkT \ln K_a^0 + ZkT \ln a_{H^+} - T \left(\frac{\partial S_m}{\partial \alpha}\right)_{\alpha f} = 0$$

$$-A_2' + ZkT \ln K_s^0 - ZkT \ln a_{C^+} - T \left(\frac{\partial S_m}{\partial \alpha f}\right)_{\alpha} = 0. \tag{7.3.7}$$

If a sufficiently detailed model has been adopted to permit evaluation of A_2, Eq. (7.3.7) could be used to obtain the potentiometric and binding properties of the polymer in terms of the electrical interactions between its charges. Adding the two Eqs. (7.3.7), we obtain after rearrangement

$$p\text{H} + \log a_{C^+} = pK_a^0 - pK_s^0 - \frac{0.434}{Zk}\left[\left(\frac{\partial S_m}{\partial \alpha}\right)_{\alpha f} + \left(\frac{\partial S_m}{\partial \alpha f}\right)_{\alpha}\right] \tag{7.3.8}$$

where the pH is given in terms of quantities which are known if the degree of binding of the polyion has been measured.

To use Eq. (7.3.8), it only remains to evaluate the entropy of mixing S_m. Since neither the un-ionized groups nor the ion pairs possess net charge, a reasonable first approximation to S_m might be achieved by considering a random distribution of un-ionized and ion paired groups. However, since the nature of the bound state (i.e., the cell partition function) is not understood in detail, it is certainly possible that all distributions of the bound pairs might not be equally probable. For random mixing

$$S_m = -Zk\left[(1-\alpha)\,\ln\left(\frac{1-\alpha}{1-\alpha+\alpha f}\right) + \alpha f\,\ln\left(\frac{\alpha f}{1-\alpha+\alpha f}\right)\right] \quad (7.3.9)$$

where the derivatives of S_m with respect to α and αf are

$$\left(\frac{\partial S_m}{\partial \alpha}\right)_{\alpha f} = Zk\,\ln\frac{1-\alpha}{1-\alpha+\alpha f}$$

$$\left(\frac{\partial S_m}{\partial \alpha f}\right)_{\alpha} = -Zk\,\ln\frac{\alpha f}{1-\alpha+\alpha f}. \quad (7.3.10)$$

Substitution of these expressions for the derivatives into Eq. (7.3.8) yields

$$pH + \log a_{C^+} + \log\frac{1-\alpha}{\alpha} - \log f = \rho K_a^0 - pK_s^0 = pK_1 \quad (7.3.11)$$

where pK_1 is defined by Eq. (7.3.11).

7.3.1 COMPARISON WITH EXPERIMENT

Equation (7.3.10) can be compared with experimental results for polymethacrylic acid whose pH at varying degrees of neutralization and counterion concentration has been measured by Oth and Doty.[15] These authors also use for binding data the more accurate measurements obtained in polyacrylic acid solutions by Huizenga et al.[16] based upon transference and diffusion measurements of radioactive sodium ions in the polymer solutions. The structures of polymethacrylic and polyacrylic acids differ only in details which, according to the theory set forth, only slightly affect the degree of binding. Confirmation of the similarity in degree of binding between polymethacrylic and polycarylic acids is available by comparison with the provisional binding data of Oth and Doty. Since the measurements of Oth and Doty were made without adding salt to the polymer solution, the concentration of sodium counterions is equal to the net charge on the polymer and may be written as the stoichiometric concentration of Na^+ ions multiplied by the factor $1 - f$.

It should be observed that the quantity actually desired for use in

Eq. (7.3.11) is the activity of the sodium ions rather than their concentration. If the ionic activity coefficients depend significantly upon other parameters than the ionic strength, defined in terms of the mobile ions, the argument leading to Eq. (7.3.4) will also need to be re-examined.

Values of pK_1, which according to the foregoing should be independent of α and c_p, were calculated from experimental data and Eq. (7.3.11) and are listed in Table 7.3.1. The original pH data vary over a range of somewhat

TABLE 7.3.1[a]

VALUES OF pK_1 FOR POLYMETHACRYLIC ACID

c_p gm./100 cc	pK_1				
	$\alpha = 0.08$	$\alpha = 0.20$	$\alpha = 0.35$	$\alpha = 0.50$	$\alpha = 0.75$
0.2166	3.90	4.42	4.55	4.90	5.51
0.1083	3.73	4.28	4.45	4.81	5.46
0.05415	3.59	4.16	4.31	4.65	5.39
0.02707	3.35	4.00	4.17	4.48	5.29
0.01353	3.20	—	4.03	4.35	5.09

[a] Reproduced from ref. 1.

more than 3 pH units, and it can be seen that the present treatment has only managed to halve this spread, leaving significant differences between results at low and at high degrees of dissociation. Since the assumptions involved in the deduction of the theory are not strongly dependent in many respects upon the form of the interaction between the various charges of the molecule, one must conclude provisionally that either some important factor governing the behavior of the system has been omitted or that the assumptions regarding the nature of the counterion binding (cell partition function) are incorrect. It should be pointed out that the magnitude of the free energy changes associated with the electrostatic interactions, ionization, and binding, do not enter into these conclusions, but that the important factors influencing the present considerations as the assumptions involved in enumerating the probable sites for the binding process and the independence of K_s^0 of α.

One factor deliberately omitted from consideration was the deviation of the activity coefficients from unity. An estimate of the order of magnitude of the change to be expected in pK_1 due to inclusion of activity coefficients may be made. If such calculations are made, it is found that the variation of pK_1 with polymer concentration can be drastically reduced, but the change in pK_1 with degree of neutralization is much too large to be accounted for in this manner.

7.3.2 Another Approximation to S_m

Another possible source of discrepancy between Eq. (7.3.11) and experimental results could reside in the assumption of random distribution of the bound counterions and the un-ionized hydrogens. If we retain the assumption that the free energy of the distribution of charges is governed by the distribution of only the net charge of the polyion, we then may introduce nonrandom mixing of the bound counterions and hydrogens by assuming, for example, that a bound counterion cannot be located adjacent to an un-ionized hydrogen. This assumption would be reasonable if the mechanism for binding required the presence of an adjacent ionized (but not necessarily unbound) site on either side of the site where the binding is to occur. Another way of phrasing the statement is to point out that our assumptions about the size and center, and hence also the potential in the cell, need modification. Although the statistics of this problem are too difficult to solve in an entirely rigorous manner, an approximate indication may be obtained which will enable us to decide whether a mechanism of this type could explain the discrepancy. We approximate by assuming the centers of net charge to be equally spaced along the polymer chain. The proposed nonrandom distribution of counterions and protons then corresponds to filling the spaces between a pair of centers of net charge all with the same species, and therefore the number of possible distributions of protons and counterions will decrease as the average number of sites between centers of net charge increases. If we calculate the ratio of the number of possible distributions with and without the added restriction, the entropy change upon going from the random mixing to the more restricted situation may be found to be

$$\Delta S_m = Zk(1 - \alpha) \left(\frac{1}{1 - \alpha + \alpha f} - 2 \right) [(f \ln f + (1 - f) \ln(1 - f)]. \quad (7.3.12)$$

The use of Eq. (7.3.12) in addition to (7.3.9) causes the additional term

$$\frac{0.434}{Zk} \left[\left(\frac{\partial \Delta S_m}{\partial \alpha} \right)_{\alpha f} + \left(\frac{\partial \Delta S_m}{\partial \alpha f} \right)_{\alpha} \right] = \left(\frac{1}{1 - \alpha + \alpha f} - 2 \right) \times$$

$$\left[\left(\frac{1 - f}{\alpha} - 1 \right) \log f - \frac{1 - {}_J}{\alpha} \log(1 - f) \right] \quad (7.3.13)$$

to be added to the left-hand side of Eq. (7.3.11).

If the correction implied by (7.3.13) is made, the values of pK_1 increase very sharply at low values of α, while remaining substantially unchanged for $\alpha > 0.4$.

If a more moderate restriction of configurations than that which led to Eq. (7.3.13) is considered to be a more appropriate approximation to the true situation, we may employ a restriction on the occupation of sites in a manner which will effect a partial prohibition of occupation of the sites which were formally totally prohibited. If such an approximation were adopted, semiquantitative calculations indicate that although pK_1 can be made substantially constant in the range $\alpha \leqslant 0.4$, no such approximation can completely account for the increase in pK_1 at higher degrees of neutralization. The methods outlined above have not led to a value of pK_1 constant to within less than 0.9 unit. The three treatments mentioned above are plotted in Fig. 7.3.1. Curve number 1 refers to the case of random mixing, curve 2 to total prohibition, and curve 3 to partial prohibition of site occupation.

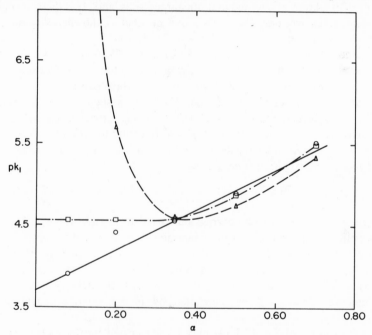

FIG. 7.3.1. A test of the accuracy of Eq. (7.3.11). [Reproduced from ref. 1.]

We, therefore, conclude that there yet remains a significant degree of difference between the model adopted here for the binding of counterions by polymeric ions, and the actual experimental situation, as observed in poly-methacrylic and polyacrylic acids. Further comments are reserved for Chapter 9.

7.3.3 THE ELECTROSTATIC ENERGY

While the discussion of the preceding section indicated the close connection between the ionization to be expected of a polyion and the binding by it of counterions, no indication was given as to the extent to which the binding should occur. To make a quantitative estimate and to fully calculate the free energy A_2 of a charged polymer chain, we can introduce forces between ionized groups on adjacent polymer sites, and calculate the entropy of the polymer arising from the distribution of its charge as a function of the net charge of the polyion. The calculations thus proposed do not depend upon the statistical treatment of the dimensions of the coiled polymer and will be valid when the interaction energy between adjacent sites is sufficiently large compared with that between sites further removed from each other. We thereupon write A_2 as a sum, $A_4 + A_5$, where A_4 is the free energy arising from the distribution of charged and uncharged sites, and A_5 is the remainder of A_2. The reader will note that we are now computing the energy in the cell model approximation.

We proceed by the method used by Kramers and Wannier for the solution of the very similar one-dimensional Ising spin lattice.[13–15] Consider the polymer as a chain consisting of Z sites, each of which can be in one of two states, ionized or un-ionized. With each site let us associate a state variable η_i, $i = 1, \cdots Z$, where the sites are numbered consecutively along the chain. The η_i may have two values, 0 and 1, $\eta = 0$ referring to an un-ionized state, and $\eta = 1$ an ionized state. We next find it convenient to introduce a function related to a grand partition function, $\Xi(\lambda, T)$,

$$\Xi(\lambda, T) = \sum_{\{\eta_i\}} \exp(-\Sigma u_{i,i+1}/kT)\lambda^{\Sigma \eta_i} \qquad (7.3.14)$$

where the summation is to be carried out over all possible values of the η_i. The quantity $u_{i,i+1}$ is the energy of interaction between sites i and $i + 1$, and is equal to χ if both sites i and $i + 1$ are ionized and zero otherwise. Obviously, $u_{i,i+1} = \chi \eta_i \eta_{i+1}$ is a satisfactory form for $u_{i,i+1}$. The quantity λ plays a role analogous to that of an absolute activity serving as a parameter whose value will be determined by the number of charges on the polymer chain.

The evaluation of Eq. (7.3.14) may be performed by the methods used for the one-dimensional Ising lattice. The summations over $\eta_2, \cdots \eta_{Z-1}$ may be regarded as the expression of the matrix product $\mathbf{V}^{Z-1}$, where the matrix $\mathbf{V}$ is defined as

$$V_{\eta\eta'} = \lambda^{(\eta+\eta')/2} e^{-\eta\eta'\chi/kT}. \qquad (7.3.15)$$

When the definition Eq. (7.3.15) is used, Eq. (7.3.14) becomes

$$\Xi(\lambda, T) = \sum_{\eta_1, \eta_Z} (\mathbf{V}^{Z-1})_{\eta, \eta_Z} \lambda^{(\eta_1 + \eta_Z)/2}. \qquad (7.3.16)$$

Since the chain is very long, end effects may be neglected, and quantities which will not produce a large change in $\ln \Xi$ may be omitted without error.

Therefore, the expression (7.3.16) for Ξ may be approximated in terms of the largest eigenvalue ϕ of the matrix $\mathbf{V}$.

$$\Xi(\lambda, T) \sim \phi^Z. \tag{7.3.17}$$

The necessary eigenvalue ϕ of $\mathbf{V}$, is readily obtained by solving the secular equation

$$\begin{vmatrix} 1 - \phi & \lambda^{1/2} \\ \lambda^{1/2} & \lambda e^{-\chi/kT} - \phi \end{vmatrix} = 0. \tag{7.3.18}$$

Since the elements of $\mathbf{V}$ are all positive, the larger eigenvalue is

$$\phi = \tfrac{1}{2}[\lambda u + 1 + \sqrt{(\lambda u + 1)^2 - 4\lambda(u - 1)}] \tag{7.3.19}$$

where

$$u = e^{-\chi/kT}. \tag{7.3.20}$$

Returning to the definition [Eq. (7.3.14)] of Ξ and using standard statistical mechanical methods, it is possible to see that

$$\left(\frac{\partial \ln \Xi}{\partial \ln \lambda}\right)_T = \alpha' Z \tag{7.3.21}$$

using the fact that the most probable value of $\sum_i \eta_i$ may be identified with $\alpha' Z$, the net number of charges of the polyion. One may also find that the free energy may be expressed in the form

$$-A(\alpha') = kT \ln \Xi - \alpha' Z kT \ln \lambda. \tag{7.3.22}$$

Differentiation of Eq. (7.3.22) with respect to T, keeping α' fixed, enables one to obtain the following equation for the distributional entropy

$$S(\alpha') = -\left(\frac{\partial A(\alpha')}{\partial T}\right)_{\alpha'} = k \ln \Xi + kT \left(\frac{\partial \ln \Xi}{\partial T}\right)_{\alpha'} - \alpha' Z \left(\frac{\partial}{\partial T}(kT \ln \lambda)\right)_{\alpha'}. \tag{7.3.23}$$

Introducing, from Eq. (7.3.17), $\ln \Xi = Z \ln \phi$, Eq. (7.3.23) becomes

$$S(\alpha') = Zk \ln \phi + \frac{ZkT}{\phi} \left(\frac{\partial \phi}{\partial T}\right)_{\alpha'} - Zk\alpha' \ln \lambda - \frac{Z\alpha' kT}{\lambda} \left(\frac{\partial \lambda}{\partial T}\right)_{\alpha'}. \tag{7.3.24}$$

The condition (7.3.21), determining λ, may be similarly transformed into

$$\frac{\lambda}{\phi} \left(\frac{\partial \phi}{\partial \lambda}\right)_T = \alpha'. \tag{7.3.25}$$

Equation (7.3.24), however, is not yet a convenient form for computation since ϕ is given by Eq. (7.3.19) in terms of the variables T and λ, whereas the differentiation of ϕ is required in (7.3.24) at constant α'. Accordingly, observing the identity

$$\left(\frac{\partial \phi}{\partial T}\right)_{\alpha'} = \left(\frac{\partial \phi}{\partial T}\right)_{\lambda} + \frac{\phi \alpha'}{\lambda} \left(\frac{\partial \lambda}{\partial T}\right)_{\alpha'} \tag{7.3.26}$$

which follows from the basic rule for partial differentiation and the use of Eq. (7.3.25), we rearrange Eq. (7.3.24) to the form

$$S(\alpha') = Zk \ln \phi - Zk\alpha' \ln \lambda + \frac{ZkT}{\phi} \left(\frac{\partial \phi}{\partial T}\right)_{\lambda}. \tag{7.3.27}$$

For the quantities in which we are interested, it is necessary to know $1/Z(\partial S(\alpha')/\partial \alpha')_T$. By suitable manipulation of Eq. (7.3.24) and the use of (7.3.25), one obtains

$$\frac{1}{Z} \left(\frac{\partial S(\alpha')}{\partial \alpha'}\right)_T = -k \ln \lambda - \frac{kT}{\lambda} \left(\frac{\partial \lambda}{\partial T}\right)_{\alpha'}. \tag{7.3.28}$$

Again there appears on the right-hand side of (7.3.28) a derivative which cannot be taken by direct means. But upon introducing the identity

$$\left(\frac{\partial \lambda}{\partial T}\right)_{\alpha'} = -\left(\frac{\partial \alpha'}{\partial T}\right)_{\lambda} \bigg/ \left(\frac{\partial \alpha'}{\partial \lambda}\right)_T \tag{7.3.29}$$

(7.3.28) may be conveniently evaluated.

The resulting equations are

$$\alpha' = \frac{\lambda[u + (\lambda u^2 - u + 2)J^{-\frac{1}{2}}]}{\lambda u + 1 + J^{\frac{1}{2}}} \tag{7.3.30}$$

$$J = (\lambda u + 1)^2 - 4\lambda(u - 1)$$

$$\frac{T}{\lambda}\left(\frac{\partial \lambda}{\partial T}\right)_{\alpha'} = \frac{-u \ln u[1 - \alpha' - \alpha'(\lambda u - 1)J^{-\frac{1}{2}} + J^{-\frac{3}{2}}\{(2\lambda u - 1)(\lambda^2 u^2 - 2\lambda u + 4\lambda + 1) - (\lambda^2 u - 1)(\lambda u^2 - u + 2)\}]}{\alpha'(\lambda u + 1 + J^{\frac{1}{2}})[(1 - \alpha')/\lambda + J^{-\frac{1}{2}}\{u - (\lambda u^2 - u + 2\}J^{-\frac{1}{2}})]}. \tag{7.3.31}$$

The free energy A_4 is to be defined as arising only from the entropy of the distribution of configurations and is, therefore, given by $-TS(\alpha')$ and not by $A(\alpha')$. The energy terms associated with the interaction between nearest neighbors would have been included in A_4 by directly using the expression of Eq. (7.3.22), but if this energy were included, it would be more difficult to adopt a simple approximation to the energy of interaction A_5 between all the charges of the polymeric chain.

Using the Eqs. (7.3.28), (7.3.30), and (7.3.31), one may proceed to evaluate the contribution of the adjacent site interaction to the entropy of polyion systems. For a random distribution, $u = 1$, $\alpha' = \lambda/(\lambda + 1)$, and

$$\frac{1}{Z}\left(\frac{\partial S(\alpha')}{\partial \alpha'}\right)_T = -k \ln \frac{\alpha'}{1 - \alpha'} \qquad \text{(random). (7.3.32)}$$

The limiting behavior for small u, corresponding to large nearest site interactions, is somewhat more complex. If u is set equal to zero, which implies absolute prohibition of the occupation of adjacent sites by charge, $(\partial S(\alpha')/\partial \alpha')_T$ tends strongly to $-\infty$ as α' approaches 0.5. This behavior is to be expected from the assumed impossibility of charging adjacent sites, in addition to which the charging of alternate sites produces an extremely well-ordered state. As will become more apparent when actual numerical values of u are computed, we shall be interested in values of u sufficiently small that the qualitative behavior of $(\partial S(\alpha')/\partial \alpha')_T$ is a great deal like the limiting case just discussed. We, therefore, consider the region $1 \gg u > 0$ in detail.

Equation (7.3.30), relating α' with λ, can be simplified for most values of α'. If λ is such that $\lambda u \ll 1$, Eq. (7.3.30) reduces essentially to

$$\alpha' = \frac{2\lambda}{1 + 4\lambda + (1 + 4\lambda)^{\frac{1}{2}}} \qquad (7.3.33)$$

and the corresponding value of $(T/\lambda)(\partial \lambda/\partial T)_{\alpha'}$ contributes negligibly to the entropy. Inspection of Eq. (7.3.33) shows that values of α' from zero up to nearly 0.5 are included in the range of applicability of Eq. (7.3.33) for sufficiently small u. When $\lambda u = 1$, Eq. (7.3.30) indicates an exact value of $\alpha' = 0.5$ and $(T/\lambda)(\partial \lambda/\partial T)_{\alpha'} = -\ln u$. On the other hand, when λ becomes sufficiently large that $(\lambda u)^2 \geqslant 0(\lambda)$, Eq. (7.3.30) simplifies to

$$\alpha' = \frac{\lambda u + [(\lambda u)^2 + 2\lambda][(\lambda u)^2 + 4\lambda]^{\frac{1}{2}}}{\lambda u + [(\lambda u)^2 + 4\lambda]^{\frac{1}{2}}} \qquad (7.3.34)$$

and the equation corresponding in this event to (7.3.31) is

$$\frac{T}{\lambda}\left(\frac{\partial \lambda}{\partial T}\right)_{\alpha'} = \frac{-\lambda u \ln u}{\alpha'(\lambda u + [(\lambda u)^2 + 4\lambda]^{\frac{1}{2}}} \times$$

$$\left[\frac{1 - \alpha' - \dfrac{\alpha'\lambda u}{\sqrt{(\lambda u) + 4\lambda}} + \dfrac{\lambda u(\lambda^2 u^2 + 6\lambda)}{((\lambda u)^2 + 4\lambda)}}{1 - \alpha' + \dfrac{(\lambda u - (\lambda^2 u^2 + 2\lambda)(\lambda^2 u^2 + 4\lambda)^{-\frac{1}{2}})}{\sqrt{(\lambda u) + 4\lambda}}}\right]. \qquad (7.3.35)$$

It is also apparent that this formulation applies for values of α' ranging downward from unity to quite close to 0.50.

It should be observed that for any value of u the behavior to be expected in the limit of low α' will approach that of a chain without interaction between charged sites. Therefore, $(\partial A_4/\partial\alpha')_T$ approaches $-T(\partial S(\alpha')/\partial\alpha')_T(\text{random})$, where the latter quantity is given by Eq. (7.3.32). In addition, $(\partial S_m/\partial\alpha)_{\alpha f}$ approaches zero since the number of bound sites goes to zero as may be seen formally from Eq. (7.3.10). The portion of $(\partial A_5/\partial\alpha')_T$ depending upon interactions between charges on the polymer will also vanish in the limit of low α' and the only contribution to A_2 remaining will be terms of $(\partial A_5/\partial\alpha)_T$ depending upon the interaction of charges of the polymer with their respective ion atmospheres. Because the binding of counterions becomes unimportant at low degrees of neutralization, α approaches α' and the two quantities may be regarded as interchangeable. In this limit Eq. (7.3.7) simplifies to the equation

$$pH = pK_a^0 + \log\frac{\alpha}{1-\alpha} + \frac{0.434}{ZkT}\left(\frac{\partial A_5}{\partial\alpha}\right)_T \qquad (7.3.36)$$

originally derived by Overbeek.[20]

7.3.4 Some Calculations

It is now necessary to investigate the range of u which will actually arise in polyelectrolyte solutions. In polymethacrylic acid (PMA), the carboxyl groups are separated by a distance of approximately 2.5 A. To complete the calculation of u, it is also necessary to know what value to assign the local dielectric constant of the region between charges on neighboring sites. If adjacent carboxylate ions are in a configuration which places them on opposite sides of the polymer skeleton, the space between them will be occupied by organic matter of dielectric constant of the order of 2.[21] On the other hand, if the charges are in a configuration such that only water could be between them, at most one water molecule would be able to squeeze into the small space between the two large negative ions, and any such water molecule would be unable to orient cooperatively with neighboring water to effect the large dielectric polarization characteristic of the pure substance. In view of this situation, we have chosen the value 5.5 for the local dielectric constant[22] because it is of the order of magnitude of the polarization which may be attributed to water molecules not free to orient in an electric field. Since the dielectric constant of the polymer skeleton is less than 5.5, the value chosen provides a useful lower bound to the energy of interaction. Furthermore, the calculations will be found to be not at all sensitive to the exact value chosen for the interaction energy when it is large compared with kT. In PMA the interaction energy will be $\chi = q^2/Dr = 39.8kT$. The corresponding value of u is 5×10^{-18}.

A few values for $(\partial S(\alpha')/\partial\alpha')_T$ calculated from Eqs. (7.3.28) and (7.3.33) to (7.3.35) are shown in Table 7.3.2. The values clearly indicate a large decrease in entropy as the fraction of charged sites approaches 0.5, and show

TABLE 7.3.2[a]

ENTROPY OF NONRANDOM DISTRIBUTION OF CHARGE SITES

α'	0.25	0.30	0.40	0.444	0.476	0.50	0.75
$-\dfrac{1}{Zk}\left(\dfrac{\partial S}{\partial\alpha'}\right)$	−0.28	+0.25	+1.80	+2.99	+4.60	+79.6	+160
$-\dfrac{1}{Zk}\left(\dfrac{\partial S(\text{random})}{\partial\alpha'}\right)$	−1.10	−0.85	−0.41	−0.22	−0.09	0	+1.10
$-\dfrac{1}{Zk}\left(\dfrac{\partial(S-S(\text{random}))}{\partial\alpha'}\right)$	0.82	1.10	2.21	3.21	4.69	79.6	159

[a] Reproduced from ref. 1.

that the reduction of net charges by binding of counterions can effect a substantial decrease in free energy from the entropy of mixing of charged and uncharged sites.

The treatment of the entropy of mixing of charged and uncharged sites described here is based on the assumption that only charges on adjacent sites can interact. The inclusion of the effect of forces between sites not immediately adjacent to each other will, of course, further increase the free energy at higher fractions of net charge but by amounts which should be much less than the corresponding quantities in Table 7.3.2 since the greater distance between sites and the increased effective dielectric constant at this distance will greatly reduce the magnitude of the interaction.

The deviations of the free energy of mixing from the values obtained in the absence of interaction between sites are also shown in Table 7.3.2. If Eq. (7.3.36) is used to proceed from experimental pH data to values of $1/Z(\partial A_5/\partial\alpha')_T$ and the latter identified with $q\psi$, where ψ is the " potential " in the solution, values of $q\psi/kT$ are obtained. Many approximations involving electrostatic interactions in the solution depend for their validity, or rapid convergence, upon the magnitude of $q\psi/kT$. We are now in a position to observe that the actual values of $q\psi/kT$, assuming that all the contributions to A_5 are electrical in nature and that the identification $1/Z(\partial A_5/\partial\alpha')_T = q\psi$ is a good approximation, will be lower than the value calculable using Eq. (7.3.36) by the deviation in the entropy of mixing of charged and uncharged sites from its ideal value. By consultation of Table 7.3.2, we can see that there will be a large difference between values of $q\psi/kT$ calculated including, or neglecting, the nonideality of the entropy of mixing.

7.4   Numerical Calculations Based on the Harris–Rice Theory

In preceding sections it has been pointed out that the use of spherical models of polyions modifies the free energy in a manner which cannot be quantitatively calculated. In addition, the models employed by Hermans and Overbeek,[6] Kimball et al.,[7] Flory,[23] and Osawa et al.[5] lead to results which are not altogether in satisfactory agreement with experiment. On the other hand, the use of a chain model in which the probability of any configuration is assumed to depend only upon its end-to-end extension is also found to lead to results which differ markedly from experiment. The two most objectionable approximations of this random chain model are, first, the inclusion with undiminished statistical weight of all configurations of the polyion with the same end-to-end distance, and, second, the assumption that it is the total electrostatic interaction energy which is capable of causing the polyion to expand. In actuality, the configurations will be weighted according to the amount of local kinking, those configurations which place large amounts of chain close together occurring with far less than their random a priori probability. When using the concept of statistical element, care must be taken that one does not think that the last monomer in a given element is independent of the first in the next element. Such a conclusion, of course, is not true since there is a constant bond angle (or at most two constant bond angles and a rotational barrier) between the two monomer units. However, it is just such interactions as these between charges situated on adjacent monomers that constitute a large part of the total electrostatic interaction energy. Therefore, far less than the total interaction energy is effective in expanding the polymer coil.

In order to remove the first of these approximations, we shall find it advantageous to proceed by introducing, step by step, the interaction between the charges of the polyion, rigorously calculating the change in the distribution of configurations at each step of the process. The method we shall employ to introduce interactions between the charges will be to consider, first, pairwise interactions between nearest neighbor links of the polymer chain, then those between the charges on next nearest neighbor links, and so forth. While we will be unable to evaluate the results of the higher approximations, such a scheme will provide, conceptually at least, a mechanism by which any desired approximation to the actual behavior of the polyion chain can be calculated. By the use of the methods of statistical mechanics, the angular correlation between a link of the polymer chain and the preceding links can be evaluated, and the theory of random flight processes with partial correlations[25] employed to calculate the resulting distribution of molecular lengths. These statistical mechanical methods will also enable us to evaluate the thermodynamic functions of the polymer coil.

When calculating the expansion of the polyion from the extension characteristic of an uncharged polymer chain, it is assumed that the actual distribution of charges on the polymeric chain can be replaced by an average distribution characterized by an equal spacing between the charges of the polyion. Although such an approximation should be relatively good for most purposes, since this uniformly charged configuration is also the most probable one, the approximation is not adequate for calculating the entropy arising from the various possible choices of charges and uncharged ionizable groups. It is, therefore, necessary when evaluating the thermodynamic functions of polyelectrolyte solutions to include in the entropy of the chain molecule the contribution arising from the actual distribution of charged and uncharged sites on the polyion. The methods by which this aim can be accomplished have been discussed in the preceding sections.

The calculations of the preceding section indicate that the entropy decreases so rapidly at high degrees of ionization that binding of counterions by the polyion must occur when a large proportion of the ionizable portions of a polyacid are removed.* The entropy involved when binding occurs is so far from that of a random distribution of ionized and un-ionized groups that there is a serious question as to the significance of comparing thermodynamic data with theoretical quantities calculated omitting a good treatment of this effect. Since such a treatment depends crucially upon the interaction energy between ions at very small separations along the polymer chain, it is not justifiable to approximate the interaction as that between point charges in a continuous dielectric.

The model for a polyion upon which the following discussion will be based consists of a chain of N " statistical elements " or links, each of length A and net number of charges ζ, connected together at the ends by universal joints. By a statistical element is meant (see Chapter 5) a number of monomer units such that the mean square length $\langle h_0^2 \rangle$ of a similar but uncharged polymer coil is correctly given by the random flight formula

$$\langle h_0^2 \rangle = NA^2. \tag{7.4.1}$$

The statistical element will differ from the real monomer unit in such a way as to take into consideration fixed bond angles, hindered rotation, etc., but not the effect of the charges. Since the idea of a statistical element has been introduced to take into account the local orientational order caused by forces common to both the charged and uncharged polymer, it would be contrary to the spirit of the present model to assume that the behavior of a statistical element when charged should affect its length. Such changes must, however, occur.

*We remind the reader that this is equivalent to the statement that the cell partition function depends on α in such a manner that the energy required to remove an ion from its cell increases as α increases.

If we consider each possible configuration of the chain as being built up by successively adding links to the chain, the probability of adding the jth link in a given relative orientation to the preceding links will be governed by the energy of interaction of the charge of the jth link with links $j - 1, j - 2, \cdots$. If the interaction between links falls off sufficiently rapidly as they become separated on the chain, the sequence of approximations to the distribution of configurations generated by successively considering interactions between sets of segments separated on the chain by increasing numbers of units will converge. The first such approximation, after that of the completely random chain, is provided when one assumes the configuration of each link to depend only upon the one added immediately before it. Such a dependence generates a Markoff[26] process, and it can be concluded that the distribution of lengths becomes Gaussian in the limit of large N with the average end-to-end distance increasing as $N^{1/2}$. The higher approximations are also Markoff processes but with the number of polymer segments per Markoff unit increasing to two, three, etc. These approximations, therefore, correspond to Markoff chains in higher dimensional vector spaces.

One must not conclude, however, that in general a sequence of approximations of the type described, each defining a Markoff process, converges to a Markoff process in the limit of increasing number of interacting links. We are assuming here that the interaction between successive links falls off sufficiently rapidly to render such a conclusion tenable. Similar considerations indicate that the limiting dependence upon the degree of polymerization of the size of chain molecules composed of links of nonzero volume cannot be ascertained by this type of approximation unless the method is known to be convergent to a Markoff process.

In view of the rapidly increasing mathematical complexity as one proceeds to higher approximations, we shall confine the more detailed discussion to the case of nearest neighbor interaction only. After finding to what extent modification of the random flight distribution occurs, one may then ask whether proceeding to higher approximations is worthwhile. It should be emphasized, however, that proceeding to higher approximations requires no fundamentally new development, but merely necessitates more tedious calculations.

In order to complete the specification of the model described here, it is necessary also to consider the interactions between the charges on monomer units of the same statistical element. These interactions will of necessity depend upon the net degree of ionization of the polymer and when there are several charges per statistical element, intra-element interactions will make the major contribution to the potential. Means of evaluating the interaction energy within a statistical element will be discussed individually for the various polymeric species concerned.

We are now ready to consider more explicitly the nearest neighbor approximation. We find it convenient to discuss separately the end-to-end extension of the polyion and its free energy.

7.4.1 EXPANSION OF POLYIONS

Since the only interaction between links of the polymer is assumed to be the electrostatic force directed along the lines connecting the charges of a pair of neighboring links, the distribution of orientations of any link will be axially symmetric with respect to the preceding link. In this case the correlation between the orientations of successive links can be defined as the average value of the cosine of the angle γ between the vectors specifying the directions of the successive elements. Writing $\langle \cos \gamma \rangle = c$ it is possible to show[25] that the average square length $\langle h_1^2 \rangle$ approaches, for large N, the value

$$\langle h_1^2 \rangle = \langle h_0^2 \rangle \left(\frac{1 - c}{1 + c} \right). \tag{7.4.2}$$

Since the distribution of lengths is Gaussian, it is given by the formula[25]

$$W(h)\, dh = \left(\frac{3}{2\langle h_1^2 \rangle} \right)^{\frac{3}{2}} \frac{4h^2}{\sqrt{\pi}}\, e^{-3h^2/2\langle h_1^2 \rangle} \tag{7.4.3}$$

where $W(h)dh$ is the probability that the end-to-end distance is in the range from h to $h + dh$. It should be emphasized that the simple formula (7.4.2) only applies when there is axial symmetry.

The evaluation of c may be accomplished by the standard methods of statistical mechanics,[27]

$$c = \langle \cos \gamma \rangle = \frac{\int \cos \gamma \, e^{-u(\gamma)/kT} \, d\Omega}{\int e^{-u(\gamma)/kT} \, d\Omega} \tag{7.4.4}$$

where $u(\gamma)$ is the potential energy of interaction between the charges on adjacent chain segments. The integral of (7.4.4) is understood to be extended over all orientations, and $d\Omega$ is the element of solid angle. If the interaction energy is written as the sum of the contributions of pair potentials, and the interactions between a pair of charges is approximated by a screened Coulomb potential

$$u(\gamma) = \sum_{i,j} \frac{q_i q_j \, e^{-\kappa r_{ij}(\gamma)}}{D r_{ij}(\gamma)} \tag{7.4.5}$$

where the summations are to be extended over all net charges i on one statistical element, and all net charges j of the other, q_i is the magnitude of charge i, D the dielectric constant of the medium, and $r_{ij}(\gamma)$ is the distance between charges i and j when the angle between their statistical elements is γ. The screening constant may be taken as the reciprocal Debye length κ and is given by[28]

$$\kappa^2 = \frac{4\pi q^2}{DkTV} \sum_i c_i^0. \tag{7.4.6}$$

Equation (7.4.5) is not in a convenient form for actual use in finding $u(\gamma)$ and, more important, the exact location of the various charges on the statistical element is not generally known. We, therefore, will approximate Eq. (7.4.5) by moving all the net charge of each statistical element to its center. In this approximation

$$u(\gamma) = \frac{\zeta^2 q^2 e^{-\kappa r_{ij}(\gamma)}}{Dr_{ij}(\gamma)}$$

$$r_{ij}(\gamma) = A\cos(\gamma/2). \tag{7.4.7}$$

The extent of the approximation in going from Eq. (7.4.5) to Eq. (7.4.7) depends upon at least two factors. The first of these resides in the approximation inherent in the use of the concept of " statistical element," and is due to the fact that the charges near the end of a statistical element cannot have their full effect as indicated by Eq. (7.4.5) in forcing orientation of the next statistical element because the actual molecule cannot bend sharply at a given point while the model can. Indeed though a major portion of the electrostatic interaction energy arises from the interactions of adjacent charges, this part of the energy is completely ineffective in expanding the polyion due to the constant bond angle between adjacent monomers. In fact, if there is a charge for each skeletal atom or rigid skeletal unit the first segment on which the interaction can have any orienting effect at all is a next nearest neighbor and this effect may be markedly reduced if there is a large potential restricting rotation about the monomer-monomer bond. A second approximation which we can qualitatively evaluate is the change in electrostatic interaction which occurs when a group of charges, situated at points of the statistical element, are replaced by the sum of their charges at their centroid. Since, as mentioned earlier, charges less than a statistical element apart cannot exert their full force toward expanding the chain, the approximation of lumping the charge should largely cancel that of overestimating the effective interaction of neighboring charges. To examine the probable size of this effect which will decrease the electrostatic interaction, we may approximate the location of the charges of each statistical element by a set of equally spaced charges on a line. The difference between $u(\gamma)$ as calculated by a single charge model and the distributed charge model is discussed in connection with the various experimental results.

7.4.2 THERMODYNAMIC FUNCTIONS OF POLYIONS

While treatment of the polyion expansion as a Markoff process enables its distribution of lengths to be determined, such considerations do not facilitate determination of the thermodynamic functions. For this purpose we will divide the free energy of the polyelectrolyte solution into a number of parts in the manner indicated in the preceding section. We shall at once specialize to polyacids, noticing that for polymeric bases an analogous development can be made.

We recall that the first contribution to the free energy, A_1, is that of the free hydrogen ions and counterions eligible to take part in binding and neutralization phenomena. The second term A_2 is the sum of the contributions of the electrostatic interactions of the polyion with itself and with the small ions. The third term A_3 is the free energy change that accompanies the processes of binding and dissociation, together with the contribution of the entropy of mixing of bound and undissociated sites.

The division of the free energy of the polymer into parts A_2 and A_3 depends upon the assumption that the electrostatic effect of binding can be regarded merely as a reduction in net charge of the polymer. In addition the evaluation of A_3 requires further assumptions regarding the binding process.

A_1 will be a function of a_{H^+}, the activity of the free hydrogen ions in the solution, and a_{C^+}, that of the counterions. A_2 will be expressible in terms of the net fraction of charge, α', of the polymer chain. α' is related to α, the degree of dissociation of the polymer, by the formula $\alpha' = \alpha (1 - f)$, where f is the fraction of the dissociable groups bound by counterions. In this notation, ζ, the number of charges per statistical element, is expressible as $\zeta = \alpha' Z / N$. A_3 will depend upon α and f separately as well as upon the dissociation constants K_a^0 and K_s^0 for the acidic groups and bound ion counterion pairs, respectively. We may summarize much of the above in the equation

$$A = A_1(a_{H^+}, a_{C^+}) + A_2(\alpha') + A_3(\alpha, f, K_a^0, K_s^0). \qquad (7.4.8)$$

Following the notation of the previous section we will further subdivide A_2 such that

$$A_2 = A_4 + A_5 = A_4 + A_6 + A_7 + A_8 \qquad (7.4.9)$$

where A_4 is the contribution to the free energy arising from the entropy of the statistical distribution of the net charge of the polymer among the sites on the chain. A_6 arises from the interaction between statistical elements and here will be approximated as being satisfactorily described by terms including only the interaction of nearest neighboring statistical elements. To calculate A_6, it is convenient to consider the set of angles

$\{\gamma_{i,i+1}\} \equiv \{\gamma_i\}$ between the orientation of the succeeding statistical elements. The $\{\gamma_i\}$ are independent, as follows immediately from the limitation of interaction to nearest neighboring elements. In addition, the $\{\gamma_i\}$ suffice to determine the energy of the system of interacting links. They are, therefore, suitable quantities to use in a statistical-mechanical ensemble. We may define a configuration partition function Q_γ by the formula

$$Q_\gamma = \int e^{-u(\gamma)/kT} \, d\Omega \qquad (7.4.10)$$

and the change in free energy A_6, resulting from the introduction of the interaction $u(\gamma)$, will be

$$A_6 = -NkT(\ln Q_\gamma - \ln Q_\gamma^0) \qquad (7.4.11)$$

where Q_γ^0 refers to the value of Q_γ when $u(\gamma)$ is zero. By direct integration we know $Q_\gamma^0 = 4\pi$, so

$$A_6 = -NkT \ln\left(\frac{Q_\gamma}{4\pi}\right). \qquad (7.4.12)$$

A_7 is the free energy change accompanying the building up of ion atmospheres about each of the ions of the system. We take this as [3, 28]

$$A_7 = -\frac{q^2\kappa\alpha'Z}{2D}. \qquad (7.4.13)$$

It must be emphasized that Eq. (7.4.13) holds when the ionic strength is due to added salt. This result arises from the use of the Debye–Hückel equation when the negative ions are immobile. Thus, though each fixed negative ion may have an atmosphere of positive ions about it, the positive ions do not have an atmosphere of negative ions about them in the manner prescribed by the Debye–Hückel theory.

Finally, it is necessary to add the free energy A_8 of the electrostatic interaction within the statistical elements. The explicit form of A_8 will be determined by the model chosen for the statistical element. To calculate A_8, we use a crude model in which the charges are assumed to be equally spaced in a straight line along the statistical element. A_8 is then given by the sum over all pairs of the pair interaction energy and may be conveniently written as

$$A_8 = \frac{Nq^2\zeta}{DA} \sum \left(\frac{\zeta-j}{j}\right) e^{-j\kappa A/\zeta}. \qquad (7.4.14)$$

To proceed to a calculation of the potentiometry or degree of binding of the polyion, we shall find it necessary to obtain derivatives of the various terms of A_2 with respect to the fraction of net charges α'. Carrying out the indicated differentiations we obtain

$$\frac{\partial A_6}{\partial \alpha'} = \frac{2Z}{\zeta} \frac{\int u(r) \, e^{-u(\gamma)/kT} \, d\Omega}{\int e^{-u(\gamma)/kT} \, d\Omega} \tag{7.4.15}$$

$$\frac{\partial A_7}{\partial \alpha'} = -\frac{q^2 \kappa Z}{2D} \tag{7.4.16}$$

$$\frac{\partial A_8}{\partial \alpha'} = \frac{q^2 Z}{DA} \sum_{j=1}^{\zeta} \left(\frac{2\zeta}{j} - 1 - \kappa A + j\frac{\kappa A}{\zeta} \right) e^{-j\kappa A/\zeta}. \tag{7.4.17}$$

It should be noted that Eqs. (7.4.14) and (7.4.17) are not well defined for nonintegral values of ζ and are to be regarded as replaced by appropriate continuous functions for nonintegral ζ.

The remaining contribution to $\partial A_2/\partial \alpha'$, namely $\partial A_4/\partial \alpha'$, is given by Eqs. (7.3.28), (7.3.30), and (7.3.31). The conditions determining the potentiometry and the binding of counterions of the polymeric electrolyte are obtained by minimizing the free energy under appropriate conditions. The results are given in Eq. (7.3.7).

$$-\frac{\partial A_2}{\partial \alpha'} + ZkT \ln K_s^0 - ZkT \ln a_{C^+} - T\left(\frac{\partial S_m}{\partial \alpha f}\right)_\alpha = 0 \tag{7.4.18}$$

$$+\frac{\partial A_2}{\partial \alpha'} - ZkT \ln K_a^0 + ZkT \ln a_{H^+} - T\left(\frac{\partial S_m}{\partial \alpha}\right)_{\alpha f} = 0.$$

The terms involving S_m, the entropy of mixing of bound ions and undissociated sites, arise from the free energy term A_3. Expressions for the derivatives indicated in Eq. (7.4.18) may be found as Eq. (7.3.10).

7.4.3 COMPARISON WITH EXPERIMENT

The experimental quantities usually compared with a theory of polyelectrolyte behavior are the expansion of the polyion as a function of ionic strength, and the potential as determined from a relation between the pH of the solution, a constant characteristic of the dissociating group, and the free energy of mixing of the ionized and un-ionized sites. The expansion of the polyions is a good test of the theory since its experimental determination is completely independent of any assumptions concerning the behavior of polyelectrolytes in solution. On the other hand, the usual comparison of theoretical and experimental potentials is not a good test.

The accepted relation among the thermodynamic functions of the polyion, the pH, and the degree of dissociation is[20]

$$pH = pK_0 - \log \frac{1-\alpha}{\alpha} + 0.434 \frac{q\psi}{kT} \qquad (7.4.19)$$

where K_0 is the ionization constant of the dissociating groups in the absence of the interference of electrostatic interactions and ψ is the electrostatic potential of the polyion. The potential as defined by (7.4.19) is related to the electrostatic free energy (i.e., A_5 of the present development) by $(\partial A_5/\partial \alpha)_\kappa = Zq\psi$. Equation (7.4.19) implies a random distribution of charged and uncharged sites, and we have shown on the basis of a model employing interactions between neighboring charge groups that the entropy of mixing of such a system, while approaching the ideal value for low degrees of ionization, deviates considerably and to an increasing extent as the degree of ionization increases. In order to calculate the potential, it is necessary to account quantitatively for the entire electrostatic free energy of the polyion. To do this correctly, the treatment must include those contributions arising from binding and other nonideal effects as well as the more obvious electrostatic terms. However, it is not suitable merely to reduce the degree of neutralization to compensate for the binding and then use Eq. (7.4.19), since the binding is symptomatic of something fundamentally wrong with the assumptions inherent in the use of Eq. (7.4.19).

The potential has been shown to be intimately related to the degree of binding of counterions by the polyion and, therefore, it suffices to compare either the pH of the solution or the degree of binding with experimentally observed values. The second test of the theory to be applied here, then, will be the agreement of the calculated values of the binding with those directly observed experimentally by transference or diffusion measurements.

The extent of binding was evaluated assuming that the entropy of mixing S_m of bound ion pairs and un-ionized sites is given by the ideal solution expression. $(\partial S_m/\partial \alpha)_{\alpha f}$ is then given by Eq. (7.3.10). Equation (7.4.18) was then solved by an iterative procedure assuming a value of 10 for K_s^0. It was found that the value of K_s^0 chosen does not seriously affect the results since $(\partial A_2/\partial \alpha')$ varies extremely rapidly with α' for the values of α' with which we are concerned. The major contributions to $(\partial A_2/\partial \alpha')$ are found to arise from the interactions within the statistical element included in A_8, and from the nonrandom mixing term A_4. For this reason the amount of binding will increase sharply as the distance between neighboring charges is decreased. One might, therefore, expect that the amount of binding in carboxymethyl-cellulose (CMC), where the sites are 5 A apart, would be less than that observed under corresponding conditions for polyacrylic acid (PAA) where the sites are separated by but 2.5 A. A serious difficulty in evaluation of the necessary free energy terms depends upon the choice to be made for the effective dielectric constant between a pair of charges only a few angstroms apart. This small space will undoubtedly be partially occupied by other portions of the polymer molecule, which as Kirkwood and Westheimer[21] have indicated, should be recognized explicitly as having a much lower dielectric constant than the bulk of the aqueous medium. It should also be remembered that even in the absence of polymer chain between charges the decrease in dielectric constant for the water molecule in such close proximity to ions would be most important.[29] For the reasons enumerated in the preceding section, we have chosen the value 5.5 for the local dielectric constant.

Binding for PAA calculated according to the above discussion is listed together with the experimental values [16] in Table 7.4.1. It will be observed

TABLE 7.4.1[a]

COMPARISON OF THE CALCULATED AND OBSERVED BINDING IN
POLYACRYLIC ACID

	$\alpha = 1$	$\alpha = 0.5$	$\alpha = 0.2$
Exptl.[16]	0.63	0.45	0.20
Calcd.	0.65	0.36	~ 0.0

[a] Reproduced from ref. 1.

that the calculated binding decreases faster than the experimental findings indicate it should. A possible explanation of this deviation is the neglect of higher order interactions between the charged groups. At high α' this omission makes little difference since there is a large decrease in the pair interaction energy on going from nearest to next nearest neighboring pairs. This decrease is due to both the increased distance between the charges and the increase in effective dielectric constant at the larger distance. On the other hand, at low α', the charges are on the average further apart, and the further neighbors contribute proportionally more of the total interaction. Since the difficulty of extending the calculations to include next nearest neighboring sites is considerable, we must temporarily be content with the eminently satisfactory qualitative agreement now obtained.

We now turn to an examination of the expansion of the polyion.

1. Sodium Carboxylmethylcellulose

Data on CMC were taken from the work of Schneider and Doty.[14]

From the observed weight average molecular weight and assuming that the polydispersity could be adequately accounted for as a Flory distribution,[30] the z average molecular weight was calculated from the relation $M_z = 1.5 M_w$*. From the z average molecular weight, the z average contour length L was obtained. The value of the unperturbed polymer size was determined by extrapolation of the light-scattering data to infinite ionic strength and was found to be 2100 A for a polymer of weight average molecular weight 435,000. The Kuhn statistical element length[31] can then be calculated by solving the simultaneous equations

$$\langle h_0^2 \rangle = NA^2$$

$$L = NA. \tag{7.4.20}$$

In this manner A was found to be 335 A. The polymer was characterized as possessing 1.1 carboxyl groups per 5.1 A and the degree of neutralization was 0.96.

The expansion factor $\langle h_1^2 \rangle^{1/2} / \langle h_0^2 \rangle^{1/2}$ was calculated by the use of Eqs. (7.4.2) and (7.4.4). The integrals necessary to evaluate (7.4.4) were computed numerically by means of a 20-point interpolation formula, and Eq. (7.4.7) was used for the function $u(\gamma)$; D was taken as 80. A comparison of the experimental values of $\langle h_1^2 \rangle^{1/2} / \langle h_0^2 \rangle^{1/2}$ with the ones calculated assuming no binding is given in Table 7.4.1. It can be seen that at low ionic strengths the calculated expansion is too large in spite of the fact that the method used would be expected to give a lower value than that actually observed. But we have already seen that the interactions between neighboring charge sites on the

TABLE 7.4.2[a]

COMPARISON OF THE CALCULATED AND EXPERIMENTAL EXPANSION
FACTORS $\langle h_1^2 \rangle^{1/2} / \langle h_0^2 \rangle^{1/2}$ FOR SODIUM CARBOXYLMETHYLCELLULOSE

$1/\kappa$	No binding	Binding 33%	Binding 67%	K.L.	H.O.	Exptl.
6	1.07	1.07	1.06	2.00	1.41	1.09
13.8	1.13	1.11	1.09	4.16	1.81	1.16
30.4	1.38	1.30	1.19	10.9	2.45	1.39
44	1.72	1.57	1.31	13.8	2.87	1.59
50	1.91					

[a] Reproduced from ref. 1.

Columns 1, 2 and 3 are calculated on the basis of the theory presented in this section. Column 4 is based on the theory of Katchalsky and Lifson and column 5 is based on the theory of Hermans and Overbeek. Column 6 presents the experimental results of Schneider and Doty.[14]

* We include this material to show how experimental data must be modified to conform to a model calculation.

polymer chain will make it impossible to remove all the sodium ions from the chain. This binding reduces the effective charge per statistical element, thereby decreasing the interaction between elements. In Table 7.4.2 we have also listed the calculated expansions for 33% and 67% binding and have plotted all the data of the table in the accompanying graph (Fig. 7.4.1).

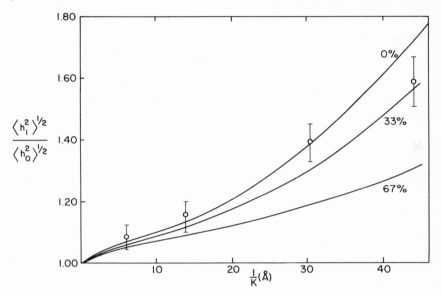

FIG. 7.4.1. Comparison of the predicted and observed expansion ratio for sodium carboxylmethylcellulose (Eqs. 7.4.2 and 7.4.4). [Reproduced from ref. 1.]

While the experimental error of approximately 10% does not permit a decisive choice at higher ionic strengths, the data at low ionic strength points to a small amount of binding, probably somewhat less than 33%. This conclusion is in accord with our discussion of binding where we suggested that the extent of binding in CMC should be less than that in PAA. The extent of binding observed for PAA at complete stiochiometric neutralization is about 63%.

It is quite clear that the present calculation treats the electrostatic interactions in a quite different way than that more random model of Kuhn, Künzle, and Katchalsky. The effect of the modifications introduced in the present theory may be seen by comparing results computed using the formula of Katchalsky and Lifson[9] with the other entries in Table 7.4.2. Also listed in the table are the expansions calculated from the spherical model of Hermans and Overbeek.[6] A substantial improvement over all these theories has been realized.

2. Polymethacrylic Acid

The only structural difference between the monomeric group in poly-methacrylic acid (PMA) and polyacrylic acid (PAA) is the presence of a methyl group on the carbon atom α to the carboxyl group in PMA. In both polymers the spacing between charged sites is 2.5 A so that to the approximation involved in the theory described here, with the possible exception of the statistical element length, there should be little difference between PMA and PAA. The major contributions to the electrostatic free energy arise from the nonrandom free energy of mixing and the interaction between nearest neighbor charge groups. The interaction between the statistical elements is a small additional term which, even if the statistical elements in PMA and PAA are not identical, cannot be very different for the polymers. Recognizing this similarity between the polymers, we have assumed that the extent of binding in PMA will be the same as the extent of binding in PAA which is 63% at complete neutralization.[16]

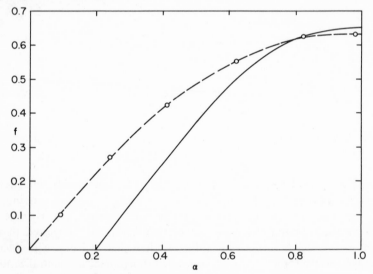

FIG. 7.4.2. Comparison of the predicted and observed counterion binding for sodium polyacrylate. [Reproduced from ref. 1.]

The calculated expansions for PMA in the absence of added salt are compared with the experimental values of Oth and Doty[15] in Table 7.4.3. It is easily seen that the calculated values, even for zero binding, fall below the experimental value.

The model employed in this theory neglects interactions between segments further separated than nearest neighbors. In CMC where the statistical

TABLE 7.4.3[a]

COMPARISON OF THE CALCULATED AND EXPERIMENTAL EXPANSION
FACTORS $\langle h_1^2 \rangle^{1/2}/\langle h_0^2 \rangle^{1/2}$ FOR SODIUM POLYMETHACRYLATE

	$A = 10$ A	
No binding	60% binding	Exptl.
3.26	1.64	6.0

[a]Reproduced from ref. 1.

element is 335 A long, the shielding radius $1/\kappa$ is only about one-eighth the element length even at the lowest ionic strength investigated. In this case the neglect of higher order interactions is not serious. In PMA the statistical element is only 10 A long and the experimental lengths listed in Table 7.4.3 were determined under conditions such that the shielding radius $1/\kappa$ was never less than 30 A. Under these circumstances it is to be expected that higher order interactions will be important and that the calculated expansions will be smaller than the experimental values.

Since the statistical element length in PMA corresponds to only four monomer units,[15] it is possible to make an estimate of the decrease in electrostatic energy when we replace the discrete ensemble of charges spaced in some manner along the element with one large charge at the centroid. In principle one may write the exact solution to the interaction between the pairs of charges on adjacent segments, but this expression is inconvenient for numerical calculations due to its great complexity when there are a large number of charges on a statistical element. For the element of four monomers, with equally spaced charges, the error involved in replacing the four charges with one of magnitude four is of the order of 25% at moderate ionic strengths and decreases at low ionic strength. It is to be expected that the error involved in overestimating the interaction energy due to the model being able to bend sharply is in the opposite direction and it is probably of the same order of magnitude.

7.5 Polyampholytes with Regularly Alternating Structure

The basic outline of the chain model for polyelectrolytes of one sign of charge, as developed in the preceding sections, is also applicable to polymers capable of bearing both positively and negatively charged groups. Since a more complete understanding of such ampholytic polymers might well prove to be valuable not only in view of the prominence of these species in biological systems, but also because techniques are beginning to be developed for the synthesis of polymapholytes, it will be of interest to extend the chain model explicitly to these systems.

While the forces which operate in polyampholyte systems are qualitatively no different than those in purely acidic or basic polyelectrolytes, significant new problems arise because of the local inhomogeneity of the polymer skeleton. In general, far from the isoelectric point (i.p.), where either the acidic or basic groups become almost totally uncharged, the problem is similar to that of a simple polyelectrolyte. However, in the nearly isoelectric region where appreciable numbers of both positive and negative charges can coexist on the chain, these complications dominate the situation.

The two physical properties with which we shall be primarily concerned are the titration curve, or relation between pH and degree of charge at given ionic strength, and the end-to-end extension of the polyampholyte as a function of charge and ionic strength. These quantities suffice to characterize the thermodynamic properties of the polyampholyte solution. Calculation of the titration curve in the neighborhood of the isoelectric point will differ from the corresponding calculation for a simple polyion in that there will be both attractive and repulsive forces present, and that it will no longer in general be permissible to assume that all of the sites capable of occupation by charge of a given sign are similar but will vary in a manner determined by the statistics of the skeletal distribution along the polymer chain. However, such a simplification is still possible in the case of a polymer of regularly alternating skeletal structure, enabling that problem to be solved in a manner similar to the simple polyion. On the other hand, polyampholytes of more general skeletal distribution will possess sites with different environments, and therefore the extent of dissociation of the various ionizable groups of the polymer will depend upon the local skeletal structure as well as upon the concentrations of the ions in the solution. We shall for this reason restrict the present discussion to polyampholytes of regularly alternating structure, reserving the more general types of copolymeric electrolytes for a separate discussion in the next section.

As in the preceding sections, it is necessary to consider the possibility of binding of the counterions by the polyampholyte. We assume here, as we did previously, that binding can be represented in terms of a thermodynamic equilibrium between bound ion–counterion complexes and free ions, with binding possible at each charge site bearing an ionizing group. The previous theory is only extended in that both cation and anion binding are here considered. The reader is again reminded that we use this descriptive language to implement what is really a one-dimensional cell model.

Calculations of the titration curves have been carried out for a hypothetical polymer of typical characteristics. The results indicate a dependence upon ionic strength qualitatively similar to the titration properties of dilute protein solutions. The effect of the nonspecific type of counterion binding assumed is small over a wide range of pH about the isoelectric point.

The second physical property to be investigated is the end-to-end extension of the polyampholyte. Calculation of the expansion or contraction also requires some additional considerations not necessary in dealing with purely acidic or basic copolymers. As in the preceding sections, we adopt a model first introduced by Kuhn,[31] representing the polymer as a chain of rigid " statistical elements," connected together by universal joints. Superimposed upon the random distribution of chain configurations is the perturbation induced by the electrostatic forces between elements which is approximately calculated in terms of point charges of appropriate magnitudes at the center of each element.* When each statistical element bears a nonvanishing average net charge, the calculation of extensions proceeds exactly as in the case of a purely acidic or basic polymer. However, when as at the isoelectric point the average net charge of each statistical element vanishes it is necessary to consider the correlations in charge between interacting elements. By using the formalism developed by Kirkwood and Shumaker,[32] it is possible to calculate the potential of average force between statistical elements of the polymer arising from fluctuations of charge. In general, there will also be a correlation between charges on adjacent elements of the chain arising from the statistical nature of the distribution of monomeric groups along the polymer skeleton. Both these effects upon the polymer extension have been considered, and their influence upon the potential of average force causing expansion of contraction evaluated.

In the present section we shall only be concerned with the fluctuation effect mentioned above since the skeletal distribution effect vanishes for a regularly alternating copolymer. We shall find the fluctuation terms to be of a very small order of magnitude indicating that, in the equimolar polymer under consideration, it is nearly impossible to move any charged groups without materially increasing the free energy since the polymer is almost completely in the zwitterion form. It might be noted that in polyampholytes not of exactly equal molar proportions, there will be at the isoelectric point a number of acidic or basic groups which remain uncharged, thus facilitating rearrangement of the charge distribution and producing correspondingly larger fluctuation energies.

7.5.1 EQUILIBRIUM CONDITIONS

We now proceed to a calculation of the titration curves for regularly alernating polyampholytes, and seek means of determining the average net charge of the polymer, $\bar{Q}$ (in units of the electronic charge), as a function of pH and counterion activities. We also require the derivative $(\partial \bar{Q}/\partial p\mathrm{H})$,

---

* We refer the reader to Chapter 5 for criticism of this model. However, for the purposes of illustrating the direction and magnitude of effects, it is more than ample.

evaluated at the isoelectric point, in order to determine the charge fluctuations whose influence upon the extension of the polymer will be treated in a later section.

As before, we divide the free energy A of the polyelectrolyte solution into three parts. The first part A_1 describes the free hydrogen ions eligible to take part in ionization and the free counterions which are eligible to take part in binding, but not the polymer itself whose contribution is totally included in A_2 and A_3. A_2 is the electrical free energy of interaction between the various charges of the polyion and includes terms arising from the entropy of distribution of the charges among the sites. The remainder of the polyion free energy A_3 therefore includes the free energy associated with the chemical changes occurring on association, dissociation, or binding, which we shall compute from a reference state in which all of the groups of the polyampholyte are uncharged. In addition, A_3 includes the free energy of mixing of un-ionized and bound groups among the uncharged sites of the chain. This definition of A_2 and A_3 is such that A_2 depends only upon the total charge of each sign on the polyion, counting sites bound by counterions as uncharged, but that A_3 will depend upon the number of un-ionized and bound states separately.

Let us consider a polymer composed of an alternation of acidic groups HA capable of dissociating into ions H^+, A^-, and basic groups B capable of adding a proton to form the ion HB^+. In addition, we shall assume that counterions may be bound to the charged sites to form complexes of the form $A^- \cdot X^+$, $HB^+ \cdot Y^-$. Setting ν_i as the average number of groups of the species i on the polymer chain, we may define the stoichiometric fractions of acid and base ionized, α_1 and α_2, respectively,

$$\alpha_1 = \frac{\nu_{A^-} + \nu_{A^- \cdot X^+}}{\nu_{HA} + \nu_{A^-} + \nu_{A^- \cdot H^+}}$$

$$\alpha_2 = \frac{\nu_{HB^+} + \nu_{HB^+ \cdot Y^-}}{\nu_B + \nu_{HB^+} + \nu_{HB^+ \cdot Y^-}}.$$

(7.5.1)

The degree of binding at acidic and basic sites, f_1 and f_2, will be defined such that

$$f_1 = \frac{\nu_{A^- \cdot X^+}}{\nu_{A^-} + \nu_{A^- \cdot X^+}}$$

$$f_2 = \frac{\nu_{HB^+ \cdot Y^-}}{\nu_{HB^+} + \nu_{HB^+ \cdot Y^-}}.$$

(7.5.2)

We further define $\alpha_1' = \alpha_1 - \alpha_1 f_1$, the fraction of the acidic sites bearing net charge, and $\alpha_2' = \alpha_2 - \alpha_2 f_2$, the corresponding quantity referring to the basic sites. We may now, following the discussion of the preceding paragraph, express the free energy in the form

$$A = A_1(a_{H^+}, a_{X^+}, a_{Y^-}) + A_2(\alpha'_1, \alpha'_2) + A_3(\alpha_1, \alpha_1 f_1, \alpha_2, \alpha_2 f_2) \quad (7.5.3)$$

where a_i is the activity of species i in the solution. Generally, a_{X^+}, a_{Y^-} will be determined by the quantities of these substances introduced into the polymer solution. It should be remarked that of the five remaining variables, a_{H^+}, α_1, α_2, f_1, f_2, only one is independent at chemical equilibrium. The relations between these quantities are given by the requirement that the free energy be a minimum. Differentiating Eq. (7.5.3) with respect to α_1, α_2, $\alpha_1 f_1$, and $\alpha_2 f_2$ we find

$$\left(\frac{\partial A_1}{\partial \alpha_1}\right)_{\alpha_1 f_1, \alpha_2 f_2} + \left(\frac{\partial A_2}{\partial \alpha'_1}\right)_{\alpha'_2} + \left(\frac{\partial A_3}{\partial \alpha_1}\right)_{\alpha_1 f_1, \alpha_2 f_2} = 0$$

$$\left(\frac{\partial A_1}{\partial \alpha_1 f_1}\right)_{\alpha_1, \alpha_2, \alpha_2 f_2} - \left(\frac{\partial A_2}{\partial \alpha'_1}\right)_{\alpha'_2} + \left(\frac{\partial A_3}{\partial \alpha_1 f_1}\right)_{\alpha_1, \alpha_2, \alpha_2 f_2} = 0 \qquad (7.5.4)$$

and similar expressions with the subscripts 1 and 2 interchanged.

Since the derivatives are to be taken in a closed system, it is possible to see that

$$\left(\frac{\partial A_1}{\partial \alpha_1}\right)_{\alpha_1 f_1, \alpha_2, \alpha_2 f_2} = -\left(\frac{\partial A_1}{\partial \alpha_2}\right)_{\alpha_1, \alpha_1 f_1, \alpha_2 f_2} = \frac{Z}{2}\mu_{H^+}$$

$$\left(\frac{\partial A_1}{\partial \alpha_1 f_1}\right)_{\alpha_1, \alpha_2, \alpha_2 f_2} = -\frac{Z}{2}\mu_{X^+} \qquad (7.5.5)$$

$$\left(\frac{\partial A_1}{\partial \alpha_2 f_2}\right)_{\alpha_1, \alpha_2, \alpha_1 f_1} = -\frac{Z}{2}\mu_{Y^-}$$

where μ_i is the chemical potential of the species i, and $Z/2$, the total number of charge sites of either sign on the polymer, is,

$$\frac{Z}{2} = \nu_{HA} + \nu_{A^-} + \nu_{A^- X^+} = \nu_B + \nu_{HB^+} + \nu_{HB^+ \cdot Y^-}. \qquad (7.5.6)$$

We may also simplify the expressions for the derivatives of A_3. Remembering that the zero of A_3 is the totally uncharged polymer, we may write

$$A_3 = \frac{Z\alpha_1}{2}(\mu^0_{A^-} - \mu^0_{HA}) + \frac{Z\alpha_1 f_1}{2}(\mu^0_{A^- X^+} - \mu^0_{A^-})$$

$$- \frac{Z\alpha_2}{2}(\mu^0_B - \mu^0_{HB^+}) + \frac{Z\alpha_2 f_2}{2}(\mu^0_{HB^+ Y^-} - \mu^0_{HB^+})$$

$$- TS_m(\alpha_1, \alpha_1 f_1, \alpha_2, \alpha_2 f_2) \qquad (7.5.7)$$

where μ_i^0 is the standard chemical potential of the species i, and $S_m(\alpha_1, \alpha_1 f_1, \alpha_2, \alpha_2 f_2)$ is the entropy of mixing of the un-ionized and bound groups among the uncharged sites of each sign on the chain. If we make the random mixing assumption for S_m we arrive at the relations

$$\left.\begin{array}{l} \left(\dfrac{\partial A_3}{\partial \alpha_1}\right)_{\alpha_1 f_1, \alpha_2, \alpha_2 f_2} = \dfrac{Z}{2}(\mu_{A^-}^0 - \mu_{HA}^0) - \dfrac{ZkT}{2}\ln\left(\dfrac{1-\alpha_1}{1-\alpha_1+\alpha f_1}\right) \\[3mm] \left(\dfrac{\partial A_3}{\partial \alpha_1 f_1}\right)_{\alpha_1, \alpha_2, \alpha_2 f_2} = \dfrac{Z}{2}(\mu_{A^-X^+}^0 - \mu_{A^-}^0) + \dfrac{ZkT}{2}\ln\left(\dfrac{\alpha_1 f_1}{1-\alpha_1+\alpha_1 f_1}\right) \\[3mm] \left(\dfrac{\partial A_3}{\partial \alpha_2}\right)_{\alpha_1, \alpha_1 f_1, \alpha_2 f_2} = -\dfrac{Z}{2}(\mu_B^0 - \mu_{HB^+}^0) - \dfrac{ZkT}{2}\ln\left(\dfrac{1-\alpha_2}{1-\alpha_2+\alpha_2 f_2}\right) \\[3mm] \left(\dfrac{\partial A_4}{\partial \alpha_2 f_2}\right)_{\alpha_1, \alpha_1 f_1, \alpha_2} = \dfrac{Z}{2}(\mu_{HB^+Y^-}^0 - \mu_{HB^+}^0) + \dfrac{ZkT}{2}\ln\left(\dfrac{\alpha_2 f_2}{1-\alpha_2+\alpha_2 f_2}\right) \end{array}\right\}$$

$$(7.5.8)$$

The μ_i^0 are related to appropriate equilibrium constants by the relations

$$\mu_{A^-}^0 + \mu_{H^+}^0 - \mu_{HA}^0 = -kT \ln K_1^0$$

$$\mu_B^0 + \mu_{H^+}^0 - \mu_{HB^+}^0 = -kT \ln K_2^0$$

$$\mu_{A^-}^0 + \mu_{X^+}^0 - \mu_{A^-X^+}^0 = -kT \ln K_3^0$$

$$\mu_{HB^+}^0 + \mu_{Y^-}^0 - \mu_{HB^+Y^-}^0 = -kT \ln K_4^0$$

$$(7.5.9)$$

where

$$\mu_{H^+}^0 = \mu_{H^+} - kT \ln a_{H^+}$$

$$\mu_{X^+}^0 = \mu_{X^+} - kT \ln a_{X^+} \qquad (7.5.10)$$

$$\mu_{Y^-}^0 = \mu_{Y^-} - kT \ln a_{Y^-}.$$

The intrinsic dissociation constants are defined to be, in the absence of electrostatic interactions,

$$K_1^0 = \frac{a_{H^+} a_{A^-}}{a_{HA}}, \qquad K_2^0 = \frac{a_{H^+} a_B}{a_{HB^+}} \qquad (7.5.11)$$

$$K_3^0 = \frac{a_{A^-} a_{X^+}}{a_{A^-X^+}}, \qquad K_4^0 = \frac{a_{HB^+} a_{Y^-}}{a_{HB^+Y^-}} \qquad (7.5.12)$$

Introducing Eqs. (7.5.9)–(7.5.11) into (7.5.8), we obtain expressions for the derivatives of A_3 which, when combined with Eqs. (7.5.4) and (7.5.5), yield

$$\ln a_{H^+} - \ln K_1^0 - \ln\left(\frac{1-\alpha_1}{1-\alpha_1+\alpha_1 f_1}\right) + \frac{2}{ZkT}\left(\frac{\partial A_2}{\partial \alpha_1'}\right)_{\alpha_2'} = 0$$

$$-\ln a_{H^+} + \ln K_2^0 - \ln\left(\frac{1-\alpha_2}{1-\alpha_2+\alpha_2 f_2}\right) + \frac{2}{ZkT}\left(\frac{\partial A_2}{\partial \alpha_2'}\right)_{\alpha_1'} = 0$$

$$\ln a_{X^+} - \ln K_3^0 - \ln\left(\frac{\alpha_1 f_1}{1-\alpha_1+\alpha_1 f_1}\right) + \frac{2}{ZkT}\left(\frac{\partial A_2}{\partial \alpha_1'}\right)_{\alpha_2'} = 0 \quad (7.5.13)$$

$$\ln a_{Y^-} - \ln K_4^0 - \ln\left(\frac{\alpha_2 f_2}{1-\alpha_2+\alpha_2 f_2}\right) + \frac{2}{ZkT}\left(\frac{\partial A_2}{\partial \alpha_2'}\right)_{\alpha_1'} = 0.$$

It now remains to evaluate the term A_2.

As has been discussed, the electrical free energy of the polyion will include the free energy of the extension of the polymer and also the free energy involved in building ion atmospheres about the charged sites as well as the free energy arising from the direct interaction of charged sites and from the distribution of the charged sites along the polymer skeleton. Following the previous notation, and assuming that the Debye–Hückel expression may be used for the free energy of ion atmospheres, we write

$$A_2 = A_4' + A_6 + A_7 \tag{7.5.14}$$

where A_6 is the free energy of expansion (or contraction), and

$$A_7 = -\frac{q^2\kappa Z(\alpha_1' + \alpha_2')}{4D} \tag{7.5.15}$$

where q is the electronic charge, D the dielectric constant of the solution, and κ the reciprocal Debye radius, defined by

$$\kappa^2 = \frac{8\pi q^2 N_0 \Gamma}{DkT10^3}. \tag{7.5.16}$$

Γ is the ionic strength in the solution. A_4', the free energy of distribution among charged and uncharged sites, differs from A_4 as defined previously in that it includes the energy terms, whereas A_4 was defined to represent the effect of only the entropy of distribution of charged and uncharged sites. It was thought to be a better approximation to define A_4 in this manner, and then to calculate A_4' including the effects of next nearest neighboring groups, rather than to estimate the effect of the groups at further distances in terms of a uniformly distributed charge model as was done for a one-component polyion.

We shall calculate A_4' in terms of a grand partition function Ξ, which will here be defined to include the effects of next nearest neighboring groups,

$$\Xi = \sum_{\{\eta_i\}} \exp\left[-\sum_{i=1}^{Z-1} \left(\frac{u_{i,i+1}}{kT}\right) - \sum_{i=1}^{Z-2} \left(\frac{u_{i,i+2}}{kT}\right) \right] \lambda_1^{\sum_{j=1}^{Z/2} \eta_{2j}} \lambda_2^{\sum_{j=1}^{Z/2} \eta_{2j-1}}$$

(7.5.17)

where η_i is the occupation number for the ith site, and can assume the values zero or one, corresponding to an uncharged or a charged group respectively at the ith site. The index i in $\{\eta_i\}$ will range from 1 to Z, even values of i referring to acidic sites, and odd values of i to basic sites. The sum is to be taken over all possible sets $\{\eta_i\}$. The $u_{i,i+1}$ and $u_{i,i+2}$ are the interaction energies between adjacent and next nearest neighboring sites respectively and may be expressed

$$u_{i,i+1} = -\frac{q^2 \eta_i \eta_{i+1}}{DR'}, \qquad u_{i,i+2} = \frac{q^2 \eta_i \eta_{i+2}}{DR''}.$$

(7.5.18)

In (7.5.18) R' and R'' are the average distances between nearest neighboring and next nearest neighboring sites respectively and will be assumed to be constants irrespective of the values of the $\{\eta_i\}$. The quantities λ_1 and λ_2 play the role of absolute activities of negative and positively charged groups respectively.

By standard methods[33] it is possible to show that A_4' is related to Ξ by

$$A_4' = -kT \ln \Xi + \frac{ZkT}{2} (\alpha_1' \ln \lambda_1 + \alpha_2' \ln \lambda_2)$$

(7.5.19)

where λ_1 and λ_2 are defined by

$$\left(\frac{\partial \ln \Xi}{\partial \ln \lambda_1}\right)_{\lambda_2, T} = \frac{Z\alpha_1'}{2}, \qquad \left(\frac{\partial \ln \Xi}{\partial \ln \lambda_2}\right)_{\lambda_1, T} = \frac{Z\alpha_2'}{2}.$$

(7.5.20)

The functional dependence of Ξ upon λ_1, λ_2, and T can be obtained by evaluation of the sum (7.5.17) in a manner to be discussed subsequently. Thus Eq. (7.5.20) may be regarded as giving α_1, α_2, in terms of λ_1 and λ_2.

Upon differentiation of Eqs. (7.5.19), (7.5.15), and (7.5.14), and incorporation of the results into Eq. (7.5.13), we may deduce the equilibrium conditions

$$\ln \lambda_1 + \ln a_{H^+} - \ln K_1^0 - \ln\left(\frac{1 - \alpha_1}{1 - \alpha_1'}\right) - \frac{q^2 \kappa}{2DkT} + \frac{2}{ZkT}\left(\frac{\partial A_6}{\partial \alpha_1'}\right)_{\alpha_2'} = 0$$

(7.5.21a)

$$\ln \lambda_2 - \ln a_{H^+} + \ln K_2^0 - \ln\left(\frac{1 - \alpha_2}{1 - \alpha_2'}\right) - \frac{q^2\kappa}{2DkT} + \frac{2}{ZkT}\left(\frac{\partial A_6}{\partial \alpha_2'}\right)_{\alpha_1'} = 0$$

(7.5.21b)

$$\ln \lambda_1 + \ln a_{X^+} - \ln K_3^0 - \ln\left(\frac{\alpha_1 f_1}{1 - \alpha_1'}\right) - \frac{q^2\kappa}{2DkT} + \frac{2}{ZkT}\left(\frac{\partial A_6}{\partial \alpha_1'}\right)_{\alpha_2'} = 0$$

(7.5.21c)

$$\ln \lambda_2 + \ln a_{Y^-} - \ln K_4^0 - \ln\left(\frac{\alpha_2 f_2}{1 - \alpha_2'}\right) - \frac{q^2\kappa}{2DkT} + \frac{2}{ZkT}\left(\frac{\partial A_6}{\partial \alpha_2'}\right)_{\alpha_1'} = 0.$$

(7.5.21d)

7.5.2 THE ISOLECTRIC POINT

We now specialize to the case when the net charge on the polyion is zero. In this event, $\alpha_1' = \alpha_2'$, necessarily $\lambda_1 = \lambda_2$. In addition, since A_6 is symmetrical between positive and negative charge, the derivatives of A_6 vanish, Taking advantage of these properties, Eqs. (7.5.21) can be solved simultaneously to yield

$$a_{H^+} = \left[\frac{K_1^0 K_2^0}{1 + \dfrac{a_{X^+}}{a_{H^+}}\dfrac{K_1^0}{K_3^0} - \dfrac{a_{Y^-}}{a_{H^+}}\dfrac{K_1^0}{K_4^0}}\right]^{1/2}.$$

(7.5.22)

It may be seen that, in the limit of no counterion binding, either by virtue of low concentration of X^+ and Y^-, or through large dissociation constants K_3^0 and K_4^0, the expression (7.5.22) reduces to the familiar form[34] valid in the absence of electrostatic forces, the result being

$$\lambda_1 = \lambda_2 = \left(\frac{K_1^0}{K_2^0}\right)^{1/2} \frac{\left(1 + \dfrac{a_{X^+}}{a_{H^+}}\dfrac{K_1^0}{K_3^0} - \dfrac{a_{Y^-}}{a_{H^+}}\dfrac{K_1^0}{K_4^0}\right)^{1/2}}{1 + \dfrac{a_{X^+}}{a_{H^+}}\dfrac{K_1^0}{K_3^0}} \exp\left(\frac{q^2\kappa}{DkT}\right).$$

(7.5.23)

Using Eq. (7.5.20) one may obtain from λ_1 and λ_2 values of α_1 and α_2, and hence Eq. (7.5.21c) or (7.5.21d) may be directly employed to calculate the fractions of the charge sites found, $\alpha_1 f_1$, and $\alpha_2 f_2$.

We shall also require the derivative of the total charge of the polymer with respect to pH at the isoelectric point. We take the derivative holding and activities of the counterions constant. In general,

$$\bar{Q} = \frac{Z}{2}(\alpha_2' - \alpha_1')$$

(7.5.24)

and therefore,

$$
\left(\frac{\partial \overline{Q}}{\partial pH}\right)_{a_{X^+},a_{Y^-}} = \frac{Z}{2}\left[\left\{\left(\frac{\partial \alpha_2'}{\partial \ln \lambda_1}\right)_{\lambda_2} - \left(\frac{\partial \alpha_1'}{\partial \ln \lambda_1}\right)_{\lambda_2}\right\} \times \left(\frac{\partial \ln \lambda_1}{\partial pH}\right)_{a_{X^+},a_{Y^-}}
$$

$$
+ \left\{\left(\frac{\partial \alpha_2'}{\partial \ln \lambda_2}\right)_{\lambda_1} - \left(\frac{\partial \alpha_1'}{\partial \ln \lambda_2}\right)_{\lambda_1}\right\}\left(\frac{\partial \ln \lambda_2}{\partial pH}\right)_{a_{X^+},a_{Y^-}}\right]. \quad (7.5.25)
$$

By virtue of Eqs. (7.5.20), $(\partial \alpha_2'/\partial \ln \lambda_1)_{\lambda_2} = (\partial \alpha_1'/\partial \ln \lambda_2)_{\lambda_1}$ and, by suitable manipulation of Eq. (7.5.21),

$$
\left(\frac{\partial \ln \lambda_1}{\partial pH}\right)_{a_{X^+},a_{Y^-}} = 2.303\left(\frac{1-\alpha_1}{1-\alpha_1'}\right)
$$

$$
\left(\frac{\partial \ln \lambda_2}{\partial pH}\right)_{a_{X^+},a_{Y^-}} = -2.303\left(\frac{1-\alpha_2}{1-\alpha_2'}\right) \quad (7.5.26)
$$

utilizing the fact that $(\partial A_6/\partial \alpha_1')_{\alpha_2}$ is an even function of $(\alpha_1' - \alpha_2')$, and that therefore its derivative with pH vanishes at the isoelectric point. Since at that point, also $(\partial \alpha_1'/\partial \ln \lambda_1)_{\lambda_2} = (\partial \alpha_2'/\partial \ln \lambda_2)_{\lambda_1}$, we may obtain the result

$$
\left(\frac{\partial \overline{Q}}{\partial pH}\right)_{a_{X^+},a_{Y^-}} = -2.303Z\left(\frac{1-\alpha_1}{1-\alpha_1'}\right)\left\{\left(\frac{\partial \alpha_1'}{\partial \ln \lambda_1}\right)_{\lambda_2} - \left(\frac{\partial \alpha_1'}{\partial \ln \lambda_2}\right)_{\lambda_1}\right\}. \quad (7.5.27)
$$

The derivatives of α_1' necessary to evalute (7.5.27) may be obtained by direct differentiation of Eq. (7.5.20).

7.5.3 MORE GENERAL CONDITIONS

When we are not at the isoelectric point, the situation becomes somewhat involved. It is then necessary to solve Eqs. (7.5.21) without the simplifying assumptions $\alpha_1' = \alpha_2'$, $\lambda_1 = \lambda_2$.

Except very near the isoelectric point, the configurational free energy A_6 depends only on the absolute magnitude of the average net charge of the polymer and not upon the values of α_1 and α_2 separately so that we may conclude

$$
-\left(\frac{\partial A_6}{\partial \alpha_1'}\right)_{\alpha_2'} = \left(\frac{\partial A_6}{\partial \alpha_2'}\right)_{\alpha_1'} \equiv A' \quad (7.5.28)
$$

where A' is defined by Eq. (7.5.28). Since we are considering only conditions not near the isoelectric point, A' may be calculated by the methods already discussed, and therefore is to be regarded as a known function of $(\alpha_1 - \alpha_2)$.

Using Eq. (7.5.28) and by suitable manipulation of Eq. (7.5.21), we may obtain expressions for λ_1 and λ_2 in terms of a_{H^+} and A',

$$\lambda_1 = \left[\frac{a_{H^+}}{K_1^0} + \frac{a_{X^+}}{K_3^0}\right]^{-1} \exp\left[\frac{1}{kT}\left(\frac{q^2\kappa}{2D} + \frac{2A'}{Z}\right)\right]$$

$$\lambda_2 = \left(\frac{K_2^0}{a_{H^+}} + \frac{a_{Y^-}}{K_4^0}\right)^{-1} \exp\left[\frac{1}{kT}\left(\frac{q^2\kappa}{2D} - \frac{2A'}{Z}\right)\right]. \qquad (7.5.29)$$

The equilibrium conditions can now be obtained without excessive labor by (1) assuming a zeroth-order value of $(\alpha_1' - \alpha_2')$ and calculating A', (2) using Eq. (7.5.29) to calculate λ_1 and λ_2; (3) using the solution of Eq. (7.5.20) to obtain new values of α_1', α_2', and $(\alpha_1' - \alpha_2')$; (4) using the new value of $(\alpha_1' - \alpha_2')$ as a basis for re-estimating A'; repeating steps (1)–(3) until a self-consistent solution is obtained.

7.5.4 EVALUATION OF Ξ

We shall proceed by the same general methods originally devised by Kramers and Wannier[35] for solution of the one-dimensional Ising lattice. We introduce into Eq. (7.5.17) for Ξ the definition

$$V(\eta_{2i-1}, \eta_{2i}; \eta_{2i+1}, \eta_{2i+2}) = \lambda_1^{\frac{1}{2}(\eta_{2i-1}+\eta_{2i+1})}\lambda_2^{\frac{1}{2}(\eta_{2i}+\eta_{2i+2})} \times$$

$$\exp(J\{\tfrac{1}{2}\eta_{2i-1}\eta_{2i} + \eta_{2i}\eta_{2i+1} + \tfrac{1}{2}\eta_{2i+1}\eta_{2i+2}\}$$

$$+ K\{\eta_{2i-1}\eta_{2i+1} + \eta_{2i}\eta_{2i+2}\}) \qquad (7.5.30)$$

where J and K are

$$J = q^2/DkTR', \qquad K = -q^2/DkTR'' \qquad (7.5.31)$$

and find

$$\Xi = \sum_{\{\eta_i\}} \prod_{j=1}^{Z/2-1} V(\eta_{2j-1}, \eta_{2j}; \eta_{2j+1}\eta_{2j+2})\lambda_1^{\frac{1}{2}(\eta_2+\eta_Z)} \times$$

$$\lambda_2^{\frac{1}{2}(\eta_1+\eta_{Z-1})}\exp\left[\frac{J}{2}(\eta_1\eta_2 + \eta_{Z-1}\eta_Z)\right]. \qquad (7.5.32)$$

If we regard $V(\eta_{2i-1}, \eta_{2i}; \eta_{2i+1}, \eta_{2i+2})$ as the elements of a matrix V,

$$\mathbf{V} = \begin{Vmatrix} 1 & \lambda_2^{\frac{1}{2}} & e^{J/2}\lambda_1^{\frac{1}{2}}\lambda_2^{\frac{1}{2}} & \lambda_1^{\frac{1}{2}} \\ \lambda_2^{\frac{1}{2}} & e^K\lambda_2 & e^{K+(3J/2)}\lambda_1^{\frac{1}{2}}\lambda_2 & e^J\lambda_1^{\frac{1}{2}}\lambda_2^{\frac{1}{2}} \\ e^{J/2}\lambda_1^{\frac{1}{2}}\lambda_2^{\frac{1}{2}} & e^{K+J/2}\lambda_1^{\frac{1}{2}}\lambda_2 & e^{2J+2K}\lambda_1\lambda_2 & e^{K+(3J/2)}\lambda_1\lambda_2^{\frac{1}{2}} \\ \lambda_1^{\frac{1}{2}} & \lambda_1^{\frac{1}{2}}\lambda_2^{\frac{1}{2}} & e^{K+J/2}\lambda_1\lambda_2^{\frac{1}{2}} & e^K\lambda_1 \end{Vmatrix}$$

$$(7.5.33)$$

Eq. (7.5.32) is equivalent to

$$\Xi = \sum_{\substack{\eta_1,\eta_2 \\ \eta_{Z-1}\eta_Z}} (V^{Z/2-1})_{\eta_1,\eta_2;\eta_{Z-1}\eta_Z} \; \lambda_1^{(\eta_2+\eta_Z-1)/2} \lambda_2^{(\eta_1+\eta_Z-1)/2} \times$$

$$\exp\left[\frac{J}{2}(\eta_1\eta_2 + \eta_{Z-1}\eta_Z)\right]. \quad (7.5.34)$$

When Z is sufficiently large compared to unity, it is possible to approximate Ξ by

$$\Xi = \phi^{Z/2} \times 0(1) \quad (7.5.35)$$

where ϕ is the largest eigenvalue of the matrix V. An iterative procedure, discussed in Appendix 7A, was used to obtain ϕ as a series in powers of λ_1, λ_2, e^J, and e^K. The result, including all terms of sufficient size to be of practical interest near the isoelectric point for typical values of the dissociation constants, is found to be

$$\phi = \lambda_1\lambda_2\, e^{2J+2K} + \lambda_1 + \lambda_2 + 2\, e^{-J-2K} + (\lambda_1^{-1} + \lambda_2^{-1})\, e^{-2J-3K}. \quad (7.5.36)$$

Equation (7.5.36) may now be used to evaluate Eq. (7.5.20) for the equilibrium degree of ionization of the polyampholyte.

$$\alpha_1' = \phi^{-1}\lambda_1(\lambda_2\, e^{2J+2K} + 1 - \lambda_1^{-2}e^{-2J-3K}). \quad (7.5.37)$$

The expression for α_2' may be obtained by permuting the subscripts 1 and 2. By differentiation of Eq. (7.5.34), Eq. (7.4.27) may also be evaluated. The final result, valid at the isoelectric point, is

$$\left(\frac{\partial \bar{Q}}{\partial \text{pH}}\right)_{a_{X^+},a_{Y^-}} = -2.303 Z\lambda_1\phi^{-1}\left(\frac{1-\alpha_1}{1-\alpha_1'}\right)(1 + \lambda_1^{-2}e^{-2J-3K}). \quad (7.5.38)$$

In addition to Eqs. (7.5.36) and (7.5.37), which are only valid near the isoelectric point where λ_1 and λ_2 are of comparable magnitude, we also need expressions for ϕ and α_2', α_2' when $\lambda_2 \gg \lambda_1$, with $\lambda_1 < 1$. By methods entirely similar in principle to those used in deriving Eqs. (7.5.36) and (7.5.37), one may find

$$\phi = e^K\lambda_2 + e^{2J+K}\lambda_1\lambda_2 + e^{-K} + e^{J-K}\lambda_1 \quad (7.5.39)$$

$$\alpha_1' = \frac{\lambda_1(\lambda_2\, e^{2J} + e^{J-2K})}{\lambda_1(\lambda_2\, e^{2J} + e^{J-2K}) + e^{-2K} + \lambda_2} \quad (7.5.40)$$

$$\alpha_2' = \frac{\lambda_2(e^{2J}\lambda_1 + 1)}{\lambda_2(e^{2J}\lambda_1 + 1) + e^{-2K} + \lambda_1 e^{J-2K}}. \quad (7.5.41)$$

Numerical values of the quantities calculated in this section will be presented in a later section for typical values of the experimental parameters.

7.5.5 END-TO-END EXTENSION

In this section are discussed the effects of the microstructure of the polymer on its mean extension in solution. We concern ourselves with all the various effects that can arise in equimolar polyampholytes in order that the discussion of expansion in polyampholytes of other than regularly alternating composition may be expedited.

In order to discuss the expansion of the polyion coil, we proceed as in Section 7.4. The formulas needed are

$$L = NA \tag{7.5.42}$$

$$\langle h_0^2 \rangle = NA^2 \tag{7.5.43}$$

$$\langle h_1^2 \rangle = \langle h_0^2 \rangle \frac{1 - c}{1 + c} \tag{7.5.44}$$

$$c = \frac{\int \cos \gamma e^{-W(\gamma)/kT} \, d\Omega}{\int e^{-W(\gamma)/kT} \, d\Omega}. \tag{7.5.45}$$

On either side of the isoelectric point, where the polymer possesses a significant average net charge, the methods previously used to discuss the expansion of one-component polyelectrolytes will be suitable to discuss the expansion of polyampholoytes. However, at the isoelectric point, even though there is no average net charge on the polyion, forces acting between the statistical elements of the polymer may arise from two distinct sources. These sources are (a) fluctuations of the charge of each statistical element about its mean value, and (b) correlations in the charge distribution of neighboring statistical elements due to the statistical distribution of acidic and basic monomer units along the polymer chain. The latter effect does not contribute a force in a regularly alternating copolymer.

The fluctuations in charge and correlations in charge distribution in neighboring statistical elements may be related to the potential of average force between the elements by the theory of Kirkwood and Shumaker.[32] As pointed out in their paper, $W(\gamma)$ may be written in terms of the potential of the force at fixed charge configuration, $U(\gamma, \{\eta_i\})$,

$$W(\gamma) = \langle U(\gamma) \rangle - \frac{1}{2kT} \langle U^2 \rangle + O(k^{-2}T^{-2}). \tag{7.5.46}$$

In Eq. (7.5.46) the averages are to be taken over all charge configurations of the two elements with the weighting factors appropriate if the effect of electrostatic interaction between the two elements is omitted. Accordingly, the averages are computed including the effect of the correlation between the composition of the chain skeletons of the two elements arising from their connectedness as well as all the effects present for each element alone. We approximate $U(\gamma, \{\eta_i\}) = q_1 q_2 \, e^{-\kappa r}/Dr$ where D is the dielectric constant of the medium between the charges, r is the distance between centers of the elements when the angle between them is γ, and q_1 and q_2 are the net charges on the two elements corresponding to $\{\eta_i\}$. Introducing the above into Eq. (7.5.46),

$$W(\gamma) = \frac{\langle q_1 q_2 \rangle \, e^{-\kappa r}}{Dr} - \frac{1}{2kT} \frac{\langle q_1^2 q_2^2 \rangle}{D^2 r^2} \, e^{-2\kappa r} + \cdots . \qquad (7.5.47)$$

To continue further with the development, we express the charge on each statistical element as a sum of a $q^{(s)}$, the average charge due to the skeletal distribution, and a $q^{(f)}$, a fluctuation contribution of average value zero. Using the fact that $q_1^{(f)}$ and $q_2^{(f)}$ are independent since they can only influence each other through the interaction term to be omitted in the present calculation, we see that

$$\langle q_1 q_2 \rangle = \langle q_1^{(s)} q_2^{(s)} \rangle \qquad (7.5.48)$$

while

$$\langle q_1^2 q_2^2 \rangle = \langle q_1^{(s)2} q_2^{(s)2} \rangle + \langle q_1^{(f)2} q_2^{(f)2} \rangle + 2\langle q_1^{(f)2} q_2^{(f)2} \rangle. \qquad (7.5.49)$$

If the skeletal distribution correlations are assumed not to alter the extent of the fluctuations, Eq. (7.5.49) reduces to

$$\langle q_1^2 q_2^2 \rangle = \langle q_1^{(s)2} q_2^{(s)2} \rangle + \langle q^{(f)2} \rangle^2 + 2\langle q^{(f)2} \rangle \langle q^{(s)2} \rangle. \qquad (7.5.50)$$

In the limit when the skeletal correlation contribution vanishes, Eqs. (7.5.48) and (7.5.50) may be substituted into (7.5.47), giving

$$W_f(\gamma) = - \frac{\langle q^{(f)2} \rangle^2}{2kTD^2 r^2} \, e^{-2\kappa r}. \qquad (7.5.51)$$

On the other hand, when the fluctuations are very small compared with the skeletal correlations, we obtain

$$W_s(\gamma) = \frac{\langle q_1^{(s)} q_2^{(s)} \rangle \, e^{-\kappa r}}{Dr} \qquad (7.5.52)$$

as a limit. In the intermediate region where both skeletal and fluctuation effects are of comparable magnitude, it may be necessary to use the unabbreviated equations (7.5.48) and (7.5.49).

The averages dependent upon skeletal correlations are to be obtained from considerations of the statistics of polymerization and will not be discussed further at this time. For the present it suffices to reiterate that the skeletal contributions to $W(\gamma)$ vanish for a regularly alternating polyampholyte.

To complete the theory for use at this time, we require the mean square charge on a statistical element of a polyampholyte at its isoelectric point. By consideration of the properties of the grand partition, it is shown in Appendix 7B that

$$\langle (q - \bar{q})^2 \rangle = - \frac{1}{2.303N} \left(\frac{1 - \alpha_1'}{1 - \alpha_1} \right) \left(\frac{\partial \bar{Q}}{\partial pH} \right)_{a_X + a_Y -} \tag{7.5.53}$$

The analogous equation for a polymer not capable of counterion binding was originally derived by Linderstrøm-Lang[36] through a detailed consideration of the successive ionization equilibria of a polyampholyte in solution. Using Eq. (7.5.38) we may reduced (7.5.53) to a computationally useful form

$$\langle q^{(f)2} \rangle = \frac{Z\lambda_1}{N\phi} (1 + \lambda_1^{-2} e^{-2J - 3K}). \tag{7.5.54}$$

7.5.6 NUMERICAL EVALUATION

Since no strictly alternating polyampholyte has as yet been studied in detail and since, as was mentioned in the introduction, significant differences in properties are to be expected if the units of the polymer are either more randomly distributed or not in exactly equal mole ratio, it is not possible to compare directly the present theory with experimental results. However, it is desirable to examine the equations set forth in the preceding sections and determine the general form of the behavior predicted.

We shall make calculations on a hypothetical copolymer of dissociation constants $K_1^0 = 10^{-5}$, $K_2^0 = 10^{-9}$. We shall assume adjacent sites of possible charge to be separated by 6 A and next nearest neighboring sites by 8 A. These distances correspond roughly to average distances in a vinyl type copolymer with the charge bearing groups on short side chains. It will be assumed that the statistical element for expansion contains 10 monomers and has an effective length of 25 A. The interaction energy between charges will be calculated assuming a dielectric constant of 50 for nearest neighboring charges, and 70 for next nearest neighbors.[37] We further assume the possibility of both anion and cation binding with binding dissociation constants $K_3^0 = K_4^0 = 10$.

At the isoelectric point the titration properties were calculated from Eqs. (7.5.22), (7.5.23), and (7.5.36), and Eqs. (7.5.44), (7.5.45), (7.5.51), and (7.5.54) were employed to calculate the end-to-end distance as a function of counterion concentrations. The integrations necessary to evaluate Eq. (7.5.45) were done numerically using a 20 point interpolation formula. The results are given in Table 7.5.1. Note that the contraction caused by charge fluctuations is extremely small due to the fact that the polymer is nearly completely in the zwitterion form, and motion of the charges among sites of equal energy is very restricted. The binding of counterions at the isoelectric point takes

TABLE 7.5.1[a]

TITRATION CURVE AND EXPANSION OF AN EQUIMOLAR POLYAMPHOLYTE AS A FUNCTION OF IONIC ACTIVITIES IN AQUEOUS SOLUTION [b,c]

Γ	a_{H^+}	$a_{X^+} = a_{Y^-}$	λ_1	λ_2	α'_1	α'_2	$\alpha_1 f_1$	$\alpha_2 f_2$	$(\langle h_1^2 \rangle / \langle h_0^2 \rangle)^{1/2}$
0	10^{-7}	0	100	100	0.998	0.998	0	0	0.999996
10^{-3}		10^{-3}	100	100	0.998	0.998	2×10^{-5}	2×10^{-5}	0.999996
10^{-1}		10^{-1}	63.5	63.5	0.997	0.997	1.4×10^{-4}	1.4×10^{-4}	0.999994
0	10^{-5}	0	1.68	5980	0.880	0.99996	0	0	1.16
10^{-3}		10^{-3}	1.65	3540	0.878	0.99994	1.2×10^{-5}	3×10^{-5}	1.16
10^{-1}		10^{-1}	1.42	105	0.860	0.998	1.4×10^{-3}	2×10^{-3}	1.16
10^{-3}	10^{-3}	0	0.0359	2.90×10^5	0.54	0.999998	0	0	1.65
1.5×10^{-3}		10^{-3}	0.0359	2.90×10^3	0.54	0.99998	4.6×10^{-7}	2×10^{-5}	1.65
10^{-1}		10^{-1}	0.0168	88	0.35	0.9998	6.5×10^{-5}	1.1×10^{-3}	1.61
10^{-1}	10^{-2}	0	0.00177	8.4×10^6	0.053	0.999998	0	0	1.94
		10^{-3}	0.00177	8400	0.053	0.99998	9.5×10^{-8}	2×10^{-5}	1.94
		10^{-1}	0.00177	84	0.053	0.9985	9.5×10^{-6}	0.0015	1.94
1.5×10^{-1}	10^{-1}	10^{-1}	0.000177	84	0.0058	0.9985	1.0×10^{-6}	0.0015	2.00

[a] Reproduced from ref. 2.

[b] Constituent properties of the hypothetical polymer used are given in the text.

[c] At the isoelectric point, the slopes of the titration curves are: at $\Gamma = 0$, $(2/Z)(d\bar{Q}/dp\mathrm{H}) = 0.0104$, at $\Gamma = 10^{-3}$, 0.0105, and at $\Gamma = 10^{-1}$, 0.0164.

place to a very small extent, insufficient to decrease by a significant amount the net fraction of charge of each sign.

Behavior at pH 5 was calculated according to the methods described, using Eq. (7.5.36) for ϕ. The expansion in (7.5.36) still converges satisfactorily at this pH. Results at pH 3, 2, and 1 were also calculated, but at these hydrogen ion activities it became necessary to use the development valid far from the isoelectric point where ϕ is given by Eq. (7.5.39). These data comprise the remainder of Table 7.5.1. Since we have chosen a system which is symmetric between acidic and basic groups in aqueous solution, the behavior on the basic side of pH 7 can be immediately inferred from the results here given.

There are several qualitative observations that may be of interest. First, we note that the counterion binding does not cause a large change in the titration curve while there are appreciable numbers of charges of both signs, an effect which is explainable in terms of the stabilizing influence of adjacent opposite charges. However, when the charge of the polyion is predominantly of one sign, the binding prevents the degree of ionization from approaching unity asymptotically. Secondly, we observe that the expansion (or contraction) of the polymer does not depend strongly upon ionic strength at a given pH. This result also can be accounted for because as the ionic strength increases, thus tending to weaken the force expanding the polymer, the charge which can be supported on it also increases, tending to counteract the first effect. The small dependence of expansion on ionic strength is shown graphically in Fig. 7.5.1.

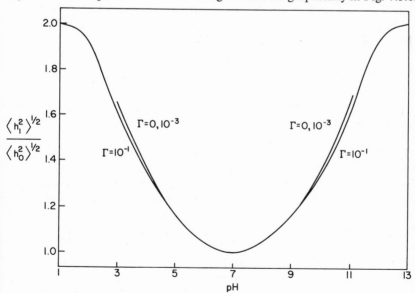

FIG. 7.5.1. The computed expansion of a regularly alternating polyampholyte as a function of pH. [Reproduced from ref. 2.]

Finally, we consider the effect of ionic strength upon the titration curve. From the data in Table 7.5.1, titration curves for no added salt, 10^{-3} molar ionic strength and 10^{-1} molar ionic strength were plotted as shown in Fig. 7.5.2. Since the binding is negligible and the concentrations of free hydrogen ions in the solutions small, the net charge of the polyion is effectively identical with the number of molecules of acid or base added to the solution. As

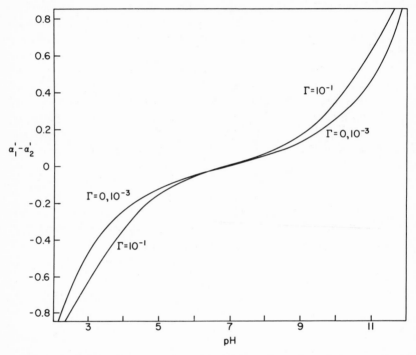

FIG. 7.5.2. The titration curve of a regularly alternating polyampholyte as a function ionic strength. [Reproduced from ref. 2.]

Fig. 7.5.2 clearly shows, the effect of added salt is to shift the pH at any given degree of neutralization towards the isoelectric point, creating a curve characteristic of weaker acidic and basic groups. Examination of the calculations leading to Table 7.5.1 reveals that the primary source of the shift in the titration curve is due to the electrostatic interaction among the segments of the polymer and not to binding phenomena. The effect of ionic strength upon the titration curve is qualitatively similar to that experimentally observed for protein solutions where it is highly likely that some of the factors here considered are significant. However, it must be borne in mind that a model such as has been developed here is an oversimplification of even the gross, nonspecific properties of real protein molecules.

Calculations have also been made for a polymer identical with that above, except that the charges are on shorter side chains such that $R' = 4$ A, $R'' = 7$ A, with local dielectric constants of 20 and 60, respectively. This corresponds more nearly to the charge spacing to be expected in synthetic polyampholytes, whereas the charge spacings of the text are more nearly characteristic of typical proteins. It is found that the buffer action of this polymer is extremely small. Over the pH range 2–12, α'_1 and α'_2 are within 1% of unity. This illustrates the large interaction to be expected in such polyions. The expansion ratio predicted at pH 1 is 1.09.

7.6 Skeletal Distribution Effects in Equimolar Polyampholytes

The preceding section of this chapter has provided a general discussion of the problems involved in the calculation of titration properties and end-to-end extension in polyampholytes of equimolar composition. At that time, the theory was developed for copolymers consisting of a regular alternation of acidic and basic groups, and it was pointed out that in addition to the effects calculated for such a polymer, in general there would be further contributions arising from the distribution of monomer units along the polymer skeleton. It is the purpose of the present section to extend the theory to consider such effects. The skeletal distribution affects the free energy changes accompanying ionization because not all sites capable of bearing charge must have the same local skeletal environment, and therefore the discussion of the previous section, which required that all the sites be essentially similar, is not applicable to polymers of general skeletal structure. In that case it is necessary to recognize explicitly that charge-bearing groups which are adjacent to groups of the same charge will have a greater tendency to remain un-ionized than will groups surrounded by oppositely charged sites. Since exact treatment of the effect of skeletal distribution is prohibitively difficult, approximate methods have been developed in which only the charge configurations of relatively favorable energies are considered. If the polymer is regarded as a succession of sequences of consecutive acid and base groups, the approximations introduced have the effect of assuming that the conditions of the various sequences are nearly independent. The physical justification for such an assumption resides in the fact that, near the isoelectric point, where this theory differs nontrivially from that of one component polyelectrolytes, the polymer is largely in the zwitterion form and the terminal groups of each sequence are almost all charged. The properties of the polymer thus depend upon the relative number of sequences of different lengths, which may be calculated by well known methods, making use of the statistical nature of the copolymerization process.

The skeletal distribution will affect not only the titration curve, but also the end-to-end distance and electrostatic energy of expansion of a polyampholyte. For, if the polymerization statistics are such that there is a tendency toward long sequences of successive monomers of the same species, adjacent statistical elements of the polymer will be, on the average, of the same sign of net charge, resulting in a repulsion between adjacent elements. Under certain conditions, when there is a tendency toward an alternating structure, it will be shown that, instead of a repulsion between adjacent elements, an attractive force is created. In the present section, the correlation in charge between successive elements of the polymer will be calculated in terms of the polymerization statistics, and use will be made of the theory developed in the preceding section to calculate the effect of skeletal correlations on the end-to-end extension and expansion energy.

7.6.1 POLYMERIZATION STATISTICS

Since the distribution of monomer units in the copolymer now under study is not assumed to be regularly alternating, it will be necessary to consider in detail the statistics of the polymerization process. Let it be assumed that the polymerization proceeds under conditions such that the probability of each of the possible chain propagating steps remains constant during the entire reaction. Then define P_{ij} as the probability that a group of type j will add to a polymeric species terminating in a group of type i. Let the indices 1, -1, refer to groups capable of bearing positive, and negative charges, respectively. The P_{ij}, by their definition, are not all independent, but must satisfy the continuity relations

$$P_{1,1} + P_{1,-1} = 1$$

$$P_{-1,1} + P_{-1,-1} = 1.$$

(7.6.1)

In addition, if the polymerization is to produce polyampholytes of equimolar proportions, it is necessary that the over-all probability of adding either type of group to the chain be equal. This requirement, together with Eqs. (7.6.1), reduces to

$$P_{1,1} = P_{-1,-1}$$

$$P_{1,-1} = P_{-1,1} = 1 - P_{1,1}.$$

(7.6.2)

The magnitudes of the P_{ij} are simply related to the rate constants for the competing polymer reactions. If the rate constant for the addition of monomer type j to a radical of type i is k_{ij}, and the reactivity ratios are defined as[30]

$$r_1 = \frac{k_{1,1}}{k_{1,-1}}, \qquad r_{-1} = \frac{k_{-1,-1}}{k_{-1,1}}$$

(7.6.3)

one may write, using Eqs. (7.6.2) and (7.6.3),

$$\frac{P_{1,1}}{P_{1,-1}} = \frac{c_1}{c_{-1}} r_1 = \frac{P_{-1,-1}}{P_{-1,1}} = \frac{c_{-1}}{c_1} r_{-1}$$

(7.6.4)

where the c_j are the concentrations of the monomer species j. Equation (7.6.4) may be solved to yield the well-known result

$$P_{1,1} = \frac{(r_1 r_{-1})^{\frac{1}{2}}}{1 + (r_1 r_{-1})^{\frac{1}{2}}}; \qquad \frac{c_1}{c_{-1}} = \left(\frac{r_{-1}}{r_1}\right)^{\frac{1}{2}}. \tag{7.6.5}$$

Values of $P_{1,1}$ in excess of $\frac{1}{2}$ are indicative of a greater tendency toward longer successions of groups of the same species than would occur in a random copolymer, becoming in the limit when $P_{1,1} = 1$, a block copolymer. Values of $P_{1,1}$ less than $\frac{1}{2}$ indicate deviation from randomness in the direction of a regularly alternating structure, the limiting case when $P_{1,1} = 0$.

In the next section, use will be made of n_j, the number of sequences of exactly j successive monomers of a given type. By standard methods, one may show that

$$n_j = \frac{Z}{2}(1 - P_{1,1})^2 P_{1,1}^{j-1} \tag{7.6.6}$$

where Z is the over-all degree of polymerization of the copolymer. Since it will turn out that sequences of length 2 cannot be treated similarly to sequences of other lengths, it is also necessary to know the numbers of isolated sequences of 2, the number of adjacent pairs of sequences of 2, etc. These quantities are exhibited and used in conjunction with the discussion of Appendix 7C.

7.6.2 TITRATION PROPERTIES

In this subsection will be considered means for calculating the titration properties of equimolar polyampholytes of general skeletal structure. Because of the close correspondence between the present methods and those developed in the discussion of regularly alternating polyampholytes, it will be possible to use much of the formalism there developed.

1. Equilibrium Conditions

The only way in which the present development differs from that of the preceding section is in the evaluation of the free energy A_4', which depends upon the charge configuration and interaction between nearby charged sites of the polymer chain. Since the skeletal structure no longer consists of regularly repeating units, an expression for a grand partition function of the form of Eq. (7.5.17) of the preceding section would not be susceptible to obvious simplification. Alternatively, it will be shown how to evaluate quantities λ_1 and λ_2 defined in such a manner that they are exactly the absolute activities λ_1 and λ_2 of the previous section. In that section, it was shown [in the steps leading from Eq. (7.8.19) to Eq. (7.5.21)] that

$$\left(\frac{\partial A_4'}{\partial \alpha_1'}\right)_{\alpha_2'} = \frac{ZkT}{2}\ln \lambda_1, \qquad \left(\frac{\partial A_4'}{\partial \alpha_2'}\right)_{\alpha_1'} = \frac{ZkT}{2}\ln \lambda_2 \qquad (7.6.7)$$

so that Eq. (7.6.7) may be regarded as a definition of λ_1 and λ_2 if alternative means are available for relating A_4' to α_1' and α_2'. Then the entire development of Section 7.5, beginning with Eq. (7.5.21), may be applied to the systems now under study. It may be seen that the quantities which must be evaluated to carry through the discussion as in Section 7.5 are λ_1, λ_2, and $(\partial \alpha_1'/\partial \ln \lambda_1)_{\lambda_2}$, $(\partial \alpha_1'/\partial \ln \lambda_2)_{\lambda_1}$.

7.6.3 THE STATISTICAL FORMULATION OF A_4'

A_4' will be determined both by the entropy associated with the multiplicity of arrangements of charge on the polymer and by the interactions among the charges. The electrostatic energy of interaction among neighboring charged sites will be computed using a nearest neighbor approximation, since the errors which will thereby be introduced are less serious than in the strictly alternating polymers previously discussed. Therefore, consider the polymer to be composed of sequences of successive groups of the same kind (i.e., acidic or basic) whose behavior will be assumed to vary independently. Although such an assumption is strictly not true, it will be satisfactory because near the isoelectric point, where the calculated results are most sensitive to the approximations used, the polymers will be largely in the zwitterion form. Since pairs of charges on adjacent sites at the point where two sequences of the polymer join (henceforth referred to as " terminal sites ") are situated in positions of optimum energy, it is reasonable to suppose that the number of such sites not occupied would be small until the over-all proportion of charges sites of one sign is quite small. The behavior of each sequence of the chain will be calculated assuming the adjacent terminal sites of the adjoining sequences to be charged a suitable fraction of the time, independently of the condition of the sequence whose properties are under study. In this way the interaction between sequences can be introduced in a manner which is consistent with the previously assumed independent behavior of each sequence. The above remarks apply primarily to sequences of length other than 2, since the charges on the two terminal sites of a sequence then do not directly interact electrostatically. For sequences of length 2, the approximation breaks down for reasons which ultimately depend upon the fact that the two terminal charges of a sequence of length 2 do directly interact as nearest neighbors. Sequences of length 2 are therefore handled separately in a manner described in Appendix 7C. The main stream of the development is best followed by temporarily ignoring this complication.

Accepting the assumptions of the preceding paragraph, A_4' can be written in terms of the properties of the individual sequences. Let $v_{ijk}^{(\sigma)}$ be the number of sequences of groups of kind (for basic groups $\sigma = 1$, for acidic groups, $\sigma = 2$), and length i, with j charges (take $j > 0$ for both positive and negatively charged sequences), of which k are on terminal sites. Then

$$\sum_{jk} v_{ijk}^{(\sigma)} = n_i^{(\sigma)} \qquad (\text{each } \sigma, i) \tag{7.6.8}$$

where $n_i^{(\sigma)}$ is the number of sequences of length i, as given by Eq. (7.6.6). The total charge of each sign on the polymer is given by

$$\sum_{ijk} j v_{ijk}^{(\sigma)} = \frac{Z}{2} \alpha_\sigma' \qquad (\sigma = 1, 2). \tag{7.6.9}$$

Let $Q_{ijk}^{(\sigma)}$ be the partition function describing possible distributions of j charges on a sequence of length i and kind σ, with k terminal charges and an energy zero at the lowest electrostatic interaction energy. The electrostatic energy of this lowest state in absolute terms will be denoted by $\mathscr{E}_{ijk}^{(\sigma)}$. In order to calculate the interaction between sequences, it is necessary to know the probability that the end groups of the adjacent sequences are occupied. It may be seen that θ_σ, the fraction of charged end groups adjacent to sequences of type σ, may be expressed as

$$\theta_\sigma = \frac{\displaystyle\sum_{ijk} \tfrac{1}{2} k v_{ijk}^{(\sigma')}}{\displaystyle\sum_{ijk} v_{ijk}^{(\sigma')}} \qquad (\sigma' \neq \sigma) \tag{7.6.10}$$

and that the interaction between a sequence and its neighbors may be approximated by including in the total energy of the sequence a therm $-\theta_\sigma \chi k$ where χ is the interaction between similar charges on adjacent sites. Although the above energy term appears to include each interaction twice, it should not be divided by 2, since the averaging process implied by the use of θ_σ would then give the energy the wrong dependence upon k. It is better simply to notice that the zero of energy does not correspond to the totally uncharged polymer.

With the above definitions, it is possible to write an expression for A_4',

$$A_4' = -kT \ln \left[\sum_{\{v_{ijk}^{(\sigma)}\}} \left(\prod_{\sigma,i} n_i^{(\sigma)}! \right) \left(\prod_{ijk\sigma} \frac{1}{v_{ijk}^{(\sigma)}!} \left\{ Q_{ijk}^{(\sigma)} \right. \right. \right.$$

$$\left. \left. \left. \times \exp\left[-\frac{1}{kT} (\mathscr{E}_{ijk}^{(\sigma)} - \chi k \theta_\sigma) \right] \right\}^{v_{ijk}^{(\sigma)}} \right) \right]. \tag{7.6.11}$$

The factorials in Eq. (7.6.11) account for the fact that the different charge states may be distributed in a number of ways among the sequences, which are distinguishable by their locations in the polymer chain. The quantities $Q_{ijk}^{(\sigma)}$ and $\mathscr{E}_{ijk}^{(\sigma)}$ are obtainable by methods described in Appendix 7C, subject to the nearest neighbor interaction approximation.

7.6.4 The Equilibrium Distribution

Equation (7.6.11) becomes a useful form if the right-hand side is evaluated. The simplest procedure is to replace the summation by its maximum term, remembering that the choices of the $v_{ijk}^{(\sigma)}$ must be made subject to the conditions (7.6.8) and (7.6.9). Applying Lagrange's method of undetermined multipliers and Stirling's formula, and treating θ_σ as a constant, one finds that the values of the $v_{ijk}^{(\sigma)}$ for the largest term are

$$v_{ijk}^{(\sigma)} = \xi_i^{(\sigma)} \eta_\sigma^j Q_{ijk}^{(\sigma)} \exp\left[-\frac{1}{kT}(\mathscr{E}_{ijk}^{(\sigma)} - \chi k \theta_\sigma)\right] \qquad (7.6.12)$$

where the Lagrange multipliers $\xi_i^{(\sigma)}$ and η_σ are to be chosen to satisfy the subsidiary conditions (7.6.8) and (7.6.9). Substitution of (7.6.12) into (7.6.11) then leads to

$$A_4' = kT \sum_{i\sigma} n_i^{(\sigma)} \ln\left(\frac{\xi_i^{(\sigma)}}{n_i^{(\sigma)}}\right) + kT \sum \frac{Z\alpha_\sigma'}{2} \ln \eta_\sigma. \qquad (7.6.13)$$

The quantities of interest are to be derived from A_4' by differentiation with respect to the α_σ', the dependence of (7.6.13) upon which is contained in the parameters η_σ and $\xi_i^{(\sigma)}$. It is therefore necessary to consider in more detail the transformation from the variables η_σ, $\xi_i^{(\sigma)}$ to the alternate set of independent variables α_σ', $n_i^{(\sigma)}$. By direct differentiation of (7.6.13)

$$\frac{ZkT}{2}\ln \lambda_1 = \left(\frac{\partial A_4'}{\partial \alpha_1'}\right)_{\alpha_2'} = \frac{ZkT}{2}\sum_\sigma \alpha_\sigma'\left(\frac{\partial \ln \eta_\sigma}{\partial \alpha_1'}\right)_{\alpha_2'}$$

$$+ kT \sum_{i\sigma} n_i^{(\sigma)}\left(\frac{\partial \ln \xi_i^{(\sigma)}}{\partial \alpha_1'}\right)_{\alpha_2'} + \frac{ZkT}{2}\ln \eta_1 \quad (7.6.14)$$

where the derivatives appearing in (7.6.14) may be obtained by applying a Jacobian transformation to (7.6.9). A typical derivative of the type in (7.6.14) is of the form

$$\left(\frac{\partial \ln \eta_1}{\partial \alpha_1'}\right)_{\alpha_2'} = \frac{D_{11}}{J}$$

$$= \frac{\begin{vmatrix} \dfrac{Z}{2} & \left(\dfrac{\partial \alpha_1' Z/2}{\partial \ln \eta_2}\right) & \left(\dfrac{\partial \alpha_1' Z/2}{\partial \ln \xi_1^{(1)}}\right) \cdots \\[2ex] 0 & \left(\dfrac{\partial \alpha_2' Z/2}{\partial \ln \eta_2}\right) & \vdots \\[2ex] 0 & \vdots \\[1ex] \vdots \\ \vdots \end{vmatrix}}{\begin{vmatrix} \left(\dfrac{\partial \alpha_1' Z/2}{\partial \ln \eta_1}\right) & \left(\dfrac{\partial \alpha_1' Z/2}{\partial \ln \eta_2}\right) \cdots \\[2ex] \left(\dfrac{\partial \alpha_2' Z/2}{\partial \ln \eta_1}\right) & \vdots \\[2ex] \left(\dfrac{\partial n_1^{(1)}}{\partial \ln \eta_1}\right) \\[1ex] \vdots \end{vmatrix}} \tag{7.6.15}$$

where the derivatives on the right-hand side of (7.6.15) may be directly computed from Eqs. (7.6.8) and (7.6.9). Since the actual values of η_σ and $\xi_i^{(\sigma)}$ may be determined when α_1' and α_2' are known, Eq. (7.6.14) provides an applicable method of determining λ_1. The presence of θ_σ in the equation should provide no essential difficulty since a consistent set of $v_{ijk}^{(\sigma)}$ may be found by successive approximations or other convenient procedure. Obviously similar remarks apply to λ_2.

The remaining quantities of interest are $(\partial \alpha_1'/\partial \ln \lambda_1)_{\lambda_2}$, $(\partial \alpha_1'/\partial \ln \lambda_2)_{\lambda_1}$. By basically simple procedures it is possible to obtain the identities.

$$\left(\frac{\partial \alpha_1'}{\partial \ln \lambda_1}\right)_{\lambda_2} = \left(\frac{\partial \ln \lambda_2}{\partial \alpha_2'}\right)_{\alpha_1'} \left[\left(\frac{\partial \ln \lambda_1}{\partial \alpha_1'}\right)_{\alpha_2'} \times \left(\frac{\partial \ln \lambda_2}{\partial \alpha_2'}\right)_{\alpha_1'} - \left(\frac{\partial \ln \lambda_2}{\partial \alpha_1'}\right)_{\alpha_2'}^2\right]^{-1} \tag{7.6.16a}$$

$$\left(\frac{\partial \alpha_1'}{\partial \ln \lambda_2}\right)_{\lambda_1} = -\left(\frac{\partial \ln \lambda_2}{\partial \alpha_1'}\right)_{\alpha_2'} \left[\left(\frac{\partial \ln \lambda_1}{\partial \alpha_1'}\right)_{\alpha_2'} \times \left(\frac{\partial \ln \lambda_2}{\partial \alpha_2'}\right)_{\alpha_1'} - \left(\frac{\partial \ln \lambda_2}{\partial \alpha_1'}\right)_{\alpha_2'}^2\right]^{-1}. \tag{7.6.16b}$$

The right-hand sides of (7.6.16a) and (7.6.16b) contain expressions all of which can be obtained by direct differentiation of the formula (7.6.14). In fact

$$
\frac{ZkT}{2}\left(\frac{\partial \ln \lambda_1}{\partial \alpha_1'}\right)_{\alpha_2'} = \frac{ZkT}{2}\sum_\sigma \alpha_\sigma'\left(\frac{\partial^2 \ln \eta_\sigma}{\partial \alpha_1'^2}\right) + ZkT\left(\frac{\partial \ln \eta_1}{\partial \alpha_1'}\right)
$$
$$
+ kT\sum_{i\sigma} n_i^{(\sigma)}\left(\frac{\partial^2 \ln \xi_i^{(\sigma)}}{\partial \alpha_1'^2}\right) \qquad (7.6.17a)
$$

$$
\frac{ZkT}{2}\left(\frac{\partial \ln \lambda_2}{\partial \alpha_1'}\right)_{\alpha_2'} = \frac{ZkT}{2}\left(\frac{\partial \ln \eta_1}{\partial \alpha_2'}\right) + \frac{ZkT}{2}\left(\frac{\partial \ln \eta_2}{\partial \alpha_1'}\right)
$$
$$
+ \frac{ZkT}{2}\sum_\sigma \alpha_\sigma'\left(\frac{\partial^2 \ln \eta_\sigma}{\partial \alpha_1'\,\partial \alpha_2'}\right) + kT\sum_{i\sigma} n_i^{(\sigma)}\left(\frac{\partial^2 \ln \xi_i^{(\sigma)}}{\partial \alpha_1'\,\partial \alpha_2'}\right). \qquad (7.6.17b)
$$

Again, the use of Jacobian transformations is required to express the second derivatives appearing in (7.6.17) in terms of known quantities. A typical result here is of the form

$$
\left(\frac{\partial^2 \ln \eta_1}{\partial \alpha_1'\,\partial \alpha_2'}\right) = \frac{1}{J}
\begin{vmatrix}
\left(\dfrac{\partial \alpha_1' Z/2}{\partial \ln \eta_1}\right) & \left(\dfrac{\partial \alpha_1' Z/2}{\partial \ln \eta_2}\right) & \left(\dfrac{\partial \alpha_1' Z/2}{\partial \ln \xi_1^{(1)}}\right) & \cdots \\[2ex]
A_1 & A_2 & A_3 & \cdots \\[2ex]
\left(\dfrac{\partial n_1^{(1)}}{\partial \ln \eta_1}\right) & \left(\dfrac{\partial n_1^{(1)}}{\partial \ln \eta_2}\right) & \cdots &
\end{vmatrix}
\qquad (7.6.18)
$$

where J is the determinant appearing in (7.6.15) and the elements of the α_2' row are of the form

$$
A_k = \frac{Z}{2}\sum_{lm}\left(\frac{\partial \ln \eta_1}{\partial x_l}\right)\left(\frac{\partial u_m}{\partial \alpha_1}\right)\left(\frac{\partial^2 x_l}{\partial u_m\,\partial u_k}\right)
$$

where u_m denotes the mth member of the ordered set of variables $\ln \eta_1$, $\ln \eta_2$, $\ln \xi_1^{(1)}$, $\cdots$, and x_m denotes the mth member of the ordered set of variables $Z\alpha_1'/2$, $Z\alpha_2'/2$, $n_1^{(1)}$, $\cdots\cdot$. Since Eq. (7.6.18) and the similar equations to be used may be directly evaluated, the computation of $(\partial \alpha_1'/\partial \ln \lambda_1)_{\lambda_2}$ and $(\partial \alpha_1'/\partial \ln \lambda_2)_{\lambda_1}$ is in principle complete.

7.6.5 END-TO-END EXTENSION

In the preceding subsection the effect of the microstructure of the polymeric ion on its mean extension in solution was discussed. It was pointed out there that in an equimolar polyampholyte near its isoelectric point, the two factors tending to affect its end-to-end extension are the nature of the skeletal distribution of acidic and basic charge sites, and the fluctuations among charged and uncharged groups at each position of the chain. In a polyampholyte of regularly alternating structure, the skeletal effect vanishes, and the fluctuation effect alone remains. In the preceding section it was found that, in a polymer where the acid and base dissociation constants are large enough to cause the polymer to be predominantly in the zwitterion form, the number of uncharged sites of either acidic or basic type was so small that the fluctuations were of negligible effect in determining the extension of the polymer chain. It may be verified that similar conclusions may be drawn for equimolar polyampholytes of general skeletal structure. Such calculations, in fact, show that the fluctuation effects will not be important except when there are an appreciable number of uncharged groups, so that sizeable local inhomogeneities in fluctuation charge are relatively probable. It appears, therefore, that fluctuations will affect the end-to-end extension only in nonequimolar polyampholytes, where, at the isoelectric point, there will of necessity be a number of uncharged groups of one type or the other.

For the purpose of calculating the end-to-end extension, the polymer chain is considered as a chain of rigid " statistical elements," with each element connected to its neighbors by universal joints. At the isoelectric point, in a polymer where skeletal correlation effects are much larger than the effect of charge correlations, it may be shown that

$$W(\gamma) = \frac{\langle q_1^{(s)} q_2^{(s)} \rangle \, e^{-\kappa r(\gamma)}}{Dr(\gamma)} \qquad (7.6.19)$$

where $W(\gamma)$ is the potential of average force between two successive statistical elements connected at an angle γ, D is the dielectric constant, $r(\gamma)$ the distance separating the centers of the elements (where the charges are assumed to be), and κ the Debye reciprocal radius; $\langle q_1^{(s)} q_2^{(s)} \rangle$ is the average of the product of the charges on the elements 1 and 2 arising from the local skeletal structure.

To simplify the following discussion $\langle q_1^{(s)} q_2^{(s)} \rangle$ will be calculated assuming the polymer to be totally in the zwitterion form. Since the actual fraction of the maximum possible charge present is large in typical copolymers (see next section), such an assumption cannot lead to seriously inaccurate results. It is therefore only necessary to consider the statistical distributions of the constituent groups along the polymer skeleton. Since the structure of the polymer is assumed to be controlled by competing reactions, with the rate of an addition governed only by the identity of the end group and the prospective addend, there will be a correlation in charge between adjacent segments determined by the value of $P_{1,1}$, the probability of adding a group similar to the group then on the end of the chain. One may proceed by considering a section of the polymer chain two statistical elements long and averaging with appropriate weight all the possible charge distributions arising from the statistical distribution of the monomeric groups.

Let each statistical element contain n monomer units. Let the total number of units be numbered in order, from 1 to $2n$, and associate with each unit i an occupation number η_i, where η_i will have the value $+1$ or -1, referring to the charge on monomer unit i. Each possible skeletal configuration of the chain then corresponds to a set of the variables $\boldsymbol{\eta} = \{\eta_1, \cdots \eta_{2n}\}$. Let $P(\boldsymbol{\eta})$ be the probability of the given configuration $\boldsymbol{\eta}$. Using the chain propagation probabilities given previously

$$P(\boldsymbol{\eta}) = \tfrac{1}{2} P_{\eta_1 \eta_2} P_{\eta_2 \eta_3} \cdots P_{\eta_{2n-1} \eta_{2n}} \tag{7.6.20}$$

where the factor $\tfrac{1}{2}$ arises from the *a priori* probability that η_1 will have a given value. Finally, let $\Xi(\theta_1, \theta_2)$ be a function defined by the equation

$$\Xi(\theta_1, \theta_2) = \sum_{\boldsymbol{\eta}} P(\boldsymbol{\eta}) \theta_1^{\sum_{i=1}^{n} \eta_i} \theta_2^{\sum_{i=n+1}^{2n} \eta_i} \tag{7.6.21}$$

where the sum is over all sets $\boldsymbol{\eta}$. Note that the exponents of θ_1 and θ_2 are, for each configuration $\boldsymbol{\eta}$, the total net charge of the skeleton in the first n, and in the second n positions, respectively. By differentiation of $\Xi(\theta_1, \theta_2)$, followed by setting $\theta_1 = \theta_2 = 1$, one may show that

$$\left(\frac{\partial^2 \Xi(\theta_1, \theta_2)}{\partial \theta_1 \, \partial \theta_2} \right)_{\theta_1 = \theta_2 = 1} = \sum_{\boldsymbol{\eta}} \left(\sum_{i=1}^{n} \eta_i \right) \left(\sum_{i=n+1}^{2n} \eta_i \right) P(\boldsymbol{\eta})$$
$$= \langle q_1^{(s)} q_2^{(s)} \rangle. \tag{7.6.22}$$

Hence, evaluation of $\Xi(\theta_1, \theta_2)$ provides a method of obtaining the quantity $\langle q_1^{(s)} q_2^{(s)} \rangle$.

Upon introducing Eq. (7.6.20) into (7.6.21), it may be seen that (7.6.21) can be rewritten in terms of a matrix product,

$$\Xi(\theta_1, \theta_2) = \sum_{\substack{\eta_1, \eta_n \\ \eta_{n+1}, \eta_{2n}}} \tfrac{1}{2} (\mathbf{V}_1^{n-1})_{\eta_1 \eta_n} (\mathbf{V}_2^{n-1})_{\eta_{n+1} \eta_{2n}} \times P_{\eta_n \eta_{n+1}} \theta_1^{\frac{1}{2}(\eta_1 + \eta_2)} \theta_2^{\frac{1}{2}(\eta_{n+1} + \eta_{2n})} \tag{7.6.23}$$

where $\mathbf{V}_s$ is a two-by-two matrix whose elements are

$$(\mathbf{V}_s)_{\eta \eta'} = P_{\eta \eta'} \theta_s^{\frac{1}{2}(\eta + \eta')}. \tag{7.6.24}$$

We now replace the matrices $\mathbf{V}_s^{n-1}$ by an expansion in terms of their eigenvalues. For any two-by-two Hermitian matrix H,[38]

$$H_{\eta \eta'} = \phi^{(1)} x^{(1)}(\eta) x^{(1)}(\eta') + \phi^{(2)} x^{(2)}(\eta) x^{(2)}(\eta') \tag{7.6.25}$$

where $\phi^{(1)}$ and $\phi^{(2)}$ are the eigenvalues of H, and $x^{(1)}(\eta)$, $x^{(2)}(\eta)$ are the corresponding eigenvectors. The eigenvectors form an orthonormal set, so that

$$\sum_{\eta} x^{(i)}(\eta) x^{(j)}(\eta) = \delta_{ij} \tag{7.6.26}$$

where δ_{ij} is unity when i and j are the same and zero otherwise. Since the eigenvalues of $\mathbf{V}_s^{n-1}$ are simply ϕ^{n-1}, where ϕ_s are the eigenvalues of $\mathbf{V}_s$, and the eigenvectors of V_s^{n-1} are identical with those of V_s, Eq. (7.6.25) may be usefully employed to simplify Eq. (7.6.23). The result is

$$
\begin{aligned}
\Xi(\theta_1, \theta_2) = \sum_{\substack{\eta_1, \eta_2 \\ \eta_{n+1}, \eta_{2n}}} & \tfrac{1}{2}\{(\phi_1^{(1)})^{n-1} x_1^{(1)}(\eta_1) x_1^{(1)}(\eta_n) \\
& + (\phi_1^{(2)})^{n-1} x_1^{(2)}(\eta_1) x_1^{(2)}(\eta_n)\} \theta_1^{\frac{1}{2}(\eta_1 + \eta_n)} P_{\eta_n \eta_{n+1}} \\
& \times \{(\phi_2^{(1)})^{n-1} x_2^{(1)}(\eta_{n+1}) x_2^{(1)}(\eta_{2n}) \\
& + (\phi_2^{(2)})^{n-1} x_2^{(2)}(\eta_{n+1}) x_2^{(2)}(\eta_{2n})\} \theta_2^{\frac{1}{2}(\eta_{n+1} + \eta_{2n})}
\end{aligned}
\tag{7.6.27}
$$

so that the matrix methods have provided a means for performing all but four of the $2n$ summations. Taking cognizance that $P_{\eta_n, \eta_{n+1}}$ is independent of the θ_s, and that the dependence upon θ_1 and θ_2 is separated, one may differentiate (7.6.27) to find

$$
\langle q_1^{(s)} q_2^{(s)} \rangle = \sum_{\substack{\eta_1, \eta_n \\ \eta_{n+1}, \eta_{2n}}} \tfrac{1}{2} P_{\eta_n, \eta_{n+1}} \frac{\partial}{\partial \theta_1} \left[\theta_1^{\frac{1}{2}(\eta_1 + \eta_2)} \{(\phi_1^{(1)})^{n-1} \cdots \} \right]_{\theta_1 = 1}
$$

$$
\times \frac{\partial}{\partial \theta_2} \left[\theta_2^{\frac{1}{2}(\eta_{n+1} + \eta_{2n})} \{(\phi_2^{(1)})^{n-1} \cdots \} \right]_{\theta_2 = 1}. \tag{7.6.28}
$$

To proceed further it is necessary to evaluate more explicitly the various quantities appearing in Eq. (7.6.28). The eigenvalues of V_s are obtained by solving the secular equation

$$
\begin{vmatrix}
P_{1,1}\theta_s - \phi_s^{(\beta)} & 1 - P_{1,1} \\
1 - P_{1,1} & P_{1,1}\theta_s^{-1} - \phi_s^{(\beta)}
\end{vmatrix} = 0. \tag{7.6.29}
$$

The result is

$$
\phi_s^{(\beta)} = \frac{P_{1,1}}{2}(\theta_s + \theta_s^{-1}) - \frac{(-1)^\beta}{2} [P_{1,1}^2(\theta_s + \theta_s^{-1})^2 - 4(2P_{1,1} - 1)]^{\frac{1}{2}}. \tag{7.6.30}
$$

The corresponding eigenvectors are

$$
\begin{aligned}
x_s^{(\beta)}(1) &= (1 - P_{1,1})\omega_s^{(\beta)} \\
x_s^{(\beta)}(-1) &= (\phi_s^{(\beta)} - P_{1,1}\theta_s)\omega_s^{(\beta)}
\end{aligned}
\tag{7.6.31}
$$

where

$$
\omega_s^{(\beta)} = [(1 - P_{1,1})^2 + (\phi_s^{(\beta)} - P_{1,1}\theta_s)^2]^{-\frac{1}{2}} \tag{7.6.32}
$$

and the superscript β, which indicates to which eigenvalue the labeled quantities refer, may assume the values 1, 2.

By suitable manipulation of Eqs. (7.6.30)–(7.6.32), it is possible to show that, when $\theta_1 = \theta_2 = 1$,

$$\left(\frac{\partial \phi_s^{(\beta)}}{\partial \theta_s}\right) = 0$$

$$\frac{\partial}{\partial \theta_s}\left[x_s^{(\beta)}(\eta)x_s^{(\beta)}(\eta')\right] = \frac{-\eta P_{1,1}}{2(1 - P_{1,1})}\,\delta_{\eta\eta'}(-1)^\beta \qquad (7.6.33)$$

$$\phi_s^{(\beta)} = P_{1,1} + (-1)^\beta(P_{1,1} - 1)$$

$$x_s^{(\beta)}(\eta) = 2^{-1/2}\eta^{\beta-1}.$$

When Eqs. (7.6.33) are introduced into Eq. (7.6.28), all the indicated summations can be performed, yielding the simple final result

$$\langle q_1^{(s)}q_2^{(s)}\rangle = \left(\frac{2P_{1,1} - 1}{4}\right)\left(\frac{1 - (2P_{1,1} - 1)^n}{1 - P_{1,1}}\right)^2. \qquad (7.6.34)$$

Let us consider in some detail the behavior implied by Eq. (7.6.34). For $P_{1,1} = \frac{1}{2}$, $\langle q_1^{(s)}q_2^{(s)}\rangle$ vanishes, as it should. When $P_{1,1}$ is greater than $\frac{1}{2}$, $\langle q_1^{(s)}q_2^{(s)}\rangle$ assumes positive values, resulting in a repulsion between neighboring segments of the chain. In the limit, $P_{1,1} \to 1$, Eq. (7.6.34) approaches n^2, the value to be expected for a block copolymer of long sequences of identical groups. When $P_{1,1}$ is less than $\frac{1}{2}$, $\langle q_2^{(s)}q_2^{(s)}\rangle$ is negative, but its magnitude strongly depends upon whether n is even or odd. In the limit $P_{1,1} = 0$, $\langle q_1^{(s)}q_2^{(s)}\rangle = 0$ for n-even, -1 for n-odd. Since the concept of statistical elements of definite length was introduced to approximately account for the incomplete and not easily described flexibility of the polymer chain, the distinction between the behavior of chains of even and odd lengths does not in fact correspond to an actual physically observable phenomenon. Since the different behavior of even and odd elements is inherent in the methods here described, the theory is not really well fitted to deal with polymers approaching alternant structure. If it is desired to consider such polymers, one might attempt to calculate $\langle q_1^{(s)}q_2^{(s)}\rangle$ by averaging the values obtained from interpolation of both odd and even values of n.

7.6.6 CORRELATION IN CHARGE BETWEEN NONADJACENT SEGMENTS

Particularly in the case of polymers for which $P_{1,1}$ is nearly unity, one may wish to know how much correlation in charge due to skeletal structure exists between nonadjacent statistical elements of the polymer. In block copolymers, the repulsion between next nearest neighboring statistical elements might be sufficiently large that it would be desirable to take it into account. The correlation in charge between a first statistical element and a kth element (numbered with respect to the first), $\langle q_1^{(s)}q_k^{(s)}\rangle$, may be calculated by methods quite similar to those used to obtain $\langle q_1^{(s)}q_2^{(s)}\rangle$.

Let us introduce a function $\Xi(\theta_1, \theta_k)$,

$$\Xi(\theta_1,\theta_k) = \sum_{\eta} P(\mathbf{\eta})\theta_1^{\sum_{i=1}^{n}\eta_i}\theta_k^{\sum_{i=(k-1)n+1}^{kn}\eta_i} \tag{7.6.35}$$

where $P(\mathbf{\eta})$ now represents the probability of a sequence of kn groups with occupation numbers $\eta_1, \cdots \eta_{kn}$. An equation analogous to (7.6.23) may now be written,

$$\Xi(\theta_1, \theta_k) = \sum_{\substack{\eta_1,\eta_n \\ \eta_{(k-1)n+1},\eta_{kn}}} \tfrac{1}{2}(\mathbf{V}_1)_{\eta_1\eta_n}^{n-1}[(\mathbf{V}_0)_{\eta_n\eta_{(k-1)n+1}}]^{(k-2)n+1} \times$$

$$[(\mathbf{V}_2)_{\eta_{(k-1n+1}\eta_{kn}}]^{n-1}\theta_1^{\frac{1}{2}(\eta_1+\eta_n)}\theta_k^{\frac{1}{2}(\eta_{(k-1)n+1}+\eta_{kn})} \tag{7.6.36}$$

where the matrix $\mathbf{V}_0$ simply has elements $(\mathbf{V}_0)_{\eta\eta'} = P_{\eta\eta'}$. Proceeding just as before, one may obtain the result

$$\langle q_1^{(s)}q_k^{(s)}\rangle = \frac{(2P_{1,1}-1)^{(k-2)n+1}}{4}\left[\frac{1-(2P_{1,1}-1)^{n-2}}{1-P_{1,1}}\right]. \tag{7.6.37}$$

Equation (7.6.37) predicts a decrease in the charge correlation as the separation of the elements involved is increased. Except when $P_{1,1}$ is nearly unity, the decrease will be quite rapid.

TABLE 7.6.1[a]

CHARGE CORRELATION BETWEEN ADJACENT SEGMENTS ARISING
FROM SKELETAL STRUCTURE

$P_{1,1}$	$\langle q_1^{(s)}q_2^{(s)}\rangle$	
	$Z/N = 10$	$Z/N = 6$
0.2	−0.23	−0.21
0.4	−0.14	−0.14
0.5	0.00	0.00
0.6	0.31	0.31
0.8	3.17	3.42

[a] Reproduced from ref. 2.

7.6.7 NUMERICAL EVALUATION AND DISCUSSION

The thermodynamic and configurational properties of polyampholytes have as yet been very incompletely investigated. At present the literature appears to contain no data for equimolar polyampholytes of any skeletal structure, and but fragmentary information concerning polymers in which the molar ratio of the monomers differs from unity. Since it is anticipated that properties of nonequimolar polyampholytes will differ considerably from those expected for equimolar polymers, it is not possible to compare the present development with experimental results. However, the general behavior predicted will be demonstrated by presenting calculations for a hypothetical copolymer.

The hypothetical copolymer chosen here is composed of the same monomeric species as used in the calculation of the preceding section. The dissociation constants are $K_1^0 = 10^{-5}$, $K_2^0 = 10^{-9}$, and the spacing between nearest neighboring groups is 6 A. The statistical element for expansion is assumed to contain 10 monomers and to possess an effective length of 25 A. The interaction energy between charges will be calculated assuming a dielectric constant of 50. Binding of anions and cations will be permitted with binding dissociation constants $K_3^0 = K_4^0 = 10$. The distribution of monomeric groups along the polymer skeleton is assumed to be random. However, to reduce the difficulties attendant upon inversion of matrices of high order, as required in the evaluation of Eq. (7.6.14), it was assumed for the purpose of calculating the titration properties that the sequences of successive monomers of length 5 and above could be replaced by an appropriate number of sequences of length 6. This approximation results in the reassignment of approximately 15% of the monomer units of the polymer in such a way that the average sequence length remains invariant, and therefore the calculated properties should not be significantly altered.

Except at the isoelectric point, the evaluation of the dependence of λ_1 and λ_2 upon pH and ionic activities is straightforward but tedious. At the isoelectric point, however, some simplification is possible because of the symmetry then prevailing between λ_1 and λ_2, and necessarily also between η_1 and η_2. The matrix to be inverted is then factorizable, and analytical forms for the inverse of the factor matrices may be obtained. The procedure followed in any case is to evaluate λ_1 and λ_2 for various values of η_1, η_2, θ_1, and θ_2, thereby permitting an interpolation of η_1 and η_2 satisfying Eqs. (7.5.21). As a rough guide to the result to be expected, it might be remarked that the values of λ_σ and η_σ are generally of the same order of magnitude, and that the final value of λ_σ differs but little from the corresponding value for a regularly alternating polyampholyte.

The titration curve for the hypothetical copolymer, calculated as described above, is indicated in Fig. 7.6.1. A most significant feature of the curve is the smaller variation near the isoelectric point in pH with changes in the net charge of the polymer, as compared with a regularly alternating copolymer of the same composition. This effect is symptomatic of the considerable number of charges situated in the interior of sequences adjacent to charges of the same sign, and therefore relatively loosely bound to the polymer. A much smaller change in chemical potential of hydrogen ion therefore results from their removal than from the removal of an equal number of the charges of a

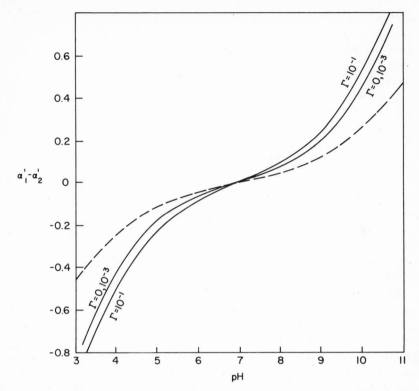

FIG. 7.6.1. A comparison of the titration curves of equimolar random and regularly alternating polyampohlytes. [Reproduced from ref. 2.]

regularly alternating copolymer, essentially all of whose charges are tightly bound by electrostatic attraction to neighboring charges of opposite sign. This effect will of course be relatively sensitive to the skeletal structure of the polymer, and a wide range of titration behavior should in principle be realizable simply by alteration of the skeletal distribution. It may also be seen that increasing the ionic strength, thereby decreasing the interaction among charge sites, both by binding and by increased electrostatic shielding, results in a shift of the titration curve in the direction of the behavior to be expected of independent acidic and basic groups. This behavior was also observed for the regularly alternating copolymer.

Finally, consider the configurational properties of general equimolar polyampholytes. As discussed in a preceding section, the only important difference between such copolymers and their regularly alternating counterpart arises from inhomogeneity in the skeletal distribution. For copolymers whose statistical elements for expansion comprise 6 and 10 monomer units, values of the correlated charge product $\langle q_1^{(s)} q_2^{(s)} \rangle$ were computed for different values of $P_{1,1}$, the parameter governing skeletal distribution. The results, shown in Table 7.6.1, indicate that, even for a relatively short statistical element, a strong deviation from random structure is needed to effect a sizeable expansion or contraction.

It should be reiterated that values of $\langle q_1^{(s)} q_2^{(s)} \rangle$ for $P_{1,1} < 0.5$ are, as discussed, of dubious validity due to the inadequacy of the model used for computing expansion.

7.7 Helix-coil Transitions in Charged Macromolecules

It has now been established that a number of macromolecules possessing helical configurations in solution may undergo a transition to a different configuration. Of the possible examples, the most thoroughly investigated case is the molecular isomerization from α-helix to random coil characteristic of the synthetic polypeptides. For these substances the elegant studies of Doty and co-workers[40] have clearly shown the dependence of the transition upon the intimate details of the interactions within one polymer molecule, as well as the interactions of the polymer with solvent molecules. There has been extensive study of the theory of the transition in uncharged polypeptides.[41] It is the purpose of the present section to study the effects of electrostatic forces. Since we shall ultimately need the difference in electrostatic energy between two molecular configurations, the methods introduced earlier in this chapter will give accurate results, even when the range of interactions is truncated and the absolute electrostatic free energy of one configuration is not too accurate.

There are two aspects of the problem considered. First, it is pertinent to investigate how suitably defined molecular parameters may be deduced from experimental data. Secondly, it is necessary to construct a detailed theory in which the existence of the isomerization is predicted from the functional dependence and numerical values of these same molecular parameters. We shall discuss these two topics in the order indicated.

7.7.1 EXPERIMENTAL EVALUATION OF PARAMETERS

The definition of appropriate molecular parameters requires specification of the system under investigation. We restrict attention to systems dilute in the polymeric component and consider only the interactions within one polymer molecule and between a polymer molecule and the various components of the solvent. We shall treat the solvent as a dielectric continuum. Each polymer consists of many identical subunits, each of which may ionize, and each of which may also hydrogen bond to the appropriate other segment of the same chain to form an α-helix. As in the preceding discussion, a system of the type indicated may be conveniently described in terms of the semigrand partition function

$$\Xi = \sum_{\{c\}} \sum_{\{\eta\}} e^{-[W(\{c\},\{\eta\})/kT]} Q^0(\{c\}) \lambda_i^{\Sigma \eta_i} \tag{7.7.1}$$

where η_i, which can be either 0 or 1, is an occupation variable specifying the charge state of site i, and c_i in the same fashion specifies the hydrogen bond state of site i. The internal partition function, $Q^0(\{c\})$ depends upon the positions of the chain subunits, hydrogen bond energies, steric hindrances, and any other relevant variables. Finally, the variable λ plays the role of the reciprocal of the absolute activity of the hydrogen ion (we take the macromolecule to be a polyacid for concreteness), and $W(\{c\}, \{\eta\})$ is the electrostatic energy corresponding to the skeletal configuration with $\{c\}$ hydrogen bonds and the charge distribution specified by $\{\eta\}$. Note that the summation of the occupation variable η_i over all sites is just the total charge on the polyion and the summation of c_i gives the total number of intact hydrogen bonds. The degree of ionization, α, is obtained from Ξ just as for the simpler case where internal structure may be neglected, i.e.,

$$\alpha = \frac{1}{Z} \frac{\partial \ln \Xi}{\partial \ln \lambda} \tag{7.7.2}$$

where there are Z charge sites on the molecule. The absolute activity λ is related to the more familiar variables of pH and dissociation constant K_a^0 by

$$0.434 \ln \lambda = pH - pK_a^0. \tag{7.7.3}$$

Let us assume that the pure helix and pure coil could exist alone. These would then be characterized by partition functions Ξ_h and Ξ_c of the form indicated in Eq. (7.7.1). The degrees of ionization of these hypothetical forms would be,

$$\alpha_h = \frac{1}{Z} \frac{\partial \ln \Xi_h}{\partial \ln \lambda}, \qquad \alpha_c = \frac{1}{Z} \frac{\partial \ln \Xi_c}{\partial \ln \lambda}. \tag{7.7.4}$$

Consider for the moment the uncharged molecule, $\alpha = 0$, and let s be the equilibrium constant for the addition to a section of helix of the appropriate segment from the adjacent section of randomly coiling chain. Then if it is assumed that when $\alpha = 0$ the polymer molecule is completely in the helical form, it is seen that

$$(\Xi)_{\alpha = 0} = s^Z \tag{7.7.5}$$

end effects being neglected. Then, by Eqs. (7.7.2) and (7.7.4),

$$\ln \mathfrak{m} - \ln s = \int_{\lambda(0)}^{\lambda(\alpha)} \alpha \, d \ln \lambda \tag{7.7.6}$$

$$\ln \mathfrak{c}_0 = \int_{\lambda(0)}^{\lambda(\alpha)} \alpha_c \, d \ln \lambda \tag{7.7.7}$$

with

$$\mathfrak{m} = \Xi^{1/Z}, \qquad \mathfrak{c}_0 = \Xi_c^{1/Z}, \qquad \mathfrak{h}_0 = \Xi_h^{1/Z} \tag{7.7.8}$$

where $\mathfrak{m}$, $\mathfrak{c}_0$ and $\mathfrak{h}_0$ are the partition functions per residue of the mixed, pure coil and pure helix forms respectively. If the fully charged state is assumed to be in the pure randomly coiled configuration,

$$(\mathfrak{m})_{\alpha=1} = (\mathfrak{c}_0)_{\alpha=1} \tag{7.7.9}$$

and thereby

$$\ln s = \int_{\lambda(0)}^{\lambda(1)} (\alpha_c - \alpha) \, d \ln \lambda \tag{7.7.10}$$

In Eqs. (7.7.6), (7.7.7), (7.7.10) and (7.7.14) use is made of the limiting conditions at $\alpha = 0$,

$$\mathfrak{c}_0 = 1, \qquad \mathfrak{h}_0 = 1$$

which follow directly from the interpretation of the partition function in terms of the relative probabilities of finding various (charge) configurations and the use of the uncharged random coil and uncharged helix as reference states for Ξ_c and Ξ_h respectively. The titration curves for the pure helix and pure coil can in principle be calculated and can in practice be determined experimentally. The necessary extrapolations past the transition region into the domain of hypothetical stability are not difficult to perform.

The integrals may be most easily evaluated by the following device. The single integral is converted to a double integral by the introduction of a new variable. The double integral may then be transformed as follows:

$$\int_{\lambda(0)}^{\lambda(\alpha)} (\alpha_c - \alpha_h) \, d \ln \lambda = \int_{\lambda(0)}^{\lambda(\alpha)} \int_{\alpha_h}^{\alpha_c} d\alpha \, d \ln \lambda$$

$$= \int_{\ln \lambda(\alpha_h) - \ln \alpha_h/(1-\alpha_h)}^{\ln \lambda(\alpha_c) - \ln \alpha_c/(1-\alpha_c)} \int_{\alpha_h}^{\alpha_c} d\alpha \, d\left(\ln \lambda - \ln \frac{\alpha}{1-\alpha} \right)$$

$$\tag{7.7.11}$$

since the Jacobian of the transformation is unity. As can be seen in Fig. 7.7.1, a plot of $\ln \lambda - \ln (\alpha/1 - \alpha)$ is much more nearly linear than a plot of α versus $\ln \lambda$ (see Fig. 7.7.2). Since the areas cut off by a line of constant λ on either plot are equal by virtue of Eq. (7.7.11) and the extrapolation of nearly linear curves is easier than that of nonlinear curves, $\ln \lambda - \ln (\alpha/1 - \alpha)$ is the more useful experimental variable. It is clear then that the determination of the titration curve of the polymer followed by interpolation of the portion characteristic

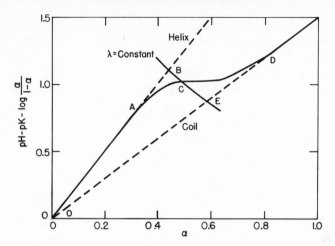

FIG. 7.7.1. Construction for the determination of the thermodynamic parameters. The experimental titration curve OACD has been drawn through the data shown in Fig. 7.7.3 for polyglutamic acid in 0.0133 M salt. According to Eqs. (7.7.10), (7.7.11), (7.7.14), and (7.7.15), the quantities $\log_{10} s$, $\log_{10} s'$ and $\log_{10} m'$ are given by the areas OACD, OACD–OBE, and CED, respectively. The curve BCE is drawn for the particular case of $\lambda = 1$, corresponding at point C on the experimental curve to $\alpha = 0.49$. [Reproduced from ref. 39].

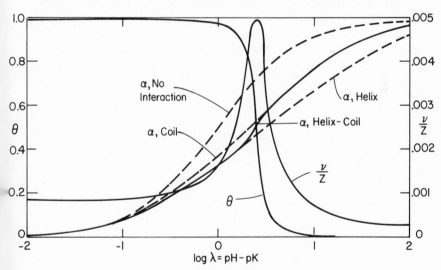

FIG. 7.7.2. Theoretical results for our model of polyglutamic acid in 0.133 M salt in a dioxane water mixture with $D = 50$. Shown are the titration curves of a monobasic acid (no interaction), a perfect helix, a perfect random coil, and the equilibrium mixture of helix and coil; also shown are the fraction of internal hydrogen bonds, ϑ, and the frequency of alternation of random and helical sections, ν/Z. Recall that $\nu/Z = \partial \ln m / \partial \ln \sigma$ with ν the number of helical sections in a chain. [Reproduced from ref. 39.]

of the pure coil and measurement of the area between the observed and the interpolated curves gives the value of the equilibrium constant s. The latter important quantity is a measure of the stability of the uncharged helix.

It is possible by similar methods to determine the fraction, ϑ, of intact hydrogen bonds in the helix. Two accessible experimental quantities are

$$s' = s\mathfrak{h}_0/\mathfrak{c}_0 \qquad (7.7.12)$$

$$\mathfrak{m}' = \mathfrak{m}/\mathfrak{c}_0. \qquad (7.7.13)$$

By methods analagous to those used in the foregoing,

$$\ln s' = \int_{\lambda(0)}^{\lambda(1)} (\alpha_c - \alpha) \, d\ln\lambda - \int_{\lambda(0)}^{\lambda(\alpha)} (\alpha_c - \alpha_h) \, d\ln\lambda \qquad (7.7.14)$$

$$\ln \mathfrak{m}' = \int_{\lambda(\alpha)}^{\lambda(1)} (\alpha_c - \alpha) \, d\ln\lambda. \qquad (7.7.15)$$

These integrals correspond to the areas shown in Fig. 7.7.1. For example, for the case illustrated in Fig. 7.7.1, $s = 1.355$, $s' = 1.02$, $\lambda = 1$, and $\mathfrak{m}' = 1.039$. The way in which these numbers can be put to use to determine the parameters characteristic of the transition will be discussed at a later point after we have introduced the necessary theoretical considerations.

Before closing this section it is pertinent to note that it is also possible to determine experimentally the fraction, ϑ, of the polymer which is in the helical configuration. For, if the junction interactions between helical and coiled sections of the macromolecule are ignored, the partition function becomes

$$\mathfrak{m}^Z = (s\mathfrak{h}_0)^k \mathfrak{c}_0^{Z-k} \qquad (7.7.16)$$

where there are k residues in helix form. Using the definition $\vartheta = k/Z$ and the definitions of Eqs. (7.7.12) and (7.7.13),

$$\ln \mathfrak{m} = \vartheta \ln s + \vartheta \ln \mathfrak{h}_0 + (1 - \vartheta) \ln \mathfrak{c}_0 \qquad (7.7.17)$$

$$\ln \mathfrak{m}' = \vartheta \ln s' \qquad (7.7.18)$$

whereby

$$\left(\frac{\partial \ln \mathfrak{m}}{\partial \ln s}\right)_\lambda = \frac{d\ln \mathfrak{m}'}{d\ln s'} = \vartheta. \qquad (7.7.19)$$

7.7.2 THE MOLECULAR MODEL

In the last section it was seen that the equilibrium constant for the addition of a residue to a helical section could be determined from experimental data. It remains to be shown how this parameter enters the molecular theory and whether or not further deductions from experiment can be made.

The examination of the consequences of Eq. (7.7.1) requires an evaluation of the internal partition function $Q^0(\{c\})$. Consider the case of a polypeptide in the helical configuration. The partition function is the sum of the weighted probabilities of finding the macromolecule with some monomeric units in a random coil configuration and all others in a helical configuration. The weighting of particular configurations is in each case a Boltzmann factor with the exponent equal to the work required to create the specified state by braking hydrogen bonds, overcoming steric hindrances, etc. In more descriptive language, the free energy of each hydrogen bond contributes to the stability of the helix. If a hydrogen bond is broken, the two amide groups which were previously severely restricted acquire an increased freedom of motion. If two adjacent hydrogen bonds are missing there is an extra increase in freedom over and above that achieved in the breaking of two separated hydrogen bonds. We can proceed in this manner to characterize the state of each hydrogen bond site in terms of interactions with other groups along the chain. The internal state of the macromolecule can thus be characterized by specifying the probabilities of occurrence of all possible arrangements of bonded and nonbonded sites.

As in our discussion of the electrostatic free energy of a polyion, practical considerations, based upon the increased mathematical complexity involved in handling large numbers of simultaneously interacting sites, force the adoption of an approximation in which only small groups of sites interact as a unit. If the energy of interaction of a given site (which shall be called the central site) with its neighbors is truncated after the terms corresponding to say the xth neighboring site, then the set of sites considered to interact forms a basic group from which the configuration of the entire polymer may be specified. The probability of finding the whole molecule in a given configuration is, in fact, the product of the probabilities of finding the required local configurations of $x + 1$ interacting groups. The essence of the approximation inherent in truncating the interactions after x groups is that the total partition function is then constructed of the subpartition functions for the interactions within the basic group. The basic groups are independent of one another. *This does not mean* that a given site is uncoupled from the rest of the molecule, but rather that interactions are reckoned groupwise, each site belonging to several groups and being respectively the central site, first neighbor, second neighbor, ... xth neighbor in the successive groups. The fact that each site belongs, in a different category, in several basic groups couples each site to all neighbors out to the truncation point. This is in the case cited the xth neighbor.

When charges are present, as in the present case, we proceed in the same way. From Eq. (7.7.1) it is seen that the weighting factor for each configuration must include both the energetic terms described above and the relevant electrostatic interactions. That is, the summation over configurations is made with a Boltzmann weighting factor of the total energy for each particular configuration. The partition function expressed in Eq. (7.7.1) thereby gives the electrical free energy of the polyion including those electrostatic effects which result from the helix-coil transition. This is, however, not the only contribution to the total change in free energy. Consideration must also be given the contribution to the Helmholtz free energy of the polyion by the free energy of the free hydrogen ions eligible to take part in neutralization of the acid groups of the polyion, by the electrical free energy due to interactions between portions of the net charge of the polyion and by the chemical free energy associated with the state of ionization of the polyion computed from a reference state in which the polyacid is completely undissociated. We have shown that the pH and degree of ionization, α, of a polyacid are related to the intrinsic acid dissociation constant, K_a^0, and the electrostatic free energy by

$$0.434 \ln \lambda = \mathrm{pH} - pK_a^0 = \frac{0.434}{ZkT} \frac{\partial A}{\partial \alpha}. \qquad (7.7.20)$$

The free energy A also contains contributions from the nonelectrostatic interactions which create the helical configuration, but only the electrostatic interactions are explicitly contained in the indicated derivative. Further progress requires the explicit evaluation of Ξ, a task to which we now turn.

Zimm and Bragg have shown[41] that the internal hydrogen bonding of nonelectrolytic polypeptides can be treated by the use of a model that includes nearest neighbor interactions only. If it is assumed that only the hydrogen bonds characteristic of the α-helix can form, the internal bonding is completely described by the specification of the bond state of the oxygen of each amide group. A digit 1 is used to represent a bonded oxygen and 0 for an unbonded oxygen. Thus a state of the chain is described, with respect to internal hydrogen bonding, by a state number $\{c\}$ constructed from the occupation variables c_i and which is a sequence of ones and zeros, one apiece for each amide group. $Q^0(\{c\})$ is represented as a product of the following factors: (1) unity for every 0 in the state number, (2) s for every 1 that follows a 1, and (3) σs for every 1 that follows one or more 0's. The equilibrium constant s was defined previously while σ is a factor less than unity representing the extra difficulty involved in the starting of a section of helix compared to the extending of it. In the case of polybenzylglutamate the initiation parameter σ was in the neighborhood of 10^{-4}.[41]

The electrostatic part of the partition function is more complicated. As in the preceding, the electrolytic state number can be written as a sequence of 0's and 1's, indicating the absence or presence of charge on the corresponding segments of the chain. The electrostatic energy may clearly be written in terms of the $\{\eta\}$ as

$$W(\{c\},\{\eta\}) = \sum_{i>j} \frac{q^2}{DR_{ij}} e^{-\kappa R_{ij}} \eta_i \eta_j. \qquad (7.7.21)$$

The summation in Eq. (7.7.21) extends over all pairs of interacting charges. Following the procedure introduced earlier in this chapter we consider electrostatic interactions between small numbers of near neighboring sites only, accepting the necessary inaccuracies that arise from the neglect of the more numerous but weaker distant interactions. In the present case we retain the interactions of each charge with eight of its neighbors, four on each side. To estimate the error involved in this truncation we compare the contributions to $W(\{c\}, \{\eta\})$ of these four nearest neighbors on a polybenzylglutamate helix with the value obtained by summing the Debye–Hückel potential out to infinity under the conditions appropriate to the experiments to be discussed later. If the solvent has a dielectric constant of 50, the four neighbor approximation accounts for 65% of $W(\{c\}, \{\eta\})$ when the salt concentration is 0.133M NaCl, 90% of the electrostatic energy is contributed when the salt concentration is 1.33 M, and at the lower concentration 0.0133 M NaCl only 37% of the electrostatic energy is contributed by the first four neighbors.

The inaccuracies generated by the truncation of the series of Eq. (7.7.21) are perhaps less important than the uncertainties arising from the choice of an appropriate dielectric constant. Although the bulk dielectric constant of a medium containing two thirds water and one third dioxane is about 50, the medium immediately surrounding the polypeptide chain may be different because of selective concentration due to adsorption or salting out effects; the saturation effects of the strong electric fields around ions are unknown, and in any case the polypeptide chain itself occupies a significant part of the space that the electric lines of force must traverse.

Although the figures provided above indicate an error of considerable magnitude in the absolute electrostatic energy, this is not the quantity of greatest interest since the transition depends upon *the difference in energy* of the two isomeric configurations. Since the errors will be in the same direction with both forms, a considerable compensation of errors is likely to result.

Thus, with some quantitative uncertainty in the fundamental formulae it is appropriate to choose models featuring simplicity rather than refinement.

For the structure of the polypeptide helix we use the paremeters proposed by Pauling and Corey[42] for a helix with 3 to 6 residues per turn. It will be assumed that the side chains of the glutamic acid residues are extended straight out along the radii of the helix because of electrostatic repulsion. The charges are then located at 6.5 A from the helix axis and at rotations of 100° apart with an axial translation of 1.5 A per residue. The distances between one charge and its first six neighbors along the chain are 10.1 A, 13.1 A, 7.9 A, 7.5 A, 14.3 A, and 14.4 A. It is clear that the first neighbors for charge interactions are the fourth and third neighbors along the skeleton. For the random form of the chain it will be assumed that the electrostatic forces favor a locally extended form of the chain with the backbone having 3.6 A per residue. It will be further assumed that the charges are arranged around the chain in a helix of three residues per turn. The distance from the charges to the axis is 5 A and the distances between one charge and each of its first three neighbors is about 10 A, and the fourth neighbor is 16 A away. Kinked configurations of the chain, although admittedly present to some extent, are nevertheless neglected by this model.

7.7.3 MATHEMATICAL METHOD

The natural method for the mathematical treatment of the problem as outlined above is the matrix analysis of the partition function which we have already exploited so much in this chapter.

We proceed by labeling the states of each segment with five binary digits. The first digit comes from the hydrogen bond occupation number and the others from the electrolytic state number corresponding to the charge on the segment in question and the three segments preceding it in the chain. The state vector then consists of the array of thirty-two statistical weights corresponding to the set of all values of the five digit binary numbers. With this labeling system, the first sixteen components of the state vector correspond to charge states of the random coil while the second sixteen components correspond to charge states of the α-helix. The interaction matrix, $\mathbf{M}$, is of order 32×32 and may be partitioned into four 16×16 submatrices as shown:

$$\mathbf{M} = \left\| \begin{array}{c:c} \mathbf{C} & \mathbf{J}_1 \\ \hdashline \sigma s\, \mathbf{J}_2 & \mathbf{H} \end{array} \right\| \tag{7.7.22}$$

where the submatrices $\mathbf{C}$ and $\mathbf{H}$ contain, respectively, only electrostatic interactions between random coil configurations or between helix configurations. The submatrices $\mathbf{J}_1$ and $\mathbf{J}_2$ contain the mixed helix-coil interactions. Since σ is small, the off diagonal matrices $\mathbf{J}_1$ and $\mathbf{J}_2$ may be considered to be perturbations. To proceed to find the dominant eigenvalue of $\mathbf{M}$, let $\mathbf{C}$ and $\mathbf{H}$ be diagonalized by similarity transformations with matrices $\mathbf{U}$ and $\mathbf{V}$,

$$\mathbf{U}^{-1}\mathbf{C}\mathbf{U} = \begin{Vmatrix} c_0 & 0 \dots \\ c & 0_1 \\ \vdots & \ddots \end{Vmatrix} \tag{7.7.23}$$

$$\mathbf{V}^{-1}\mathbf{H}\mathbf{V} = \begin{Vmatrix} \mathfrak{h}_0 & 0 \dots \\ 0 & \mathfrak{h}_1 \\ \vdots & \ddots \end{Vmatrix} \tag{7.7.24}$$

with the eigenvalues so ordered that c_0 and $\mathfrak{h}_0$ are the largest in their respective matrices. $\mathbf{M}$ may now be transformed by a similarity transformation with the matrices

$$\begin{Vmatrix} \mathbf{U}^{-1} & 0 \\ 0 & \mathbf{V}^{-1} \end{Vmatrix}, \qquad \begin{Vmatrix} \mathbf{U} & 0 \\ 0 & \mathbf{V} \end{Vmatrix} \tag{7.7.25}$$

and the new matrix has the same eigenvalues as $\mathbf{M}$. The dominant eigenvalue of $\mathbf{M}$ is denoted $\mathfrak{m}$ and is obtained from the secular equation of the transformed matrix by neglecting elements containing σ in all places except where they occur in parallel with the small but important elements $c_0 - \mathfrak{m}$ and $\mathfrak{h}_0 - \mathfrak{m}$. The secular equation reduces to

$$(c_1 - \mathfrak{m})(c_2 - \mathfrak{m}) \cdots (\mathfrak{h}_1 - \mathfrak{m})(\mathfrak{h}_2 - \mathfrak{m}) \cdots \begin{vmatrix} c_0 - \mathfrak{m} & \mathfrak{l}_1 \\ \sigma s \mathfrak{l}_2 & s \mathfrak{h}_0 - \mathfrak{m} \end{vmatrix} = 0 \tag{7.7.26}$$

where $\mathfrak{l}_1$ and $\mathfrak{l}_2$ are the upper left-hand corner elements of the transforms of $\mathbf{J}_1$ and $\mathbf{J}_2$. $\mathfrak{l}_1$ and $\mathfrak{l}_2$ are related to the vectors $\mathbf{u}$, $\mathbf{v}$ consisting of the first columns of $\mathbf{U}$ and $\mathbf{V}$, and to $\mathbf{u}^{-1}$ and $\mathbf{v}^{-1}$, the first rows of $\mathbf{U}^{-1}$ and $\mathbf{V}^{-1}$, by

$$\mathfrak{l}_1 = \mathbf{u}^{-1}\mathbf{J}_1\mathbf{v}, \qquad \mathfrak{l}_2 = \mathbf{v}^{-1}\mathbf{J}_2\mathbf{u}. \tag{7.7.27}$$

The elements of $\mathbf{C}$, $\mathbf{H}$, $\mathbf{J}_1$, and $\mathbf{J}_2$ depend upon the distance between segments in the indicated molecular configurations. For a given difference of number of segments, the intersegment distance is smallest in the helix and largest in the coil. Thus the elements of the interaction matrices are intermediate in size between the corresponding elements of $\mathbf{C}$ and $\mathbf{H}$. Since all nonzero elements are positive and not very different from unity, it is consistent to approximate $\mathbf{J}_1$ and $\mathbf{J}_2$ by the average of $\mathbf{C}$ and $\mathbf{H}$. This leads to the approximate formulas

$$\mathfrak{l}_1 = \tfrac{1}{2}(c_0 + \mathfrak{h}_0)(\mathbf{u}^{-1}\mathbf{v}) \tag{7.7.28}$$

$$\mathfrak{l}_2 = \tfrac{1}{2}(c_0 + \mathfrak{h}_0)(\mathbf{v}^{-1}\mathbf{u}).$$

The principal eigenvalue of $\mathbf{M}$ is the largest root of Eq. (7.7.26) which is the large root of the 2×2 determinant. With the approximation of Eq. (7.7.28) this is the larger value of

$$\mathfrak{m} = \tfrac{1}{2}[\mathfrak{c}_0 + s\mathfrak{h}_0 \pm \{(\mathfrak{c}_0 - s\mathfrak{h}_0)^2 + \sigma s(\mathfrak{c}_0 + \mathfrak{h}_0)^2 (\mathbf{u}^{-1}\mathbf{v})(\mathbf{v}^{-1}\mathbf{u})\}^{\frac{1}{2}}].$$

$$(7.7.29)$$

As usual, the partition function is given by the Zth power of the dominant eigenvalue. In addition to Eq. (7.7.2) defining the degree of ionization, the fraction ϑ of segments whose oxygen atoms are involved in hydrogen bonds is

$$\vartheta = \frac{1}{Z}\frac{\partial \ln \Xi}{\partial \ln s} = \frac{\partial \ln \mathfrak{m}}{\partial \ln s}. \tag{7.7.30}$$

Neither $\mathfrak{c}_0$, $\mathfrak{h}_0$ nor the vectors $\mathbf{u}$, $\mathbf{v}$ depend upon s, so that the differentiation of $\mathfrak{m}$ with respect to s presents no difficulties. The differention displayed in Eq. (7.7.2) requires the derivatives of $\mathfrak{c}_0$ and $\mathfrak{h}_0$ with respect to λ. These can be obtained from the matrix equations

$$\frac{\partial \mathfrak{c}_0}{\partial \lambda} = \mathbf{u}^{-1}\left(\frac{\partial \mathbf{C}}{\partial \lambda}\right)\mathbf{u} \tag{7.7.31}$$

$$\frac{\partial \mathfrak{h}_0}{\partial \lambda} = \mathbf{v}^{-1}\left(\frac{\partial \mathbf{H}}{\partial \lambda}\right)\mathbf{v}. \tag{7.7.32}$$

7.7.4 CALCULATIONS AND RESULTS

The only elements of $\mathbf{C}$ and $\mathbf{H}$ that are nonzero are those whose indices in binary notation are of the form $C_{abcx,\ yabc}$ where x and y may be the same or different. The nonzero elements of $\mathbf{C}$ and $\mathbf{H}$ are the factors which enter the corresponding term of the partition function when a segment in charge state x is added to a chain ending in four successive segments of charge states $y, a, b. c$. Included in these factors are the Boltzmann weights of the corresponding terms in the energy expression. For the purposes of numerical calculation we have taken $T = 298°K$, $D = 50$. From Eq. (7.7.16) the reciprocal shielding length κ is 0.50, 0.158, and 0.050 A^{-1} for salt concentrations of 1.33 M, 0.133 M, and 0.0133 M respectively.

As an example of the calculation of a typical element consider the case of 0.133 M salt. We shall construct $H_{1011,0101}$. This matrix element contains the following factors: (1) λ for the missing hydrogen ion on the added segment ($x = 1$), (2) the factor 0.8, the rounded off value of the Boltzmann weight for the electrostatic interaction between the added charge and the charge on the next preceding segment at distance 10.1 A, and (3) the factor 0.7, the Boltzmann weight for the interaction of the added charge and the charge on the third preceding segment at a distance of 7.9 A. All other factors are unity since there are no charges on the second and fourth preceding segments and interactions beyond the fourth neighbor are neglected.

The actual Boltzmann weights used are given in Table 7.7.1. Each of the entries represents the interaction between a charge on the added segment and the charge on the indicated one of the four preceding segments. The principal eigenvalue and the left and right eigenvectors were found numerically for **C** and **H** for various values of λ by iteration on an arbitrary initial vector. The numerical work was carried out on an IBM 650 automatic computer. From the results obtained, the degrees of ionization were computed using Eq. (7.7.4), also by the use of the IBM 650. The results may be extended by the theorem of Lifson, Kaufman, and Lifson[43] which states that the titration curve is symmetrical about its midpoint. Selected examples of the results are shown in Table 7.7.2.

It should be noted that the course of the transition depends upon the value of the parameter s which expresses the stability of the internal hydrogen bonds. If s is either too large or too small the chain will remain in one or the other of its two forms at all degrees of ionization. To obtain the transition between configurations, s must be selected so that the helix and coil forms are of equal stability somewhere near the midpoint of the titration curve, i.e., $c_0 = s\mathfrak{h}_0$ at this point. Once s has chosen α, ϑ and all other quantities of interest are easily computed by Eqs. (7.7.2), (7.7.29) and (7.7.30).

It is convenient to separate the discussion of the results into several categories.

1. *Form of the transition.* The formulas of the preceding paragraphs show that the system under investigation can be analyzed by separation into two parts: the calculation of the titration curve of a polyelectrolyte, and the calculation of the transition behavior between two forms of the molecule. The transition is determined by the eigenvalue $\mathfrak{m}$, which is the solution of an equation identical in form to the corresponding equation in the case of the uncharged polypeptide. This can be shown by rewriting Eq. (7.7.29) as follows:

$$\mathfrak{m}' = \tfrac{1}{2}[1 + s' \pm \{(1 - s')^2 + 4\sigma's'\}^{\frac{1}{2}}] \tag{7.7.33}$$

where $\mathfrak{m}'$ and s' are defined by Eqs. (7.7.12) and (7.7.13) and

$$\sigma' = \sigma(\mathbf{u}^{-1}\mathbf{v})(\mathbf{v}^{-1}\mathbf{u})(c_0 + \mathfrak{h}_0)^2/4c_0^2\mathfrak{h}_0^2. \tag{7.7.34}$$

Equation (7.7.33) is identical with Eq. (22) of ref. 41. Thus the fact that the molecule is a polyelectrolyte introduces only one novel feature into the description of the transition. This is the presence of a new independent variable which may be taken either as the fraction of charged groups, α, or as λ, which is effectively the reciprocal of the hydrogen ion activity. This variable, however, affects the transition only in so far as it alters the canonical variable, s'.

The variable s' has a significance similar to that attached to s in the non-electrolyte case; s' can be thought of as the equilibrium constant for the addition of a segment, initially in the random form, to the end of a helical section, the molecule being allowed to ionize as it will. From the temperature variation of s' one can compute the net heat of such an addition, ΔH, by means of the equation

$$d \ln s'/dT = \Delta H/RT^2 \qquad (7.7.35)$$

Data of Doty et al.[44] on polyglutamic acid show that the variation of s' with temperature is very slight. If σ is assumed to be approximately the same as the value for polybenzylglutamate,[41] Eq. (7.7.35) leads to an estimate of only -50 cal/mol for the heat of addition of a segment to a helical section for polyglutamic acid in the mixture of water, dioxane, and NaCl studied by Doty et al.[44]

2. *The transition and the titration curve.* In Fig. 7.7.2 are shown the results of the calculation for 0.133 M salt, with σ assumed to be 10^{-4} and s adjusted to 1.114, a value that causes the transition to take place in the neighborhood of fifty percent ionization. The transition causes a small anomaly in the titration curve. The frequency of alternation of helical and random regions goes through a maximum at the midpoint of the transition, just as in the case of a nonelectrolyte; on the average, the maximum has a value of nearly one helical section in every two hundred segments.

Since the titration anomaly is rather small when plotted in the usual manner, it is helpful to adopt other representations of the curve that emphasize the changes produced by the transition. Such a representation is shown in Fig. 7.7.3, where the deviation of the pH from the ideal titration curve of a monobasic acid is plotted against the degree of ionization. Here the transition introduces a sigmoid region into the otherwise smoothly rising curve. The fact that the middle of the curve is nearly horizontal is an accidental result of the choice of parameters, and has no special significance.

Also shown in Fig. 7.7.3 are the titration data taken by Wada[45] on polyglutamic acid in various concentrations of salt. The parameter s has been adjusted to give the best agreement with these data. When one remembers that this is the *only* parameter that has been adjusted, the value of $\sigma = 10^{-4}$ having been taken from experiments on polybenzylglutamate in nonaqueous solvents,[41] the agreement of the theory and experiment must be considered astonishingly good.

3. *Titration curves of the individual forms.* The agreement just noted between theory and experiment is the more remarkable in view of the approximations in the model. Inspection of Fig. 7.7.3 shows that we have succeeded

in reproducing the experimental titration curves for the helical form almost exactly at the two lower salt concentrations, with some modest discrepancies at the highest salt concentration where the non-ideality is small anyway. Though the titration curve of the random form is reproduced somewhat less well, the discrepancies are still minor. It should be remembered that no adjustable parameters have entered these calculations, which are preliminary to and distinct from any discussion of the transition behavior.

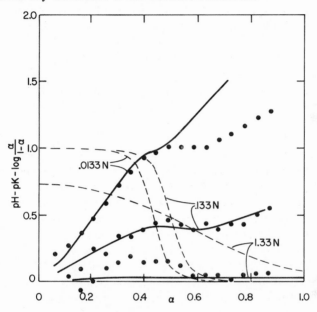

FIG. 7.7.3. Titration curves plotted as the deviation from the monobasic acid curve. The solid lines are theoretical curves and the dashed curves represent the theoretical values of ϑ as a function of α. The data comes from the work of Professor A. Wada. [Reproduced from ref. 39.]

Let us recall the two principal approximations that have been made. These are truncation of the sum over the electrostatic interactions to four members, and the use of the linearized Debye–Hückel theory to represent the interactions. The effect of the first approximation has already been estimated (Section 7.7.1); the electrostatic interaction energy should be too small by about a factor of three in the case of 0.0133 M salt. The titration curve should therefore rise about three times more steeply in Fig. 7.7.3 than our calculations. On the other hand the linearization of the Poisson–Boltzmann equation is also most serious at the lowest salt concentration, and might be expected to cause deviations in the opposite direction. It is known that in the neighborhood of a highly charged macro-ion the electrostatic potential greatly exceeds kT, with the result that a dense atmosphere of counter ions surrounds the polymer. Therefore the internal ionic strength exceeds that of the bulk solution and the screening of the charges by the counterions should be greater than calculated from the Debye–Hückel theory using the over-all concentration of counterions. The magnitude of the counterion association is dependent upon the geometry of the macromolecule. The calculations of Fuoss et al.[46] for a rod lead to a predicted surface potential of 26 kT/q for the case of no added external electrolyte. In contrast, the uniform sphere model of Wall and Berkowitz[47] leads to a maximum potential of 1.30 kT/q in the presence of 0.133 M NaCl and 3.53 kT/q in the presence of 0.0133 M NaCl. The Wall and Berkowitz model leads to a domain binding of 40–50% whereas the rod geometry would have much more extensive counter ion clustering close to the macro-ion. In both cases the local clustering is large enough that our neglect of the electrostatic interactions beyond the fourth neighbor may have providentially approximated more closely to the actual screening effects at low salt concentrations than our attempted use of the Debye–Hückel screening constant.

4. *A simplified approximation.* Since sixteenth order matrices are cumbersome, and, as we have seen, only the principal eigenvalue and its derivatives are important to the results, it becomes interesting to examine the adequacy of simpler approximations. One such approximation consists of lumping the interactions of a charge with all of its neighbors together in one term, as if all the neighboring charges were combined into a single charge at a single point. The distance of this point is then chosen so that the total electrostatic energy of the completely charged molecule is the same as without the lumping. This problem can then be described by a matrix of order two:

$$\left\| \begin{matrix} 1 & 1 \\ \lambda & \lambda \hat{u} \end{matrix} \right\| \tag{7.7.36}$$

where

$$\hat{u} = \exp\left[-\frac{1}{kT} \sum_{i \neq j} \frac{q^2}{DR_{ij}} e^{-\kappa R_{ij}} \right] \tag{7.7.37}$$

This mathematical problem has been treated extensively in Section 7.3. The degree of ionization, α, is given by

$$\alpha = \frac{\lambda[\hat{u} + (\lambda \hat{u}^2 - \hat{u} + 2)J^{-\frac{1}{2}}]}{\lambda \hat{u} + 1 + J^{\frac{1}{2}}} \tag{7.7.38}$$

$$J = (\lambda \hat{u} - 1)^2 + 4\lambda$$

We have calculated a few values from this equation, cutting off the sum in Eq. (7.7.37) at the fourth term and using the values of Table 7.7.1. The results,

shown in Table 7.7.3, are in remarkable agreement with the previous calculations, which used the matrix of sixteenth order.

Part of the explanation of this rather surprising agreement is probably to
be found in Table 7.7.4, where the distribution of charges derived from the
sixteenth order matrix, is compared with the random distribution having the
same average charge. The matrix distribution was calculated from the
products of the corresponding elements in the left and right eigenvectors of
the matrix **H**, which were automatically generated in the course of finding the
eigenvalue. The numbers in the random column are simply the products of
the probabilities of finding 0 or 1, 0.3355, and 0.6645, respectively. One is
immediately struck by the degree to which the two distributions are in
agreement. We may conclude that consideration of the detailed features of
the charge distribution is not essential to obtaining reasonably accurate
results, at least in this case where the individual potentials of interactions are
not much larger than kT.

5. *The determination of σ' from experimental data.* From Eq. (7.7.33)
we can derive the approximate relation

$$\frac{1}{\sigma'} = 16\left(\frac{\partial^2 m'}{\partial s'^2}\right)^2_{s'=1} \qquad (7.7.39)$$

at $s' = 1$. By examination of Fig. 7.7.1 and Eq. (7.7.19) it is seen that
$(\partial \ln m'/\partial \ln s')$ is determined by the ratio of the line segments CE and BE.
A plot of this ratio versus s' yields the desired second derivative as the slope
at the point $s' = 1$. Figure 7.7.4 shows this method applied to the data
represented in Fig. 7.7.1 for the case of 0.0133 M NaCl; we find $\sigma' = 1.4 \times$
10^{-3}. This is to be compared with the value $\sigma = 10^{-4}$ assumed in the theoretical calculations and which was successful when the external salt concentration
was 0.133 M NaCl. Uncertainties in the graphical work are probably too
small to account for this discrepancy. We have no ready explanation for the
disagreement.

6. *Concluding remarks.* In the calculations presented there is a direct
analogy between the electrolytic and nonelectrolytic transitions which has
been exploited at several points. The comparison may be completed by noting
that a constant temperature, the pH plays the role for the electrolytic case that
the temperature plays for the uncharged polymer. As can be seen clearly
from the results, the breadth of the transition depends upon the ionic strength
of the medium. This solvent effect may also in principle be mimicked for
uncharged polypeptides by the use of media in which the net heat of formation
of the hydrogen bond differs with solvent composition.

This latter observation serves as an introduction to one of the subtler
approximations of the theory. It has been assumed in the foregoing that the
energy of a hydrogen bond is independent of both the charge state of the

polyion and the ionic strength. It is not difficult to imagine that a hydrogen bond adjacent to a charge may differ in energy from one far from a charge. It is to be anticipated that such an effect is of secondary importance compared to our ignorance of the local dielectric constant and other pertinent variables. (See the general discussion of Chapter 5).

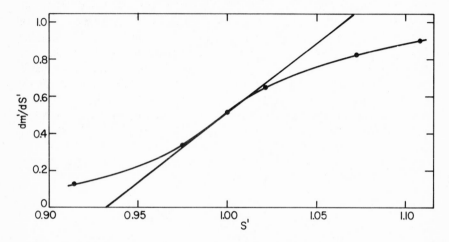

FIG. 7.7.4. The evaluation of σ' from the slope of dm/ds' versus s'. See Fig. 7.7.1. [Reproduced from ref. 39.]

The statistical theory of the transition offers no difficulties. It is the calculation of the electrostatic free energy of the polyion which is complex. This is due to the non-linearity of the Poisson–Boltzmann equation and the difficulties in handling the coordinate representation of the random coil configurations as well as to basic uncertainties in fundamental parameters. Good agreement between experiment and theory was obtained. Even if this agreement is regarded as fortuitous, the combinations of experimental analysis and statistical theory makes the titration curve a useful way of studying the energetics of the transition.

We would like to call attention to another treatment of this problem by Peller.[48]

TABLE 7.7.1[a]

BOLTZMANN FACTORS FOR VARIOUS NEIGHBORS AT SEVERAL
SALT CONCENTRATIONS

Neighbor No.	1.33 M NaCl	0.133 M NaCl	0.0133 M NaCl
Helix			
1	0.99	0.8	0.51
2	1.00	0.9	0.64
3	0.97	0.7	0.39
4	0.97	0.6	0.36
Coil			
1, 2, 3	0.99	0.8	0.5
4	1.00	0.95	0.75

[a] Reproduced from ref. 39.

TABLE 7.7.2[a]

DEGREE OF DISSOCIATION, α, AS A FUNCTION OF THE
RECIPROCAL ACTIVITY OF HYDROGEN ION, $\lambda = 10^{(pH - pK)}$

λ	1.33 M Salt		0.133 M Salt		0.0133 M Salt	
	α coil	α helix	α coil	α helix	α coil	α helix
0.01	0.009894	0.009886	0.00980	0.00971	0.00957	0.00951
0.1	0.09046	0.08986	0.0823	0.0783	0.0710	0.0682
0.4	0.2822	0.2778	0.2263	0.2035	0.1672	0.1548
1.0	0.4926	0.4829	0.3711	0.3229	0.2510	0.2278
2.0	—	—	0.4949	0.4241	0.3210	0.2869
2.5	0.7055	0.6937	0.5354	0.4576	—	—
5.0	0.8263	0.8166	0.6575	0.5625	0.4180	0.3679
10.0	—	—	0.7660	0.6645	0.4930	0.4300
20.0	0.9497	0.9459	0.8524	0.7583	0.5682	0.4922
40.0	—	—	0.9139	0.8384	0.6422	0.5544
80.0	0.9869	0.9858	0.9527	0.9002	—	—
100.0	—	—	0.9614	0.9160	0.7361	0.6364
150.0	0.9929	0.9924	—	—	—	—
400.0	0.9974	0.9971	—	—	—	—

[a] Reproduced from ref. 39.

TABLE 7.7.3[a]

COMPARISON OF 2ND AND 16TH ORDER MATRIX CALCULATIONS

λ	α coil		α helix	
	Eq. (7.7.32)	16th order matrix	Eq. (7.7.32)	16th order matrix
		0.133 M		
2.057	0.500	0.495	—	—
3.304	—	—	0.500	0.50
10	0.772	0.766	0.652	0.664
		0.0133 M		
10.2	0.500	0.495	—	—
22.2	—	—	0.500	0.50

TABLE 7.7.4[a]

COMPARISON OF EXACT CHARGE DISTRIBUTION ON HELIX WITH A
RANDOM DISTRIBUTION (0.133 M SALT, $\lambda = 10$, $\alpha = 0.6645$)

Charge configuration[b]	Probability, matrix calculation	Probability, random distribution	Difference
0000	0.00844	0.01267	−0.00423
0001	0.02401	0.02509	−0.00108
0010	0.02124	0.02509	−0.00385
0011	0.05229	0.04970	0.00259
0100	0.02124	0.02509	−0.00385
0101	0.05593	0.04970	0.00623
0110	0.04641	0.04970	−0.00329
0111	0.10600	0.09844	0.00756
1000	0.02401	0.02509	−0.00108
1001	0.04953	0.04970	−0.00017
1010	0.05593	0.04970	0.00623
1011	0.10012	0.09844	0.00168
1100	0.05231	0.04970	0.00261
1101	0.10012	0.09844	0.00168
1110	0.10600	0.09844	0.00756
1111	0.17642	0.19498	−0.01856

[a] Reproduced from ref. 39.

[b] 0 means no charge, 1 means a charge on one of four successive contiguous groups on one turn of the helical polypeptide.

Appendix 7A. Maximum Eigenvalue of the Matrix **V**

In Section 7.5 we saw that the calculation of certain thermodynamic properties of an alternating polyampholyte was reduced to the problem of finding the maximum eigenvalue ϕ of the matrix **V** given by Eq. (7.5.33). The procedure we used to obtain ϕ near the isoelectric point was as follows.[38] An initial approximation $y^{(1)}$ to the eigenvector corresponding to ϕ is obtained by setting equal to zero all terms smaller than e^J in the exponentials of **V** and solving the resultant simpler eigenvalue problem exactly. For the purpose of estimating the relative magnitude of various exponentials involving K as well as J, it was assumed that K was of the order of magnitude of $-J/2$; in typical cases this would be an upper limit to $|K|$. Using the approximate eigenvector $y^{(1)}$ in the equation

$$Ay^{(1)} = \phi^{(1)} y^{(2)} \tag{7A.1}$$

we obtain a first-order eigenvalue $\phi^{(1)}$ and a second-order eigenvector $y^{(2)}$. The eigenvalue $\phi^{(1)}$ is chosen so that the largest term of $y^{(1)}$ is equal to the corresponding term of $y^{(2)}$. The second-order eigenvector $y^{(2)}$ is then put into the equation

$$Ay^{(2)} = \phi^{(2)} y^{(3)} \tag{7A.2}$$

and the procedure continued until any desired degree of accuracy is attained. It can be shown that, for any real matrix, the iterative procedure described here converges to the dominant eigenvalue when the eigenvalue is real. Near the isoelectric point the initial approximate eigenvector used was

$$y^{(1)} = \begin{Bmatrix} e^{-K} \\ \lambda_2^{1/2} e^J \\ \lambda_1^{1/2}\lambda_2^{1/2} e^{(3J/2)+K} \\ \lambda_1^{1/2} \end{Bmatrix} \tag{7A.3}$$

and ϕ converges to the expression given in Eq. (7.5.36).

Further from the isoelectric point, a procedure similar to the above was adopted, but it was assumed that $\lambda_2 \gg \lambda_1$, $\lambda_1 < 1$, and $\lambda_1\lambda_2 = 0(1)$. Therefore, the initial approximation to the eigenvector was determined by suppressing all terms of **V** except those with the largest powers of λ_2 (estimating $\lambda_1 \sim \lambda_2^{-1}$). Solving the resultant eigenvalue problem to first order, it is found that

$$y^{(1)} = \begin{Bmatrix} \lambda_2^{1/2} \\ e^K \lambda_2 \\ e^{K+(J/2)}\lambda_1^{1/2}\lambda_2 \\ \lambda_1^{1/2}\lambda_2^{1/2} \end{Bmatrix} \tag{7A.4}$$

and under these conditions converges to Eq. (7.5.39).

Appendix 7B. Charge Fluctuations

Fluctuations of the charge on a segment of a polyion may be calculated in terms of the grand partition function as given in Eq. (7.5.17). We need concern ourselves only with the isoelectric point. We consider a statistical element of the polymer with Z/N charge sites for which the grand partition function Ξ_s is, therefore, given by (7.5.17), but with Z replaced by Z/N. We assume that the statistical elements are sufficiently long that end effects may be neglected in calculating the grand partition function, and therefore Ξ_s is related to Ξ, the grand partition function for the entire polymer, by

$$\ln \Xi_s = \frac{1}{N} \ln \Xi. \tag{7B.1}$$

Differentiation of the logarithm of Ξ_s with respect in $\ln \lambda_1$ and $\ln \lambda_2$ leads to, as may be seen by inspection of Eq. (7.5.17),

$$\left(\frac{\partial^{l+l'} \ln \Xi_s}{\partial \ln \lambda_1^l \, \partial \ln \lambda_2^{l'}} \right) = \frac{1}{N} \left(\frac{\partial^{l+l'} \ln \Xi}{\partial \ln \lambda_1^l \, \partial \ln \lambda_2^{l'}} \right)$$

$$= \left\langle \left(\frac{Z\alpha_1'}{2} \right)^l \left(\frac{Z\alpha_2'}{2} \right)^{l'} \right\rangle. \tag{7B.2}$$

The charge fluctuations of a statistical element of the polymer at the isoelectric point may be written

$$\langle q^2 \rangle = \left\langle \left[\frac{Z}{2N} (\alpha_2' - \alpha_1') \right]^2 \right\rangle$$

$$= \left\langle \left(\frac{Z\alpha_2'}{2N} \right)^2 \right\rangle + \left\langle \left(\frac{Z\alpha_1'}{2N} \right)^2 \right\rangle - 2 \left\langle \left(\frac{Z\alpha_1'}{2N} \right) \left(\frac{Z\alpha_2'}{2N} \right) \right\rangle \tag{7B.3}$$

which, using Eq. (7B.2), is equivalent to

$$\langle q^2 \rangle = \frac{1}{N} \left[\left(\frac{\partial^2 \ln \Xi}{\partial \ln \lambda_1^2} \right)_{\lambda_2} + \left(\frac{\partial^2 \ln \Xi}{\partial \ln \lambda_2^2} \right)_{\lambda_1} - 2 \left(\frac{\partial^2 \ln \Xi}{\partial \ln \lambda_1 \, \partial \ln \lambda_2} \right) \right]. \tag{7B.4}$$

Introducing expressions (7B.2) for the first derivatives of $\ln \Xi$, and remembering that we are at the isoelectric point,

$$\langle q^2 \rangle = \frac{Z}{N} \left[\left(\frac{\partial \langle \alpha_1' \rangle}{\partial \ln \lambda_1} \right)_{\lambda_2} - \left(\frac{\partial \langle \alpha_1' \rangle}{\partial \ln \lambda_2} \right)_{\lambda_1} \right]. \tag{7B.5}$$

Since the macroscopic value of α_1' is exactly what we mean by $\langle \alpha_1' \rangle$, Eq. (7B.5) is equivalent to Eq. (7.5.53) of the main text, when Eq. (7.5.27) for $(\partial \bar{Q}/\partial pH)_{a_{X^+}, a_{Y^-}}$ is employed.

Appendix 7C.   Details of the Calculation of Titration Properties

In this appendix will be calculated $Q_{ijk}^{(\sigma)}$ and $\mathscr{E}_{ijk}^{(\alpha)}$, the partition function and minimum energy of a sequence of i successive groups of type σ, on which there are j charges, including k on terminal groups. We shall also discuss the modifications to the theory presented in the main text, so as to make it applicable to polymers containing adjacent sequences of two successive groups.

$Q_{ijk}^{(\sigma)}$ and $\mathscr{E}_{ijk}^{(\alpha)}$ may be obtained simply by summing, with a Boltzmann weighting factor, the ways the required numbers of charges can be arranged in a sequence of length i, with k charges on terminal groups. No attempt was made to write down the general form of $Q_{ijk}^{(\sigma)}$ or $\mathscr{E}_{ijk}^{(\alpha)}$. In calculating the energies of the various configurations, only nearest neighbor interactions were considered, and the interaction energy was taken to be χ, the parameter already introduced for that purpose in the main text. Table 7.C.1 shows some typical values of Q_{ijk} and $\mathscr{E}_{ijk}$, for $i = 4$.

TABLE 7.C.1 [a]

PARTITION FUNCTIONS AND ENERGIES FOR SEQUENCES OF LENGTH 4

i	j	k	Q_{ijk}	$\mathscr{E}_{ijk}$
4	0	0	1	0
4	1	0	2	0
4	1	1	2	0
4	2	0	1	χ
4	2	1	$2 + 2e^{-\chi/kT}$	0
4	2	2	1	0
4	2	2	1	0
4	3	1	2	2χ
4	3	2	2	χ
4	4	2	1	3χ

[a] Reproduced from ref. 2.

Sequences of length 2 cannot be treated just as are the other sequences because in adjacent sequences of length 2 the behavior of the terminal sites differs greatly from the average over terminal sites on all sequences. This is because the charging of both terminal groups of a sequence of 2 necessarily requires the introduction of an increase in energy due to the interaction of these groups. In successions of sequences of 2, this increase in energy becomes larger than the decrease caused by adjacent opposite charges at the ends of the sequences. Therefore we consider separately isolated sequences of 2, which are adjacent only to sequences of length other than 2, consecutive pairs of sequences of 2, etc. Therefore, let $n_2^{(\sigma)}$, $n_2^{(\sigma,\sigma')}$, $n_2^{(\sigma,\sigma',\sigma)}$, $\cdots$, be the numbers of isolated sequences of 2 of type σ; the numbers of pairs of sequences of 2; the numbers of triples of sequences of 2, 2 of which are of type σ and the other of which is of type σ'; $\cdots$. By the methods used in deriving Eq. (7.6.6), one may show that

$$n_2^{(\sigma\cdots)} = \frac{Z}{2}(1 - P_{1,1} + P_{1,1}^2)^2(1 - P_{1,1})^{r+1}P_{1,1}^r \qquad (7C.1)$$

where r is the number of consecutive sequences of 2 in a group of sequences whose number is $n_2^{(\sigma,\cdots)}$.

TABLE 7.C.2 [a]

Partition Functions and Energies for Sequences of Length 2

A. Isolated				
	j	k	Q_{ijk}	$\mathscr{E}_{ijk}$
	0	0	1	0
	1	1	2	0
	2	2	1	x

B. Pairs				
j	j'	k	Q_{ijk}	$\mathscr{E}_{ijj'k}$
0	0	0	1	0
1	0	0	1	0
1	0	1	1	0
0	1	0	1	0
0	1	1	1	0
1	1	0	1	$-x$
1	1	1	2	0
1	1	2	1	0
2	0	0	0	x
0	2	1	1	x
2	1	1	1	0
2	1	2	1	x
1	2	1	1	0
1	2	2	1	x
2	2	2	1	x

[a] Reproduced from ref. 2.

Now let $v_{2j_1\cdots j_r k}^{(\sigma,\cdots)}$ be a charge distribution among a group of r consecutive sequences, with k terminal charges. The terminal charges now refer to the ends of the entire group of sequences. The development leading to (7.6.12) is extended so that wherever a sum or product over $v_{ijk}^{(\sigma)}$ appears, it is understood to include contributions from all groups of sequences of length 2. The only essential complication arises from the fact that the additional Eq. (7.6.11)

$$v^{(\sigma,\cdots)}_{2j_1\cdots j_r k} = \zeta^{(\sigma\cdots)}_2 \eta^{j_1+j_3+\cdots}_\sigma \eta^{j_2+j_4+\cdots}_{\sigma'} Q^{(\sigma\cdots)}_{2j_1\cdots j_r k} \exp\left[-\frac{1}{kT}(\mathscr{E}^{(\sigma,\cdots)}_{2j_1\cdots j_r k} - \theta k x\right]$$

(7C.2)

introduces a dependence of the same term on both η_σ and $\eta_{\sigma'}$, thus preventing, in general, the factorizing of the Jacobian matrix appearing in (7.6.15).

The values of Q and $\mathscr{E}$ for isolated and double sequences of 2 are shown in Table 7.C.2.

REFERENCES

1. F. E. HARRIS AND S. A. RICE, *J. Phys. Chem.* **58,** 725, 733 (1954).
2. S. A. RICE AND F. E. HARRIS, *J. Chem. Phys.* **24,** 326, 336 (1956).
3. F. E. HARRIS AND S. A. RICE, *J. Chem. Phys.* **25,** 955 (1956).
4. P. DEBYE AND E. HÜCKEL, *Physik. Z.* **24,** 305 (1923).
5. F. OOSAWA, N. IMAI, AND I. KAGAWA, *J. Polymer Sci.* **13,** 93 (1954).
6. J. J. HERMANS AND J. T. G. OVERBEEK, *Rec. trav. chim.* **67,** 761 (1948).
7. G. E. KIMBALL, M. CUTLER, AND H. SAMELSON, *J. Phys. Chem.* **56,** 57 (1952).
8. R. A. MARCUS, *J. Chem. Phys.* **23,** 1057 (1955).
9. W. KUHN, O. KÜNZLE, AND A. KATCHALSKY, *Helv. Chim. Acta* **31,** 1994 (1948); A. KATCHALSKY, O. KÜNZLE, AND W. KUHN, *J. Polymer Sci.* **5,** 283 (1950); A. KATCHALSKY AND S. LIFSON, *Ibid.* **11,** 409 (1953).
10. L. ONSAGER, *Chem. Rev.* **13,** 73 (1933); J. G. KIRKWOOD, *J. Chem. Phys.* **2,** 767 (1934).
11. N. BJERRUM, *Kgl. Danske Videnskab. Selskab, Mat.-fys. Medd.* **7,** No. 9 (1926).
12. R. H. FOWLER AND E. A. GUGGENHEIM, " Statistical Thermodynamics," p. 407, Cambridge Univ. Press, London and New York, 1950.
13. L. I. SCHIFF, " Quantum Mechanics," pp. 76–79 (McGraw-Hill, New York, 1949.
14. N. S. SCHNEIDER AND P. DOTY, *J. Phys. Chem.* **58,** 762 (1954).
15. A. OTH AND P. DOTY, *J. Phys. Chem.* **56,** 43 (1952).
16. J. R. HUIZENGA, P. F. GRIEGER, AND F. T. WALL, *J. Am. Chem. Soc.* **72,** 2636, 4228 (1950).
17. H. A. KRAMERS AND G. H. WANNIER, *Phys. Rev.* **60,** 252 (1941).
18. L. ONSAGER, *Phys. Rev.* **65,** 117 (1944).
19. G. F. NEWELL AND E. W. MONTROLL, *Revs. Modern Phys.* **25,** 353 (1953).
20. J. T. G. OVERBEEK, *Bull soc. chim.* **57,** 252 (1948).
21. J. G. KIRKWOOD AND F. H. WESTHEIMER, *J. Chem. Phys.* **6,** 506, 513 (1938).
22. J. B. HASTED, D. M. RITSON, AND C. H. COLLIE, *J. Chem. Phys.* **16,** 1 (1948).
23. P. J. FLORY, *J. Chem. Phys.* **21,** 162 (1953).
24. F. E. HARRIS AND S. RICE, *J. Polymer Sci.* **15,** 151 (1954).
25. C. M. TCHEN, *J. Chem. Phys.* **20,** 214 (1952).
26. J. DOOB, " Stochastic Processes." Wiley, New York, 1953.
27. J. E. MAYER AND M. G. MAYER, " Statistical Mechanics." Wiley, New York, 1940.
28. See, for example, H. FALKENHAGEN, "Electrolytes." Oxford Univ. Press, London and New York, 1934.
29. J. G. KIRKWOOD AND F. H. WESTHEIMER, *J. Chem. Phys.* **6,** 506 (1938).

30. P. M. FLORY, " Principles of Polymer Chemistry." Cornell Univ. Press, Ithaca, New York, 1953.
31. W. KUHN, *Kolloid-Z.* **26,** 258 (1936); *Ibid.* **87,** 3 (1939).
32. J. G. KIRKWOOD AND J. B. SHUMAKER, *Proc. Natl. Acad. Sci.* **38,** 863 (1952).
33. G. S. RUSHBROOKE, " Introduction to Statistical Mechanics." Oxford Univ. Press, London and New York, 1949.
34. E. J. COHN AND J. T. EDSALL, " Proteins, Amino Acids and Peptides." Reinhold, New York, 1943.
35. H. A. KRAMERS AND G. H. WANNIER, *Phys. Rev.* **60,** 252 (1941).
36. K. LINDESTRØM-LANG, *Rec. trav. Lab. Carlesberg* **15,** No. 7 (1924).
37. J. B. HASTED, D. M. RITSON, AND C. H. COLLIE, *J. Chem. Phys.* **16,** 1 (1948).
38. See, for example, F. B. HILDEBRAND, " Methods of Applied Mathematics." Prentice Hall, Englewood Cliffs, New Jersey, 1952.
39. B. H. ZIMM AND S. A. RICE, *J. Mol.* **3,** 391 (1960).
40. P. M. DOTY, A. M. HOLTZER, A. BRADBURY, AND E. BLOUT, *J. Am. Chem. Soc.* **76,** 4493 (1954); P. M. DOTY AND J. T. YANG, *J. Am. Chem. Soc.* **78,** 498 (1956).
41. B. H. ZIMM AND J. K. BRAGG, *J. Chem. Phys.* **31,** 526 (1959).
42. L. PAULING, R. COREY, AND R. BRANSON, *Proc. Natl. Acad. Sci.* **37,** 205 (1951).
43. S. LIFSON, B. KAUFMAN, AND H. LIFSON, *J. Chem. Phys.* **27,** 1356 (1957).
44. P. M. DOTY, A. WADA, J. T. YANG, AND E. BLOUT, *J. Polymer Sci.* **23,** 851 (1957).
45. A. WADA, to be published.
46. R. M. FUOSS, A. KATCHALSKY, AND S. LIFSON, *Proc. Natl. Acad. Sci. U.S.* **37,** 579 (1951).
47. F. T. WALL AND J. BERKOWITZ, *J. Chem. Phys.* **26,** 114 (1957).
48. L. PELLER, *J. Phys. Chem.* **63,** 1194, 1199 (1959).

8. Characterization of the Nonideal Polyelectrolyte Solution

8.0 Introduction

There are many equivalent ways of representing the deviations from ideal behavior of a solution. For dilute solutions of polymeric solutes, by far the most convenient formulation is in terms of the virial coefficients introduced in Chapter 1. On the other hand, for solutions of molecules of approximately equal size, the activity coefficient formulation introduced by Lewis has the sanction of usage. In general, the two different representations may be interchanged. To establish the translation, consider the thermodynamic relation defining the osmotic pressure π,

$$\mu_1 - \mu_1^0 = -\pi \bar{v}_1 \tag{8.0.1}$$

with μ_1, μ_1^0, and $\bar{v}_1$ being the chemical potential, standard state chemical potential, and partial molal volume of the solvent. By straightforward differentiation it is found that

$$\frac{\partial \mu_1}{\partial c_2} = -\bar{v}_1 \frac{\partial \pi}{\partial c_2}. \tag{8.0.2}$$

It was shown in Chapter 1 that the osmotic pressure can also be expressed in terms of the virial coefficients and powers of the weight concentration of the solute c_2 as

$$\pi = RT \left[\frac{c_2}{M_2} + \sum_{n \geqslant 2} (n-1) B_n c_2^n \right] \tag{8.0.3}$$

where M_2 is the molecular weight of the solute. The use of the Gibbs–Duhem relation at constant temperature and pressure,

$$N_1 \, d\mu_1 + N_2 \, d\mu_2 = 0 \tag{8.0.4}$$

leads to

$$\frac{\partial \mu_2}{\partial c_2} = -\frac{N_1}{N_2} \frac{\partial \mu_1}{\partial c_2} \tag{8.0.5}$$

and thereby

$$\frac{\partial \pi}{\partial c_2} = \frac{N_2}{N_1 \bar{v}_1 c_2} \frac{\partial \mu_2}{\partial \ln c_2}. \tag{8.0.6}$$

In the usual way, the activity coefficient γ_2 is defined in terms of the chemical potential μ_2 by

$$\mu_2 = \mu_2^0 + RT \ln \gamma_2 c_2 \tag{8.0.7}$$

and thereby

$$\frac{\partial \mu_2}{\partial \ln c_2} = RT\left[1 + \frac{\partial \ln \gamma_2}{\partial \ln c_2}\right]. \tag{8.0.8}$$

In dilute solution, $N_1\bar{v}_1 \gg N_2\bar{v}_2$ and the concentration is $c_2 = (N_2 M_2)/V$ whereby $(N_2/N_1\bar{v}_1 c_2) = (1/M_2)$. Introducing this latter expression into (8.0.6) and equating (8.0.6) and the derivative of (8.0.3) leads to

$$\frac{1}{M_2}\frac{\partial \ln \gamma_2}{\partial \ln c_2} = \sum_{n\geqslant 2} n(n-1)B_n c_2^{n-1} \tag{8.0.9}$$

which for sufficiently dilute solutions reduces to the form

$$\ln \gamma_2 = 2M_2 B_2 c_2. \tag{8.0.10}$$

Note that this relationship is for the activity coefficient of the solute species and will be most useful in the nonelectrolyte case. In the case of polyelectrolyte solutions, Eq. (8.0.10) is not valid in the form displayed.

Corresponding to the two representations of nonideality discussed above, we shall divide our discussion into two parts. In the first part there will be considered measurements of activity coefficients by standard thermodynamic techniques, whereas in the second part emphasis will be placed on the dilute solution region and the corresponding measurements of virial coefficients. This division is convenient because the use of activity coefficients is advantageous in describing the behavior of the counterions and byions, whereas the use of virial coefficients focuses attention on polyion-polyion interactions.

8.1 Measurement of the Activity Coefficient

Owing to the magnitude of the unit electric charge and our inability to change independently the concentration of a single ion, the activity coefficients of single ionic species have not thermodynamic basis, anhed only mean activity coefficient of the salt is operationally definable. The determination of individual ionic activity coefficients is therefore always subject to some ambiguity, this ambiguity usually arising from liquid junction potentials in the cells constructed to measure them. If cells without liquid junctions are used, only the mean activity coefficient of the salt is measurable. However, these observations should not be taken to mean that the activity coefficient of an individual ionic species does not "exist" or that it need not be defined. There are many cases where the use of an individual ionic activity coefficient is convenient, the most familiar example being the definition of pH. Of greater pertinence is the observation, from studies of Donnan membrane equilibrium in polyion solutions, that the difference between the nominal activity coefficients of counterions and byions is of primary importance in discussing the equilibrium ionic distribution.

Single ion activity coefficients are conventionally defined by the following hypothesis of Lewis.

(1) The activity coefficients of K^+ and Cl^- are the same in solutions of KCl. Presumably this is a consequence of the near equality of the ionic radii of the two species in solution.

(2) An ion has the same activity coefficient in all solutions having the same ionic strength. This is an extrapolation of the limiting law behavior of ionic solutions. With these two assumptions, the activity coefficients of single ionic species may be deduced from experimental results.

In general, the procedures used to measure activity coefficients in solutions of polyelectrolytes are the same as those used for simple salt solutions. To date, however, techniques based on the Gibbs–Duhem relationship and cyroscopic and osmotic pressure measurements have not been employed. Experimental emphasis has fallen instead on electrochemical methods such as (1) electromotive force measurements in cells with liquid junctions, (2) the measurement of membrane potentials, and (3) the measurement of Donnan equilibria. The first method is direct and gives reliable results, but the latter two methods are experimentally easier. Typical examples of the use of membrane potentials are the measurement of pH with the glass electrode[1] and the measurement of the activities of alkali-ions using ion exchange membranes. The third method has frequently been used to determine the activity coefficients of ions internal to resins,[2] and in only a few cases for polyelectrolyte solutions.[3,4]

Consider the following cell:

$$M\,|\,M^+\ (\text{solution})\,|\,\text{saturated KCl}\,|\,\text{standard electrode.}$$

The electromotive force of the cell is related to the activity of the metal ion by

$$\mathscr{E} = \mathscr{E}_0 + \frac{RT}{\mathscr{F}} \ln a_{M^+} \qquad (8.1.1)$$

where $\mathscr{E}_0$ is a constant containing the potential of the standard electrode, the electrode potential and the liquid junction potential between M^+ (solution) and saturated KCl. Studies of diffusion potentials in polyelectrolyte solutions[5] suggest that the ambiguities resulting from the liquid junction potential are likely to be small. Moreover, the mean activity coefficient, as obtained from ionic activity coefficients, is probably free of most of the error resulting from the ambiguity in the liquid junction potential. To date, the following systems have been examined using this technique:

$Na^{+6,7,8}$	Na (amalgam) | Na^+ (soln.) | sat. KCl | calomel electrode
Ag^{+9}	Ag–AgCl | Ag^+ (soln.) | sat. KNO_3 | calomel electrode
H^{+1}	glass electrode | H^+ (soln.) | sat. KCl | calomel electrode
$Cl^{-8,10}$	Ag–AgCl | Cl^- (soln.) | sat. KNO_3 | calomel electrode

In all the experiments, values of the constant $\mathscr{E}_0$ were obtained by using solutions with known ionic activities as standard samples, i.e., solutions of NaCl for the Na^+ measurements, solutions of $AgNO_3$ for the Ag^+ measurements, and solutions of KCl for the Cl^- measurements.

By far the most common counterion encountered is Na^+ since most polyelectrolytes are examined as the sodium salt. The measurement of the activity coefficient of Na^+ in polyelectrolyte solutions was first reported by Kern[6] and later more precise data were presented by Nagasawa et al.[7,8] In most of these studies a sodium amalgam electrode was used. The results are very reliable at high salt concentrations, the data in dilute solution being less reliable because of the reaction of sodium with water. For example, when the salt concentration is less than $0.02N$, $\mathscr{E}$ deviates from linearity in ln a_{Na^+} and to obtain a_{Na^+} graphical corrections must be applied. When the salt concentration is less than $0.005N$, even graphical correction becomes impossible. For such dilute solutions the sodium amalgam electrode must be abandoned. In lieu of such an electrode, Nagasawa and Kagawa[7] employed an ion-exchange membrane method. As is well known, cation exchange membranes are permeable to cations and sensibly impermeable to anions. Therefore, if an ion exchange membrane is converted to the Na^+ form, the membrane can be used to determine Na^+ activities in pure polyelectrolyte solutions in exactly the same manner as the glass electrode is used for measurements of pH. Consider, then, the following cell wherein the ion exchange membrane is in the Na^+ form:

calomel electrode	standard soln. of NaCl $a^I_{Na^+}$	sample soln. of polyelectrolyte $a^{II}_{Na^+}$	calomel electrode

membrane

The membrane potential is

$$\mathscr{E} = \mathscr{E}_0 + \alpha_0 \frac{RT}{\mathscr{F}} \ln a^{II}_{Na^+} \tag{8.1.2}$$

where $\mathscr{E}_0$ and α_0 are constants depending upon the NaCl concentration and the characteristics of the polyelectrolyte and the membrane. To calibrate the cell, the values of $\mathscr{E}_0$ and α_0 are determined in concentrated solution where the value of $a^{II}_{Na^+}$ is known from the sodium-amalgam electrode measurements. Having thus determined $\mathscr{E}_0$ and α_0 the activity of Na^+ in very dilute solution may be obtained directly from the measured membrane potential.

Methods for measuring the activities of the other ions in the solution are exactly the same as the techniques ordinarily used to study solutions of simple electrolytes.

The activity coefficients of Na^+ and Ag^+ in the absence of added salt are shown in Figs. 8.1.1 and 8.1.2. The effect of the polyion charge density upon

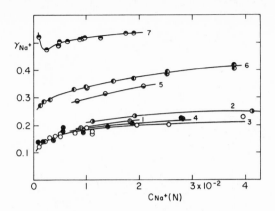

FIG. 8.1.1a. Dependence of Na^+ activity coefficient upon sample concentration, using samples of different charge density and without added salt. Sample: sodium polyvinyl alcohol sulphate. [Reproduced from ref. 7.]

No.	Degree of Polymerization	Degree of Esterification
1	1400	0.726
2	400	0.692
3	470	0.740
4	1440	0.711
5	—	0.494
6	680	0.431
7	1100	0.301

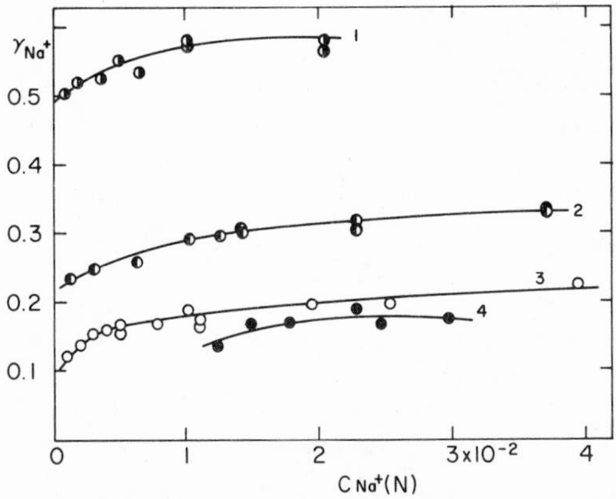

FIG. 8.1.1b. Dependence of Na^+ activity coefficient upon sample concentration, using samples of different polymer skeleton and without added salt. [Reproduced from ref. 7.]

No.	Polymer	Degree of Polymerization	Degree of Esterification
1	Na-carboxymethyl cellulose	—	0.736
2	Na-polystyrene sulfuric acid	340	0.824
3	Na-polyvinyl alcohol sulphate	470	0.740
4	Na-cellulose sulphate	—	2.32

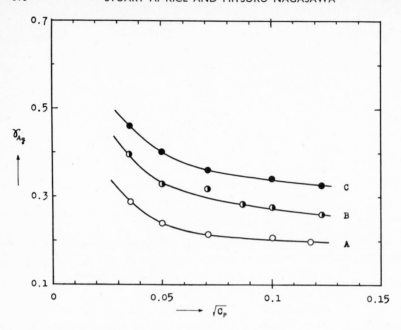

FIG. 8.1.2. Dependence of Ag⁺ activity coefficient upon sample concentration, without added salt. Sample: silver salt of carboxymethyl cellulose; degree of esterification: curve A—1.33; curve B—1.15; curve C—0.847; degree of neutralization: curve A—1.00; curve B—0.975; curve C—0.951. [Reproduced from ref. 9.]

the counterion activity coefficient is shown in Fig. 8.1.3 while in Figs. 8.1.4 and 8.1.5 the activity coefficients of Na⁺ and Cl⁻ in polyelectrolyte solutions

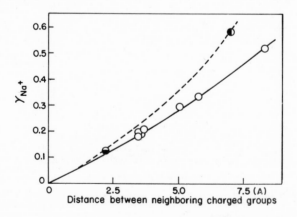

FIG. 8.1.3. Variation of the counterion activity coefficient with linear charge density.

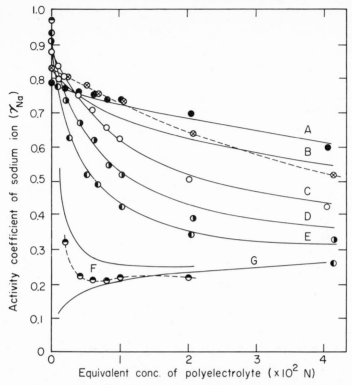

Fig. 8.1.4. Variation of sodium ion activity coefficient, in polyvinyl alcohol sulphate solutions with NaCl. Degree of polymerization = 1600; degree of Esterification = 0.65. Solid line curves calculated from additivity hypothesis. C_{NaCl} values: curve A—$1.00 \times 10^{-1}N$; curve B—$5.00 \times 10^{-2}N$; curve C—$2.01 \times 10^{-2}N$; curve D—$1.01 \times 10^{-2}N$; curve E—$5.00 \times 10^{-3}N$; curve F—$0.991 \times 10^{-3}N$; curve G—0, estimated from other data. [Reproduced from ref. 8.]

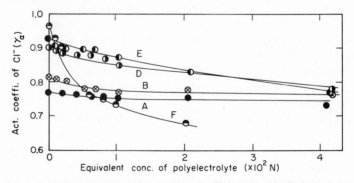

FIG. 8.1.5. Variation of chloride ion activity coefficient in polyvinyl alcohol sulphate solutions with NaCl. Samples are the same as those in Fig. 8.1.4. [Reproduced from ref 8.]

with added NaCl are shown. Finally, in Fig. 8.1.6 is depicted the mean activity coefficient of NaCl as calculated from the same data represented in Figs 8.1.4 and 8.1.5.

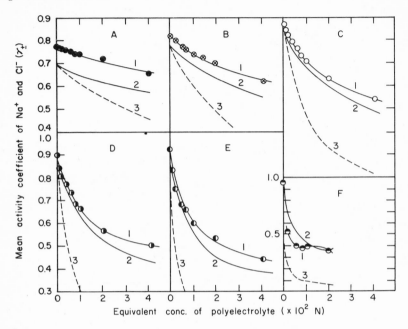

FIG. 8.1.6. Concentration dependence of the mean activity coefficient of NaCl in polyvinyl alcohol sulphate solutions with NaCl. Samples are the same as those in Fig. 8.1.4. ....., observed values; ——, values calculated from Katchalsky and Lifson theory, using calculated values of $\langle h^2 \rangle^{\frac{1}{2}}$; - - - -, values calculated from K.L. theory, using the $\langle h^2 \rangle^{\frac{1}{2}}$ determined from viscosity. [Reproduced from ref. 8.]

From these studies, the following conclusions may be drawn.

(1) The activity coefficient of the counterion in a polyelectrolyte solution is very small relative to the activity coefficient of the same ion in an equivalent solution of a simple salt.

(2) The activity coefficient of the byion (such as Cl$^-$ in a Na polyelectrolyte) does not show a correspondingly large decrease relative to the value in an equivalent simple salt solution.

(3) The activity coefficient of the counterion, in the absence of added salt, tends to zero as the charge density of the polyion increases at a constant concentration of polyelectrolyte.[7]

(4) The concentration dependence of the activity coefficient of the counterion shows a species dependence indicating that nonelectrostatic forces must also be considered.[7]

Note that the characteristic concentration dependence of the counterion activity is such that the activity coefficient of the counterion *decreases as the concentration decreases*. This tendency is in remarkable contrast with the behavior of solutions of simple salts, where the reverse is observed. However, the activity coefficient of Ag^+ does not show this peculiar concentration dependence.[9]

(5) It is possible to construct a rule of additivity for activities. That is, the Na^+ activity in mixed solutions of Na polyvinyl alcohol sulfate and NaCl is almost equal to the concentration weighted sum of the activities in pure water solutions of each salt. The solid lines in Fig. 8.1.3 show the calculated activities using the additivity assumption in the form

$$(\gamma_{Na^+})_{obs.} = \frac{\gamma_{Na^+}^P c_{NaP} + \gamma_{Na^+}^S c_{NaCl}}{c_{NaP} + c_{NaCl}}$$

where $\gamma_{Na^+}^P$ is the activity coefficient of the sodium ion in the pure water solution of the polyelectrolyte, and c_{NaP} and c_{NaCl} are the equivalent concentration of polyelectrolyte and NaCl in the solution, respectively.

It should be noted that the additivity of the counterion acvitities was first shown in the elegant work of Mock and Marshall.[1] The system these investigators studied was quite different from the one considered above, namely poly *p*-styrene sulfonic acid and HCl. The fact that additivity is valid for two different ions in solutions with different polyions, suggests a broad measure of validity for the hypothesis.

(6) The activity coefficient of the counterion is independent of molecular weight and dimension provided only that the molecular weight is not too small. The results depicted in Fig. 8.1.3 are for several polymer samples with differing degrees of polymerization.[7, 8] (See Appendix 8A.)

8.2 Analysis of Activity Coefficient Data

It is clear that the large electrostatic field generated by the charge on the polyion will influence the behavior of the neighboring counterions. In a crude picture, the medium surrounding a polyion may be imagined to be composed of successive " phases " of varying concentration. At equilibrium, the concentration of counterions near the polyion is large, this concentration falls off with distance, and the bulk concentration is attained in the limit of large distance from the polyion. Although it has not been demonstrated theoretically that the ions adjacent to the polyion have lower activity coefficients than the average value, it seems natural to assume that this is indeed the case. It is the large increase in counterion concentration adjacent to the polyino which suggests that the counterion activity coefficient is lower than that in the bulk solution.

A convenient, but nonrigorous, manner of handling this phenomenon is to divide the counterions into two classes: bound ions and free ions.* The success of such an approximation depends on the spatial extension of the region wherein the counterion concentration is larger than the average. If this spatial extension is small, then the external region comprises almost all of the solvent and an external ion sees a small net potential from the polyion. On the other hand, the ions internal to the region see a very large electric potential. Under these circumstances the fraction of ions falling in neither extreme case is small and the approximation of regarding the ions of the internal region as bound is probably quite good. This is the physical genesis of the counterion binding proposed in the Harris–Rice[11] and Imai–Oosawa–Kagawa[12] theories. In a certain sense, this interpretation of the concept of ion binding in polyelectrolyte solutions corresponds to the concept of Bjerrum ion-pair formation in solutions of ordinary electrolytes. However, no consistent theory of the activity coefficient based on the ion localization model has been reported. As might be expected it is very difficult to calculate the amount of counterion binding in a theoretically satisfying manner. We shall return to the theory of ion localization after making some comments about the theory of Katchalsky and Lifson,[3] which theory is the only one for the activity coefficient which does not utilize adjustable parameters.

Katchalsky and Lifson have calculated the mean activity coefficient in polyelectrolyte solutions containing simple electrolytes but neglecting polyion-polyion interactions. The procedure is as follows.

(a) The macromolecules are uniformly stretched from the uncharged configuration with root mean square end-to-end distance $\langle h_0^2 \rangle^{1/2}$ to a new configuration with rms end-to-end distance $\langle h_1^2 \rangle^{1/2}$. The free energy is thereby changed an amount A_1.

(b) Keeping the end-to-end distance fixed, ion atmospheres are constructed about each charge on the polymer and in the solution. This causes a free energy change A_2 which does not include intrapolymer repulsions.

(c) The fixed charges on the polymer are allowed to interact increasing the free energy by an amount A_3.

For the first stage,

$$A_1 = \frac{3kT}{\langle h_0^2 \rangle} \int_{\langle h_0^2 \rangle^{1/2}}^{\langle h_1^2 \rangle^{1/2}} \lambda h \, dh \qquad (8.2.1)$$

$$\lambda = 1 + 0.6 \frac{(h/h_{max})^2}{1 - (h/h_{max})^2} \qquad (8.2.2)$$

* We again remind the reader that these bound ions may be reclassified in different terms. The basic interpretation we place on " bound " in terms of a cell model is given in Chapter 7.

with h_{max} the contour length of the molecule. The second stage is calculated in the Debye–Hückel approximation with the result that

$$A_2 = -\frac{q^2\kappa}{3D}\left(\sum N_i + \alpha Z\right)$$

$$\kappa^2 = \frac{4\pi q^2}{DVkT}\sum N_i \tag{8.2.3}$$

with αZ the number of charges affixed to polymer chains and N_i the number of simple ions of species i. The calculation of A_3 is the most difficult part of the analysis. To sum the pair interaction energy over all interacting groups the charge is uniformly smeared along the polymer chain and the distribution of chain configurations is taken to be the same as in the absence of charges. Neglecting the size of the counterions,

$$A_3 = \frac{\alpha^2 Z^2 q^2}{D\langle h_1^2\rangle^{1/2}} \ln\left(1 + \frac{6\langle h_1^2\rangle^{1/2}}{\kappa\langle h_0^2\rangle}\right). \tag{8.2.4}$$

By addition of Eqs. (8.2.1), (8.2.3), and (8.2.4) and differentiation with respect to the number of ions of the species concerned, it is found that

$$-kT \ln \gamma_i = \frac{3\alpha^2 Z^2 q^2}{D\sum N_i}\frac{1}{\kappa\langle h_0^2\rangle + 6\langle h_1^2\rangle^{1/2}} + \frac{\kappa q^2}{2D}. \tag{8.2.5}$$

Further, by utilizing the condition

$$\left(\frac{\partial A}{\partial h}\right)_{h=h_1^*} = 0 \tag{8.2.6}$$

we find for the equilibrium end-to-end separation of the chain ends

$$\frac{3h_1^* kT}{\langle h_0^2\rangle}\lambda = \frac{\alpha^2 Z^2 q^2}{Dh_1^{*2}}\left[\ln\left(1 + \frac{6h_1^*}{\kappa\langle h_0^2\rangle}\right) - \frac{6h_1^*/\kappa\langle h_0^2\rangle}{1 + 6h_1^*/\kappa\langle h_0^2\rangle}\right]. \tag{8.2.7}$$

In marked contrast with the experimental observations, Eq. (8.2.5) gives the same value for the activity coefficients of counterions and byions. Eq. (8.2.5) is to be interpreted as the mean activity coefficient of both species.

As is depicted in Fig. 8.1.1, the predicted concentration dependence of γ_i is parallel to but displaced from the experimental data. The disagreement in the absolute magnitude is due to the failure of the Debye–Hückel limiting law. From the data exhibited it may be concluded that the Katchalsky–Lifson theory is in agreement with experiments, insofar as the mean activity coefficient is concerned. Note however that in order to obtain this result the theoretical values of h_1^* had to be used. These calculated end-to-end distances are known to be much too large. (See Chapter 10). Moreover, the use of experimental values of h_1^* destroys the agreement between the calculated and observed values of γ_i. (See Fig. 8.1.6.) It is clear that the Katchalsky–Lifson theory is not entirely adequate with respect to three observations:

(a) the computed root mean square end-to-end distances are much too large;

(b) the marked difference in behavior between counterions and byions is not accounted for;

(c) the theory cannot be used to describe the thermodynamic properties of the counterion in the absence of added salt.

To remove these defects, particularly (b), it may prove adequate to graft onto the theory of form of ion localization. For:

(1) The counterion activity coefficient is independent of molecular weight and dimension. It is clear that a simple domain association model cannot provide an adequate explanation of the data since the volume occupied by a polymer molecule is not a linear function of the molecular weight and the fraction of free ions trapped within the polymer domain would therefore not be independent of molecular weight.

(2) The species dependence of the counterion activity indicates that non-electrostatic forces must be considered. The simplest forces of this nature are those giving rise to different ionic size and thereby different distances of closest approach. The possibility of covalent binding (as in the case of Ag^+ and carboxymethyl cellulose$^-$) or monopole-dipole interactions (as in polyvinyl pyridinium$^+$ and Br^-) cannot be eliminated.

(3) The additivity of activities suggests that the ion localization model is substantially correct since, if the extent of ion localization is relatively insensitive to ionic strength, additivity of activities would be predicted.

In the ion localization approach to the theory of the activity coefficient, the bound ions are presumed to make no contribution to the thermodynamic properties of the solution. The nonpaired counterions are, of course, never completely free, since there is to be considered an electrostatic interaction arising from the residual charge of the polyion as well as the interactions between the small ions.

A completely consistent theory would calculate both the electrostatic interactions between all charges in the solution, and the extent of ion pairing caused by such interactions. No such consistent theory of polyelectrolyte solutions has been presented. For example, the Harris–Rice[11] theory proceeds by two linked but largely independent procedures. If it is assumed that some ion binding exists so that the net charge on the polyion is less than the stoichiometric charge, then the corresponding Poisson–Boltzmann equation may be solved and the electrostatic potential calculated. Using this electrostatic potential and the additional assumption that the ion-pair formation, whatever its genesis, is describable in terms of an equilibrium between bound and free ions,[11,13] the number of bound ions may be calculated.* The extent of the inconsistency may be estimated as follows. If the nonpaired counterions are regarded as noninteracting, then all deviations from ideality are ascribable to incomplete dissociation. For solutions of simple salts, it is well known that such a description is inferior to the Bjerrum theory, which treats the unassociated ions by ordinary Debye–Hückel techniques, and calculates the number of ion pairs and hence also the pair dissociation constant from the size of the ions and the electrostatic interaction. Optimistically, the errors in the polyelectrolyte case will not exceed those in the simple salt case. Indeed if the binding is strong it may be hoped to achieve the same order of success as in the theory of ideal weak electrolytes.

To correct the binding model discussed above, it is necessary to account for interactions between unbound ions and the polyion. This is difficult to do exactly because of the complicated chain structure of the polyion. In the spherical approximation such a treatment has been carried out by Harris and Rice. When there is a large excess of added salt, the Harris–Rice treatment leads to

$$\Delta A_{elec} = -\frac{\kappa}{3D} \sum_\alpha L_\alpha q_\alpha^2 + \frac{N}{2} \sum_{i \neq j} \frac{q^2 \exp(-\kappa'|\mathbf{r}_i - \mathbf{r}_j|)}{D_0 |\mathbf{r}_i - \mathbf{r}_j|} - \frac{\kappa n N q^2}{2D} - \frac{\kappa' n N q^2}{2D_0}$$

$$(8.2.8)$$

where the primed symbols refer to the interior of the domain occupied by the polymer. The second term is the electrostatic free energy of an isolated polyion with n free charges in a medium of ionic strength characterized by κ'. From Eq. (8.2.8) the activity coefficient may be derived by standard techniques. Since no comparisons of this formalism with experiment have been made we shall make no further comment. In closing this section, the reader should be referred to the literature[12–15] for further discussion of counterion activity coefficients.

8.3   Thermodynamics of the Donnan Membrane Equilibrium†

As described in Section 8.0, the second virial coefficient of a polyelectrolyte is very different from that of a nonelectrolyte polymer. One of the major differences is that the second virial coefficient of a polyion is determined primarily by the behavior of the simple ions in the solution and therefore may be large even if all the particles in the solution behave in ideal fashion. In fact, B_2 is a maximum (in a certain sense) when the solution is an ideal solution.

* Note here our use of the language of mass action theory. In view of our analysis in Chapter 7, we could use the language of the cell model and come to the same conclusion about the state of the system.

† For recent work see the papers by Z. ALEXANDROWICZ, J. Polymer Sci. **18**, 325, 337 (1960).

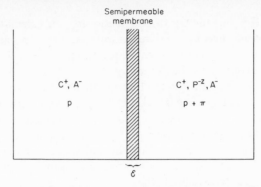

FIG. 8.3.1. Schematic representation of a Donnan equilibrium.

Suppose that a solution of a colloidal species having charges on the surface of each particle is placed in contact with a simple salt solution such as $C^+ A^-$ through a semipermeable membrane, and further suppose that the counterion to the colloidal electrolyte is the same as the positive ion of the salt. Finally, let all small particles but not the colloidal particle be permeable through the membrane. In the initial state described, this system is not at equilibrium. To attain equilibrium, some particles diffuse through the membrane, while to maintain electroneutrality in both phases, the diffusing species must be neutral salts and solvent. Thus, the number of cations transferred must be equal to the number of anions transferred. Let us suppose that the initial concentration of the colloidal electrolyte is C (equivalent concentration per unit charge), that the initial equivalent concentration of the simple salt is c_0 and that the amount of salt which diffused from phase II to phase I is y. Then the ionic distribution at equilibrium can be expressed in the form

$$\frac{C + y}{c_0 - y} = \frac{c_0 - y}{y} \tag{8.3.1}$$

or,

$$\frac{c_{C+}^{I}}{c_{C+}^{II}} = \frac{c_{A-}^{II}}{c_{A-}^{I}}. \tag{8.3.2}$$

In general, if there are many kinds of diffusible ions in both phases, the equilibrium condition is

$$\left(\frac{c_{1+}^{I}}{c_{1+}^{II}}\right)^{1/z_{1+}} = \left(\frac{c_{2+}^{I}}{c_{2+}^{II}}\right)^{1/z_{2+}} = \cdots = \left(\frac{c_{1-}^{II}}{c_{1-}^{I}}\right)^{1/z_{1-}} = \left(\frac{c_{2-}^{II}}{\frac{1}{2}c_{-}^{I}}\right)^{1/z_2} = \cdots \tag{8.3.3}$$

where the c's are the concentrations of the ions in the equilibrium state and the z's are the charges of the ions. Because of the distribution of these ions, a pressure difference as well as an electrostatic potential difference appears across the membrane. The thermodynamics of this phenomenon was studied by Donnan.[17,18]

As always, the equilibrium of the system described is determined by the equality of the electrochemical potentials of the diffusible species, so that

$$\tau_{C^+}^I = \tau_{C^+}^{II}$$
$$\tau_{A^-}^I = \tau_{A^-}^{II}$$
$$\tau_{H_2O}^I = \tau_{H_2O}^{II} = \mu_{H_2O} \tag{8.3.4}$$

where the τ's are the electrochemical potentials

$$\tau_i = \mu_i^0 + z_i^0 q\psi + \bar{v}_i p + kT \ln x_i \tag{8.3.5}$$

for each simple ion, and

$$\mu_{H_2O} = \mu_{H_2O}^0 + \bar{v}_{H_2O} p + kT \ln x_{H_2O} \tag{8.3.6}$$

for the solvent. Of the remaining symbols, $\bar{v}_i$ is the partial molar volume of species i, x_i the mole fraction, and μ_i^0 the standard chemical potential of the ith species. Substituting Eqs. (8.3.5) and (8.3.6) into Eqs. (8.3.4) and replacing $\psi^I - \psi^{II}$ by $\mathscr{E}$ and $p^I - p^{II}$ by π, we obtain the following relations for an ideal system,

$$q\mathscr{E} + \bar{v}_{C^+}\pi = kT \ln \frac{x_{C^+}^{II}}{x_{C^+}^I}$$

$$-q\mathscr{E} + \bar{v}_{A^-}\pi = kT \ln \frac{x_{A^-}^{II}}{x_{A^-}^I} \tag{8.3.7}$$

$$\bar{v}_{H_2O}\pi = kT \ln \frac{x_{H_2O}^{II}}{x_{H_2O}^I}.$$

There are three equations for the three unknown quantities, $\mathscr{E}$, π, and y and we can therefore solve the set of equations for this ideal system without ambiguity if we assume that the partial molar volumes are independent of π,

$$\frac{x_{C^+}^I x_{A^-}^I}{(x_{H_2O}^I)^{(\bar{v}_{C^+} + \bar{v}_{A^-})/\bar{v}_{H_2O}}} = \frac{x_{C^+}^{II} x_{A^-}^{II}}{(x_{H_2O}^{II})^{(\bar{v}_{C^+} + \bar{v}_{A^-})/\bar{v}_{H_2O}}}. \tag{8.3.8}$$

If the solutions are sufficiently dilute, $x_{H_2O}^I \approx x_{H_2O}^{II} \approx 1$. Let $c_0 - y = \hat{C}$ ($\hat{C}$ is usually taken as constant), so that

$$\frac{x_{C^+}^I}{x_{C^+}^{II}} = \frac{x_{A^-}^{II}}{x_{A^-}^I}; \qquad \frac{C + y}{\hat{C}} = \frac{\hat{C}}{y} \tag{8.3.9}$$

which relations are the same as Eqs. (8.3.1) and (8.3.2). $\mathscr{E}$ is given by

$$\mathscr{E} = \frac{kT}{q} \ln \frac{C + \sqrt{C^2 + 4\hat{C}^2}}{2\hat{C}} \tag{8.3.10}$$

if $\bar{v}\pi \ll q\mathscr{E}$, which is the case for the usual experimental conditions. The osmotic pressure π is given by

$$\pi = \frac{kT}{\bar{v}_{H_2O}} \ln \frac{1 - x_{C^+}^{II} - x_{A^-}^{II}}{1 - x_{C^+}^I - x_{A^-}^I - x_p^I} \tag{8.3.11}$$

where x_p^{I} is the mole fraction of the colloidal species. After the expansion of the logarithm, and the replacement of the x's with the weight concentration of the colloidal particle c_p and the molar concentrations of the simple salts c_s, and using Eq. (8.3.9) to calculate y, we have [19]

$$\frac{\pi}{c_p} = RT \left[\frac{1}{M_2} + \frac{c_p}{4a^2 c_s} + \cdots \right] \tag{8.3.12}$$

or

$$B_2 = (1/4a^2 c_s) \tag{8.3.13}$$

with M_2 the molecular weight of the colloidal particle, and a the molecular weight per unit charge,

$$a = (M_2/Z).$$

If the colloidal particle is a linear polyion, the contribution due to the polymer-polymer interactions, B_p, may be added,

$$B_2 = B_p + (1/4a^2 c_s). \tag{8.3.14}$$

The second virial coefficient may be determined by light scattering measurements from which Edsall et al.[20] have shown that the second term of Eq. (8.3.14) is dominant in solutions of rigid colloidal electrolytes.

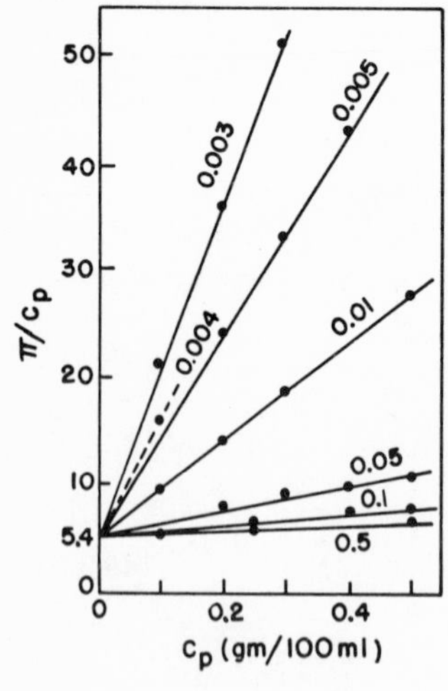

FIG. 8.3.2. An example of the linear relationship between π/c_p and c_p [Reproduced from ref. 19.]

The first application of this theory to linear polyelectrolyte solutions was carried out by Pals and Hermans.[19] The results of these investigators show that the theory agrees with the experimental results qualitatively, but there are large quantitative deviations. One example of the linear relationship between π/c_p and c_p is shown in Fig. 8.3.2. The values of B_2 obtained from the figure are plotted against $1/c_s$ and $1/a^2$ in Figs. 8.3.2 and 8.3.3. It is clear that there is good linearity in all cases, as predicted by the theory. Moreover, Eq. (8.3.13) correctly predicts the independence of the second virial coefficient of the molecular weight of the sample.[19,21] However, the slopes of Figs. 8.3.3 and 8.3.4 are much smaller than the calculated slopes. The ratio of the theoretical second virial coefficient to the observed one is almost 2. Moreover, when a vinyl compound or polyphosphate is used, $B_{\mathrm{theor.}}/B_{\mathrm{obs.}}$ often becomes greater than 10, and sometimes reaches 300.[16,19,21–30]

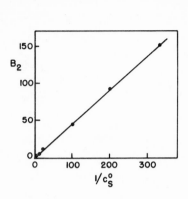

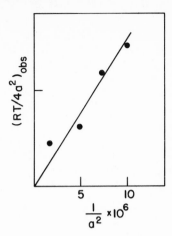

FIG. 8.3.3 The second virial coefficient as a function of added salt concentration. [Reproduced from ref.
19.]

FIG. 8.3.4. A test of Eq. (8.3.14).

Several explanations have been advanced for this quantitative disagreement. Schneider and Doty[23] reported that the analysis of the second virial
coefficient of a polyelectrolyte can best be based on Flory's theory for nonelectrolyte polymers,[31] inasmuch as the polyion is very effectively shielded by
the counterion atmosphere. This conclusion is very instructive, but cannot be
taken to mean that the polyion can be considered to be electrically neutral. If
the polyion had no effective charge, the concentration of the extraneous
electrolyte $(C^+ A^-)$ would be the same on both sides of the membrane.
Actually, however, there is a large difference between the concentrations of
the extraneous salt on the two sides of the membrane (an example is shown in
Fig. 8.3.5). The osmotic pressure calculated from the concentration of the
polyelectrolyte and the extraneous salt is much larger than the observed
osmotic pressure. The problem which we must solve is why the second virial
coefficient of a polyelectrolyte is so small that it may be computed from the
theory of nonionic polymers, in spite of the large difference of the ionic
concentrations in the two phases. Recently, Orofino and Flory[30] have
extended the theory of Flory to polyelectrolyte solutions and calculated the
second virial coefficient. The calculated second virial coefficient was found to
be much larger than the experimental one leading Flory and Orofino to
conclude that the discrepancy is due to the neglect of the ionic activity coefficients. We will return to the theory of Orofino and Flory after examining
the effect of the introduction of ionic activity coefficients on the Donnan
membrane equilibrium.

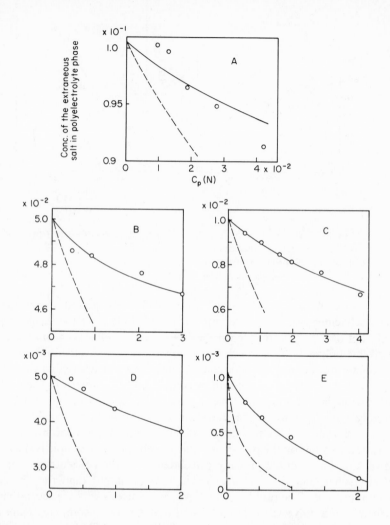

FIG. 8.3.5. Analytical concentration of the extraneous salt in polyelectrolyte phase. Polymer—Na-polyvinyl alcohol sulphate; salt—NaCl. The concentration of the phase with polyion $= 1.004 \times 10^{-2}$ N. [Reproduced from ref 7.]

The theoretical treatment of the Donnan equilibrium of such an extremely nonideal system as a polyelectrolyte solution has not yet been completed. In particular, a most difficult point arises in the treatment of the activity coefficient of the solvent. If we replace the concentrations of the various species by the corresponding activities, Eq. (8.3.7) becomes:

$$q\mathscr{E} + \bar{v}_{C^+}\pi = kT \ln \frac{\gamma_{C^+}^{II} x_{C^+}^{II}}{\gamma_{C^+}^{I} x_{C^+}^{I}}$$

$$-q\mathscr{E} + \bar{v}_{A^-}\pi = kT \ln \frac{\gamma_{A^-}^{II} x_{A^-}^{II}}{\gamma_{A^-}^{I} x_{A^-}^{I}}$$

$$\bar{v}_{H_2O}\pi = kT \ln \frac{\gamma_{H_2O}^{II} x_{H_2O}^{II}}{\gamma_{H_2O}^{I} x_{H_2O}^{I}}. \tag{8.3.15}$$

In problems in which the effective activity coefficient of the solvent is close to unity, Eq. (8.3.15) works very well. For example, the membrane potential $\mathscr{E}$ is given by

$$\mathscr{E} = \frac{RT}{2\mathscr{F}} \left[\ln \frac{\gamma_{C^+}^{II} c_{C^+}^{II}}{\gamma_{C^+}^{I} c_{C^+}^{I}} + \ln \frac{\gamma_{A^-}^{I} c_{A^-}^{I}}{\gamma_{A^-}^{II} c_{A^-}^{II}} \right] \tag{8.3.16}$$

if $\bar{v}_{C^+} = \bar{v}_{A^-}$. One example of the comparison between the membrane potential and the value of the right-hand side of Eq. (8.3.16) as determined from activity measurements is shown in Fig. 8.3.6. It is clear that satisfactory agreement is obtained. Moreover, from Eq. (8.3.15), we have an equation corresponding to Eq. (8.3.8):

$$\frac{\gamma_{C^+}^{I} x_{C^+}^{I} \gamma_{A^-}^{I} x_{A^-}^{I}}{\gamma_{C^+}^{II} x_{C^+}^{II} \gamma_{A^-}^{II} x_{A^-}^{II}} = \left(\frac{\gamma_{H_2O}^{I} x_{H_2O}^{I}}{\gamma_{H_2O}^{II} x_{H_2O}^{II}} \right)^{(\bar{v}_{C^+} + \bar{v}_{A^-})/\bar{v}_{H_2O}} = 1$$

under the usual experimental conditions.[16] That is,

$$\frac{\gamma_{C^+}^{I} c_{C^+}^{I} \gamma_{A^-}^{I} c_{A^-}^{I}}{\gamma_{C^+}^{II} c_{C^+}^{II} \gamma_{A^-}^{II} c_{A^-}^{II}} = 1. \tag{8.3.17}$$

The right-hand side of Eq. (8.3.16) can be calculated from Eq. (8.3.17) and the following empirical rule for the additivity of ionic activities (see Section 8.1)

$$\gamma_{C^+}^{I} c_{C^+}^{I} = \gamma_{A^-}^{I} c_{A^-}^{I} + \gamma_{C^+}^{0} c_p \tag{8.3.18}$$

where $\gamma_{C^+}^{0}$ is the activity coefficient of the counterion in a salt-free solution having the same equivalent concentration of polymer c_p. The calculated values are plotted in Fig. 8.3.6. Again we find satisfactory agreement. However, the osmotic pressure of a polyelectrolyte solution, which is determined by the activity of the solvent, has not yet been predicted. Donnan suggested an expansion of π in powers of concentration by transforming the activity coefficient of the solvent into the osmotic coefficient g,

$$\ln \gamma_{H_2O} x_{H_2O} = g \ln x_{H_2O}. \tag{8.3.19}$$

This substitution leads to[32]

$$\pi = RT[g^{I}(N_p^{I} + N_{C^+}^{I} + N_{A^-}^{I}) - g^{II}(N_{C^+}^{II} + N_{A^-}^{II})]$$

or

$$\frac{\pi}{c_p} = RT \left[\frac{g^{II}}{M} + \frac{g^{I}}{4a^2} \cdot \frac{\gamma_{CA}^{I}}{\gamma_{CA}^{II}} \cdot \frac{1}{c_s} + \cdots \right] \tag{8.3.20}$$

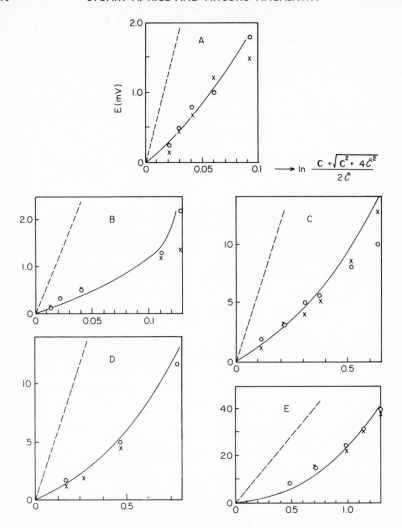

FIG. 8.3.6. Membrane potential. Experimental conditions are the same as in FIG. 8.3.5. ----, theoretical line for an ideal solution; ———, values calculated from Eqs. 8.3.16 or 17 assuming additivity of counterion activity Eq. 8.3.18. O—membrane potential;

$$\times \frac{1}{m} \frac{RT}{2F} \left[\ln \frac{\gamma_{c^+}^{II} c_{c^+}^{II}}{\gamma_{c^+}^{I} c_{c^+}^{I}} + \ln \frac{\gamma_A^{I} c_{A^-}^{I}}{\gamma_A^{II} c_{A^-}^{II}} \right]$$

determined experimentally. A—$C_{NaCl} = 1.005 \times 10^{-1}N$; B—$C_{NaCl} = 5.00 \times 10^{-2}N$; C—$C_{NaCl} = 1.004 \times 10^{-2}N$; D—$C_{NaCl} = 5.00 \times 10^{-3}N$; E—$C_{NaCl} = 1.022 \times 10^{-3}N$.

which corresponds to Eq. (8.3.12) for ideal solutions. By comparison with experiment, it is found that this equation does not account for the abnormalities in osmotic pressure in polyelectrolyte solutions. The origin of the discrepancy probably lies in the fact that the osmotic coefficient does not distinguish between the contributions of the counterions, the polyelectrolyte, and the added salt.

A very important thermodynamic theory of the osmotic pressure of a nonideal system was presented by Scatchard.[33] He avoided the difficulty described above by using the Gibbs–Duhem relationship. In his original paper a general case is described, but herein we shall treat again the example of Fig. 8.3.1. The nondiffusible and diffusible components which are used in Scatchard's theory are not ions but salts, i.e., C_ZP, CA, and water. That is, the chemical potential μ of each component is taken to be

$$\frac{\mu_2}{RT} = \ln m_2 + Z \ln (m_k + Zm_2) + \beta_2 + \frac{\mu_2^{00}}{RT}$$

$$\frac{\mu_2}{RT} = \frac{Z\mu_{C^+}}{RT} + \frac{\mu_{P^{-z}}}{RT} \tag{8.3.21}$$

$$\frac{\mu_k}{RT} = \ln(m_k + Zm_2) + \ln m_k + \beta_k + \frac{\mu_k^{00}}{RT}$$

$$= \frac{\mu_{C^+}}{RT} + \frac{\mu_{A^-}}{RT} \tag{8.3.22}$$

in which m is the molal concentration of the component, the subscript 2 indicates the polyelectrolyte (C_ZP), k denotes the extraneous salt (CA), μ^{00} is the value of the chemical potential in the standard state at the same temperature and pressure, and β is the excess chemical potential divided by RT,

$$\beta_2 = \ln \gamma_2 = \ln \gamma_p \gamma_{C^+}^Z$$

$$\beta_k = \ln \gamma_k = \ln \gamma_{C^+} \gamma_{A^-} \tag{8.3.23}$$

if γ_2 and γ_k are the mean activity coefficients of the components. Therefore,

$$\frac{1}{RT} \frac{\partial \mu_2}{\partial m_2} = \frac{1}{m_2} + \frac{Z^2}{m_k + Zm_2} + \beta_{22} \tag{8.3.24}$$

$$\frac{1}{RT} \frac{\partial \mu_2}{\partial m_k} = \frac{1}{RT} \frac{\partial \mu_k}{\partial m_2} = \frac{Z}{m_k + Zm_2} + \beta_{2k}$$

with

$$\beta_{22} = \frac{\partial \beta_2}{\partial m_2}$$

$$\beta_{2k} = \frac{\partial \beta_2}{\partial m_k} = \frac{\partial \beta_k}{\partial m_2}. \tag{8.3.25}$$

Because of the definition of the components chosen, the electrostatic potential does not appear in the chemical potential. Now, the equilibrium condition is

$$\frac{\partial \mu_k}{\partial m_2} + \frac{\partial \mu_k}{\partial m_1}\frac{dm_1}{dm_2} + \frac{\partial \mu_k}{\partial m_k}\frac{dm_k}{dm_2} = \frac{\partial \mu_k'}{\partial p'}\frac{dp'}{dm_2} + \frac{\partial \mu_k'}{\partial m_1'}\frac{dm_1'}{dm_2} + \frac{\partial \mu_k'}{\partial m_k'}\frac{dm_k'}{dm_2} \quad (8.3.26)$$

which is a differential form of Eq. (8.3.4). In Eq. (8.3.26) p' is equal to $-\pi$. Multiplying Eq. (8.3.26) by m_k and summing for all diffusible components, including 1, we have

$$\sum_k m_k \frac{\partial \mu_k}{\partial m_2} + \sum_j \frac{dm_j}{dm_2}\sum_k m_k \frac{\partial \mu_k}{\partial m_j} = \frac{dp'}{dm_2}\sum_k m_k \frac{\partial \mu_k'}{\partial p'} + \sum_j \frac{dm_j'}{dm_2}\sum_k m_k \frac{\partial \mu_k'}{\partial m_j'}.$$

$$(8.3.27)$$

By the Gibbs–Duhem equation (at constant T and p)

$$\sum_k m_k \frac{\partial \mu_k}{\partial m_2} = -m_2 \frac{\partial \mu_2}{\partial m_2} \quad (8.3.28)$$

whereupon,

$$m_2 \frac{\partial \mu_2}{\partial m_2} + m_2 \sum_j \frac{dm_j}{dm_2}\frac{\partial \mu_2}{\partial m_j} = -\frac{dp'}{dm_2}\sum_k m_k \frac{\partial \mu_k'}{\partial p'} - \sum_j \frac{dm_j'}{dm_2}\sum_k m_k \frac{\partial \mu_k'}{\partial m_j'}. \quad (8.3.29)$$

After further differentiation of Eq. (8.3.29), taking the limit as m_2 approaches zero and omitting two negligibly small terms, we have

$$\frac{Z^2}{2m_k} + \beta_{22} - \left(\frac{dm_k}{dm_2} - \frac{dm_k'}{dm_2}\right)^2 \left(\frac{2}{m_k} + \beta_{kk}\right) = \frac{d^2\pi}{dm_2^2}\frac{V_m^0}{RT} \quad (8.3.30)$$

where V_m^0 is the volume of the solution which would contain one kilogram of component 1. The second virial coefficient is given by [see Eq. (8.0.3)]

$$B_2 = \frac{d^2\pi}{dm_2^2}\frac{V_m^0}{RT} \cdot \frac{1}{2m_2}$$

so that

$$B_2 = \frac{1}{2M_2}\left[\frac{Z^2}{2m_k} + \beta_{22} - \frac{\beta_{2k}^0 m_k}{2 + \beta_{kk}^0 m_k}\right] \quad (8.3.31)$$

which is to be compared with Eq. (8.3.13) or Eq. (8.3.20). The first term of Eq. (8.3.31) is equal to the ideal Donnan term Eq. (8.3.13). The second and third terms are complementary contributions. The third term, which can be determined from ionic distribution experiments[34,16] and also calculated from the theory of Katchalsky and Lifson,[3,24,16] is usually small compared with the first and second terms. The second term can be estimated from the activity coefficients. The comparison of Eq. (8.3.31) with experiment was carried out by Nagasawa et al.[16] The results of this comparison are shown in Table 8.3.1.

TABLE 8.3.1[a]

COMPARISON OF THE EXPERIMENTAL SECOND VIRIAL COEFFICIENT OF SODIUM
POLYVINYL ALCOHOL SULFATE WITH THE VALUES CALCULATED FROM
EQ. (8.3.31), USING THE ACTIVITY COEFFICIENTS OF THE IONS

c_{NaCl}	B_2 obs./B_{ideal}	$\dfrac{1}{2M^2}\,\beta_{22}/B_{\text{ideal}}$
$1.005 \times 10^{-1}\,N$	0.13	-0.96
$5.00\ \times 10^{-2}$	(0.007)	-1.06
1.004×10^{-2}	0.013	-0.83
$5.00\ \times 10^{-3}$	(0.016)	-1.15
1.029×10^{-3}	0.018	-1.35

[a] Reproduced from ref. 16.

It is clear that the second term β_{22} is so large that it may almost cancel the first (ideal) term to give a very low second virial coefficient. Unfortunately, a quantitative comparison of terms is almost impossible because to determine the difference accurately requires very great precision in the determination of each term of Eq. (8.3.31).

To explain quantitatively such extremely low values of the second virial coefficient, a product of the ideal Donnan term and a correction factor is to be preferred over additive correction terms such as are presented in Eq. (8.3.31). An example of a relation of this type is the following equation, suggested by Nagasawa et al.[16]

$$B_{\text{nonideal}} = B_{\text{ideal}}\frac{(\gamma_{\text{C}+}^0)^2}{\gamma_{\text{CA}}}. \tag{8.3.32}$$

Equation (8.3.32) was derived from the assumption that the osmotic pressure is proportional to the difference between the sum of all solute activities on each side of the membrane.

From the investigations cited it may be concluded that the abnormally low second virial coefficient of a polyelectrolyte is due to the abnormally low activity coefficient of the counterions. However, no rigorous and quantitatively satisfactory theory has as yet been advanced.

8.4 Statistical Thermodynamics of the Second Virial Coefficient of Polyelectrolytes

In thermodynamic theories, it is always assumed that there is a uniform distribution of the polymer units throughout the volume. Since the second virial coefficient of a polyelectrolyte is only slightly dependent on the molecular weight, a uniform distribution of polymer units is not necessarily a poor assumption. However, in dilute solutions it is undoubtedly necessary to account for the discontinuity of the polymer distribution.

From the theory described in Chapter 1, we see that the second virial coefficient is determined by the potential of average force between two particles (W_{12}),

$$B_2 = -\frac{N_0}{2VM_2^2} \int^V \int (e^{-W_{12}/kT} - 1) \, d(1) \, d(2)$$

(8.4.1)

$$= -\frac{2\pi N_0}{VM_2^2} \int_V r^2 (e^{-W_{12}/kT} - 1) \, dr$$

where N_0 is Avogadro's number. If W_{12} is given as a function of the distance r between two particles, B_2 may be calculated from Eq. (8.4.1) and the effects of the ionic distribution (as discussed in Section 8.3) are automatically taken into account.

Calculations of the second virial coefficient of a colloidal electrolyte have been made by Hill and Stigter.[35,36] However, the theory is applicable only to spherical polyions having a small amount of charge on the surface. This limitation arises from the use of the Debye–Hückel potential for W_{12}

$$W_{12}(r) = Db^2 \psi_b^2 \frac{e^{-\kappa(r-2b)}}{r} f$$

(8.4.2)

where f is a correction term equal to unity if $\kappa(r - 2b) \gg 1$, and κ is the Debye–Hückel reciprocal shielding length. Since most polyions have high charge densities, the use of Eq. (8.4.2) is too restrictive and the Hill–Stigter theory will be omitted from this chapter.

For flexible polyelectrolytes, it is quite difficult to calculate W_{12} because of the complicated distribution of chain segments and the ionic atmospheres around these segments. Recently, however, an attempt to overcome this difficulty has been presented by Orofino and Flory.[30] Their theory is an extension of the theory of the expansion of a polyion and of the theory of the second virial coefficient of nonionic polymers as presented by Flory and others.

Orofino and Flory start with a single polyion immersed in an infinite bath of solution of strong electrolyte $M_{\nu+}^{z+}A_{\nu-}^{z-}$, at a concentration c_s moles/liter. It is assumed that the polyion is substantially neutral, and that the concentration of ions inside the polyion domain is determined by the Donnan membrane equilibrium. The free energy required to assemble such a system can be calculated from the mixing free energy of the segments of the polyion and the solvent (including small ions). The assumption that the polyion is substantially neutral permits the neglect of electrical interactions between the polyions. The distribution of the segments of each molecule is assumed to be Gaussian. W_{12}, which is interpreted as the free energy associated with the process of bringing together (to the separation r) two identical polyion molecules, is calculated as the difference between the free energies of two polyions in the infinitely separated configuration and in the partially over-lapped configuration. The process is the same as that used for uncharged polymers except that the mixing of the movable ions is taken into account. The result of the calculation is

$$W_{12} = kT \sum_{i=1}^{2} X_i e^{-\sigma_i a^2} \tag{8.4.3}$$

where

$$X_1 = 1000 \left(\frac{3^{3/2}}{4\pi^{3/2}}\right)\left(\frac{1}{Nv_1}\right)\left(\frac{Mv_u}{M_u}\right)^2 (\langle R_G^2 \rangle)^{-3/2}\left(\tfrac{1}{2} - \chi_1 + \frac{v_1\alpha^2}{4v_u^2\Gamma}\right) \tag{8.4.4}$$

$$\sigma_1 = \tfrac{4}{3}(\langle R_G^2 \rangle)^{-1}$$

$$X_2 = 10^6 \left(\frac{3^{5/2}}{8\pi^3}\right)\left(\frac{1}{N^2 v_1}\right)\left(\frac{Mv_u}{M_u}\right)^3 (\langle R_G^2 \rangle)^{-3}\left(\tfrac{1}{3} - \chi_2 + v_1(z_- - z_+)\frac{\alpha^3}{12v_u^3\Gamma^2}\right) \tag{8.4.5}$$

$$\sigma_2 = (\langle R_G^2 \rangle)^{-1}$$

with $\langle R_G^2 \rangle$ the square of the radius of gyration, v_1, v_u the molar volumes of solvent and segments respectively, M, M_u, the molecular weights of the polymer and its repeating unit, Γ the ionic strength of the external solution, and χ_1 and χ_2 the parameters determining the heat of mixing.

Substitution of this result into Eq. (8.4.3) gives

$$B_2 = \frac{16\pi N_0}{3^{3/2}}\left(\frac{(\langle R_G^2 \rangle)^{3/2}}{M^2}\right) I_2(X_1, X_2) \tag{8·4.6}$$

where

$$I_2(X_1, X_2) = \int_0^\infty t^2[1 - \exp(-X_1 e^{-t^2} - X_2 e^{-4t^{3/2}})]\, dt. \tag{8.4.7}$$

If α were set equal to zero, or Γ equal to ∞, Eq. (8.4.6) would reduce to the second virial coefficient of an uncharged polymer. Equation (8.4.6) can be simplified as follows:

$$B_2 \cong \frac{16\pi N_0}{3^{3/2}}\left(\frac{(\langle R_G^2 \rangle)^{3/2}}{M^2}\right)\ln\left(1 + \frac{\pi^{1/2}}{4}X_1 + \frac{\pi^{1/2}3^{3/2}}{32}X_2\right) \qquad X_1 < 100 \tag{8.4.8}$$

and

$$B_2 \cong \frac{16\pi N_0}{3^{3/2}}\left(\frac{(\langle R_G^2 \rangle)^{3/2}}{M^2}\right)(\ln(X_1 + X_2^{3/4}) + 0.577)^{3/2} \qquad X_1 > 35. \tag{8.4.9}$$

Orofino and Flory also derived the following relation, which combines the inter- and intramolecular theories, by comparing Eq. (8.4.6) with the equation for the expansion of a polyion [Eq. (10.1.22)].

$$B_2 \cong \frac{16\pi N_0}{3^{3/2}}\left(\frac{(\langle R_G^2 \rangle)^{3/2}}{M^2}\right)\ln\left(1 + \frac{\pi^{1/2}}{2}(\alpha_E^2 - 1)\right) \tag{8.4.10}$$

where α_E is the ratio of a linear dimension of the expanded polyion to that of the unperturbed molecule. Equation (8.4.9) is the same as the equation used by Schneider and Doty.[23]

A comparison of this theory with experimental determinations of B_2 for sodium polyacrylate was carried out by Orofino and Flory. Very large differences between theory and experiment were observed. The discrepancy may be expressed in terms of a parameter defined by

$$\hat{p} = \left[\frac{X_1(\text{obs.})}{X_1(\text{theor.})}\right]^{\frac{1}{2}} = \frac{\alpha(\text{obs.})}{\alpha(\text{theor.})}.$$

The values of $\hat{p}$ reported by Orofino and Flory are shown in Table 8.4.1. It is clear that $\alpha\hat{p}$ is the effective charge of the polyion and $\hat{p}$ plays the same role as the activity coefficient in Eq. (8.3.32).

TABLE 8.4.1[a]

THE VALUES OF $\hat{p}$ FOR POLYACRYLIC ACID IN NaCl SOLUTIONS

α	Γ	$\hat{p}$ (from B_2 data)	$\hat{p}$ (from viscosity data)
0.102	0.10	0.38	0.44
0.335	0.10	0.24	0.30
0.344	0.01	0.13	0.30
0.947	1.00	0.17	0.15
0.950	0.10	0.13	0.15
0.994	0.01	0.09	—

[a] Reproduced from ref. 30.

Appendix 8A

We consider in this Appendix some properties of polyethylene imine. This substance is of interest for two reasons. First, when charged it corresponds very closely to theoretical models since there are no side groups, the charge is on the chain, all bond angles are the same and, aside from the small difference in CN and CC bond lengths, the polymer skeleton is microscopically uniform. Second, very low molecular weight analogues may be obtained and the transition from simple electrolyte to polyelectrolyte studied. Lapanje et al,[37] reported studies of counterion activity coefficients and counterion binding in the hydrochlorides of polyethylene imine, ethylene diamine, diethylene triamine, triethylene tetramine and tetraethyelene pentamine, as well as a theoretical analysis of the results obtained.

Using the cell

$$\text{Hg, Hg}_2\text{Cl}_2 \mid \text{saturated KCl} \mid \text{sample} \mid \text{AgCl, Ag}$$

and neglecting the liquid junction potential, the following results were obtained for the chloride ion activity coefficients (Table 8A.1):

TABLE 8A.1

MOLAL ACTIVITY COEFFICIENTS OF CHLORIDE IONS IN SOLUTIONS
OF AMINE HYDROCHLORIDES AND PEI(HCl)

	Molality	ED(HCl)$_2$	DT(HCl)$_3$	TT(HCl)$_4$	TP(HCl)$_5$	PEI(HCl)
1	1.0	0.437	0.414	0.382	0.365	0.298
2	0.8	0.453	0.437	0.401	0.382	0.300
3	0.6	0.479	0.465	0.426	0.408	0.302
4	0.5	0.497	0.481	0.442	0.423	0.304
5	0.4	0.525	0.502	0.464	0.441	0.308
6	0.3	0.565	0.531	0.492	0.479	0.314
7	0.2	0.610	0.566	0.538	0.528	0.321
8	0.1	0.683	0.649	0.607	0.589	0.337
9	0.08	0.703	0.675	0.634	0.610	0.339
10	0.06	0.736	0.707	0.667	0.638	0.343
11	0.05	0.754	0.727	0.685	0.655	0.351
12	0.04	0.774	0.750	0.710	0.676	0.360
13	0.03	0.796	0.779	0.744	0.702	0.374
14	0.02	0.820	0.813	0.786	0.758	0.391
15	0.01	0.873	0.866	0.844	0.823	0.428
16	0.009	0.875	0.870	0.849	0.828	0.433
17	0.008	0.878	0.876	0.859	0.836	0.440
18	0.007	0.882	0.886	0.869	0.848	0.448
19	0.006	0.891	0.898	0.883	0.864	0.457
20	0.005	0.904	0.908	0.901	0.883	0.467
21	0.004	—	—	—	—	0.479
22	0.003	—	—	—	—	0.496
23	0.002	—	—	—	—	0.516
24	0.001	—	—	—	—	0.562

Counterion binding measurements were also made (see Chapter 9) with the results in Table 8A.2:

TABLE 8A.2

FRACTION OF COUNTERIONS BOUND, f, TO AMINE HYDROCHLORIDES

Compound	f
ED(HCl)$_2$	0.00 ± 0.01
DT(HCl)$_3$	0.00 ± 0.01
TT(HCl)$_4$	0.00 ± 0.01
TP(HCl)$_5$	0.10 ± 0.01
PEI(HCl)	0.54 ± 0.02

The ionic molal activity coefficients of Cl^- for the dihydrochlorides were given in Appendix 6C. The most characteristic feature of the tabulated values is that below a certain concentration, (0.03m) the sequence is such that γ_{Cl^-} for $ED(HCl)_2$ is the smallest and for $TP(HCl)_2$ the largest. If we assume that the two end nitrogen atoms are charged, an assumption which seems justified from the point of view of interaction energy considerations and which seems to be confirmed by the form of the titration curves,[37] we may conclude that, at the same concentration, activity coefficients of Cl^- increase with increasing distance between the charges on the cation. A theoretical discussion of this case was given in Appendix 6C and a simple model shown to account quantitatively for the change in γ_{Cl^-} with charge separation.

The ionic molal activity coefficients of Cl^- in solutions of fully neutralized amines and PEI(HCl) are given in Table 8A.1. Here, the general trend is in accord with expectation. With increasing number of charges on the cation, the counterion activity coefficients at the same concentration become smaller. The differences between the activity coefficients of Cl^- in the solutions of the several amines are seen to be rather small. In the case of $DT(HCl)_3$ an anomaly is encountered: in the concentration range below $0.7 \times 10^{-3}m$ the activity coefficients of Cl^- are seen to be slightly larger than those of $DT(HCl)_2$. This we believe can be attributed to experimental error.

Though in solutions of $TP(HCl)_5$ the cation already has 5 charges its influence on the activity coefficients of Cl^- is not nearly so large as in the case of the polyion. In other words, the differences between the respective counterion activity coefficients are still large A plot of the activity coefficients of Cl^- in PEI(HCl) and $TP(HCl)_5$ solutions versus m_{Cl^-} is given in Fig. 8A.1. It may be seen that the values of activity coefficients increase rather slowly and that even at a concentration of 1×10^{-3} m the counterion activity is still low. There is another feature which should be noted. We have previously commented on the fact that in some cases the activity coefficients of counterions to polyelectrolytes decrease with decreasing concentration.[7] For the compounds described in this Appendix, such behavior is not observed.

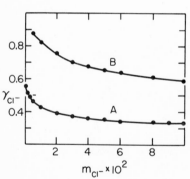

FIG. 8A.1 Activity coefficients of Cl^- in solutions of neutralized amines. Curve A: PEI(HCl), and curve B: $TP(HCl)_5$.

But, the case of PEI(HCl) is somewhat different from the others. Whereas the amines are moderately strong bases and hydrolysis in solutions of their respective hydrochlorides may be neglected, PEI, on the other hand, is a very weak base and a non-negligible amount of hydrolysis occurs. The degree of hydrolysis can be estimated by measuring the pH of single solutions. These values are given in Table 8A.3. We may compare this situation with that of a polyelectrolyte solution with added simple electrolyte, in which case behavior similar to that described above has been observed.[8]

TABLE 8A.3

pH AND DEGREE OF HYDROLYSIS OF SOLUTIONS OF PEI(HCl)

	Molarity	pH	Degree of hydrolysis
1	1.0	0.79	0.16
2	0.8	0.82	0.19
3	0.6	0.90	0.21
4	0.5	0.99	0.21
5	0.4	1.09	0.21
6	0.2	1.34	0.23
7	0.1	1.61	0.25
8	0.08	1.68	0.26
9	0.06	1.79	0.27
10	0.05	1.89	0.26
11	0.04	1.99	0.26
12	0.02	2.28	0.26
13	0.01	2.59	0.26
14	0.008	2.69	0.26
15	0.006	2.78	0.28
16	0.005	2.86	0.28
17	0.004	2.93	0.29
18	0.002	3.18	0.33
19	0.001	3.44	0.36

We turn now to a discussion of the results of the counterion binding experiments as listed in Table 8A.2. It is readily seen that the degree of binding parallels the behavior of the counterion activity coefficient, i.e., there is a slow increase as the number of charges on the polyion increases but the values for $TP(HCl)_5$ are very far from those of PEI(HCl). This observation confirms and extends the results of Nagasawa and Rice[38] which showed that the amount of counterion binding was not determined solely by the local charge density (see Chapter 9 for discussion). The present data indicate that a charged chain of length greater than 5 is needed for any considerable amount of counterion binding. In our opinion these results completely rule out any interpretation of the cell model in terms of discrete ion pair formation with a one to one correspondence between ion pairs and charged sites along the polymer skeleton. Note however that the basic premises of the cell model are valid. The data do not preclude an interpretation of binding in terms of loose association of counterions with long segments of charged chain. We have already used this interpretation, i.e. in the formulation of the cell model the distribution of ions about the charged polymer skeleton is arbitrarily cut off at a given distance and ions within said radius are considered bound while those outside are considered substantially free.

It is pertinent to make two remarks at this point of the argument. First, because the PEI(HCl) is partially hydrolyzed, the amount of binding should be compared with that of say sodium polyacrylate at 75% neutralization. For this compound, at three-fourths full charge, f is approximately 0.5 (see Chapter 9), in agreement with the data cited in Table 8A.2. Since the linear charge densities of PEI(HCl) and sodium polyacrylate are almost the same, this agreement is as expected. Second, we wish to point out that the extent of binding is not totally independent of the local charge density. In a series of experiments, Wall and Doremus[39] determined the binding of sodium ion to various polyphosphates. They found that f was approximately 0.05, 0.21, 0.29, and 0.71 for Na_2HPO_4, $Na_4P_2O_7$, $Na_5P_3O_{10}$ and polymer, respectively. In this case the local charge density is inhomogeneous and the over-all linear charge density rather higher than in PEI(HCl). In view of the small increase in binding with considerable increase in linear charge density for sodium polyacrylate ($f = 0.5$ at $\alpha = 0.75$, $f = 0.6$ at $\alpha = 1$, see Chapter 9) we attribute the larger binding of the polymer and the larger binding of the low molecular weight analogues to the charge inhomogeneity.

In Appendix 6C we considered a simple model of the bolion and showed it to be in good agreement with experiment. Can the same be done for the polymer?

It is clear that the considerations of Appendix 6C may be extended by formal techniques to more highly charged ions. However, it is also clear that such an extension runs afoul of very complex algebra. In view of this situation we consider a different and very over simplified model of linear multiply charged ions. The principal simplifying assumption which we shall use is that end effects may be neglected. With this assumption, each charge may be treated as being exactly equivalent to every other charge.

Consider a linear array of $Z + 1$ charges. As depicted in Fig. 8A.2 we represent the coordinates of each charge with respect to an arbitrary point by the variables ϑ_i and r_i. Let there be $Z/2$ charges to each side of the charge labeled zero. Then the distance of the jth charge from an arbitrary point P is

$$r_j^2 = r_0^2 + j^2 R^2 - 2jr_0R \cos \vartheta_0 \qquad (8A.1)$$

where γ_0, ϑ_0 represent the distance of charge zero from P and the angle between the molecular axis and the vector $\mathbf{r}_0$. The distance between any pair

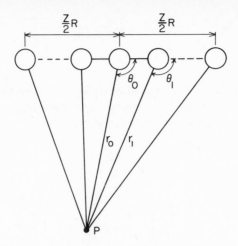

FIG. 8A.2. The multiply charged linear model ion.

of adjacent charges is taken to be the constant R. Thus,

$$\cos \vartheta_j = \frac{jR - r_0 \cos \vartheta_0}{r_j} \tag{8A.2}$$

As in Appendix 6C we solve the linearized Poisson–Boltzmann equation outside the ion, and the Laplace equation inside. In particular, let the polyion be represented as a linear array of charged spheres of radius a and near pair separation, R. The solution to both equations may be written in terms of Legendre polynomials. Let the coefficients of the expansion be C_i' outside and A_{ij}' inside, with i the order of the polynominal and j the ion label. Using the boundary condition $\psi'(a) = \psi(a)$ on the surface of each charge and neglecting end effects leads to

$$C_0' \frac{e^{-\kappa Rx}}{Rx} + C_1' \frac{1 + \kappa Rx}{(Rx)^2} \cos \vartheta_0 + 2C_0' \sum_{j=1}^{Z/2} \frac{e^{-j\kappa R}}{jR} \left[1 + \tfrac{1}{2}(1 + j\kappa R)y_j\right]$$

$$+ 2C_1' \sum_{j=1}^{Z/2} \frac{(1 + y_j)}{(jR)^2} + 2C_1' \kappa \sum_{j=1}^{Z/2} \frac{(1 + \tfrac{1}{2}y_j)}{jR}$$

$$= \frac{zq}{DRx} + (Z + 1)A_{00}' + A_{10}' Rx \cos \vartheta_0 \tag{8A.3}$$

$$+ \frac{2zq}{D} \sum_{j=1}^{Z/2} \frac{(1 + \tfrac{1}{2}y_j)}{jR} + 2A_{10}' \sum_{j=1}^{Z/2} jR\left(1 - \frac{y_j}{2}\right)$$

where

$$y_j = (2x/j)\cos \vartheta_0$$

$$x = r_0/R$$

(8A.4)

and we have truncated the series after terms in $P_1^{(0)}$ and have further assumed x^2 is negligible relative to x. If we now apply the second boundary condition, namely $\nabla\psi'(a) = \nabla\psi(a)$ then it is found that

$$C_0' = \frac{zq}{D}\frac{e^{\kappa a}}{1 + \kappa a}$$

$$C_1' = 0$$

(8A.5)

$$A_{00}' = \frac{zq}{(Z+1)D}\left\{\frac{e^{\kappa a}}{1+\kappa a}\left[\frac{e^{-\kappa a}}{a} + 2\sum_{j=1}^{Z/2}\frac{e^{-j\kappa R}}{jR}\right] - \frac{1}{a} - 2\sum_{j=1}^{Z/2}\frac{1}{jR}\right.$$

$$\left. - \frac{Z\left(\frac{Z}{2}+1\right)}{1-Z}\left[\frac{e^{\kappa a}}{1+\kappa a}\sum_{j=1}^{Z/2}\frac{e^{-j\kappa R}}{j^2 R}(1+\kappa jR) - \sum_{j=1}^{Z/2}\frac{1}{j^2 R}\right]\right\}$$

(8A.6)

$$A_{01}' = \frac{2zq}{(1-Z)RD}\left\{\frac{e^{\kappa a}}{1+\kappa a}\left[\sum_{j=1}^{Z/2}\frac{e^{-j\kappa R}}{j^2 R}(1+j\kappa R)\right] - \sum_{j=1}^{Z/2}\frac{1}{j^2 R}\right\}$$

Using the same method of calculation as detailed in Appendix 6C we find the activity coefficient to be

$$kT\ln\gamma_{Cl^-}^{(Z+1)} = -\frac{z^2 q^2}{2D}\cdot\frac{\kappa}{1+\kappa a_{Cl^-}} + \frac{z^2 q^2}{4D}\cdot\frac{Z+1}{Z}\left\{-\frac{a}{(1+\kappa a)^2}\right.$$

$$\times\left[\frac{1}{a} + 2\sum_{j=1}^{Z/2}\frac{e^{-\kappa(jR-a)}}{jR} - \frac{Z(Z/2+1)}{1-Z}\sum_{j=1}^{Z/2}\frac{e^{-\kappa(jR-a)}}{j^2 R}(1+\kappa jR)\right]$$

$$+ \frac{1}{1+\kappa a}\left[-2\sum_{j=1}^{Z/2}\frac{jR-a}{jR}e^{-\kappa(jR-a)}\right.$$

(8A.7)

$$- \frac{Z(Z/2+1)}{1-Z}\sum_{j=1}^{Z/2}\frac{e^{-\kappa(jR-a)}}{j}$$

$$\left.\left.+ \frac{Z(Z/2+1)}{1-Z}\sum_{j=1}^{Z/2}\frac{jR-a}{j^2 R}e^{-\kappa(jR-a)}(1+\kappa jR)\right]\right\}$$

for the linear array of charged spherical beads. For the cases $Z + 1 = 3$ and $Z + 1 = 5$ considered in this Appendix, the general formula reduces to

$$kT \ln \gamma_{Cl^-}^{(3)} = -\frac{z^2 q^2}{2D} \frac{\kappa}{1 + \kappa a_{Cl^-}} - \frac{3}{8} \frac{z^2 q^2}{D} \left\{ \frac{\kappa a}{(1 + \kappa a)^2} \left[\frac{1}{a} \right. \right.$$

$$\left. + \left(\frac{6}{R} + 4\kappa \right) e^{-\kappa(R - a)} \right]$$

$$+ \frac{\kappa e^{-\kappa(R - a)}}{1 + \kappa a} \left[2 + 6\kappa(R - a) - \frac{6a}{R} \right] \right\} \qquad (8A.8)$$

and

$$kT \ln \gamma_{Cl^-}^{(5)} = -\frac{z^2 q^2}{2D} \frac{\kappa}{1 + \kappa a_{Cl^-}} - \frac{5}{16} \frac{z^2 q^2}{D} \left\{ \frac{\kappa a}{(1 + \kappa a)^2} \left[\frac{1}{a} \right. \right.$$

$$\left. + \left(\frac{6}{R} + 4\kappa \right) e^{-\kappa(R - a)} + \left(\frac{2}{R} + 2\kappa \right) e^{-\kappa(2R - a)} \right]$$

$$+ \frac{\kappa}{1 + \kappa a} \left[\left(\frac{6(R - a)}{R} - 4 + 4(R - a)\kappa \right) e^{-\kappa(R - a)} \right. $$

$$\left. + \left(\frac{2(2R - a)}{R} - 2 + 2(2R - a)\kappa \right) e^{-\kappa(2R - a)} \right] \right\}$$

$$(8A.9)$$

Consider now the variation of γ_{Cl^-} with charge. As seen in Fig. 8A.3, the expected increase in concentration dependence with increasing charge is

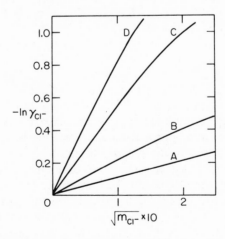

FIG. 8A.3. Dependence of $\ln \gamma_{Cl^-}$ on the $\sqrt{m_{Cl^-}}$ in the presence of: curve A—spherical univalent cation; curve B—bolaform ion; curve C—linear trivalent ion; curve D—linear pentavalent ion.

obtained. Of more interest is the approach to the polymeric limit, i.e. the dependence of on the number of charges. Here, the predicted variation is much more rapid than that observed (Fig. 8A.4) although, of course, the direction predicted is correct. It should be noted that the multiple charge linear model with the neglect of end effects overestimates the electrostatic potential considerably for short chains. For, each charge is assumed to be

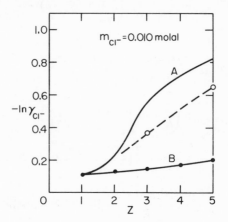

FIG. 8A.4. Dependence of $\ln \gamma_{Cl^-}$ on number of charges on " polymer " at fixed concentration: curve A—calculated; curve B—observed; – – – corrected.

symmetrically surrounded by Z other charges. In a very rough estimate we may say that the computed potential is a factor of 3/2 too high at the end charges of the triple ion. For the ion of charge five, both the end and next to end charges have real potentials lower than the computed potential. In a very rough approximation, the computed activity coefficient should be (6/5) larger for the triple ion and (15/12.8) larger for the quintuple ion. These crude estimates are shown as circles in Fig. 8A.4. While the variation with Z is appreciably reduced, it is still too large when compared with experiment. We have no ready explanation of the remaining discrepancy.

We now turn briefly to the possible extension of the models proposed herein to the case of coiled polyions. The results of Nagasawa and Rice[38] indicated that the mean electrostatic field at a charged site inside a polyion had major contributions from both the near neighbor and the rest of the polyion. This suggests that we take the multiply charged linear model and imbed it inside a polyion. Such a model would correspond to choosing an electrostatic Kuhn element and surrounding this element with a more or less spherical charge distribution. Qualitatively, we expect the potential due to the rest of the ion to be approximately constant throughout the polymer domain.

Thus, we would anticipate that the activity coefficient of the counterion would be further depressed (over that calculated herein for the small compounds) and that its value would be independent of molecular weight. While these qualitative features are easily seen to follow from the physical nature of the model, the mathematical formulation of this problem is fraught with difficulties.

REFERENCES

1. R. A. MOCK AND C. A. MARSHALL, *J. Polymer Sci.* **13,** 263 (1954).
2. For example: H. P. GREGOR AND M. H. GOTTLIEB, *J. Am. Chem. Soc.* **69,** 2830 (1953); K. KIMURA, M. NAGASAWA, AND K. NAKAMURA, *J. Chem. Soc. Japan, Ind. Chem. Sect.* **56,** 435 (1953).
3. A. KATCHALSKY AND S. LIFSON, *J. Polymer Sci.* **11,** 409 (1953).
4. A. M. LIQUORI, F. ASCOLI, C. BOTRÉ, *J. Polymer Sci.* **15,** 169 (1959).
5. M. NAGASAWA, S. OZAWA, AND K. KIMURA, *J. Chem. Soc. Japan, Ind. Chem. Sect.* **59,** 1201 (1956) (in Japanese); M. NAGASAWA, S. OZAWA, K. KIMURA, AND I. KAGAWA, *Mem. Fac. Eng., Nagoya Univ.* **8,** 50 (1956) (in English).
6. W. KERN, *Makromol. Chem.* **2,** 279 (1948).
7. M. NAGASAWA AND I. KAGAWA, *J. Polymer Sci.* **25,** 61 (1957); errata **31,** 256 (1958).
8. M. NAGASAWA, M. IZUMI, AND I. KAGAWA, *J. Polymer Sci.* **37,** 375 (1959).
9. I. KAGAWA AND K. KATSUURA, *J. Polymer Sci.* **17,** 365 (1955).
10. I. KAGAWA AND K. KATSUURA, *J. Polymer Sci.* **9,** 405 (1952).
11. F. E. HARRIS AND S. A. RICE, *J. Phys. Chem.* **58,** 725 (1954).
12. F. OOSAWA, N. IMAI, AND I. KAGAWA, *J. Polymer Sci.* **11,** 409 (1954).
13. F. OOSAWA, *J. Polymer Sci.* **23,** 421 (1957).
14. R. A. MARCUS, *J. Chem. Phys.* **23,** 1057 (1955).
15. I. KAGAWA AND M. NAGASAWA, *J. Polymer Sci.* **16,** 299 (1955).
16. M. NAGASAWA, T. TAKAHASHI, M. IZUMI, AND I. KAGAWA, *J. Polymer Sci.* **38,** 213 (1959).
17. F. G. DONNAN AND E. A. GUGGENHEIM, *Z. physik. Chem.* **162,** 346 (1932).
18. F. G. DONNAN, *Z. physik. Chem.* **168,** 369 (1934).
19. D. T. F. PALS AND J. J. HERMANS, *Rec. trav. chim.* **71,** 469 (1952).
20. J. T. EDSALL, H. EDELHOCK, R. LONTIE, AND P. R. MORRISON, *J. Am. Chem. Soc.* **72.** 4641 (1950).
21. I. KAGAWA AND M. FUKUDA, *J. Chem. Soc. Japan, Ind. Chem. Sect.* **54,** 97 (1951).
22. I. KAGAWA AND A. TAKAHASHI, *J. Chem. Soc. Japan, Ind. Chem. Sect.* **56,** 252 (1953).
23. N. S. SCHNEIDER AND P. DOTY, *J. Phys. Chem.* **58,** 762 (1954).
24. H. J. L. TRAP AND J. J. HERMANS, *J. Phys. Chem.* **58,** 757 (1954).
25. H. INAGAKI AND T. ODA, *Makromol. Chem.* **21,** 1 (1956).
26. H. INAGAKI, S. HOTTA, AND M. HIRAMI, *Makromol. Chem.* **23,** 1 (1957).
27. G. SAINI AND L. TROSSARELLI, *J. Polymer Sci.* **23,** 563 (1957).
28. U. P. STRAUSS AND P. L. WINEMAN, *J. Am. Chem. Soc.* **80,** 2366 (1958).
29. H. TERAYAMA, *J. Polymer Sci.* **19,** 181 (1956).
30. T. A. OROFINO AND P. J. FLORY, *J. Phys. Chem.* **63,** 283 (1959).
31. P. J. FLORY, " Principles of Polymer Chemistry." Cornell Univ. Press, Ithaca, New York, 1953.
32. M. NAGASAWA, H. NAKOJI, AND I. KAGAWA, *J. Chem. Soc. Japan, Ind. Chem. Sect.* **57,** 9 (1954).

33. G. SCATCHARD, *J. Am. Chem. Soc.* **68**, 2315 (1946).
34. G. SCATCHARD, A. C. BATCHELDER, AND A. BROWN, *J. Am. Chem. Soc.* **68**, 2320 (1946).
35. T. L. HILL, *Discussions Faraday Soc.* **21**, 31 (1956); *J. Phys. Chem.* **61**, 458 (1957).
36. D. STIGTER AND T. L. HILL *J. Phys. Chem.* **63**, 551 (1959).
37. S. LAPANJE, J. HAEBIG, H. T. DAVIS, AND S. A. RICE, *J. Am. Chem. Soc.* **83**, 1590 (1961).
38. M. NAGASAWA AND S. A. RICE, *J. Am. Chem. Soc.* **82**, 5207 (1960).
39. F. T. WALL AND R. H. DOREMUS, *J. Am. Chem. Soc.* **76**, 868 (1954).

9. Ion Binding

9.0 Introduction

In 1938, Kern[1] published the first modern physico-chemical study of linear polyelectrolytes. In his paper he used the words " elektrostatische Inaktivierung " to indicate that some of the counterions in the polyelectrolyte solution did not seem to contribute to the osmotic pressure of the solution, nor to other colligative properties. In Fig. 9.0.1, the osmotic pressure of a poly-

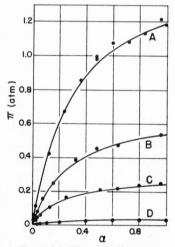

FIG. 9.0.1 The osmotic pressure as a function of degree of neutralization for a typical polyelectrolyte. $c_p = 0.25$ (A), 0.125 (B), 0.625 (C), 0.01 (D). [Reproduced from ref. 1.]

acrylate solution is plotted against the degree of neutralization of the polyion. It is observed that the osmotic pressure is much lower than the ideal value calculated from the number of particles in the solution. Moreover, the rate of increase with α (of the osmotic pressure) decreases with increasing degree of neutralization. It is as if the counterions were inactivated by the charge of the polyion with the degree of inactivation increasing with the charge on the polyion. Similar phenomena are observed when other properties of the counterions are examined, i.e., the conductivity, activity coefficient. . . . Our understanding of these observations was greatly advanced in 1950 when Wall and his co-workers reported measurements of the transference number of a polyacrylate, by which it was shown that a significant fraction of the counterions move together with the polyion against an applied electric field. This observation has stimulated much of the subsequent experimental and theoretical study of polyelectrolyte solutions.

Kern did not present any detailed molecular interpretation of his observations. In 1944, Kagawa[2] suggested by the use of the words " Ionkoeti " (ion fixation) that the counterions were in part immobilized by some interaction with the charge on the polyion. The first attempt to understand ion fixation on the basis of a simple model was presented by

Harris and Rice in 1954. In this model the counterions are attracted to the neighborhood of fixed ions by the strong electrostatic fields present and they form a kind of ion pair.* In contrast, Fuoss[3] used a model based solely on the occlusion of ions within the polymer domain. Presaging some later work. Oosawa et al.[5] divided the " Ion-kotei " into two groups: the ions occluded inside the polymer domain (first discussed by Fuoss[3]) and the site-bound ions introduced by Harris and Rice. A classification scheme does not, however, inevitably solve a scientific problem and there remains much to be done before a quantitative understanding of the original observation of Kern can be reconciled with molecular models of the polyion. It is the purpose of this chapter to discuss several aspects of the phenomenon of ion binding from a molecular point of view. This chapter is intended to supplement the discussion of the counterion distribution presented in Chapter 5.

9.1 Specific Ion Binding with Bond Formation[6,7]

It is well known that most transition metal ions form complexes with other ions or molecules, some well-known examples being:

$$[Co(NH_3)_4]^{+3}, [Cu(NH_3)_4]^{+2}, [Cu(COO)_4]^{-2}. \qquad (9.1.1)$$

It has been observed that most of the complex forming metal ions have a coordination number of 6, but the coordination numbers of Cu^{++}, Cd^{++}, and Zn^{++} are 4 and that of Ag^+ is 2. If a ligand has in the same molecule several complexing groups, these may coordinate simultaneously with the central metal ion to make a chelate of the type shown in Eq. (9.1.2)

$$
\begin{array}{ccc}
& \overset{H_2}{} & \\
OC{-}O & N{-}CH_2 & \\
| & \diagdown \nearrow & | \\
& Cu & \\
| & \nearrow \diagdown & | \\
H_2C{-}NH_2 & O{-}CO &
\end{array}
\qquad (9.1.2)
$$

Chelated complexes are usually more stable than the corresponding non-chelated complexes of similar chemical nature.

The bond between a metal ion and a ligand may be conveniently represented (though not necessarily correctly) as partialy ionic and partialy covalent. The contribution of each extreme bond type to the bond of course differs with the species of the metal ion and the ligand, i.e., hydrates of the alkali ions Na, K, ··· are probably purely ionic in character, whereas complexes such as those shown in Eq. (9.1.1) have much greater covalent character. The study of the nature of the ligand-metal bond has been one of the most active fields in theoretical chemistry in recent years. Several theories

---

* The reader is referred to Chapter 7 for an interpretation of these ion pairs in terms of a cell model of the counterion polyelectrolyte system.

of bond formation have been presented, detailed descriptions of which are unnecessary for the purposes of this book. (See reference 7.)

Examination of the theories mentioned shows that all metal ions should have some tendency to form complexes with other groups, the magnitude of the complexing tendency depending upon the ionization potential and the electronegativity of the metal. If the divalent metal ions are ordered according to the strength of the complexes formed, it is found that[7]

$$Pt > Pd > Hg > Be > Cu > Ni > Co > Pb > Zn > Cd > Fe > Mn > Cu > Sr > Ba.$$

Univalent and tervalent ions have not been extensively studied, but data on the systems univalent ions—dibenzoylmethanate ion show[8] the binding sequence

$$Ag > Tl > Li > Na > K > Rb > Cs,$$

and for trivalent ions—acetylacetonate ion[9] it is found that

$$Fe > Ca > Al > Se > In > Y > Pr > Ce > La.$$

The complexes of trivalent cobalt and chromium are generally much more stable than those of Fe^{+++}. In general the stabilities of metal complexes increase with increases of the charge on the central ion, with decreasing radius of the central ion, and with increasing electron affinity of the metal ion. It is clear then that the alkali metal ions will have the least tendency to form complexes and the transition metals will have the greatest tendency to form complexes. In aqueous solution only exceptionally strong chelating agents are able to make complexes with alkali metal ions, i.e., the complexes of alkali metal ions with ethylenediaminetetraacetic acid,[10] and uranyl diacetic acid[11] are known.

The detection of a complex compound in solution is not always easy. Complex compounds of the transition metals can be studied by the use of absorption spectroscopy in the visible region. The absorption of light is in these cases due to an electronic transition between the d orbitals, and hence cannot be observed for most metals. Often two absorption bands are observed, one being the d absorption mentioned above, and the other an absorption due to an electron transfer within the molecular complex, in most cases from the metal to the ligand. This latter transition gives rise to the so-called " charge transfer spectrum," and can be observed in all complexes including those of the alkali metal ions. The observation of a charge transfer spectrum is one of the few methods which permit the detection of certain classes of complex compounds in aqueous solution. Unfortunately, charge transfer spectra usually appear in the ultraviolet region where interfering absorptions make the experiment difficult. Two other spectroscopic techniques, the Raman effect, and nuclear magnetic resonance have also been used for the detection of complex compounds.

The strength of the complex between a metal ion and a ligand in solution is usually expressed in terms of a dissociation constant. Let us denote the concentration of the metal ion by [M] and that of the ligand by [A], with obvious extensions to the complex species. The equilibrium constants are defined by

$$M + A \rightleftharpoons MA \qquad K_1 = \frac{[MA]}{[M][A]}$$

$$MA + A \rightleftharpoons MA_2 \qquad K_2 = \frac{[MA_2]}{[MA][A]} \qquad (9.1.3)$$

$$\vdots$$

$$MA_{n-1} + A \rightleftharpoons MA_n \qquad K_n = \frac{[MA_n]}{[MA_{n-1}][A]}.$$

Note that the possibility of defining a meaningful equilibrium constant presupposes a unique configuration for the complex. To determine the complex formation constants, the spectroscopic methods described above are useful. These give directly the amount of complex compound in solution. In addition to the spectroscopic methods, a few indirect techniques can be employed.

(1) Bjerrum's titration method: This method depends upon the competition of the cation M and hydrogen ion for the ligand A defining the equilibrium

$$[HA] + [HA_{i-1}] \rightleftharpoons [H] + [HA_i]$$

The titration curve of HA will then be displaced to lower pH in the presence of the complex forming cation M. Writing C_A for the stoichiometric ligand concentration, α for the degree of neutralization of its conjugate acid, and $[X] = \sum_i [MA_i]$ for the concentration of the ligand in the various complex ions formed,

$$[A] = C_A\alpha + [H] - [OH] - [X]$$

$$C_A = [HA] + [A] + [X] \qquad (9.1.4)$$

$$[H][A]/[HA] = K_a$$

thereby permitting the calculation of [X]. The average number of ligands complexed with a cation present at a stoichiometric concentration C_M is then

$\bar{n} = [X]/C_M$, with the relationship between $\bar{n}$ and [A] given by

$$\bar{n} = \frac{\sum_{1}^{n} \prod_{1}^{i} K_j[A]^i}{1 + \sum_{1}^{n} \prod_{1}^{i} K_j[A]^i}. \tag{9.1.5}$$

Therefore, provided that the successive formation constants are far apart, $[A] = (1/K_1)$ for $\bar{n} = 0.5$ and the other constants can be obtained successively. The treatment for the general case, where the successive formation constants may be close to one another, is given by Bjerrum[12] and by Martell and Calvin.[13]

Bjerrum's method gives the concentration constants rather than thermodynamic equilibrium constants. If the complex has a large dissociation constant, the nonideality of the solution may not be neglected. Moreover, it is known both theoretically and experimentally that the addition of a neutral salt, such as NaCl, will give a shift of pH in polyelectrolyte solutions, even when there is little possibility of complex formation. In the particularly interesting case when the complexing tendency is very small, it is difficult to distinguish between the pH shift due to true complex formation and that resulting from such an electrostatic effect. This error may usually be neglected if the experiment is conducted in a medium with a sufficient amount of neutral salt.

(2) Potentiometric measurements employing an electrode of the metal whose complexes are being studied provide a convenient method for the direct determination of the free cation activity. The method is, however, limited by difficulties in obtaining reversible electrodes of some of the metals which are most interesting in their complexation behavior. In polyelectrolyte solutions, moreover, it is almost impossible to distinguish between the activity coefficient depression due to the electrostatic interaction and the effective depression due to complex formation.

(3) The activity of the free cation may be fixed by employing a second phase acting as a " cation buffer." This buffer may be a sparingly soluble salt or an ion exchange resin.[14] If a radioactive isotope of the cation is used, it may be employed at extremely low concentrations so that the amount of ligand present in the various complexes is negligible compared to the total amount of ligand in solution. If the stoichiometric concentration of M in solution is C_M and C_M^0 in the presence and the absence of the ligand, respectively, then

$$C_M - C_M^0 = \sum [MA_i] \tag{9.1.6}$$

and since $C_M^0 = [M]$ we may use the relations (9.1.3) to obtain

$$\sum [MA_i] = [M]\{K_1[A] + K_1K_2[A]^2 + \cdots\}. \tag{9.1.7}$$

Setting $[A] = C_A$, one finds

$$\frac{C_M - C_M^0}{C_M^0} = K_1 C_A + K_1 K_2 C_A^2 + \cdots \qquad (9.1.8)$$

yielding the successive formation constants from determination of the cation concentration as a function of C_A at constant ionic strength. However, as in the preceding method, it may be very difficult to differentiate between an effect due to electrostatic interaction and that due to complex formation if a polyelectrolyte is used.

(4) Ordinary chemical analysis, for example, polarography or dialysis by which all free ions are removed and their concentrations determined, may occasionally be of use.

In summary, the direct methods of observation can show the existence of complex compounds under a wide variety of conditions, whereas the indirect methods give the equilibrium constant only when the complex formation constant is so large that the electrostatic interaction between the free ions may be neglected. In polyelectrolyte solutions, the electrostatic interaction is generally very large and therefore great care must be taken when studying complex formation by weakly complexing ions. The complexes of the transition metal ions are easily studied and the complex formation constants simply determined as will be seen in the following.

The complexes of linear polyions with strongly complexing ions have been studied by Morawetz, Gregor, Wall, and their co-workers. Morawetz et al.[15] determined the complex formation constant for alkaline earth metal ions with a maleic acid copolymer, for which it could be assumed that a chelate involving two neighboring carboxyl groups is the only type of complex formed, and the titration shift could be interpreted in terms of the extent of counterion binding. These authors used Bjerrum's titration method. To apply this method to a polyelectrolyte system, account must be taken of the fact that the apparent ionization constant K_a of a polyelectrolyte varies with the degree of ionization. Morawetz et al. neglected the specific effects of the counterion chelation on the polymer configuration and the equilibrium constant K_a and assumed that K_a depends only on the net charge density, ρ, along the polymeric chain. Equation (9.1.4) must then be replaced, for the binding of bivalent cations, by the relations

$$[A] = C_A \alpha + [H] - [OH] - [X]$$
$$C_A = [HA] + [A] + [X]$$
$$\frac{[H][A]}{[HA]} = K_a(\rho) \qquad (9.1.9)$$
$$\rho = [A] + \frac{\bar{n} - 2}{\bar{n}} \cdot \frac{[X]}{C_A}$$

which may be combined to give the relation

$$\frac{2K_a(\rho)}{\bar{n}\rho C_A + (2 - \bar{n})[\alpha C_A + [H] - [OH]]} = \frac{[H]}{C_A(1 - \alpha) - [H] + [OH]}.$$

$$(9.1.10)$$

The function $K_a(\rho)$ is obtained from the titration curve of the polymeric acid in the absence of complexing ions and [X] is calculated from Eq. (9.1.10). Thus, the chelate formation constant

$$K_f = \frac{[X]}{[M][A]} \qquad (9.1.11)$$

can be calculated, since the free metal concentration [M] is the difference between the stoichiometric metal concentration and the chelate concentration. Table 9.1.1. gives a comparison between the values of K_f for alkaline earth

TABLE 9.1.1[a]

COMPARISON OF CHELATE FORMATION CONSTANTS OF ALKALINE EARTH IONS,
WITH *VEE/MA*, *ST/MA*, AND SUCCINIC ACID

	log K_f			
	Ca^{++}	Mg^{++}	Ba^{++}	S^{++}
VEE/MA ($\rho = 1.3$)	2.45	2.30	2.00	1.96
ST/MA ($\rho = 1.3$)	2.11	1.74	1.36	1.46
Succinic acid [13]	1.16	1.02	0.97	0.75

[a] Reproduced from ref. 15.

complexes with the copolymers of maleic acid-styrene and of maleic acid-vinyl ethyl ether at an arbitrarily chosen value of $\rho = 1.3$ and the analogous complexes with succinic acid.

The binding of Cu^{++} to polyacrylic and polymethacrylic acids, which has been investigated by Kotliar and Morawetz,[16] Wall and Gill,[17] and Gregor et al.,[18] presents more difficult problems. The high affinity of the polyanions for Cu^{++} leaves no doubt that chelate formation must be involved. The cooperation of two neighboring carboxylates would, however, result in eight-membered rings which are known to form with great difficulty and larger rings involving carboxylates at greater distances along the polymeric chain must account for most of the cation binding. Moreover, Fronaeus' data[19] on the binding of acetate by Cu^{++} also makes it clear that the possibility of association with more than two carboxylate groups cannot be disregarded when a Cu^{++} ion finds itself surrounded by the high local concentration of carboxylate in the region occupied by the polyanion.

In analyzing the factors which determine the equilibrium between a polyacid and a complex forming cation, we have problems more difficult than in analyzing the ionization constants of a polybasic acid, not only because the complex formation constant depends on the charge density of the polyacid, but also because the metal ion has a coordination number greater than 2. Gregor et al.,[18] and Morawetz and Kotliar[16] have suggested very interesting (and similar) methods to resolve the dependence of the complex formation constant on the charge density of polyion. Both replaced the charge density dependent part in the complex formation constant by the ionization constant of the polyacid, a quantity which can be easily determined by titration. Moreover, Morawetz and Kotliar simplified the discussion by distinguishing between the first step, in which the free cation combines with a single ligand of the polyion, and subsequent steps in which the bound cation forms chelates by coordination with further ligand groups.

The equilibrium established in the first step will depend on the stoichiometric concentration of ligand groups in solution and the electrostatic field of the polyion

$$\frac{[Cu\,A]^+}{[Cu^{++}][A]} = K_1^0 \exp\left(\frac{-\Delta A'_{elec}}{RT}\right) \tag{9.1.12}$$

where $\Delta A'_{elec}$ is the electrostatic free energy change corresponding to the first association step and K_1^0 is the first complex formation constant of a suitable monodentate. Assuming that electrostatic factors account for all of the variation of the apparent ionization constant of polymeric acids with the degree of ionization ΔA_{elec} is related to the electrostatic free energy of ionization by

$$\Delta A'_{elec} = -2\Delta A_{elec} \tag{9.1.13}$$

and we obtain

$$\frac{[Cu\,A]^+}{[Cu^{++}][A]} = K_1^0 \left(\frac{K_a^0}{K_a}\right)^2 \tag{9.1.14}$$

where K_a and K_a^0 are the apparent and the intrinsic ionization constants, respectively.

In dilute solutions of polymeric acids and with a large excess of carboxyl groups over Cu^{++}, the probability of chelation with two carboxylate groups of the same chain will be heavily favored over the association with groups belonging to different polymers. Such chelation equilibria depend, therefore, only on the state of the isolated macromolecular coils, being independent of their number as long as interpenetration of the polymers can be neglected. Under these conditions the equilibrium equations are

$$\frac{[Cu\ A_2]}{[Cu\ A]^+} = K_2' \ ; \quad \frac{[Cu\ A_3]^-}{[Cu\ A_2]} = K_3' \ ; \cdots \qquad (9.1.15)$$

where K_2', K_3', etc., depend only on the degree of ionization of the polyion, provided the number of ligand groups is very large compared to the number of chelated cations so that interference between different parts of the polymer engaged in chelate formation may be disregarded. At relatively higher " loading " of the polyion with bound cations, such interference will result in a severe restraint on possible chain configurations and the magnitudes of K_2', K_3', $\cdots$ would be expected to decline.

In view of the factors described above, the calculation of " chelate formation constants," in the conventional sense, has no meaning with respect to the chelation of ions by polyelectrolytes. The best approach to a study of such systems requires the use of at least two experimental methods, one to determine the concentration of the free ions and the other to provide information about the type of complex formed. Using for the concentration of all species of bound copper

$$Cu_b = \sum_i [Cu\ A_i] \qquad (9.1.16)$$

Eqs. (9.1.14) and (9.1.15) may be combined to yield

$$\frac{Cu_b}{[Cu^{++}]} = \bar{K}[A] \qquad (9.1.17)$$

$$\bar{K} = K_1^0 \left(\frac{K_a^0}{K_a}\right)^2 [1 + K_2' + K_2' \cdot K_3' + \cdots].$$

The ratio $Cu_b/[Cu^{++}]$ was determined polarographically by Wall and Gill and by dialysis equilibrium by Kotliar and Morawetz. With polyacrylic acid, both methods gave $Cu_b/[Cu^{++}]$ proportional to the first power of [A], as the theory requires. With polymethacrylic acid, studies of dialysis equilibria indicated a decrease in the affinity of the polyion for Cu^{++} with increasing polyelectrolyte concentration. According to Morawetz,[6] this may be due to molecular association of the polyanions.

Spectroscopic studies of Cu^{++} complexes were carried out by Kotliar and Morawetz (visible region),[16] Wall and Gill (ultraviolet),[17] Morawetz (ultraviolet),[20] and Morawetz and Sammak.[21] The intensity of the absorption is a function of the degree of neutralization of the polyion but the location of the absorption maximum is independent of the degree of neutralization. Therefore, it is likely that only a single type of complex is formed. By comparing the location of the absorption maximum with the absorption spectra of cupric acetate complexes, Morawetz and Kotliar came to the conclusion that $[Cu(COO)_4]^{--}$ is the complex formed.

Studies of the complexes of metal ions other than Cu^{++} are few in number. Aside from the work of Morawetz, Kotliar, and Mark mentioned above, Gregor et al.[18] determined the complex formation constants of Mg^{++}, Ca^{++}, Mn^{++}, Co^{++}, and Zn^{++} with polyacrylic acid. The formation constants for these ions are much smaller than that for Cu^{++}, and those of the alkaline earth metals are smaller than those of the transition metals, as expected. Ag^+ has a considerably higher propensity towards complex formation. In view of this observation, it is possible that the activity coefficient of Ag^+ in the presence of a polyion, which was measured by Kagawa and Katsuura[22] and discussed by Kagawa and Nagasawa,[23] should be reinterpreted in terms of a complex formation constant.

Several studies of the precipitation of a polymeric acid by the addition of divalent cations have been published. However, since precipitation is not always related to complex formation, we shall not discuss these experiments.

The investigation of the complexes of alkali metal ions with linear polyelectrolytes is particularly interesting from the point of view of theory. Recently Strauss and his co-workers [25-30] have carried out a series of studies of the binding of alkali metal ions by polyphosphate ions. It is not at present clear whether this binding is due to the formation of a covalent bond or simply due to Coulombic interactions (ion pair formation). Although there may be complex formation between alkali metal ions and the phosphate ion, it seems more probable that the binding is due to Coulombic interactions. The work of Strauss will be discussed in detail in the following section.

A very interesting observation was reported by Strauss et al.,[24] who found that quaternized polyvinylpyridinium bromide acquired a negative charge in solutions of high bromide ion concentration. This proves that the cationic group of the polymer can bind more than an equivalent amount of counterions and that other than Coulombic interactions must be involved. The work cited is by no means the only experiment reporting the binding of anions to neutral polymers or proteins; for example, there is anion binding to polyvinylpyrrolidone,[31] and to polyvinylalcohol.[32] In general, the amount of anion binding in these cases is very small, and the nature of the binding remains unknown.

9.2 Ion-Pair Formation in Simple Electrolyte Solutions[34,35]

A dissociation constant (or complex formation constant) can be successfully used to describe the properties of strongly complexed ions since these are, after all, completely analogous to weak electrolytes. It is well known, however, that a dissociation constant description of the properties of strong electrolytes is unsuccessful (see Chapter 2). Moreover, no detectable quantity of undissociated molecules has ever been found in aqueous solutions of strong electrolytes, such as the alkali halides. The Debye–Hückel theory is, however, inadequate to describe the behavior of concentrated solutions of strong electrolytes. One convenient method of describing such solutions proceeds by the definition of a complex (ion pair). It must be emphasized that the defined ion pair is entirely different from an undissociated molecule in the sense originally used by Arrhenius.

The postulation of ion pairs in solutions of strong electrolytes was first presented by Bjerrum.[33] He suggested that the failure of the Debye–Hückel theory in concentrated solutions was due to the breakdown of the assumption that no ions are close to other ions, in which region the electrostatic potential is very large. To account for the interactions in the region of high potential Bjerrum divided the ions into two classes; free ions which obey the Debye–Hückel theory and associated ion pairs composed of two oppositely charged ions sufficiently close to each other that screening of the interaction by other ions cannot occur. The associated ion pairs are not true undissociated molecules but are nevertheless assumed to behave in a manner similar to undissociated molecules, and to have no electrostatic effect on the remaining free ions. It is clear that the division of the ions into two such groups is arbitrary.

Bjerrum's specification of an ion pair included the assumption that two oppositely charged ions are paired if they come closer to each other than a distance $r_{min} = [(z_+ z_- q^2)/2DkT]$. The distance r_{min} was chosen because the probability of finding an oppositely charged ion at a distance r from a given ion,

$$\mathscr{P}(r)\, dr = \frac{N_0 c}{1000} [\exp(-z_+ z_- q^2/DrkT)]4\pi r^2\, dr \qquad (9.2.1)$$

has a minimum at r_{min}. Note that the free ions in the solution are assumed to have no effect on the interaction, i.e., the Coulomb potential is unscreened. From (9.2.1) it is easily seen that the degree of association, $1 - \alpha$, is given by

$$1 - \alpha = \frac{N_0 c}{1000} \int_a^{r_{min}} [\exp(-z_+ z_- q^2/DrkT)]4\pi r^2\, dr \qquad (9.2.2)$$

where α is the " apparent " degree of dissociation and a is the distance of closest approach of the two ions. For 1–1 type electrolytes in water at 18°C, r_{min} has the value 3.52 A and, consequently, electrolytes of 1–1 type which have diameters less than 3.52 A will more or less form ion pairs, while other electrolytes having diameters greater than 3.52 A ought to follow the Debye–Hückel theory.

If it is assumed that the ion-pair formation can be expressed in terms of an equilibrium constant K^{-1}, it follows that

$$K^{-1} = \frac{1 - \alpha}{\alpha^2 c} \frac{\gamma_{\text{ion pair}}}{\gamma_+ \gamma_-} \tag{9.2.3}$$

for z–z type electrolytes, and where the γ's are the activity coefficients of the respective species. For most $1 - 1$ electrolytes in water, it is sufficient to write

$$K^{-1} = \frac{1 - \alpha}{C} \cdot \frac{1}{\gamma_\pm} \tag{9.2.4}$$

since α is nearly unity and $\gamma_{\text{ion pair}}$ can be assumed to be close to unity. As usual $\gamma_\pm$ is given by the Debye–Hückel theory,

$$\ln \gamma_\pm = -\frac{z^2 q^2}{DkT} \cdot \frac{\kappa}{1 + \kappa r_{\text{min}}}$$

$$\kappa = \left[\frac{8\pi N_0 c(1 - \alpha)z^2 q^2}{1000 DkT} \right]^{\frac{1}{2}} \tag{9.2.5}$$

since the distance of closest approach of the free ions must be, by definition, r_{min}. From Eqs. (9.2.2) and (9.2.4),

$$K^{-1} = \frac{4\pi N_0}{1000} \int_a^{r_{\text{min}}} \exp(z^2 q^2 / DrkT) r^2 \, dr \tag{9.2.6}$$

so that by substitution of

$$Y = \frac{z^2 q^2}{DrkT}$$

$$b = \frac{z^2 q^2}{DakT} \tag{9.2.7}$$

into Eq. (9.2.5) we obtain

$$K^{-1} = \frac{4\pi N_0}{1000} \left(\frac{z^2 q^2}{DkT} \right) \int_2^b e^Y Y^{-4} \, dY$$

$$= \frac{4\pi N_0}{1000} \left(\frac{z^2 q^2}{DkT} \right) Q(b) \tag{9.2.8}$$

where $Q(b)$ is defined by

$$Q(b) = \int_2^q e^Y Y^{-4} \, dY. \tag{9.2.9}$$

Values of $Q(b)$ were tabulated as a function of b by Bjerrum and by Fuoss and Kraus.[36]

An experimental determination of α is possible by any of several methods. If we know the value of a, we can determine the best value of α by use of Eqs. (9.2.3) and (9.2.8), and the observed activity coefficient as calculated from

$$(\gamma_\pm)_{obs} = \alpha\gamma_\pm. \tag{9.2.10}$$

An example given by Bjerrum for the case of 1—1 ions having $a = 2.82 \times 10^{-8}$ cm in water at 18°C is shown in Table 9.2.1.

TABLE 9.2.1[a]

DEGREE OF ASSOCIATION, $1 - \alpha$, IN WATER AT 18°C [35]

$a = 2.82 \times 10^{-8}$ cm

(mol /liter)	0.002	0.005	0.01	0.05	0.1	0.2
$1 - \alpha$	0.002	0.002	0.005	0.017	0.029	0.048

[a] Reproduced from ref. 35.

Precise conductance measurements have been used by Fuoss, Kraus, and their co-workers to determine α. The conductivity of the free ions is analyzed on the basis of the Onsager theory and a detailed description of all the factors of importance has been given by Fuoss.[37] It is easier to study ion association in nonaqueous solution having a lower dielectric constant than that of water for the rather obvious reason that more association occurs in such a medium.† Despite the larger amount of ion association it is pertinent to remark, as do Sadek and Fuoss, that there is ample evidence that strong electrolytes are completely dissociated into ions even in nonaqueous solutions.[38]

The results of many years of research may be summarized with the statement that the equilibrium constant of Eq. (9.3.8) is adequate to describe the association phenomenon. Thus, while the application of the Debye–Hückel theory gives unreasonable—sometimes negative—ionic diameters, the introduction of the ion-pair formation constant with a determined from Eq. (9.2.8) results in ionic diameters of suitable magnitude. These ionic diameters are almost constant with large variations of the solvent and consequently of the dielectric constant. An example is given in Table 9.2.2.

† Nevertheless, analysis of conductance data (R. L. KAY, *J. Am. Chem. Soc.* **82**, 2099 (1960)) gives the following approximate association constants for the alkali halides in aqueous solution at 25°C: LiCl (0.0), NaCl (0.2), KCl (0.4), RbCl (0.6), CsCl (0.8), NaBr (0.2), NaI (0.0), KBr (0.4), KI (0.3).

TABLE 9.2.2ᵃ

CONSTANTS FOR TETRAISOAMYLAMMONIUM NITRATE IN DIOXANE–WATER
MIXTURES [36]

Wt % H_2O	D	$-\log K$	K	$a \times 10^8$
0.60	2.38	15.7	2×10^{-16}	6.01
1.26	2.56	14.0	1×10^{-16}	6.23
2.35	2.90	12.0	1×10^{-12}	6.36
4.01	3.48	9.6	2.5×10^{-10}	6.57
6.37	4.42	7.53	3.0×10^{-8}	6.65
9.50	5.84	5.78	1.65×10^{-6}	6.45
14.95	8.5	4.00	1.00×10^{-4}	6.50
20.2	11.9	3.05	9.0×10^{-4}	6.70
53.0	38.0	0.60	0.25	6.15

ᵃ Reproduced from ref. 36.

It may be surmised that the concentrations of water in the mixtures used are high enough to maintain constant solvation of the ions.

Bjerrum's theory has a few obvious defects,[39] some mathematical and some physical. First, note that Bjerrum's probability distribution function diverges at the origin. This difficulty was surmounted by arbitrarily cutting the integral off at the point where the integral has a minimum and more detailed calculations justify the procedure.[39] Second, the theory counts as associated pairs ions which are not in physical contact thereby giving the impression that the definition of an ion pair is arbitrary. Third, the cutoff leads to the prediction that association should cease abruptly at a critical value of the dielectric constant D and/or temperature T for a given electrolyte. In Fig. 9.2.1, the change of K calculated using an averaged value of a is depicted by the solid line. The critical dielectric constant beyond which no ion pairs could exist is clearly shown in the figure. Fourth, dependent upon the solvent, different values of a are obtained from the theory, a manifestation of the fact that the ion pair can contain solvent molecules between the two ions.

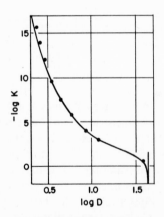

FIG. 9.2.1. A comparison of the calculated and observed dependence of the ion pair dissociation constant on the macroscopic dielectric constant. [Reproduced from ref. 36.]

There have been some recent attempts to overcome these defects of the Bjerrum theory. Denison and Ramsey[40] used a Born cycle to show that the dissociation constant K should be a continuous function of D, obtaining the relation

$$K = e^{-b} \tag{9.2.11}$$

for 1–1 electrolytes, where

$$b = q^2/DakT. \tag{9.2.12}$$

At a different level, Gilkerson[41] used the cell model partition function of Kirkwood[42] to construct a detailed molecular model of the ion pair. According to the zeroth approximation of the Kirkwood free volume theory,[42] the partition function for a particle in solution is given by

$$\mathscr{Q}_i = \left(\frac{2\pi m_i kT}{h^2}\right)^{3/2} g_i v_{f_i} \bar{\sigma}_i e^{-E_{0i}/kT} \tag{9.2.13}$$

where v_f is the free volume available to the particle, and $\bar{\sigma}$ is a factor varying between unity for solids and e for gases and related to the "communal entropy." The factor g is the internal rotational and vibrational contribution to the molecular partition function, while E_0 is the potential energy difference between the particle in the liquid and in the gaseous states. The Gibbs free energy for the dissociation of one mole of ion pairs

$$A^+, B^- \rightleftharpoons A^+ + B^-$$

may be shown to be

$$\Delta G^0 = -RT \ln(\mathscr{Q}_+\mathscr{Q}_-/\mathscr{Q}) + W \tag{9.2.14}$$

where W is the work necessary to remove the ions from a distance of closest approach, a, to infinite separation. It has been shown by Denison and Ramsey that

$$W = N_0 q^2/Da. \tag{9.2.15}$$

Therefore,

$$K = \left(\frac{2\pi\mu kT}{h^2}\right)^{3/2} g v_f \bar{\sigma} e^{-E_s/kT} e^{-N_0 q^2/DakT} \tag{9.2.16}$$

where $\mu = [m_+m_-/(m_+ + m_-)]$, $E_s = E_- + E_+ - E_\pm$ and the factor $g v_f \bar{\sigma}$ is $g_+ g_- v_{f+} v_{f-} \bar{\sigma}_+ \bar{\sigma}_-/g_\pm v_{f\pm} \bar{\sigma}_\pm$. The introduction of the free volume and E_s renders K specifically dependent upon the solvent. E_s is the difference between solvent-ion and solvent-ion-pair interaction energies. To a first approximation it should be proportional to the dipole moment μ_s of the solvent. Then

$$E_s = A\mu_s \tag{9.2.17}$$

and Eq. (9.2.16) becomes

$$K = \left(\frac{2\pi\mu kT}{h^2}\right)^{3/2} gv_f\bar{\sigma}e^{-A\mu_s/kT}e^{-N_0q^2/DakT}. \tag{9.2.18}$$

If the dielectric constants of two solvents are almost identical and their chemical constitutions are not too dissimilar, $(gv_f\bar{\sigma})$ might possibly be the same in the two media and the values of A should not be too different. Under these conditions, the following relationship should exist among three different solvents having almost equal dielectric constants,

$$\frac{\ln(K_1/K_2)}{\ln(K_1/K_3)} = \frac{\mu_1 - \mu_2}{\mu_1 - \mu_3} \tag{9.2.19}$$

where 1, 2, 3 denote the different solvents. This relationship was compared with the experimental results of Accasina et al.[43] for several salts in the three solvents ethylene chloride (solvent 1), ethylidene chloride (solvent 2), and ortho-dichlorobenzene (solvent 3), having dielectric constants of 10.23, 10.00, and 9.93, respectively. (See Table 9.2.3.) The very good agreement observed shows that the distance of closest approach of the two ions, a, is independent of the solvent. On the other hand, if Bjerrum's theory is used to describe the ion association, three different values of a are obtained.

Equation (9.2.18) can be rearranged to the form

$$F = \log K_a - \tfrac{3}{2}\log T + E_s/2.303\ RT \tag{9.2.20}$$

$$= \log\left[\left(\frac{2\pi\mu k}{h^2}\right)^{3/2}(gv_f\bar{\sigma})\right] - N_0q^2/2.303DaRT.$$

TABLE 9.2.3[a]

RATIOS OF DISSOCIATION CONSTANTS

Salt	$\dfrac{\log(K_1/K_2)}{\log(K_2/K_3)}$	$\dfrac{\log(K_1/K_3)}{\log(K_1/K_2)}$
Et$_4$ N Pi	2.36	1.73
Pr$_4$ N Pi	2.44	1.69
Bu$_4$ N Pi	2.65	1.60
Am$_4$ N Pi	2.56	1.64
Octd Me$_3$ N Pi	2.20	1.86
Calc. from		
Eq. (9.2.19)†	2.20	1.83

[a] Reproduced from ref. 41.

† Dipole moments used for the calculation: $\mu_{s_1} = 1.75$ D; $\mu_{s_2} = 2.05$ D; $\mu_{s_3} = 2.30$ D.

If E_s is adjusted until a plot of F versus $1/DT$ for a given electrolyte in several different solvents yields a straight line (see Fig. 9.2.2), then the value of a

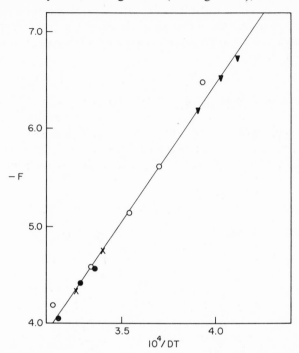

FIG. 9.2.2. Plot of $-F$ versus $1/DT$ for Bu₄NPi in ethylene chloride (●), ethylidene chloride (×), propylene chloride (▼, $-F + 1.45$), and anisole (○, $-F - 10.48 \times 10^4/DT - 4.30$). [Reproduced from ref. 41.]

determined from the slope of the line is much lower than that obtained using the Bjerrum theory.

In general, the association constant may be written in the form,

$$K_a = K_a^0 e^b \tag{9.2.21}$$

$$b = e^2/DakT$$

a relationship shown by Fuoss and Kraus to be consistent with data of the highest precision. It should be noted that Fuoss has derived Eq. (9.2.21) by a method different from that of Denison and Ramsey and Gilkerson and finds for the pre-exponential term

$$K_a^0 = (4\pi N_0 a^3)/3000. \tag{9.2.22}$$

We note that the ion pairs defined in Eqs. (9.2.11)–(9.2.22) consist of two oppositely charged ions which are in physical contact, a model basically different from that of Bjerrum.

We have previously remarked that in water at room temperature, no ion pair formation is observed for the case of 1 – 1 electrolytes.† On the other hand, if the valence of the ion is greater than or equal to three and moreover the ion is large, a charge transfer absorption due to the presence of ion pairs is often detected even in solvents such as water. An example of such behavior is provided by the salt

$$Co(NH_3)_6^{3+} \cdot I^- \rightleftharpoons Co(NH_3)_6^{3+} + I^-.$$

The bond between $Co(NH_3)_6^{3+}$ and I^- is different from that in an ordinary complex (i.e., Co—NH₃) and is presumed to be electrostatic in origin. The descriptive designation " outer sphere complex," has been suggested for such ion pairs.[7] The equilibrium association constants can be determined by spectroscopic measurements, and conductance measurements. Some examples are given in Table 9.2.4.‡

TABLE 9.2.4ᵃ

ASSOCIATION CONSTANTS FOR OUTER SPHERE
COMPLEX IONS AT 25°C[44]

Ion pair	K_a
$Co(NH_3)_6^{3+} \cdot Cl^-$	74 ± 4
$Co(NH_3)_6^{3+} \cdot Br^-$	46 ± 2
$Co(en)_3^{3+} \cdot Br^-$	$21 \pm 0 \cdot 5$
$Co(en)_3^{3+} \cdot I^-$	$8 \cdot 6 \pm 0 \cdot 4$

en: ethylenediamine

ᵃ Reproduced from ref. 44

Recently, Jardetsky and Wertz[45] have demonstrated the existence of weak complexes of the sodium ion in some aqueous solutions of strong electrolytes by nuclear spin resonance. Although no chemical shifts were observed in solutions of a number of inorganic sodium salts (the halides, NaN_3, Na_2SO_3 $NaClO_4$, Na_3PO_4, Na_2SO_4, NaOH), the widths of the resonances showed an increase with concentration taken relative to $3N$ NaCl in the cases of the hydroxide, phosphate, iodide, thiocyanate, versenate, and citrate of sodium. Similar line broadening was observed in solutions of many other sodium salts of hydroxy-organic acids. Broadening of the absorption curve was not observed for NaCl, NaF, NaBr, NaN_3, $NaCH_3CO_2$ or $NaHCO_2$. The line amplitude taken relative to the line amplitude of $3N$ NaCl increases linearly with concentration in the case of the latter electrolytes, but in the former

† See however the footnote on page 439.

‡ Outer sphere complexes to polyelectrolytes have also been observed. See, for example, M. MATHEWS, *Biochim. Biophys. Acta.* **37**, 288 (1960).

electrolytes, the dependence on concentration is not linear. From these experiments, Jardetsky and Wertz concluded that when the broadening of the absorption curve occurs only in highly concentrated solutions, it can be attributed to ion pair formation enhanced by a relative unavailability of solvent molecules, and that its presence in more dilute solutions probably signifies an actual preference on the part of the sodium ion for the given complexing agent over a solvent molecule. Sodium hydroxide, sodium iodide, and sodium thiocyanate would fall into the first category, with sodium pyrophosphate, sodium versenate, and sodium citrate representing the second class. With this assignment of results, there are no ion pairs observed in solutions of the sodium halides, sodium acetate, or sodium formate.*

A much more clear-cut case is presented in the determination of the true dissociation constant of very strong acids such as nitric acid and sulfuric acid. Hood et al.[46] and others [47,48] have used nuclear magnetic resonance measurements to supplement studies using Raman intensities. These methods are free of the problem of ionic interaction which is inherent in the classical methods discussed previously. The observations indicate very clear chemical shifts of the proton resonance in these cases, so that true undissociated acid must exist in these solutions. Some examples of dissociation constants of strong acids determined by this method are: $K = 22$ for nitric acid, 38 for perchloric acid, and 0.18 for iodic acid.

Despite the extensive work since Bjerrum's original proposal, it appears to the authors that there remain unsatisfactory features in the treatment of ion pairing. The new definition, by virtue of which only two ions in physical contact are called ion pairs has a plausibility which hides the fact that the classification is still basically arbitrary. In the terminology of Fuoss[39]: " two free ions have each three degrees of translational freedom and the ion pair they form has three degrees each of translational and rotational freedom. By insisting on contact in pairs, the vibrational degree of freedom is excluded." Moreover, it is unlikely that the new definition could account for the behavior of concentrated solutions of very simple electrolytes such as NaCl, in which ion pairs in physical contact cannot be detected even by nuclear magnetic resonance. The Bjerrum definition or an even more arbitrary definition may still be useful in the high concentration region. The difference between the ion pairs defined by Bjerrum and those defined by physical contact, was the subject of a paper by Sadek and Fuoss[38] published prior to the theory of Denison, Ramsey, Gilkerson, and Fuoss. Sadek and Fuoss divided the ion pairing process into two parts: the first step is considered to be ion pairing in the Bjerrum sense,

$$A^+ + B^- (+ S) \rightleftharpoons A^+SB^- \tag{9.2.23}$$

* See, however, footnote on page 439.

and the second step is the process of desolvation

$$A^+SB^- \rightleftharpoons A^+B^- + S \qquad (9.2.24)$$

where S denotes solvent molecules. The equilibrium constant for process (9.2.23) is given by the Bjerrum theory as

$$[A^+][B^-] = K_d[A^+SB^-]$$

$$K_d^{-1} = (4\pi N_0/1000)(q^2/DkT)^3 Q(b) \qquad (9.2.25)$$

$$b = q^2/DakT$$

and that for process (9.2.24) is

$$[A^+B^-][S] = k[A^+SB^-]. \qquad (9.2.26)$$

If [S], which is the local concentration of the polar solvent near the ion pair, can be assumed to be a constant independent of the bulk composition of the solvent mixture, then

$$k/[S] = k_S = \text{const} \qquad (9.2.27)$$

and the equilibrium condition for the two stage process may be represented in the form

$$\frac{[A^+][B^-]}{[A^+B^-] + [A^+SB^-]} = K_d/(1 + k_S). \qquad (9.2.28)$$

Sadek and Fuoss used this relation to explain the differences between the association constants of tetrabutyl ammonium bromide in nitrobenzene-carbon tetrachloride, methanol-carbon tetrachloride, and methanol-benzene mixtures. They deduce that the solvent molecules in CCl_4—ϕNO_2 are easily expelled from the ion pair and thereby that the association constant in this mixture is much smaller than that in the other mixtures cited. It is concluded that, if the ion is strongly solvated and if k_S is negligible relative to unity, then Bjerrum's theory is valid. However, if k_S is very large, the new definition of Denison, Ramsey, Gilkerson, and Fuoss must be used.

9.3   Counterion Binding by Linear Polyions

We have already noted that the osmotic coefficient of the solvent, the activity coefficient of the counterion, the conductivity of the polyion, as well as many other properties of polyelectrolyte solutions are anomalously low when compared to the properties of simple electrolyte solutions. All of these observations can be qualitatively explained if we assume that some of the counterions of the polyelectrolyte are bound on or inside the polyion. Although some of the experimental results cited may be due to specific complex formation in the sense defined in Section 9.1, such cases will be excluded from consideration in this section. Moreover, most phenomena which appear to require explanation in terms of ion binding have been observed with sodium, potassium, or bromide ions, none of which are believed to form complexes with ordinary acid groups.

Two different molecular models have been presented to account for ion-binding phenomena. Fuoss[3], Kagawa,[5,22] Wall,[49] and their co-workers suggested that ion binding is due to occlusion of counterions by the coiled polyion, whereas Harris and Rice,[4] Strauss,[24-30] and Inagaki[50,51] suggested that it is due to binding of counterions by localization in the vicinity of the fixed charges of the polyion.

The first model recognizes the fact that a polymer in a salt solution has a more or less coiled configuration, and therefore some of the counterions are contained inside the polymeric domain. The distribution of counterions inside and outside the polymeric domain has been studied by Hermans and Overbeek,[52] Kimball et al.,[53] Oosawa et al.,[5] Wall and Berkowitz,[54] Lifson,[55] and Nagasawa and Kagawa.[56] A most extreme case is considered in the theory of Kimball et al. who assumed that the polymeric domain is neutral and that a Donnan membrane equilibrium is established between the "inside" and "outside" of the polymeric domain. If the concentration of the 1-1 type added salt is c_s^0 and a polyion having a fixed charge density $\rho_f = Zq/\frac{4}{3}\pi R_s^3$ is in the solution, the distribution of simple ions is determined by

$$c_+ = c_s^0 \exp[-q\psi(r)/kT]$$
$$c_- = c_s^0 \exp[+q\psi(r)/kT] \tag{9.3.1}$$

where $\psi(r)$ is the electrostatic potential to be determined from the Poisson equation

$$\nabla^2 \psi = -\frac{4\pi}{D}\rho(r). \tag{9.3.2}$$

As usual, $\rho(r)$ is the net charge density,

$$\rho(r) = \rho_f + qc_+(r) - qc_-(r) \tag{9.3.3}$$

and if the polymeric domain is assumed to be electrically neutral,

$$\rho(r) = 0 \tag{9.3.4}$$

and thereby

$$c_+(r) = \frac{1}{2}\left\{\frac{\rho_f}{q} + \left(\left(\frac{\rho_f}{q}\right)^2 + 4c_s^{02}\right)^{\frac{1}{2}}\right\}$$
$$c_-(r) = \frac{1}{2}\left\{-\frac{\rho_f}{q} + \left(\left(\frac{\rho_f}{q}\right)^2 + 4c_s^{02}\right)^{\frac{1}{2}}\right\}. \tag{9.3.5}$$

A moment's reflection serves to convince us that the condition of electro-neutrality is overly restrictive. Inasmuch as a potential difference exists between the inside and outside of the polymeric domain, some free charges must be present at the boundary to maintain the potential difference and the charge distribution. The quantity of charge at the boundary $\tilde{Q}$ is related to the potential difference E_D between the polyion domain and the bulk solution by

$$- \tilde{Q} = \Re E_D \tag{9.3.6}$$

with $\Re$ the capacity of the polyion–counterion system. If the double layer at the boundary is assumed to be of the Helmholtz type, that is of the non-diffusive type, the capacity is

$$\Re = D \frac{R_s^2 - (\delta^2/4)}{\delta} \cong \frac{R_s^2}{\delta} = \frac{D(\kappa R_s)^2}{2\kappa} \tag{9.3.7}$$

where the second form follows when the thickness of the double layer is replaced by $2/\kappa$, an approximation valid if R_s is sufficiently large. The charge $\tilde{Q}$ is then

$$\tilde{Q} = -D \frac{(\kappa R_s)^2}{2\kappa} E_D. \tag{9.3.8}$$

This may be considered to be the surface charge of the polyion sphere, envisaged in the model of Kimball et al.

The first detailed calculation of the surface charge of a spherical polyion was reported by Oosawa et al. [5] and was discussed in Chapter 5. The ratio of the net charge to the total fixed charge, $\langle \lambda \rangle$, is given by

$$\langle \lambda \rangle = \frac{\tilde{Q}}{Zq} = \frac{3DkT}{4\pi q^2 R_s^2(\rho_f/q)} \ln(f + \sqrt{1+f^2}) \tag{9.3.9}$$

where $f = \rho_f/2qc_s^0$.

Equation (9.3.9) may be rewritten as follows:

$$\langle \lambda \rangle = \frac{1}{Zq} DR_s E_D \tag{9.3.10}$$

from which it is clear that the capacity $\Re$ defined in Eq. (9.3.6) is taken to be DR_s, which value corresponds to the capacity of a sphere without an outer ionic atmosphere. The physical origin of this error lies in the assumption that the effect of the charge distribution outside the polymeric domain can be neglected.

The effect of the ionic distribution outside the polymer coil on the surface charge is calculable from the Hermans–Overbeek theory. $\tilde{Q}$ is then defined by

$$\tilde{Q} = \int_0^{R_s} \rho 4\pi r^2 \, dr = -\int_{R_s}^{\infty} \rho 4\pi r^2 \, dr \tag{9.3.11}$$

and the use of the charge density calculated in the Hermans–Overbeek theory gives

$$\tilde{Q} = \frac{4\pi\rho_f R_s^3}{p^2}(1 + p)e^{-p}\left[\cosh p - \frac{\sinh p}{p}\right]$$

$$= D\frac{E_D^0}{2\kappa}(1 + \kappa R_s)e^{-\kappa R_s}[(1 + \kappa R_s)e^{-\kappa R_s} - (1 - \kappa R_s)e^{\kappa R_s}]$$

(9.3.12)

where $p = \kappa R_s$, and E_D^0 is the value of E_D when $c_s^0 \gg \rho_f/q$. If the concentration of added salt, c_s^0, is very small or ρ_f is very large, the computed surface charge is fallaciously large. In view of this situation, Nagasawa and Kagawa

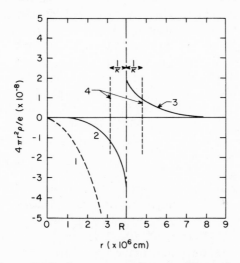

FIG. 9.3.1. Charge distribution at one surface of polymer sphere. 1: ρ_f = constant, 2, 3: Eq. (9.3.13), 4: Donnan–Helmholtz model. [Reproduced from ref. 56.]

suggested that the most practical relation for the computation of the surface charge of a polyion is represented by the approximation

$$\tilde{Q} = D\frac{E_D}{2\kappa}(1 + \kappa R_s)e^{-\kappa R_s}[(1 + \kappa R_s)e^{-\kappa R_s} - (1 - \kappa R_s)e^{\kappa R_s}].$$ (9.3.13)

Now, an exact solution of Eqs. (9.3.1)–(9.3.3) has been obtained by Wall and Berkowitz[54] by numerical methods using a high speed electronic computer. This work was discussed in Chapter 5 and Fig. 5.2.5 shows $\tilde{Q}$ as a function of f for various values of κR_s. Inasmuch as the perturbation calculations of Lifson[55] are in good agreement with the computed potential distribution of Wall and Berkowitz, Eq. (9.3.13) may be expected to be a good approximation. If the charge distribution at the surface as calculated from

(9.3.13) is plotted as a function of the distance from the center of the sphere, it is easily seen that most of the free charge accumulates at the boundary of the polymeric domain (Fig. 9.3.1).

We believe it to be very doubtful that the nonzero surface charge is a major source of counterion binding although it is likely that the surface charge plays a role in determining the interpolymer interactions. Our reasons for coming to this conclusion are as follows: (a) The properties of poly-electrolyte solutions which are determined by the behavior of the counterions (activity coefficient, osmotic pressure, conductivity) are generally independent of the degree of polymerization. However, $\tilde{Q}$(or $1 - \langle \lambda \rangle$) as defined above explicitly depends upon the degree of polymerization. (b) In many experiments the fundamental phenomena are observed in solutions of sufficient concentration that the equivalent polymeric domains occupy the entire volume. (c) The activity coefficients of the byions cannot be explained even qualitatively on this model. The concentration of the byions outside the polymeric domain is much larger than the average concentration. Nevertheless, the activity coefficient of the byion (for example, Cl^-) in Na-polyvinyl alcohol sulfate is less than unity.[57,58] (d) The model is inadequate to explain the behavior of the polyion under the influence of an applied electric field. In contrast to the basic assumption of the domain binding model, it follows from application of the Kirkwood–Riseman or Debye and Bueche theories that the polyion is almost freely drained when $\kappa R_s \gg 1$.[59] (See Chapter 11.) (e) The effect of the species of counterion on the thermodynamic and configurational properties of the polymer cannot be explained.

A different model of ion binding, which we call the site-binding hypothesis, was first presented by Harris and Rice.[4] In this model, counterions are supposed to be localized on or near to the fixed charges of the polyion (or on the skeleton of the polyion) due to the high electrostatic potential originating in the fixed charges. In a certain sense the site-localization model is an extension of the ion pair model for simple electrolyte solutions. It must be determined by examination whether this model is free from many of the defects of the domain-binding model discussed above.

It is clear that site localized ion pairs differ from undissociated molecules. One of the best examples of the difference between ion pairs and undissociated molecules in the case of polyelectrolytes is provided by the conductmetric titration curves of polyacrylic acid and polyethylene sulfonic acid.[60] Poly-ethylene sulfonic acid and polystyrene sulfonic acid have very low "apparent" degrees of dissociation in aqueous solution. The apparent degree of dissociation is in fact almost the same as that of the polycarboxylic

* Recent nuclear resonance investigations show that the concentration dependence of the proton resonance in polystyrene sulfonic acid is only in accord with the hypothesis of *complete dissociation* [M. NAGASAWA AND L. KOTIN, *J. Am. Chem. Soc.* **83**, 1026 (1961)].

acids (polyacrylic acid). However, a polysulfonic acid is not a weak acid but rather a strong acid in which almost all acidic hydrogens are dissociated.* The low " apparent " degree of dissociation is attributable to the fact that the dissociated hydrogen ions have a low activity coefficient due to the strong electrostatic interactions with the polyion, or in other words, due to ion localization. That this interpretation is valid can be demonstrated by comparing the conductimetric titration curve of a polysulfonic acid with that of polyacrylic acid. Typical curves are shown in Fig. 9.3.2.

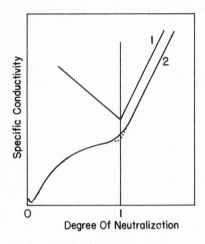

FIG. 9.3.2. The conductimetric titration curves of a strong and a weak polyelectrolyte. Curve 1—polyvinyl sulfonic acid; curve 2—polyacrylic acid.

Now, the hydrogen atom of —COOH has zero mobility whereas the sodium ion which replaces the hydrogen atom by titration has finite mobility. Therefore, the specific conductance of a solution of a polycarboxylic acid increases monotonically with increasing neutralization. On the other hand, a hydrogen ion localized on a polysulfonic acid residue (or residues) has a non-zero mobility, though the mobility would be expected to be lower than that of the free ion. Moreover, the mobility of a localized proton is lower than the mobility of a localized sodium ion. Thus, the specific conductance of a polysulfonic acid solution decreases monotonically as the degree of neutralization increases (NaOH as titrant). It is clear that there is a marked and observable difference between undissociated acid residues and localized (or paired) ions even in polyelectrolyte solutions. The site-localization theory which will be discussed here deals with dissociated but localized (or paired) ions.

With the assumption that the ion pairing could be approximated as a dissociation equilibrium, Harris and Rice[4] were able to semiquantitatively account for the data of Wall et al.[62,63] This theory has been presented in detail in Chapter 7 to which the reader is referred.

It is not at all clear but that the Harris and Rice treatment of ion-pair formation to polyions exaggerates too greatly the similarity between ion pairs and undissociated molecules. The ion pair, at least in the model due to Bjerrum, is not an undissociated molecule, but rather is composed of completely dissociated ions which are paired by virtue of being very near each other. To form a Bjerrum ion pair in sodium acetate solution, for example, the sodium ion need not necessarily have the same molecular configuration as would the hydrogen in CH_3COO—H. Any of the ions depicted in Fig. 9.3.3 can be considered to be ion pairs. It is interesting, therefore, to note that there have been no reports of ion-pair equilibria satisfying the Bjerrum relationship for the ratio of dissociation constants of a difunctional molecule except for the case of bolaform electrolytes in which there is a large separation of the two charged groups. For example, maleic acid has dissociation constants $pK_1 = 1.921$, $pK_2 = 6.225$ and the calculated interaction energy between the two charged groups, χ, is 8.22 RT after correction for the statist-

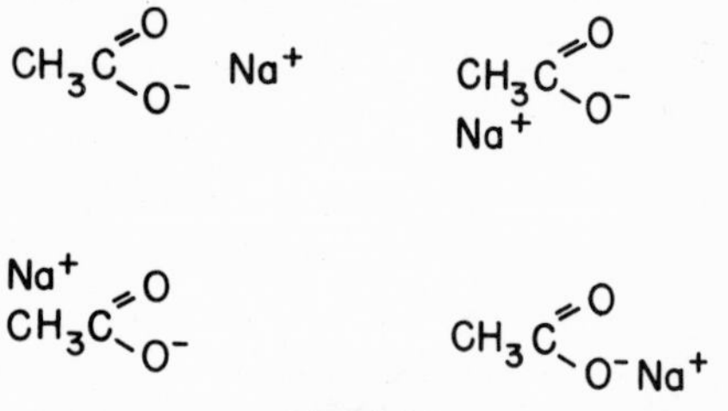

$$CH_3C \overset{\displaystyle =O}{\underset{\displaystyle O^-}{\Big\backslash}} \quad Na^+ \qquad\qquad CH_3C \overset{\displaystyle =O}{\underset{\displaystyle O^-}{\Big\backslash}} \\ Na^+$$

$$Na^+ \\ CH_3C \overset{\displaystyle =O}{\underset{\displaystyle O^-}{\Big\backslash}} \qquad\qquad CH_3C \overset{\displaystyle =O}{\underset{\displaystyle O^- Na^+}{\Big\backslash}}$$

Fig. 9.3.3

ical factor. If it is assumed that the mechanism of dissociation of a carboxyl group is the same as the mechanism of association of a counterion, the ratio of the dissociation constants of the primary and secondary ion pairs would be

$$K_{s2}^0/K_{s1}^0 = 2.5 \times 10^{-4}.$$

Therefore, even if the dissociation constant of the primary ion pair is large (as large as in the case of 1–1 electrolytes $\sim 10^2 - 10^3$), the dissociation constant of the secondary ion pair would be quite small. If K_{s1}^0 were 10^2, K_{s2}^0 would be 0.025 and there would be appreciable ion pairing. This deduction is not in agreement with experiment, the extent of ion-pair formation in a dilute sodium maleate solution being negligible.[65] Moreover, Nagasawa and Rice[65] made several copolymers of maleic acid and nonelectrolyte monomers having different ratios of ionizable to nonionizable monomers to study the effect of a nonrandom local charge density on the dissociation of carboxyl groups and sodium carboxylate ion pairs (if they exist). It was found that the effect of the nonrandom distribution of ionizable groups is not as large for sodium ion binding as for the dissociation of the carboxyl acid. This result indicates that a discussion of ion binding based on the same mechanism as the dissociation of weak acids overly exaggerates the " ion pair " feature of the site binding.

In view of the preceding remarks it is pertinent to briefly review some of the evidence prejudicial to localized binding rather than volume binding. The experiments to be cited cannot distinguish between binding to only one charged site and localization in the region of a few charged sites, but we believe they do exclude over-all volume binding as the dominant source of ion association.

The elegant experiments of Wall et al.,[62,63] leave little doubt that the counterions of a polyion are intimately associated with it. However, these experiments do not distinguish between counterions localized as ion pairs and counterions that are merely trapped in the region where $|q\psi/kT| > 1$. There are two pieces of evidence that tend to support the former point of view. Howard and Jordan[64a] studied the sedimentation of polymethacrylic acid as a function of the degree of neutralization of the polymer and the ionic strength of the medium. Kraut[64b] has succeeded in interpreting their data on the basis that the sedimenting polyion has a net charge much smaller than the stiochiometric degree of neutralization. The net charge needed to fit the data is in quantitative agreement with the transference and diffusion measurements of Wall *et al.* Since the rate with which the localized ions exchanged with those in solution is large, the experiments of Howard and Jordan largely eliminate the possibility that counterions external to the polymer coil are dragged along with it. A second set of experiments is far more clear-cut. Strauss and co-workers[64c] found that it was possible to change the sign of the charge on polyvinylpyridinium bromide and several related polyelectrolytes if the bromide ion concentration was made sufficiently large. That is, the polyion starts out positive, then becomes negative, and moves to the opposite electrode as the bromide ion concentration increases. These results cannot be interpreted on the basis of electrostatic dragging of counterions in the region when $|q\psi/kT| > 1$, since once the polyion had no net charge there would be no incentive for the counterions to further cluster in the vicinity of the polymer. It is probable, therefore, that the binding observed by Strauss and co-workers is occurring at specific sites. Moreover, these experiments indicate that there are other forces (i.e., ion-dipole) than those commonly considered which may play a significant role in determining the thermodynamic and configurational properties of polyelectrolytes.

Finally we note that an immediate implication of the proposed site localization is that the extent of ion-pair formation should be essentially independent of the electric field strength applied, for example, in an electrophoresis experiment. In contrast, if the ion association were due to electrostatic dragging in the region where $|q\psi/kT| > 1$, the amount of association should decrease continuously with increasing electric field strength. It should be noted that these statements are meant to be applied only in the region of low field strengths. At very high field strengths, where a Wien effect may occur, both types of ion binding will respond in the same way, i.e., decrease with increasing field strength. However, by the time fields sufficiently strong are reached, the amount of $q\psi/kT$ binding, assuming both exist in the absence of any field, should be negligible small, and only the site localized counterions would give the Wien effect.

Recent experiments by Wall[64d] have shown that the binding of sodium ion to polyacrylic acid is independent of the applied field strength above a minimum value. The associated counterions may therefore be classed as loosely or strongly bound. The amount of nonspecific loosely bound counterion association is of the order of 8 % of the total amount of binding when the degree of neutralization is approximately unity. Theoretical considerations previously cited (See Chapter 7) indicate that in addition to the site bound ions there will be a number of loosely bound ions within the volume occupied by the macroion. The loosely bound counterions found in Wall's experiments may be identified with these. It should be noted that all of Wall's experiments are carried out at low field strengths (0.1 to 1.5 volts per cm), and therefore are in agreement with the implications of the proposed model.

Bailey et al.[64e] have examined the Wien effect in a strong polyelectrolyte, finding an increase in conductivity of the order of 13 % when the field was 10,000 volts per cm, with larger increases at still larger field strengths. The order of magnitude of the observed Wien effect was comparable to that observed in solutions of tribenzylammonium picrate in a solvent of very low dielectric constant. Since the localized ion pairs ought to be similar to ion pairs formed by salts in media of very low dielectric constant, the observed similarity of Wien effect is not unexpected. Further, Bailey et al., noted that, though the Wien effect was reproducible, even when the pulses followed one another at very short intervals, the total conductivity of the polyelectrolyte solution did not reattain its equilibrium value after the high-voltage pulse for a period of time much longer than either the pulse length or the interval between pulses. If the Wien effect observed is due only to the site localized counterions, as suggested, then a possible explanation of this anomaly might be as follows. Due to the fact that the loosely bound counterions can be removed by very low fields, they will be much further from the polyion after a high-voltage pulse than the site bound ions that have also been affected. There will therefore be a relatively longer relaxation time for the return of the loosely bound counterions than for the strongly bound counterions. That is, loosely bound counterions will be found mostly outside the polyion and will have to diffuse to and into it, whereas strongly bound counterions, though in many cases removed from their charge sites, will remain largely within the domain occupied by the polyion. Since it is postulated that the Wien effect derives only from the strongly bound counterions, whereas the low field conductivity depends upon the numbers of both kind, the relaxation time for the recovery of low field conductivity should be longer than that for the Wien effect. This is in the observed direction.

A real polyion, particularly in salt solution, is undoubtedly in a configurational state intermediate between the coil with counterions uniformly smeared inside a spherical domain and the " strictly defined " site-binding model. We visualize the polyion as a more or less coiled structure having a cylindrical ionic atmosphere around the polymer skeleton (see Fig. 9.3.4). The domain binding model refers to the over-all volume of the coiled structure, and the site-binding model refers to the cylindrical ionic atmosphere around the polyion skeleton. It seems to us that the latter is to be preferred as a basis for the explanation of the behavior of the counterions. We again refer the reader to Chapter 7 for a discussion of the ionic distribution portrayed in Fig. 9.3.7 in terms of a cell model.

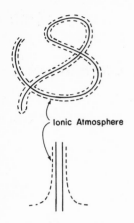

Ionic Atmosphere

FIG. 9.3.4. A schematic representation of a flexible polyion and counterions.

Consider now, in brief summary, the methods of determining the extent of counterion binding.

(1) The best known method of determining the extent of ion binding is the measurement of the transference numbers of the polyion and counterion using radioactive tracers. This technique was developed by Huizenga *et al.*[63] who studied the binding of sodium ions to polyacrylate polyions. The same solution of sodium polyacrylate containing Na^{22} is placed in both parts of the cell shown in Fig. 11.5.1. After passage of a known amount of electricity, the change in weight of the solute is determined by evaporating to dryness the solutions from each compartment. The change in the concentration of the sodium ions is measured by standard counting methods. It is necessary to know the specific conductance of the original solution.

Let C, C_{Na}, C_H and C_{OH} be, respectively, the stoichiometric concentrations of polyacrylic acid, free sodium ion, hydrogen ion and hydroxyde ion, all expressed in equivalents per liter. Also, let κ, κ_p, κ_{Na}, κ_H and κ_{OH} be, respectively, the total specific conductance minus the solvent conductance, the contribution to the specific conductance due to polymer ion, sodium ion, hydrogen ion, and hydroxyl ion. If v is the number of charges on the polymer ion, j the number of sodium ions associated with each polymer ion, Z the degree of polymerization, f' the fraction of sodium ions not associated with the polymer, Q_1, Q_2 the increases in number of equivalents of sodium ion and polymeric ion in the anode compartment, N_e the total number of equivalents of electricity passed, and r the stoichiometric concentration of sodium hydroxyde, then

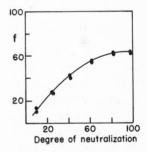

FIG. 9.3.5. The degree of ion binding to a polyion, f, as a function of the degree of neutralization. [Reproduced from ref. 63.]

$$\kappa = \kappa_p + \kappa_{Na} + \kappa_H + \kappa_{OH} \qquad (9.3.14)$$

where

$$\kappa_p = C\Lambda'_p v/Z$$

$$\kappa_{Na} = rf'\Lambda'_{Na} \qquad (9.3.15)$$

$$\kappa_H = C_H\Lambda'_H$$

and Λ' is written for $10^{-3}\Lambda$. The fraction of the current carried by the polymer ion t_p is given by

$$t_p = \frac{Q_2 v}{ZN_e} = \frac{\kappa_p}{\kappa} \qquad (9.3.16)$$

while the condition of electroneutrality requires that

$$\frac{Cv}{Z} = C_H + rf' - C_{OH}. \qquad (9.3.17)$$

On combining these relations, there is obtained

$$\kappa - C_H\left(\Lambda'_H + \frac{\kappa Q_2}{CN_e}\right) - C_{OH}\left(\Lambda'_{OH} - \frac{\kappa Q_2}{CN_e}\right) = f'r\left(\Lambda_{Na} + \frac{\kappa Q_2}{CN_e}\right) \quad (9.3.18)$$

from which f' can be computed. Also from the sodium balance equation we see that

$$\frac{j}{Z} = \frac{r(1-f')}{C} \qquad (9.3.19)$$

The choice of Λ_{Na} effects the computed values of f'. Wall and co-workers recommend the use of conductance data for aqueous NaCl. The fraction of associated sodium ion, $f = (1 - f')$, is shown in Fig. 9.3.6 as a function of the degree of neutralization.

(2) Huizenga, Grieger and Wall[63] also developed a diffusion method for determining the amount of counterion binding. If we assume that the mobility of the free ions is the same as in simple electrolyte solutions and that the bound ions have no mobility, we can calculate the amount of ion binding from the lowering of the apparent self-diffusion constant. Thus,

$$f' = \frac{D_{Na}}{D^*_{Na}} \qquad (9.3.20)$$

with D^*_{Na}, D_{Na} the self-diffusion constants of sodium ion in a simple salt solution and the polyelectrolyte solution, respectively. The diffusion constant is conveniently measured by the use of Na^{22}. In one compartment of a diffusion cell which has a glass frit at the center is placed a solution having no isotope, and in the other compartment is placed a solution having the same concentration but containing Na^{22}. The apparent diffusion constant can be calculated from the amount of Na^{22} transported across the glass frit in a given time by using Fick's law. As can be seen in Fig. 9.3.6, the transference and diffusion measurements are in good agreement with respect to the extent of ion binding.

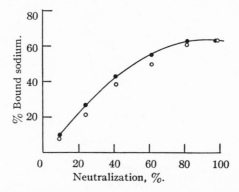

FIG. 9.3.6. Sodium ion association as a function of neutralization for 0.0151 N polyacrylic acid: O—diffusion measurements; ●—transference measurements. [Reproduced from ref. 63.]

(3) Measurements of the ionic activity coefficient may be used to calculate the amount of ion binding, if it is assumed that the bound ions do not contribute to the thermodynamic properties (i.e., colligative properties) of the solution. That is, if the activity coefficient of the free ions is assumed to be unity, the measured activity coefficient is simply the degree of dissociation of the counterions. The same argument may be used to interpret the osmotic coefficient. It is observed that the degrees of dissociation calculated in this manner from the activity coefficient and the osmotic coefficient are not in agreement with each other[1,66] (see Fig. 9.3.8). From the pH measurements of Mock and Marshall[67] and the sodium activity measurements of Nagasawa et al.,[58] it has been found that the counterion activities are additive in systems to which salt has been added. Thus, the degree of ion binding is independent of the amount of extraneous electrolyte, a fact which can also be used to account for the Donnan membrane equilibrium[68] and other observed phenomena. Osmotic pressure measurements of counterion dissociation can be used only in the absence of extraneous salt.

(4) An effective charge, defined as $(1 - f)\alpha Z$, is often used to fit experimental data and thereby calculate f'. Fujita et al.[69] and Inagaki et al.[51] determined the effective charge density of polyacrylic acid and carboxymethyl

cellulose by fitting the observed viscosity to the theory of Hermans *et al.*[70] The degree of association thereby obtained is in agreement with expectation. Terayama and Wall[71] determined the amount of binding by isoionic viscosity measurements. The method consists of diluting the polymer solution with a simple salt solution, whose concentration C_1 is selected by trial and error to make the reduced viscosity a linear function of concentration. It is assumed that under these conditions there is a constant polymer configuration and

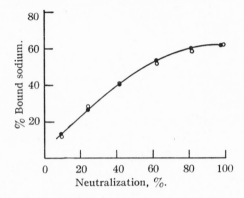

FIG. 9.3.7. Sodium ion association as a function of neutralization for 0.0378 N poly-acrylic acid: open circles, diffusion measurements; solid circles, transference measurements. [Reproduced from ref. 63.]

charge. The ratio $f' = C_1/C_0$, where C_0 is the concentration of the counter-ions, is considered to be the degree of dissociation of the polyion. The values of $1 - f'$ obtained in this manner for sodium polyacrylate were compared with the extents of counterion binding determined by transference and diffusion measurements and semiquantitative agreement was found between the two sets of determinations (Fig. 9.3.8).

Flory and Osterheld[72] and Orofino and Flory[73] determined the effective concentration of counterions inside the polyion domain by fitting theoretical second virial coefficients to the experimental data. As in the studies of Inagaki, quite reasonable effective charges are deduced. (See Chapter 8).

(5) Strauss and his co-workers[24–30] have calculated the amount of alkali ion binding to polyphosphates from electrophoresis and Donnan membrane equilibrium measurements. According to Hermans and Fujita[74] and Overbeek and Stigter,[75] the polyion is freely drained in the limit of infinite ionic strength and the electrophoretic velocity of the polyion (U_p^0) becomes equal to the mobility of a segment (see Chapter 11)

$$U_p^0 = \frac{vq}{Z\Upsilon} \qquad (9.3.21)$$

where vq/Z is the charge on a segment and Υ is the frictional constant of the segment. When counterion binding exists, v must be replaced by $f'v$ with f' the degree of dissociation. Therefore, from the mobility of a segment $vq/Z\Upsilon$ and the mobility of the polyion, f' can be calculated in the limit of infinite ionic strength. However, it is quite difficult to determine the value of $vq/\Upsilon Z$. Strauss and Bluestone[28] from electrical conductance measurements obtained $vq/Z\Upsilon = 5.2 \times 10^{-4}$ for polyphosphates, but the fact that they used the

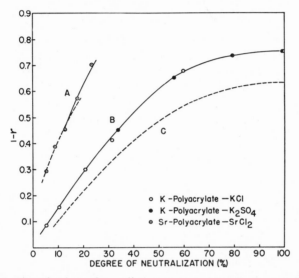

FIG. 9.3.8. Plot of values of $1 - f'$ that cause η_{sp}/C to be linear versus degree of neutralization of polyacrylic acid. Solid lines represent viscosity data and dotted lines represent bound ion fraction obtained from electrolysis experiments. [Reproduced from ref. 71.]

measured electrophoretic velocity instead of the limiting mobility (which is unknown) casts doubt upon the reliability of this value. Notwithstanding the unreliability in the value of the mobility of the monomer, it is not unlikely that the binding of the alkali metal ions to polyphosphates increases with diminishing radius of the anhydrous cation in the order $Li^+ > Na^+ > K^+ > Cs^+$. Such a series would be expected if the PO_3^- group penetrates the hydration layer surrounding the cation. The method just discussed is inapplicable if a constant value of the electrophoretic velocity cannot be obtained at high salt concentrations. (See the experiments of Strauss et al.[25]) Under these conditions the theory developed by Hermans, Overbeek, and co-workers is not valid. Finally, one cannot dismiss the possibility that the effect of ion pairing on the electrophoretic velocity of the polyion is negligible. (See Chapter 11.)

(6) Strauss and his co-workers[24-30] have also used measurements of Donnan membrane equilibria to calculate the extent of ion binding to polyphosphates. Scatchard[76] (see Chapter 8) has derived the following equation for the ionic distribution of simple electrolytes between a phase containing colloidal ions and a phase containing no colloidal particles:

$$\lim_{m_2 \to 0} \frac{\partial \ln (m_3'/m_3)}{\partial c_2} = \frac{1}{M_2} \frac{(v/m_3') + \beta_{23}^0}{2 + \beta_{33}^0 m_3'} . \tag{9.3.22}$$

In Eq. (9.3.22) M_2 is the molecular weight of the polymer, $v = \alpha Z$ the number of charges on a polymer, m the molarity, c_2 the colloidal concentration in grams per 1000 ml; 1, 2, and 3 denote, respectively, the water, the polyelectrolyte, and the simple electrolyte; the primes indicate the solution without colloidal particles; and $\beta_{jk} = \partial \ln \gamma_j / \partial m_k$, γ_j being the activity coefficient. If β_{23}^0 and β_{33}^0 are assumed to be zero, i.e., all the deviations from ideality are lumped into an effective degree of ionization, Eq. (9.3.22) can be simplified to read:

$$f' = \frac{v_{eff}}{v} = \frac{2M_2 m_3}{v} \lim_{c_2 \to 0} \frac{d \ln (m_3'/m_3)}{dc_2}. \tag{9.3.23}$$

Strauss and his co-workers used instead of Eq. (9.3.23) the simpler relation

$$f' = \frac{2M_2}{v} \frac{m_3' - m_3}{c_2}. \tag{9.3.24}$$

Some of the results reported by Strauss and Ander[27] are shown in Table 9.3.1.

TABLE 9.3.1[a]

Electrolyte	m_3'	f'
LiBr	0.02013	0.18
	0.3968	0.28
NaBr	0.02043	0.15
	0.3991	0.23
KBr	0.01965	0.16
	0.2377	0.23
TMABr	0.02019	0.26
(Tetramethyl-	0.19306	0.43
ammonium bromide)	0.9692	1.02

[a] Reproduced from ref. 27.

Note that the value of f' increases with increasing concentration and the value of f' for TMABr is unreasonably large. In view of this Strauss and Ander concluded that calculations based on Eq. (9.3.24) are unsound. It would be interesting to know how consistent the results from different methods can be, but sufficient measurements have as yet not been reported. With the information now available it is found that kinetic and thermodynamic methods do not always give concordant results. Such discrepancies may arise from the existence of a spectrum of binding stages, some stages being more susceptible to alteration by external fields than others, a suggestion consistent with Wien effect measurements.[77-79]

As a final note we remark that the assumption that the free ions are completely " free " or as free as in a corresponding simple salt solution, cannot be rigorously valid. Whereas the Bjerrum theory builds upon the Debye–Hückel analysis of the " free " ions, simple application of the law of mass action without account of the nonideality of the unpaired ions does not give consistent results. It is likely that theories based on ion binding and theories where no ion binding is taken into account are complementary rather than competitive with each other.

9.4 Ion Exchange Phenomena and Ion Binding[80]

In this section we return to a different use of the cell model to describe systems containing polyelectrolytes. We propose to develop a simple model for ion-exchange resins, based on relevant ideas from the theories of linear and lightly cross-linked polyelectrolytes. A resin, then, will be considered as a cross-linked polyelectrolyte gel, whose polymer chains contain groups capable of specific interaction with the ions which exhibit exchange properties. The polyelectrolyte gel will be able to dilate or contract to maintain mechanical equilibrium, just as does any network of polymer chains.[81] The groups responsible for the exchange properties usually posses a charge opposite in sign to that of the exchangeable ions. We therefore treat the specific interaction between the exchange groups and the exchangeable ions by again considering the possibility of formation of ion pairs similar to, but not identical with the kind first introduced by Bjerrum in the treatment of ordinary ionic solutions. The ion pair equilibria, which in this model are the only quantities which distinguish among different kinds of ions of the same charge, comprise here the sole source of selectivity in the ion-exchange process.

We shall show that a most important characteristic of ion exchange resins is the interaction between neighboring exchange sites. The experimental evidence in support of this contention includes the strong dependence of the exchange properties on the distances between neighboring sites, and the vast difference in behavior between the exchange groups in a resin and the same groups in unpolymerized compounds. It has apparently not been altogether clear, however, to what extent the interaction is due to electrostatic forces and to what extent it is a secondary effect dependent upon spatial requirements or other more subtle changes in environment. Calculations for groups at typical separations in exchange resins lead to the conclusion that the electrostatic forces are indeed large, and that they must be a dominant factor in the understanding of the resin equilibria. In computing the electrostatic forces in a resin, it is necessary to realize that the effective dielectric constant of the internal medium will be far less than that of water. This perhaps suggests why this effect has not heretofore received the attention its importance merits, and also shows why effects occur in resins which are almost completely negligible in ordinary ionic solutions.

The electrostatic interactions discussed in the preceding paragraph are sufficiently strong that considerable ion pair formation will take place so as to reduce the electrostatic potential. The extent of ion pair formation will thus depend, not only on the intrinsic properties of the ion pairs, but upon the density of exchange sites and the concentrations of eligible ions. This is the basis of the dependence of exchange selectivity upon cross-linking and exchange capacity, since only the ions bound into pairs have been selectively treated. Increased cross-linking has the same effect as increased exchange capacity, because the cross-linking makes it more difficult for the resin to decrease its site concentration by swelling. The stoichiometry relating the numbers of bound ions of each type and the numbers of unbound ions within the resin to the exchange equilibrium quotients can be shown to lead to the experimentally observed dependence of the selectivity upon the exchange state. We shall therefore find that the ion-binding concept in its simplest form is adequate to explain the gross features of resin selectivity.

The introduction of ion binding in no way implies that at any time a macroscopic quantity of resin should possess an appreciable net charge. The theory shows, as must be expected, that the charges of the resin not neutralized in ion-pair formation are compensated for by the presence of counterions within the interior of, but not closely bound to, the polymer network. This concept has a parallel in linear polyelectrolytes where an experimental distinction can be made between ions localized near specific sites on the polymer network, and unbound ions freely moving within the interior of the polymeric ion.

The electrostatic interactions also play an important role in the swelling of the resin gel. The equilibrium volume of the cross-linked gel depends upon a balance between the expansion force resulting from electrostatic repulsions, the contractile force resulting from the natural elasticity of the polymer segments, and the forces arising from the free energy of mixing of the gel with the external solution. The magnitude of the electrostatic repulsions will depend upon the extent of ion pair formation and the exchange site density. The theory predicts larger swellings at high concentrations of unpaired sites. The elasticity of the gel depends on its cross-linking, with, as already mentioned, the less cross-linked polymers being more easily stretched.

The nature of the ion pair equilibria is a most difficult question in the formulation of the theory. Some approaches have been made to this problem, both from the point of view of understanding processes in ordinary solutions and as part of earlier attempts to explain ion-exchange phenomena.[82] The idea common to all these studies has been to compute the electrostatic interaction between the ions concerned at a suitable distance of closest approach. The answers obtained depend quite markedly upon the extent to which hydration of the ions is considered to influence the distance of closest approach. It is also apparent, as has been pointed out, that drastic changes in the nature of the solution composition may well have an important effect upon the ion hydration and, in turn, upon the ion-exchange equilibria. We have assumed here that the intrinsic quantities of the equilibria remain invariant, and that the apparent changes in the binding constants are solely the result of the superposition of the binding phenomena and the electrostatic interactions between exchange sites. The actual values chosen here for ion-pair equilibria were estimated from available data and from computations of the type referred to above. The development presented here, then, should apply to typical ion-exchange systems, but cannot be expected to validly represent systems in which changes in ion hydration or a more specific kind of chemical interactions are important.

We will present later a comprehensive discussion of the successes and limitations of the model described in the foregoing. However, we feel it is important from the outset to notice that the model represents an attempt to discuss the ion-exchange phenomena on a molecular scale. It is therefore necessary to account for all the actual interactions between parts of the system, but it is equally necessary to avoid the introduction of " forces " which actually represent the result of the consideration of the numbers of states of various macroscopic properties. As discussed at appropriate points in more detail, it is not necessary to include explicitly in the model any terms descriptive of osmotic interactions between the regions external and internal to the resin. This removes the conceptual difficulty attendant upon consideration of resin of decreasing degrees of cross-linking and correspondingly increasing lack of definition of internal and external regions.

9.4.1 FREE ENERGY

We shall begin by computing the free energy of a resin system in which the resin occupies a volume v and contains $\{n_\sigma\}$ resin sites in occupation states σ. (We choose $\sigma = 0$ to refer to unbound sites and other values of σ to refer to ion pairs of the various kinds.) By minimizing the free energy with respect to v and $\{n_\sigma\}$, we may then determine their equilibrium values for any specified values of the independent parameters governing the system. Many of the details of the free energy computation have been omitted from the main text. For these details, and a more precise definition of the assumptions and approximations introduced, the reader is referred to Appendices 9A—9C. More specific reference will be made to the Appendices where appropriate.

The free energy may be written as a sum of terms, the first of which, denoted by A_{chem}, is the chemical contribution to the over-all free energy associated with the formation of $\{n_\sigma\}$ ion pairs involving ions of each kind σ, $\sigma \neq 0$. As shown in Appendix 9A, A_{chem} may be written in the form

$$A_{chem}(\{n_\sigma\}) = \sum_{\sigma \neq 0} n_\sigma kT[\ln K_\sigma - \ln c_\sigma] \qquad (9.4.1)$$

where K_σ is the intrinsic equilibrium constant for the reaction

$$(\text{ion pair})_\sigma = (\text{free site}) + (\text{free ion})_\sigma \qquad (\sigma \neq 0)$$

and c_σ is the concentration of ions of the kind σ in the external solution. As is apparent from Eq. (9.4.1), A_{chem} is a function of $\{n_\sigma\}$ alone, and not of v. The zero of A_{chem} has been chosen to be a resin in which no sites have bound ions. The assignment of different values to K_σ for different kinds of ions σ is the source of exchange specificity in this model.

A second contribution to the free energy, A_{conf}, arises from the multiplicity of configurations available to the polymer chains. We introduce the approximation that, for a specified volume, the charge state of the resin does not affect the configurational free energy. This approximation is to a certain extent in error because the relative probabilities of occurrence of various configurations must be altered by the presence of electrostatic forces. However, in appreciably cross-linked resins, the chain lengths between cross-links are too short to permit great independence to the positions of neighboring charges, and most configurations of the same volume will not differ greatly in energy. In this connection, it should be noticed that in a three-dimensional structure, most motions of a charge-bearing group separate it from some neighboring charges while simultaneously causing it to approach more closely other neighboring charges.

Having made the approximation discussed in the preceding paragraph, we may legitimately regard A_{conf} as a separate, additive contribution to the total free energy depending only upon v. Moreover, A_{conf} may then be computed by the methods already devised for the swelling of nonelectrolytic gels. With the aid of the Flory theory of network swelling,[81] one may obtain, as shown in detail in Appendix 9B,

$$A_{conf}(v) = \frac{3}{2} kT\mathfrak{b}\left[0.4\left(\frac{v}{v_0}\right)^{2/3} - 1 - 0.6\left(\frac{v_m}{v_0}\right)^{2/3} \ln\left(1 - \left(\frac{v}{v_m}\right)^{2/3}\right)\right]$$
$$- \frac{\mathfrak{b}kT}{2} \ln\left(\frac{v}{v_0}\right) + \left(\frac{v - v_0}{v_s}\right)kT\left[\ln\left(\frac{v - v_0}{v}\right) + \chi_1 \frac{v_0}{v}\right]. \qquad (9.4.2)$$

In Eq. (9.4.2), b is the effective number of cross-links in the network, v_s is the volume of a solvent molecule, and v_0 is the volume of the unswollen gel, which is taken as the reference state from which A_{conf} is computed. The volume v_m is the maximum volume to which the gel may swell, and is governed by the contour lengths of its constituent chains. The parameter χ_1 is descriptive of the difference between the nearest neighbor interaction energy of adjacent polymer segments and the interaction energy of a polymer segment and an adjacent solvent molecule.

Equation (9.4.2) describes the excess of the configurational free energy of the solvent-gel system over that of separate solvent and gel. In the systems of interest here the solvent is actually a solution of small molecules, and the reference state is for the solvent in its mixed form. The derivation of Eq. (9.4.2) is also based on the premise that the polymer-solvent interactions do not include the free energies associated with the redistribution of the mobile ions under the influence of the electric field of the polymer charges. These contributions are to be included in the electrostatic terms to be discussed later.

The remainder of the free energy, which we shall refer to as A_{int}, will include the effects of all electrostatic interactions, and the free energy associated with the distributions of unbound sites and ion pairs of each kind among the exchange sites. These quantities are interrelated because the various microscopically different ways of placing $\{n_\sigma\}$ groups of kinds σ on the resin are not all of the same interaction energy. The details of the calculation of A_{int}, set forth in Appendix 9C, depend upon statistical-mechanical considerations not unlike those employed in the lattice theory for simple liquids.[84] It is convenient to assume that only nearest neighboring exchange sites interact electrostatically, and that all site arrangements with the same sites unbound are equally probable. Screened Coulomb potentials are used to represent the interaction potential. Use of the screened potential, together with appropriate ion atmosphere free energy contributions, includes in the electrostatic free energy the effects of redistribution of the mobile ions. Therefore, it is consistent to regard the total free energy of a system in any configuration as the sum of all nonelectrostatic contributions plus an electrostatic term of the form used here. The final result for A_{int} is

$$\frac{A_{int}}{Z} = -q^2\alpha'\frac{\kappa'}{D_0} + \frac{q^2\alpha'ze^{-\kappa'r_0}}{2D_E r_0} + kT[\alpha'\ln\alpha' + (1-\alpha')\ln(1-\alpha')]$$

$$+ \frac{zkT}{2}\left[\alpha'\ln\left(\frac{\beta-1+2\alpha'}{\alpha'(\beta+1)}\right) + (1-\alpha')\ln\left(\frac{\beta+1-2\alpha'}{(1-\alpha')(\beta+1)}\right)\right]$$

$$+ \frac{kT}{Z}\sum_{\sigma\neq0} n_\sigma\ln\left(\frac{n_\sigma}{Z-n_0}\right) \tag{9.4.3}$$

where

$$\alpha' = \frac{n_0}{Z} \tag{9.4.4a}$$

$$\beta^2 = 1 - 4\alpha'(1 - \alpha')\left(1 - \exp\left\{-\frac{q^2 e^{-\kappa' r_0}}{D_E r_0 kT}\right\}\right). \tag{9.4.4b}$$

Z is the total number of exchange sites, q is the absolute value of the electronic charge, D is the dielectric constant of the external solution, D_0 the average dielectric constant within the resin, and D_E the effective dielectric constant at a distance r_0 from an ion. The distance r_0 is the average distance between nearest neighboring sites, and z is the average number of nearest neighbors. The dependence of r_0 on the resin volume v is assumed to be that corresponding to isotropic swelling, so that $r_0 = (v/v_0)^{1/3} r_0^0$, where r_0^0 is the distance between adjacent sites in a dry resin. The quantities κ and κ' are screening constants of the type appearing in the Debye–Hückel theory, referring to the regions outside and within the resin, respectively. They are given by the equations

$$\kappa^2 = \frac{4\pi}{DkT} \sum_i c_i q^2 \tag{9.4.5a}$$

$$\kappa'^2 = \left(\frac{D\phi}{D_0} e^{-\chi/kT}\right)\kappa^2 \tag{9.4.5b}$$

where c_i is the concentration of the ith kind of ions in the external solution, ϕ is the volume fraction of the resin available to occupancy by mobile ions, and the parameter χ is defined as

$$\chi = \frac{q^2}{2}\left(\frac{\kappa}{D} - \frac{\kappa'}{D_0}\right). \tag{9.4.5c}$$

The zero of A_{int} refers to an uncharged resin with the same kind of ion bound on every site. The above development has been carried out for a system containing only univalent ions. Minor changes would be required in several places if the formalism was to be applied to a more general electrolyte. A_{int} depends not only on $\{n_\sigma\}$, but also, implicitly, on v, through r_0, D_0, D_E, and κ'.

Combining the results of this section, we obtain the total free energy in the form

$$A(\{n_\sigma\}, v) = A_{\text{chem}}(\{n_\sigma\}) + A_{\text{conf}}(v) + A_{\text{int}}(\{n_\sigma\}, \text{Conf} v). \tag{9.4.6}$$

9.4.2 EQUILIBRIUM CONDITIONS

The methods described in the preceding paragraphs have led to an expression for the free energy of the system for specified resin volume and numbers of ion pairs of each kind. The equilibrium state will be that of minimum free energy with respect to v and $\{n_\sigma\}$ consistent with the chosen values of the various independent parameters. By differentiation of the expression for A with respect to the resin volume, one may find

$$\left(\frac{\partial A}{\partial v}\right)_{\{n_\sigma\}} = \frac{Zzq^2 e^{-\kappa' r_0}}{2D_E r_0}\left(\frac{2-2\alpha'}{\beta+1}-1\right)\left[\frac{1+\kappa' r_0}{3v}+\frac{1}{D_E}\left(\frac{\partial D_E}{\partial v}\right)+r_0\left(\frac{\partial \kappa}{\partial v}\right)\right]$$

$$+\frac{n_0 q^2 \kappa'}{D_0}\left[\frac{1}{D_0}\left(\frac{\partial D_0}{\partial v}\right)-\frac{1}{\kappa'}\left(\frac{\partial \kappa'}{\partial v}\right)\right]$$

$$+\frac{bkT}{v}\left[\left(\frac{v}{v_0}\right)^{\frac{2}{3}}\left(\frac{1-0.4(v/v_m)^{\frac{2}{3}}}{1-(v/v_m)^{\frac{2}{3}}}\right)-\frac{1}{2}\right]$$

$$+\frac{kT}{v_s}\left[\ln\left(\frac{v-v_0}{v}\right)+\frac{v_0}{v}+\chi_1\left(\frac{v_0}{v}\right)^2\right]=0 \qquad (9.4.7)$$

the condition for mechanical equilibrium of the swollen resin. In obtaining Eq. (9.4.7), we have made use of the proportionality of r_0^3 to v, so that $(dr_0/dv) = r_0/3v$.

Minimization of A with respect to the $\{n_\sigma\}$ determines the exchange state of the resin. Since $\sum_\sigma n_\sigma = Z$, we may regard n_0 as a function of the other n_σ, and obtain the following independent equations

$$\left(\frac{\partial A}{\partial n_\sigma}\right)_{n_{\sigma'},\,v} = \frac{q^2\kappa'}{D}-\frac{q^2 ze^{-\kappa' r_0}}{2D_E r_0}-\frac{zkT}{2}\ln\left(\frac{\beta-1+2\alpha'}{\beta+1-2\alpha'}\right)+kT\left(\frac{z}{2}-1\right)\ln\left(\frac{\alpha'}{1-\alpha'}\right)$$

$$+kT\ln\left(\frac{n_\sigma}{Z-n_0}\right)+kT(\ln K_\sigma-\ln c_\sigma)=0 \quad (\sigma,\sigma'\neq 0). \quad (9.4.8)$$

Simultaneous solution of Eqs. (9.4.7) and (9.4.8) yields the sought-for equilibrium values of v and $\{n_\sigma\}$. The set of Eqs. (9.4.8) can be combined so as to yield one nontrivial Eq. (9.4.9) to be solved simultaneously with (9.4.7), along with equations among the $\{n_\sigma\}$ only. By algebraic manipulation of the set (9.4.8), one obtains

$$\frac{q^2\kappa'}{kTD_0}-\frac{q^2 ze^{-\kappa' r_0}}{2D_E r_0 kT}-\frac{z}{2}\ln\left(\frac{\beta-1+2\alpha'}{\beta+1-2\alpha'}\right)+\left(\frac{z}{2}-1\right)\ln\left(\frac{\alpha'}{1-\alpha'}\right)-\ln\left(\sum_{\sigma\neq 0}\frac{c_\sigma}{K_\sigma}\right)=0$$

$$(9.4.9)$$

and

$$\frac{n_\sigma}{n_{\sigma'}}=\frac{K_{\sigma'}c_\sigma}{K_\sigma c_{\sigma'}}. \qquad (\sigma,\sigma'\neq 0) \qquad (9.4.10)$$

From Eq. (9.4.10) we see that the electrostatic interactions do not alter the specificity of the ion binding, for the ratios of the bound ions of the various kinds depend only on the external concentrations of those ions. The effect of the interaction is only to increase the total amount of ion binding of all kinds. This total amount of ion binding is determined by the solution to Eqs. (9.4.7) and (9.4.9) for n_0.

Having obtained values of the resin volume and of the numbers of bound ions of each kind, we may inquire as to how these are related to the usual quantities appearing in ion exchange studies. The swelling of the resins may be characterized simply by v/v_0. However, the ion exchange equilibrium quotient for the reaction

$$(\text{ion})_\sigma + (\text{ion in resin})_{\sigma'} = (\text{ion in resin})_\sigma + (\text{ion})_{\sigma'}$$

requires somewhat more careful treatment. Ions " in the resin," as determined by any presently available technique, include not only ions site-bound to the resin, but also include any unbound ions simply occupying free space within the resin volume. These unbound ions are analogous to the " volume-bound " ions referred to in the study of linear polyelectrolytes. Since in any easily realizable physical situation the resin volume is electrically neutral, it is necessary to have appreciable numbers of unbound mobile ions except when essentially all the sites are occupied by bound ions. It is found that under all conditions of interest, many unbound sites still exist, and it is therefore evident that n_σ and $n_{\sigma'}$, together with only the external ion concentrations, will not suffice to determine the ion exchange equilibrium quotient. It is also evident, since the unbound ions are not chosen selectively, that the ratios of the unbound ions of each kind will be those of the corresponding ions in the external solution, so that the equilibrium quotient, which describes the combination of two processes with different equilibrium constants, will not remain constant as the various $\{n_\sigma\}$ are varied.

We therefore turn our attention to the quantitative prediction of the numbers of unbound ions within the resin volume. The problem is actually quite simple when the concentration of exchange sites in the resin is much greater than the total concentration of mobile electrolyte in the external solution. For then, the great majority of the mobile ions with the resin are just those exchangeable ions necessary to maintain electrical neutrality, and it can be assumed that there is essentially no additional salt present.

In that case the total number of mobile exchangeable ions within the resin is just equal to the number of unbound sites n_0, and we may, as a result of the proportionality to the external concentrations, write

$$(\text{mobile ions in resin})_\sigma = \frac{c_\sigma n_0}{\sum_{\sigma \neq 0} c_\sigma}. \tag{9.4.11}$$

This completes the preparation necessary to enable us to write the exchange equilibrium quotient for reaction (II),

$$K_{\sigma'}^{\sigma} = \frac{c_{\sigma'}}{c_{\sigma}} \left(\frac{n_{\sigma} + \left(n_0 c_{\sigma} / \sum_{\sigma \neq 0} c_{\sigma}\right)}{n_{\sigma'} + \left(n_0 c_{\sigma'} / \sum_{\sigma \neq 0} c_{\sigma}\right)} \right). \tag{9.4.12}$$

Equation (9.4.12) shows that $K_{\sigma'}^{\sigma}$ depends upon n_0, as expected, and also reveals a hitherto unnoticed dependence of $K_{\sigma'}^{\sigma}$ upon c_{σ} and $c_{\sigma'}$.

To determine the way in which $K_{\sigma'}^{\sigma}$ can be expected to vary with changes in n_0 and the external concentrations, let us consider a resin in equilibrium with an external solution in which the only exchangeable ions are of kinds σ or σ'. To do so, we substitute Eq. (9.4.10) into (9.4.12). Simplifying, and using the fact that $\sum_{\sigma} n_{\sigma} = Z$, one may obtain

$$K_{\sigma'}^{\sigma} = \frac{(K_{\sigma'}/K_{\sigma}) + \left(\dfrac{n_0}{Z + n_0}\right)\left[\dfrac{(K_{\sigma'}c_{\sigma}/K_{\sigma}c_{\sigma'}) + 1}{(c_{\sigma}/c_{\sigma'}) + 1}\right]}{1 + \left(\dfrac{n_0}{Z + n_0}\right)\left[\dfrac{(K_{\sigma'}c_{\sigma}/K_{\sigma}c_{\sigma'}) + 1}{(c_{\sigma}/c_{\sigma'}) + 1}\right]}. \tag{9.4.12'}$$

From Eq. (9.4.12'), we see that $K_{\sigma'}^{\sigma}$ differs more from unity the smaller is n_0, at constant $c_{\sigma}/c_{\sigma'}$, indicating that increased selectivity accompanies increased binding. Such a conclusion might also be expected on intuitive grounds. We may also see from (9.4.12') that $K_{\sigma'}$ will increase as $c_{\sigma}/c_{\sigma'}$ decreases, at constant n_0 irrespective of the relative magnitudes of K_{σ} and $K_{\sigma'}$. For, if $(K_{\sigma'}/K_{\sigma}) < 1$, decreasing $c_{\sigma}/c_{\sigma'}$ increases the second term of both numerator and denominator, carrying $K_{\sigma'}^{\sigma}$ toward unity, which in this case is an increase. If $(K_{\sigma'}^{\sigma}/K_{\sigma}) > 1$, the second terms decrease, causing $K_{\sigma'}^{\sigma}$ to depart from unity, or, in this case, to increase also.

The behavior of $K_{\sigma'}^{\sigma}$ with respect to external concentrations provides an understanding of a phenomenon experimentally observed in many exchange studies.[83] The extent to which the model agrees with experiment can not be determined, however, until calculations are made which consider the variation of n_0 with concentration. We defer a more quantitative discussion of the variation in selectivity to a later section of this chapter.

9.4.3 FURTHER SPECIFICATION OF THE MODEL

In order to obtain even a qualitative evaluation of the predictions of the model developed here, we must first consider the factors governing the values of the parameters entering the theory. We discuss, in turn, factors influencing the electrostatic forces, the configuration parameters, and the intrinsic specificity of the exchange sites.

1. *Electrostatic Forces*

The electrostatic forces between neighboring charges in the resin depend upon the effective screening constant κ', the inside dielectric constants D_0 and D_E, and the distance between nearest neighbors and number of nearest neighboring exchange sites. The screening constant in turn depends also upon the fraction of the resin volume available to small ions. The volume available to small ions is to some extent related to the swelling an originally dry resin undergoes when immersed in an aqueous solution. If we regard the resin as always excluding from small ion occupancy its dry volume, we find, by examination of typical swelling data, that in moderately cross-linked (4–8 %) resins of the order of half the volume is available. That the volume increase on wetting is a reasonable measure of the available volume is corroborated by measurements on the concentrations of nonexchange electrolyte occluded in resins; such concentrations are lower than those of the external solution by of the order of the fraction of volume excluded.

Given a rough measure of the excluded volume and its probable range of values, consider next the appropriate values for the dielectric constant of the internal medium. Pure water, with $D = 80$, is a liquid characterized by a high degree of cooperative orientation among neighboring molecules. At room temperature, the neighbors of a given water molecule on the average contribute about 1.6 times the molecular dipole moment in the direction of the dipole of the given molecule.[85] The internal medium, consisting of water diluted by organic matter (of $D \sim 2$) and containing a number of ions (near which water molecules have little orientational freedom), will have a lower dielectric constant than will pure water not only from proportional dilution effects, but also from a partial destruction of the orientation correlations. For the purpose of computing D_0, and thence κ', we are concerned with the effective dielectric constant of the region between charges at distances such that the discontinuous nature of the internal medium can probably to a first approximation be neglected. We postpone quantitative estimates to a following section, but remark here that in a typical resin for which $\phi = 0.5$. the dielectric constant D_0 will certainly be less than $\frac{1}{2}(80 + 2) = 41$, but will probably be of the order of 30. When considering D_E, the effective dielectric constant for the interaction of a pair of nearest neighboring charge sites, it is necessary to consider, in addition to the above factors, the local structure of the resin system. If adjacent charge sites are within approximately 5 A, the effective dielectric constant of the region between the charged groups will be the order of that of the ion (~ 2). As the distance between charge sites increases, D_E will rise gradually to D_0, not reaching the limiting value until the separation is probably larger than 10 A. Estimates of the dielectric constant of pure water as a function of distance from an ion have been given by Hasted, *et al.*[86] Examination of their values (for half the separation, since we have an ion centered at each end of the separation distance) suggests that the estimates made here are reasonable.

To complete the evaluation of the electrostatic forces we need an estimate of the actual distances to be expected between neighboring groups. The theory is too crude to justify great expenditure of effort in calculation for different arrangements of nearest neighbors. For a resin of 5 equivalents/liter capacity, each exchange site has of the order of 300 cubic angstroms, which corresponds to a cube less than 7 A across. Each site might have of the order of 10 other sites near enough to be worth counting as nearest neighbors.

The considerations of the above paragraphs clearly indicate that electro-static forces will be more important than in ordinary aqueous solutions. The numbers quoted as examples, which are not unreasonable for ion exchange resins, lead to an interaction energy between nearest neighboring sites at room temperature of the order of several times kT. This strong interaction suggests, just as it does in linear polyelectrolyte theory, that ion localization will be greatly enhanced, and that the degree of the ion pair formation will depend, among other things, upon the geometrical and charge properties of the resin. In particular, resins with a high concentration of exchange sites will tend toward much ion pair formation. Moreover, resins which are highly cross-linked and cannot expand easily will be forced to assume configu-rations of higher electrostatic energy than will less cross-linked resins with the result that the more cross-linked resins will tend to contain more ion pairs. We shall soon see how these observations aid in forming a semi-quantitative understanding of the systematic behavior of many ion exchange systems.

2. *Configuration Parameters*

The configurational free energy has been assumed to depend upon some parameters which require further specification. The effective number of cross-links, b, differs from the actual number of cross-links in that b is supposed to count only those which actually contribute to the network and not those which only produce dangling side chains. For a resin composed of a mixture of di- and tetra-functional groups (as in a styrene-divinylbenzene copolymer) it is probably reasonable to assume that b and the number of tetra-functional groups are comparable.

The parameters v_s and χ_1 are, as pointed out elsewhere in this section, suitable composites of the properties of the individual species of the solvent. The limitations of the present model make undesirable any course which involves great labor; we therefore shall adopt for v_s and χ_1 values appropriate to pure water. A study of the interaction of hydrocarbon polymers and water, and an estimate of the interaction between the exchange groups and solvent, permit a value of χ_1 to be approximated on the basis of the relative proportions of the nonpolar and polar constituents of the resin[85,87]. A value of χ_1 of the order of $+1$ would appear to be a reasonable choice.

The volume of the resin when totally stretched, v_m, is not capable of direct measurement. This quantity can be estimated, however, from the contour lengths of the chains between cross-links, and the number of cross-links in the resin. The effective contour lengths must be estimated in a way which allows for the imperfect flexibility of the chains. For a fuller discussion we refer to the extensive literature.[81]

3. Exchange Specificity

The specificity of ion pair formation at the exchange sites is the only source of ion-exchange selectivity provided by the model here presented. The specificity enters through intrinsic equilibrium constants K_σ for ion-pair dissociation. As is well known, such constants must be sufficiently large that little ion pair formation will occur in ordinary solutions until the concentrations reach several molar. However, as has already been pointed out here, even fairly large dissociation constants can account for selective ion-exchange behavior when the role of electrostatic forces is considered.

Approximate quantitative values for ion pair dissociation constants have been estimated by several workers [36-41] and the approximate range in which the constants fall is indicated by a small amount of experimental work. More indirect evidence is provided by the values necessary to bring experimental measurements on linear polyelectrolytes and resins into accord with simple models for these substances. These criteria suggest that dissociation constants of the order of 10–1000 are to be expected.

The model depends as crucially upon the differences in K_σ for different kinds of exchangeable ions as it depends upon the range of magnitude in which they all lie. Here experimental data are particularly scanty, and we shall rely for the most part upon some very simple calculations. For definiteness, consider a cation resin with sites which we shall represent as A^-, in contact with two kinds of cations, X^+ and Y^+. We require information concerning the reaction

$$A^-X^+ + Y^+ = A^-Y^+ + X^+. \qquad \text{(III)}$$

If we assume the main factor governing the free energy change of this reaction is the difference in electrostatic energy of the ion pairs A^-X^+ and A^-Y^+, the free energy can be estimated from the size of the hydrated X^+ and Y^+ ions, together with the radius of the site ion A^-. In making such a calculation, it must be remembered that the dielectric constant is much lower than its value in pure water. The calculation is crude, not only because the electrostatic energy cannot be accurately measured, but also because entropy differences, which must exist, are not taken into consideration.

Tentatively, we conclude from considerations such as those of the above paragraph, that, among ions of the same charge, the larger hydrated ion will possess the greater dissociation constant. For the alkali metal ions, this causes dissociation constants to decrease in the direction Li^+, Na^+, K^+, in conformity with the bulk of ion exchange experiment. The dissociation constant for Na^+ might be expected to be from 2 to 10 times as large as that for K^+.

9.4.4 NUMERICAL RESULTS FOR SAMPLE HYPOTHETICAL RESINS

To give insight into the behavior to be expected of the model described in the preceding sections, we present calculations of the ion-exchange equilibria and swelling in several hypothetical resins. Calculations were made for resins of 5%, 15%, and 25% cross-linking, and, for the 15% cross-linked resin, at two densities of exchange sites. The parameters governing the configurational properties of the resins were chosen so as to approximate those of sulfonated polystyrene-divinyl-benzene copolymers. The lattice parameters correspond to a somewhat imperfect closest packed lattice. The bulk dielectric constant in the resin was taken as $D_0 = 40$, a value estimated from the average volume fraction of water in the swollen resins. As the theory is almost completely insensitive to D_0, its change with the swelling of the resin was neglected, even though large swellings occurred, particularly in the 5% cross-linked resin. The effective dielectric constant for nearest neighbor interaction, D_E, was taken to be a linear function of the distance between adjacent sites. The screening parameter κ' will at first be taken to be equal to κ, the value in the external solution.* This simplification will later be removed and accurate values of κ' computed. Using the molecular weight and density of the monomer, the dry volume per exchange group could be estimated, as could be r_0^0, the distance between adjacent sites in dry resin. The exchange groups were assumed to have little affinity for hydrogen ion. Binding constants for Na^+ and K^+ were chosen to be consistent with observed properties of sodium p-toluene sulfonate.[87] These input data are summarized in Table 9.4.1.

In the first series of calculations, exchange equilibria and swelling were computed for resins in contact with an external solution of total ionic strength 0.01, with concentration ratios $(Na^+)/(K^+) = 99$, 1, and 1/99. Some remarks indicating the methods of computation are provided in Appendix 9D. The

---

* Note that this simplifies the arithmetic but is a gross misrepresentation of the real system for which $\kappa' \gg \kappa$. We later remove this simplification.

TABLE 9.4.1[a]

ASSUMED CHARACTERISTICS OF SAMPLE HYPOTHETICAL RESINS

A. Properties common to all resins considered

$z\ = 10$	$\chi_1 = 1$	$K_K = 10$
$D_0 = 40$	$\kappa\ = 1/30$ (ionic strength $= 0.01$)	
$D_E = 5(r_0 - 4),\quad r_0 < 12$	$\kappa\ = 1/9.5$ (ionic strength $= 0.1$)	$K_{Na} = 40$
$= D_0,\qquad r_0 > 12$	$\kappa' = \kappa$	

B. Properties differing for different resins

Resin	I	II	III	IV
$r_0^0(A)$	7.5	7.5	9.4	7.5
v_0/Zv_s	13.9	13.9	27.8	13.9
b/Z	0.25	0.15	0.30	0.05
$(v_m/v)^{2/3}$	1.44	2.36	2.36	9.00

C. Summary

The above is equivalent to polystyrene-divinylbenzene resins

I: 25% cross-linked, fully sulfonated
II: 15% cross-linked, fully sulfonated
III: 15% cross-linked, 50% sulfonated
IV: 5% cross-linked, fully sulfonated

[a] Reproduced from S. A. RICE and F. E. HARRIS, *Z. physik. Chem.* [N.F.] **8,** 207 (1956).

results are shown in Table 9.4.2. Let us first compare the ion-exchange specificity of resins of different cross-linking. We find good qualitative confirmation of the trend, predicted earlier, toward higher specificity in more highly cross-linked resins. The order of magnitude of the equilibrium constants compares favorably with that of experimental data, some of which are listed for comparison in Table 9.4.3. The crudeness of the model and the lack of precise knowledge of the input parameters make more than a qualitative comparison inadvisable so long as we choose parameters for a hypothetical model and not a real system. Moreover, no attempt was made to match any theoretical and experimental point, and therefore, the theoretical and experimental exchange constants are not strictly comparable. Note also that the partially desulfonated resin (III) shows a lower specificity than does the fully sulfonated resin (II) of the same cross-linking, in accord with experiment.

TABLE 9.4.2 [a]

PROPERTIES CALCULATED FOR HYPOTHETICAL ION EXCHANGE RESINS
OF CHARACTERISTICS LISTED IN TABLE 9.4.1 [b]

External (Na⁺)			
External (Na^+)	0.0009 M	0.0050 M	0.0001 M
(K^+)	0.0001 M	0.0050 M	0.0099 M
Resin I. r_0	8.65 A	8.60 A	8.55 A
α'	0.46	0.43	0.42
v/v_0	1.53	1.51	1.48
K_K^{Na}	0.379	0.490	0.563
Resin II. r_0	109.5 A	10.80 A	10.73 A
α'	0.70	0.64	0.61
v/v_0	3.12	2.98	2.93
K_K^{Na}	0.522	0.645	0.706
Resin III. r_0	11.50 A	11.33 A	11.20 A
α'	0.79	0.70	0.64
v/v_0	1.83	1.75	1.70
K_K^{Na}	0.622	0.695	0.729
Resin IV. r_0	18.3 A	18.2 A	18.1 A
α'	0.987	0.970	0.954
v/v_0	14.6	14.3	14.1
K_K^{Na}	0.960	0.964	0.970

[a] Reproduced from S. A. RICE and F. E. HARRIS, *Z. physik. Chem.* [N.F.] **8**, 207 (1956).

[b] r_0 is the distance, in angstroms, between adjacent exchange sites, α' is the fraction of exchange sites not bound to an exchangeable ion, v/v_0 is the ratio of resin volume to that of the same amount of dry resin, K_K^{Na} is the exchange equilibrium constant (K^+) $(Na^+$ in resin)/(Na^+) $(K^+$ in resin).

A most gratifying feature of the results is the trend in K_K^{Na} with the external concentration ratios. This shows that changes in the degree of binding, as the concentration ratio changes, are small enough that the direction of variation of the ion exchange equilibrium quotient is governed by its direct dependence on $(Na^+)/(K^+)$. As a result, in the high sodium systems the resin has, relatively, a greater affinity for K^+, while in the high potassium systems it has a relatively greater affinity for Na^+. In agreement with experiment, we find this trend to be more pronounced for the more selective exchangers. Thus, for the first time a simple mechanism is found to be capable of predicting the major features of ion exchange equilibria on a molecular basis.

The swelling predicted by the model is also in qualitative accord with experiment. The volume increases when the resin suffers replacement of favored by less favored ions.

TABLE 9.4.3[a]

EXPERIMENTAL RESIN PARAMETERS

A. Ion-exchange equilibrium constants K_K^{Na} for divinylbenzene-polystyrene sulfonated resins[83]

External conc. ratio $(K^+)/(Na^+)$	1/99	1	99
25% Cross-linked	0.36	0.52	0.63
15% Cross-linked	0.41	0.52	0.57
$5\frac{1}{2}$% Cross-linked	0.54	0.68	0.12

B. Swelling ratios v/v_0 for divinylbenzene-polystyrene sulfonated resins in potassium form[88]

25% Cross-linked	1.33
15% Cross-linked	1.75
5% Cross-linked	2.05

[a] Reproduced from S. A. RICE and F. E. HARRIS, *Z. physik. Chem.* [N.F.] **8**, 207 (1956).

TABLE 9.4.4[a]

DEPENDENCE OF CALCULATED RESIN PROPERTIES ON IONIC STRENGTH OF THE EXTERNAL SOLUTION. CALCULATION FOR $(Na^+)/(K^+) = 99$

Ionic strength		0.01	0.1
Resin I.	r_0	8.65 A	8.64 A
	α'	0.46	0.53
	v/v_0	1.53	1.52
	K_K^{Na}	0.379	0.420
Resin II.	r_0	10.95 A	10.22 A
	α'	0.70	0.79
	v/v_0	3.12	2.54
	K_K^{Na}	0.522	0.622
Resin III.	r_0	11.50 A	11.10 A
	α'	0.79	0.89
	v/v_0	1.83	1.65
	K_K^{Na}	0.622	0.755

[a] Reproduced from S. A. RICE AND F. E. HARRIS, *Z. physik. Chem.* [N.F.] **8**, 207 (1956).

A second series of calculations was performed to study the effect of the external electrolyte concentration upon the swelling and exchange equilibria. The resins were assumed to be in contact with 99 % Na^+, 1 % K^+ solutions of ionic strengths 0.01 and 0.1, and the degree of binding and the volume were determined, as described in Appendix 9D. The results are shown in Table 9.4.4. The calculations were only carried out for relatively small ionic strengths as the equations used were derived assuming that the resin could contain only negligible amounts of nonexchange electrolyte.

The increased shielding occurring at the higher ionic strength is seen to affect both the swelling and the exchange equilibria. The smaller swelling in the more concentrated solution is to be expected because the electrostatic interaction has been reduced. The decrease in binding indicates that the decrease in interaction influences the binding more than does the mass-action effect of the higher concentration of exchangeable ions. As a result of the decrease in binding, the exchange equilibrium quotients tend toward unity as the concentration increases. The available data indicate that both these phenomena predicted by the model have been observed. Both calculation and experiment indicate that the change in swelling with concentration is greater for the more sparsely cross-linked resin.

Despite the inadvisability of attempting quantitative comparisons with experiment, it is tempting to see just how large the discrepancy between model and experiment is, and whether any deviations can be explained. In Table 9.4.5 can be found a comparison of the calculated and observed exchange constants. The 25 % cross-linked resin exchange constants are within 15 % of the experimental values and for the 15 % cross-linked resin, there is a probably fortuitous quantitative agreement over the entire concentration range. Note that only the 5 % cross-linked resin shows large deviations from experiment. It should be reiterated that no attempt was made to fit the computations at any experimental point. If we were to modify the values of K_K and K_{Na}, not only would the absolute magnitudes of the exchange constant K_K^{Na} change, but also the dependence of K_K^{Na} on concentration would be modified. It is therefore probable that some values of K_K and K_{Na} could be found that would provide a better fit to experiment.

Over and above this, however, the quantitative discrepancies evidenced in Table 9.4.5 are indicative of a slight failure of the model used. We note that in all three resins, the observed change of K_K^{Na} with concentration is greater than the computed change. (We must admit some surprise in noting that the rate of change, and the absolute magnitudes of the exchange constants for the 5 % cross-linked resin are so close to those of the 15 % cross-linked resin.) An examination of the computed and experimental volume changes, recorded in Tables 9.4.2 and 9.4.3, show that the model discussed expands more than a real resin of the same degree of cross-linking. The amount by which the model

TABLE 9.4.5[a]

COMPARISON OF CALCULATED AND EXPERIMENTAL EXCHANGE EQUILIBRIUM
CONSTANTS K_K^{Na}[87-90]

The Quantity tabulated is the ratio of K_K^{Na} to its value for the same resin when
$(K^+)/(Na^+)$ is 1/99

External conc. ratio $(K^+)/(Na^+)$	1/99	1	99
25% Cross-linking (Resin I)			
exptl.	1.00	1.44	1.75
calc.	1.00	1.29	1.49
15% Cross-linking (Resin II)			
exptl.	1.00	1.27	1.39
calc.	1.00	1.24	1.35
5% Cross-linking (Resin II)			
exptl.	1.00	1.26	1.33
calc.	1.00	1.00	1.01

[a] Reproducedfrom S. A. RICE AND F. E. HARRIS, Z. physik. Chem. [N.F.] 8, 207 (1956).

overexpands is indicative of and partially responsible for the insufficient
computed change of selectivity. The flaw in the molecular model responsible
for this excess expansion is easy to discern. We have assumed that all the
electrostatic energy is available for the expansion of the pseudolattice. Let us
consider some one charge surrounded by its ten nearest neighbors. Of these
ten nearest neighbors, two are almost certain to be on the same chain as the
central charge. As pointed out in Chapter 10, due to the fixed valence angle
between atoms and the hindered rotation about bonds, adjacent charges on a
polymer chain are largely ineffective in expanding the molecule. Now, in the
25% cross-linked resin, some 20% (2/10) of the electrostatic energy, due to
interaction of the central charge with its neighbors, is of just this type. While
there is not a linear relationship between the total electrostatic energy available for expansion and the volume, it is obvious that the smaller the effective
repulsive energy, the less the resin swells. Note that the experimental and
theoretical volumes differ by about 15%.

Of course, as the amount of cross-linking decreases, the fraction of the
energy unavailable for expansion increases. This occurs because nearest
neighbors on the same chain as the central charge cannot get further away

than a few bond distances and angles will permit, whereas charges on other chains are not so restricted in their movements. Therefore, the fraction of interaction energy due to groups on the same chain increases. By the time the cross-linking is as low as 5% we should really be using a model in which the interactions between chains are neglected and only interactions along one chain used. Except for his neglect of ion binding, this corresponds to the model for dilute polyelectrolyte gels developed by Katchalsky and co-workers.[89] The neglect of small ion binding makes the Katchalsky theory inapplicable to ion-exchange phenomena.

Finally, we note that the above considerations explain both the deviations in volume and selectivity. For, the selectivity depends directly upon the amount of binding which in turn depends upon the existence of a large repulsive potential energy between neighboring sites. The energy which may be unavailable for expansion is just the force responsible for ion binding. Thus we anticipate that the 5% cross-linked resin should have greater selectivity than computed on the basis of the liquid-like model due to the combined effects of larger binding and diminished expansion. It should be recalled that these latter two quantities are also interdependent and have been separated only for convenience of discussion.

We conclude then that the model described in this section is in very reasonable agreement with experiment for large degrees of cross-linking. The deviations from experiment at low degrees of cross-linking seem to be inherent primarily in the approximations used to compute the electrostatic energy. We require no new concepts or unusual processes to explain the observations and the deviations between model and experiment.

9.4.5 GENERAL DISCUSSION

The only previous attempt at a completely molecular description of polyelectrolyte gels is due to Katchalsky and co-workers.[89] Their considerations were intended to apply only to lightly cross-linked gels and exclude any discussion of ion-exchange phenomena. Katchalsky did point out the qualitative relation between electrostatic forces and gel volume, and considered the role of elastic deformation and gel solvent mixing. However, he did not consider ion binding, and calculated the electrostatic interactions in a way which can only be justified when interactions between charges on different chains can be neglected relative to interactions between charges on the same chain. Katchalsky's treatment, except for the omission of ion binding, is a very valuable discussion of very dilute gels, to which part of the development of this chapter does not apply.

Most of the theoretical discussion of ion exchange resins, has, in the past, been thermodynamic in nature. It would be desirable to consider the relation between the usual thermodynamic formalism and the molecular picture presented in this chapter. In the thermodynamic development, as presented by Gregor[88,90] for example, the resin is represented as a deformable membrane impermeable to and enclosing the fixed ions of the resin. The membrane is to be permeable to mobile ions and solvent. All components of the system except the membrane are to be regarded as incompressible, and the region inside the membrane is maintained electrically neutral. The condition of thermodynamic equilibrium with respect to the distribution of solvent across the membrane leads to the relation

$$\pi\bar{v}_s = RT \ln (a_{so}/a_{si}) \tag{9.4.13}$$

where $\bar{v}_s$ is the partial molal volume of solvent, a_{so} and a_{si} are the activities of solvent outside and inside the membrane, and $p + \pi$ is the pressure inside the membrane when p is the pressure outside the membrane. For definiteness, we now consider a cation exchanger. Thermodynamic equilibrium with respect to the distribution of exchangeable cations of types σ and σ' requires that

$$\pi(\bar{v}_\sigma - \bar{v}_{\sigma'}) = RT \ln \left(\frac{a_{\sigma o} a_{\sigma' i}}{a_{\sigma i} a_{\sigma' o}}\right) \tag{9.4.14}$$

where $\bar{v}_\sigma$ and $\bar{v}_{\sigma'}$ are the partial molal volumes of species σ and σ'. The volumes $\bar{v}_\sigma$ and the activities a_σ must of course refer to the same species, e.g., if the volumes of the hydrated ions are used, then so must also be the corresponding activities for hydrated ions. The cationic activities do not satisfy equations corresponding to Eq. (9.4.13) because of the imposition of the condition of electrical neutrality. Finally, thermodynamic equilibrium with respect to the distribution of mobile anions (of kinds τ) leads to

$$\pi(\bar{v}_\sigma + \bar{v}_\tau) = RT \ln \left(\frac{a_{\sigma o} \, a_{\tau o}}{a_{\sigma i} \, a_{\tau i}}\right). \tag{9.4.15}$$

The pressure π is determined by the elastic properties of the membrane, and is defined thermodynamically by

$$\pi = \frac{\partial A_{\text{membrane}}}{\partial v} \tag{9.4.16}$$

where A_{membrane} is the free energy of deformation of the membrane. The activities a_s, a_σ, and a_τ are referred to, as a standard state, the hypothetical ideal one molal solution. As in any osmotic equilibrium, it is to be understood that the activities appearing in Eqs. (9.4.13)–(9.4.16) are all computed at the pressure p.

In order to evaluate the properties of a resin in terms of Eqs. (9.4.13) to (9.4.16), it is necessary to know the volume dependence of A_{membrane}, and the activity coefficients of all species on both sides of the membrane. It is also necessary to apply the condition of electrical neutrality to the concentrations of the ions within the membrane.

The thermodynamic theory is a completely consistent and valid formulation, irrespective of the actual properties of the exchange material, assuming only that the resin volume is electrically neutral. However, its usefulness will depend upon our ability to evaluate π and the activities by independent means. If the membrane model is not an accurate physical representation of the resin, the activities necessary to satisfy experimental data will deviate widely from those of ordinary solutions of the same composition. Measurements of activity coefficients of ions in resins indeed indicate that there is a marked qualitative difference from the behavior of ions in ordinary electrolyte solutions.[90] Ion-exchange equilibria also show that there is not even the expected correlation between the activity coefficients of various ions under similar conditions.[80,83] Moreover, the thermodynamic theory is incapable of suggesting the form of the volume dependence of the membrane stretching free energy, and an empirical relation must be adopted. Considerations such as these indicate the essential limitations of a thermodynamic theory which is not based on a sufficiently detailed representation of the system.

Let us now turn to a discussion of the implications of the molecular theory presented in this section. We shall discuss, in order, the reason for the abnormal depression of the activity coefficients of the mobile ions,[90] the volume changes on swelling, the effects of stripping of the hydration shells of mobile ions in very concentrated solutions, and the diffusion of both charged and neutral particles through the resin.

The considerations of this chapter lead us to believe that the activity coefficients of the counterions of the resin should behave in a manner qualitatively similar to that of counterions to an ordinary polyelectrolyte in solutions. That is, the activity coefficient is depressed by the combined effects of ion localization and electrostatic interaction with the other charges. If the concentration of counterions in the solution is increased at constant polymer concentration, the fraction of ions bound to the polymer represents an increasingly smaller fraction of the total number of ions in solution. Moreover, the increased electrolyte concentration provides more efficient shielding of the fixed charges. Both of these effects tend to markedly increase the activity coefficient of the counterions towards unity. This behavior has been observed in both cationic and anionic exchange resins.[90]

The behavior of the activity coefficients, cited in the last paragraph, is direct evidence of the strong interactions operative between the charge groups on the resin. It should be noted that the additivity of integrated average exchange constants does not imply a lack of interaction between the exchange groups.

We have already pointed out that the functional form of the free energy of stretching the membrane cannot be predicted by the thermodynamic theory. On the other hand, the molecular picture adopted in this section has led to a prediction of the volume changes to be expected as a function of the exchange reaction constants and the concentration which is in good agreement with experiment. It should be pointed out that the only force responsible for the swelling of the resin was the electrostatic interaction between the charged groups fixed on the polymer network. At no time was it necessary to consider the space filling properties of the ions or to ascribe any portion of the volume change to the differences in volumes of the ions involved in the exchange reaction. Further, it has not been necessary to introduce a distribution of sites to account for the experimental data.[83] It is one of the principal achievements of the theory presented that the volume changes are explained rather than used to correlate the data. The fit of the data which can be obtained with the use of only one kind of site does not preclude a distribution of sites but it does indicate that the major features of the ion exchange equilibria are determined by the electrostatic forces operative between fixed charges of the polymer network and not by some other and more subtle force.

The development has been primarily concerned with those properties which should be characteristic of all resin systems. A variety of experimental evidence suggests that in particular experiments other more specialized factors come into play. Probably among the more important of these are processes involving changes in the hydration states of exchangeable ions. This can be responsible for reversing the order of selectivity, notably in systems where essentially all the water is in the hydration spheres of the various ions. It is possible that, in extremely concentrated resins, a deficiency of water will be experienced by ions within the resin, even when the external concentrations are small. A more extensive discussion of the possible effects of changes in hydration may be found in the work of Diamond.[91]

Let us now turn to a discussion of the diffusion of mobile ions[92] and nonelectrolyte molecules[90] through the resin. We may immediately note that the rate of diffusion of anions through the resin should be slower than in an ordinary solution since they are repelled from it by the fixed charges on the polymer chains. On the other hand, the rate of transfer of the cations should be of the same order of magnitude as in an ionic solution of comparable concentration. Moreover, if the mechanism of cation transfer is similar to the mechanism of the transfer of protons through the water lattice, (i.e., transfer from site to site), then it might be anticipated that the rate of diffusion will be greater than in the open solution. However, the rate of diffusion will then be controlled by the rates of ionization processes. Neutral molecules will diffuse at a slower rate within the resin than in solution for three reasons. Firstly, there is electrostriction of the solvent which tends to exclude from the resin any material of low dielectric constant. Secondly, the volume fraction of solvent in the resin matrix is much smaller than outside, leading to a small free volume. Finally, there is the possibility of the adsorption of the non-electrolyte onto the hdrocarbon residues of the fixed plymer chains. As is easily seen, all these effects tend to reduce the rate at which a nonelectrolyte molecule may move through the resin.

Finally, we may mention that this model indicates what must be considered to predict the behavior of ion-exchange resins in mixed solvents.[93] A most important factor is the dielectric constant, which, for most mixtures containing organic solvents, is less than that of water. The local effective dielectric constants within the resin will be correspondingly still further reduced. In addition, it is possible that the dissociation constants of the various kinds of ion pairs may be different due to differences in solvation. It is likely that under most circumstances the factors just considered will lead to an increased swelling of the resin. It is not advisable to make particularly general statements with regard to the exchange equilibria in mixed solvents.

9.4.6 Comparison with Experimental Studies of Real Resins

In the preceding section we made calculations for hypothetical resins with the intention of providing some insight into otherwise formal calculations. A comparison of the theory with experimental studies of the swelling of ion exchange resins was made by Lapanje and Dolar,[94a] who carried out an extensive program of measurements of water absorption isotherms and heats of absorption for variously cross-linked ion exchange resins. The free energy change accompanying the swelling can be calculated from the water absorption isotherm by the following relation:[95,96]

$$\frac{\Delta G}{n_r} = \frac{n_w}{n_r} \cdot RT \ln x - \int_0^x \frac{n_w}{n_r} \frac{RT}{x} dx \qquad (9.4.17)$$

where $x = p/p^0$ is the relative water vapor pressure and n_w/n_r is the number of moles of water per equivalent of dry resin.* Lapanje and Dolar used isopiestic measurements to obtain n_w/n_r at various values of p/p^0, and calculated $\Delta G/n_r$ to compare with Eq. (9.4.3). The comparison between calculated values of Eq. (9.4.3) and experimental values is shown in Table 9.4.6, and the parameters used for the calculation are shown in the same table. The value of a was computed from the data for sodium p-toluenesulfonate published by Bonner and Holland.[87] The values of D_0 and D_E in Eq. (9.4.3) were equated and computed from the expression $D_0 = 5(r_0 - 4)$. The intrinsic binding constant K was determined by the degree of dissociation

* In condensed systems, $A \approx G$, so that all the equations derived for the Helmholtz free energy may be used for the Gibbs free energy with very small error.

TABLE 9.4.6[a]

COMPARISON OF THE CALCULATED $-\Delta G_{swell}$ WITH THE EXPERIMENTAL
VALUES (cal)

1. Assumed characteristics of resins
   A. Properties common to all resins considered $z = 10$, $a = 3.5$ A, $D_0 = 5(r_0 - 4)$
   B. Properties differing for different resins.

Resin	IR	IIR	IIIR	IVR
α'	0.7	0.8	0.9	0.95
K_{Na}	7.0	11	14	22
b/Z	0.15	0.08	0.03	0.01
v_m/v_0	4.0	4.3	5.3	6.5

2. Properties calculated from the above characteristics and experimental data.

	IR	IIR	IIIR	IVR
r_0 (A)	7.1	7.9	9.5	11.1
χ_1	0.90	0.76	0.64	0.59
β	0.96	0.99	0.99	0.99
$v_0(ml)$	141	140	138	140
$v(ml)$	229	295	514	824

3. Values of $-\Delta G_{swell}$ (cal)

Exp.	Calc.	Exp.	Calc.	Exp.	Calc.	Exp.	Calc.
4490	3850	5000	3890	5130	3470	5460	3190

Resin IR: 15% cross-linked; IIR: 8% cross-linked; IIIR: 3% cross-linked; IVR: 1% cross-linked.

[a] Reproduced by permission of Prof. Lapanje.

α', which was guessed from an analogy with concentrated simple electrolyte solutions. The value at maximal swelling v_m and the volume of the dry resin v_0 were determined by experiment.

Although the theory is semiquantitative in nature and some of the parameters used in the calculations described were chosen arbitrarily, the agreement of the calculated values with experiment is satisfactory, especially for large degrees of cross-linking.

An even more stringent test of the theory has been made by Lapanje[94b] who has compared theoretical and experimental swelling pressures. This, of course, is a direct test of the derivative of the theoretical free energy. Lapanje uses the following expressions derivable from the theory presented:

$$p_{mix} = \frac{kT}{v_s} \ln\left(\frac{v - v_0}{v}\right) + \frac{v_0}{v} + \chi_1\left(\frac{v_0}{v}\right)^2, \tag{9.4.18}$$

$$p_{osm} = \frac{\alpha' N_0 kT}{v}$$

$$p_{str} = \frac{bkT}{v} \left[\left(\frac{v}{v_0}\right)^{2/3}\left(\frac{1 - 0.4(v/v_m)^{2/3}}{1 - (v/v_m)^{2/3}}\right) - \frac{1}{2}\right] \tag{9.4.19}$$

$$p_{el} = -\frac{\alpha' \kappa' N_0 Z q^2 e^{-\kappa' r_0}}{4 D_0 v}\left(\frac{2 - 2\alpha}{\beta + 1} - 1\right) \tag{9.4.20}$$

$$+ \frac{\alpha' N_0 q^2 \kappa'}{2 v D_0 (1 + \kappa' a)}\left(1 - \frac{\kappa' a}{1 + \kappa' a}\right)$$

$$p_{mech} = \frac{\alpha' N_0 q^2 e^{-\kappa' r_0}}{2 D_0 r_0}\left(\frac{2 - 2\alpha}{\beta + 1} - 1\right)\frac{1 + \kappa' r_0}{3v} \tag{9.4.21}$$

$$p_D = \frac{\alpha' N_0 z q^2 e^{-\kappa' r_0}}{2 D_0 r_0}\left(\frac{2 - 2\alpha}{\beta + 1)} - 1\right)\frac{1}{D_0}\left(\frac{dD_0}{dv}\right)$$

$$+ \frac{\alpha' N_0 q^2 \kappa'}{D_0 (1 + \kappa' a)}\frac{1}{D_0}\left(\frac{dD_0}{dv}\right) \tag{9.4.22}$$

Note that Lapanje has introduced the slight modification of finite ion size (Eq. 9.4.20). The term p_{el} represents the decrease of osmotic pressure due to the interaction of fixed ions and counterions, while the term p_{mech} reflects the fact that the electrostatic repulsions of the fixed ions are reduced by an increase in volume. Nothing is known of the dependence of D_0 on v in systems as complicated as ion exchange resins, but for simple systems this contribution to the pressure may be neglected. The condition of mechanical equilibrium requires that

$$p_{mix} + p_{osm} + p_{str} + p_{el} + p_{mech} + p_D = 0 \tag{9.4.23}$$

Using the data quoted in Table 9.4.6, Lapanje computed the pressures displayed in Table 9.4.7.

TABLE 9.4.7

PRESSURE OPERATIVE IN SWOLLEN ION-EXCHANGE RESINS
(ALL PRESSURES EXPRESSED AS DYNES/CM^2)

Negative terms				
p_{mix}	3.6×10^7	0.2×10^7	0.1×10^7	0.1×10^7
p_{osm}	7.5×10^7	6.7×10^7	4.4×10^7	2.9×10^7
p_{mech}	4.2×10^7	2.2×10^7	1.3×10^7	0.7×10^7
Positive terms				
p_{el}	14.0×10^7	8.9×10^7	4.6×10^7	2.4×10^7
p_{str}	3.1×10^7	1.9×10^7	1.1×10^7	0.9×10^7
Total deviation from zero (in percent of the sum of the magnitudes of the contributions)	5%	8%	1%	12%

Reproduced by permission of Prof. Lapanje.

Within an accuracy of approximately 10%, the cell model (site binding) theory presented herein is verified. The authors believe that this demonstrates that the model contains much of the physically relevant information. Of course, much more quantitative work remains to be done.

REFERENCES

1. W. KERN, *Z. physik. Chem.* **A181**, 240, 283 (1938); **A184**, 201 (1939).
2. I. KAGAWA, *J. Chem. Soc. Japan, Ind. Chem. Sect.* **47**, 574 (1944).
3. R. M. FUOSS, *J. Polymer Sci.* **12**, 185 (1954).
4. F. HARRIS and S. A. RICE, *J. Phys. Chem.* **58**, 725, 733 (1954).
5. F. OOSAWA, N. IMAI, and I. KAGAWA, *J. Polymer Sci.* **13**, 93 (1954).
6. H. MORAWETZ, *Fortschr. Hochpolym. Forsch.* **1**, 1 (1958).
7. F. BASOLO AND R. C. PEARSON, " Mechanisms of Inorganic Reactions—A Study of Metal Complexes in Solution." Wiley, New York, 1958.
8. W. C. FERNELIUS AND L. G. VAN UITERT, *Acta Chem. Scand.* **8**, 1726 (1954).
9. R. M. IZATT, W. C. FERNELIUS, C. G. HAAS, AND B. P. BLOCK, *J. Phys. Chem.* **59**, 170 (1955).
10. G. SCHWARZENBACH AND H. ACKERMANN, *Helv. Chim. Acta* **30**, 1798 (1947).
11. G. SCHWARZENBACH, E. KAMPITSCH, AND R. STEINER, *Helv. Chim. Acta* **29**, 364 (1946).
12. J. BJERRUM, " Metal Ammine Formation in Aqueous Solution." Haase, Copenhagen, 1941.
13. A. E. MARTELL AND M. CALVIN, " The Chemistry of Metal Chelate Compounds." Prentice-Hall, Englewood Cliffs, New Jersey, 1952.

14. J. SCHUBERT, *J. Phys. Chem.* **56**, 113 (1952).
15. H. MORAWETZ, A. M. KOTLIAR, AND H. MARK, *J. Phys. Chem.* **58**, 19 (1954).
16. A. M. KOTLIAR AND H. MORAWETZ, *J. Am. Chem. Soc.* **77**, 3692 (1955).
17. F. T. WALL AND S. J. GILL, *J. Phys. Chem.* **58**, 1128 (1954).
18. H. P. GREGOR, L. B. LUTTINGER, AND E. M. LOEBL, *J. Phys. Chem.* **59**, 34 (1955).
19. S. FRONAEUS, *Acta Chem. Scand.* **5**, 859 (1951).
20. H. MORAWETZ, *J. Polymer Sci.* **17**, 442 (1955).
21. H. MORAWETZ AND E. SAMMAK, *J. Phys. Chem.* **61**, 1357 (1957).
22. I. KAGAWA AND K. KATSUURA, *J. Polymer Sci.* **17**, 365 (1955).
23. I. KAGAWA AND M. NAGASAWA, *J. Polymer Sci.* **16**, 299 (1955).
24. U. P. STRAUSS, N. L. GERSHFELD, AND H. SPIERA, *J. Am. Chem. Soc.* **76**, 5909 (1954).
25. U. P. STRAUSS, D. WOODSIDE, AND P. WINEMAN, *J. Phys. Chem.* **61**, 1353 (1957).
26. U. P. STRAUSS AND P. L. WINEMAN, *J. Am. Chem. Soc.* **80**, 2366 (1958).
27. U. P. STRAUSS AND P. ANDER, *J. Am. Chem. Soc.* **80**, 6494 (1958).
28. U. P. STRAUSS AND S. BLUESTONE, *J. Am. Chem. Soc.* **81**, 5292 (1959).
29. U. P. STRAUSS AND P. D. ROSS, *J. Am. Chem. Soc.* **81**, 5295 (1959).
30. U. P. STRAUSS AND P. D. ROSS, *J. Am. Chem. Soc.* **81**, 5299 (1959).
31. H. P. FRANK, S. BARKIN, AND F. R. EIRICH, *J. Phys. Chem.* **61**, 1375 (1957).
32. T. ISEMURA AND A. IMANISHI, *J. Polymer Sci.* **33**, 337 (1958).
33. N. BJERRUM, *Kgl. Danske Videnskab. Selskab., Mat.-fys. Medd.* **7**, No. 9 (1926).
34. H. S. HARNED AND B. B. OWEN, " The Physical Chemistry of Electrolytic Solutions," 2nd ed. Reinhold, New York, 1950.
35. R. H. FOWLER AND E. A. GUGGENHEIM, " Statistical Thermodynamics." Cambridge Univ. Press, London and New York, 1956.
36. R. M. FUOSS AND C. A. KRAUS, *J. Am. Chem. Soc.* **55**, 1019 (1933).
37. R. M. FUOSS, *J. Am. Chem. Soc.* **81**, 2659 (1959).
38. H. SADEK AND R. M. FUOSS, *J. Am. Chem. Soc.* **76**, 5905 (1954).
39. H. REISS, *J. Chem. Phys.* **25**, 400, 408 (1956); R. M. FUOSS, *J. Am. Chem. Soc.* **80**, 5059 (1958).
40. J. T. DENISON AND J. B. RAMSEY, *J. Am. Chem. Soc.* **77**, 2615 (1955).
41. W. R. GILKERSON, *J. Chem. Phys.* **25**, 1199 (1956).
42. J. G. KIRKWOOD, *J. Chem. Phys.* **18**, 380 (1950).
43. F. ACCASCINA, E. L. SWARTS, P. L. MERCIER, AND C. A. KRAUS, *Proc. Natl. Acad. Sci. U.S.* **39**, 917 (1953).
44. M. G. EVANS AND G. H. NANCOLLAS, *Trans. Faraday Soc.* **49**, 363 (1953).
45. O. JARDETSKY AND J. E. WERTZ, *J. Am. Chem. Soc.* **82**, 318 (1960).
46. G. C. HOOD, C. A. REILLY, AND O. REDLICH, *J. Chem. Phys.* **22**, 2067 (1954); **23**, 2229 (1955); **27**, 1126 (1957); **28**, 329 (1958); *J. Phys. Chem.* **63**, 101 (1959); *Discussions Faraday Soc.* **24**, 87 (1957).
47. H. S. GUTOWSKY AND A. SAIKA, *J. Chem. Phys.* **21**, 1688 (1953).
48. Y. MASUDA AND T. KANDA, *J. Phys. Soc. Japan* **8**, 432 (1953).
49. F. T. WALL, *J. Phys. Chem.* **61**, 1344 (1957).
50. H. INAGAKI AND T. ODA, *Makromol. Chem.* **21**, 1 (1956).
51. H. INAGAKI, S. HOTTA, AND M. HIRAMI, *Makromol. Chem.* **23**, 1 (1957).
52. J. J. HERMANS AND J. T. G. OVERBEEK, *Rec. trav. chim.* **67**, 761 (1948).
53. G. E. KIMBALL, M. CUTLER, AND H. SAMELSON, *J. Phys. Chem.* **56**, 57 (1952).
54. F. T. WALL AND J. BERKOWITZ, *J. Chem. Phys.* **26**, 114 (1957).
55. S. LIFSON, *J. Chem. Phys.* **27**, 700 (1957).
56. M. NAGASAWA AND I. KAGAWA, *Bull. Chem. Soc. Japan* **30**, 961 (1957).
57. I. KAGAWA AND K. KATSUURA, *J. Polymer Sci.* **9**, 405 (1952).

58. M. NAGASAWA, M. IZUMI, AND I. KAGAWA, *J. Polymer Sci.* **37**, 375 (1959).
59. P. J. NAPJUS AND J. J. HERMANS, *J. Colloid Sci.* **14**, 252 (1959).
60. H. EISENBERG AND G. R. MOHAN, *J. Phys. Chem.* **63**, 671 (1959).
61. S. LIFSON, *J. Chem. Phys.* **26**, 727 (1957).
62. J. R. HUIZENGA, P. F. GRIEGER, AND F. T. WALL, *J. Am. Chem. Soc.* **72**, 2636 (1950).
63. J. R. HUIZENGA, P. F. GRIEGER, AND F. T. WALL, *J. Am. Chem. Soc.* **72**, 4228 (1950).
64a. G. HOWARD AND D. O. JORDAN, *J. Polymer Sci.* **12**, 209 (1954).
64b. J. KRAUT, *J. Polymer Sci.* **14**, 222 (1954).
64c. U. P. STRAUSS, N. GERSHFELD, AND H. SPIERA, *J. Am. Chem. Soc.* **76**, 5909 (1954).
64d. F. T. WALL, H. TERAYAMA, AND S. TECHAKUMPUCH, *J. Polymer Sci.* **20**, 477 (1956).
64e. F. E. BAILEY, JR., A. PATTERSON, JR., AND R. M. FUOSS, *J. Am. Chem. Soc.* **74**, 1845 (1952).
65. M. NAGASAWA AND S. A. RICE, *J. Am. Chem. Soc.* **82**, 5207 (1960).
66. M. NAGASAWA, S. OZAWA, K. KIMURA, I. KAGAWA, *Mem. Fac. Eng., Nagoya Univ.* **8**, 50 (1956).
67. R. A. MOCK AND C. A. MARSHALL, *J. Polymer Sci.* **13**, 263 (1954).
68. M. NAGASAWA, A. TAKAHASHI, M. IZUMI, AND I. KAGAWA, *J. Polymer Sci.* **38**, 213 (1959).
69. H. FUJITA, K. MITSUHASHI, AND T. HOMMA, *J. Colloid Sci.* **9**, 466 (1954).
70. D. T. F. PALS AND J. J. HERMANS, *Rec. trav. chim.* **71**, 434 (1954).
71. H. TERAYAMA AND F. T. WALL, *J. Polymer Sci.* **16**, 357 (1955).
72. P. J. FLORY AND J. E. OSTERHELD, *J. Chem. Phys.* **26**, 114 (1957).
73. T. A. OROFINO AND P. J. FLORY, *J. Phys. Chem.* **63**, 283 (1959).
74. J. J. HERMANS AND H. FUJITA, *Koninkl. Ned. Akad. Wetenschap., Proc.* **B58**, 182 (1955).
75. J. TH. G. OVERBEEK AND D. STIGTER, *Rec. trav. chim.* **75**, 543 (1956).
76. G. SCATCHARD, *J. Am. Chem. Soc.* **68**, 2315 (1946).
77. F. E. BAILEY, JR., AND A. PATTERSON, JR., *J. Polymer Sci.* **9**, 285 (1952).
78. F. E. BAILEY, A. PATTERSON, JR., AND R. M. FUOSS, *J. Am. Chem. Soc.* **74**, 1845 (1952).
79. K. F. WISSBRUN AND A. PATTERSON, JR., *J. Polymer Sci.* **33**, 235 (1958).
80. See, for example, R. KUNIN AND R. J. MYERS, " Ion Exchange Resins." Wiley, New York, 1950; H. P. GREGOR, *J. Colloid. Sci.* **6**, 20, 245, 304, 323 (1951); **7**, 511 (1952); *J. Am. Chem. Soc.* **70**, 1293 (1948); **73**, 642, 3537 (1951); G. E. BOYD, *Ann. Rev. Phys. Chem.* **2**, 320 (1951); C. W. DAVIES AND G. D. YEOMAN, *Trans. Faraday Soc.* **49**, 968 975 (1953); E. GLUECKAUF, *Proc. Roy. Soc.* **A214**, 207 (1952).
81. See, for example, P. J. FLORY, " Principles of Polymer Chemistry." Cornell Univ. Press, Ithaca, New York, 1953.
82. See, for example, J. L. PAULEY, *J. Am. Chem. Soc.* **76**, 1422 (1954) and references contained therein.
83. D. REICHENBERG AND D. J. MCCAULEY, *J. Chem. Soc.* p. 2741 (1955).
84. See Chap. I of this Volume.
85. J. G. KIRKWOOD, *J. Chem. Phys.* **7**, 911 (1939).
86. J. B. HASTED, D. M. RITSON, AND C. H. COLLIE, *J. Chem. Phys.* **16**, 1 (1948).
87. W. F. MCDEVIT AND F. A. LONG, *J. Am. Chem. Soc.* **74**, 1090 (1952); O. D. BONNER AND V. F. HOLLAND, *J. Am. Chem. Soc.* **77**, 5828 (1955).
88. H. P. GREGOR, F. GUTOFF, AND J. I. BREGMAN, *J. Colloid Sci.* **6**, 245 (1951).
89. A. KATCHALSKY, S. LIFSON, AND H. EISENBERG, *J. Polymer Sci.* **7**, 57 (1951); **8**, 476 (1952); *Prog. in Biophys. and Biophys. Chem.* **4**, 1 (1954).
90. H. P. GREGOR AND M. H. GOTTLIEB, *J. Am. Chem. Soc.* **75**, 3539 (1953); M. H. GOTTLIEB AND H. P. GREGOR, *J. Am. Chem. Soc.* **76**, 4639 (1954).
91. R. M. DIAMOND, *J. Am. Chem. Soc.* **77**, 2978 (1955).
92. G. BOYD, *J. Am. Chem. Soc.* **69**, 2836 (1947).

93. H. P. Gregor, D. Nogel, and M. H. Gottlieb, *J. Phys. Chem.* **59**, 10 (1955).
94a. S. Lapanje and D. Dolar, *Z. physik. Chem.* [N.F.] **18**, 11 (1958); **21**, 376 (1959).
94b. S. Lapanje, *Vestnik. Slovensk. Kem. Drustva*, **6**, 53 (1959).
95. M. H. Waxman, B. R. Sundheim, and H. P. Gregor, *J. Phys. Chem.* **57**, 969, 974 (1954).
96. E. Glueckauf and G. P. Kitt, *Proc. Roy. Soc.* **A228**, 322 (1955).

Appendix 9A. The Free Energy of Ion Binding

Equation (9.4.1) of the main text describes the chemical contribution to the free energy when $\{n_\sigma\}$ ions of types σ are bound to resin sites. Electrostatic contributions to the free energy are to be omitted from A_{chem}, so that the activity coefficients of the ions taking part in the binding may be assumed to be nearly unity. Furthermore, the sites to which various ions will bind must be regarded as completely determined, since A_{chem} is not to include any free energy resulting from the multiplicity of ways the bound ions can be arranged in the resin. The free energy for binding an ion, then, will involve the standard state chemical potentials for the exchange site in the free state, μ_0^0, and in a bound state involving an ion σ, μ_σ^0. The chemical potential for the free ion μ_σ may be taken as the ideal solution value.

$$\mu_\sigma = \mu_\sigma^0 + kT \ln c_\sigma. \tag{9A.1}$$

(See the main text for definitions of symbols used there.)

For the reasons discussed above, A_{chem} will be of the form

$$A_{\text{chem}} = \sum_{\sigma \neq 0} n_\sigma(\tilde{\mu}_\sigma^0 - \mu_\sigma - \tilde{\mu}_0^0). \tag{9A.2}$$

Introducing the intrinsic binding constants K_σ, related to the chemical potentials by

$$\tilde{\mu}_0^0 + \mu_\sigma^0 - \tilde{\mu}_\sigma^0 = -kT \ln K_\sigma \qquad (\sigma \neq 0) \tag{9A.3}$$

Equation (9.4.1) of the main text may be obtained.

Appendix 9B. Configurational Free Energy in Highly Swollen Networks

The distribution of configurations at a given resin volume has been assumed to be independent of the presence of charged groups, so that the configurational free energy may be evaluated by methods appropriate to uncharged gels. Network swelling in such structures has been considered by Flory and others. Although in our opinion the details of the theory and its application to specific systems continue to raise serious questions, the approach developed by Flory will certainly lead to a qualitative description of the swelling effect. The networks with which we are concerned are sufficiently stretched, however, that Flory's theory must be modified to allow for the finite contour lengths of the polymer chains. Such a modification has been considered briefly by Katchalsky, Künzle, and Kuhn. Unfortunately they do not provide sufficient detail that we can follow their argument, so we have made our own crude analysis, which yields somewhat different results.

According to Flory's theory, the configurational free energy can be written as the sum of an elastic contribution A_{str} and a mixing contribution calculable from the Flory–Huggins solution theory.

$$A_{conf}(v) = A_{str}(v) + \frac{v - v_0}{v_s} kT \left[\ln \frac{v - v_0}{v} + \chi_1 \frac{v_0}{v} \right]. \tag{9B.1}$$

(Throughout the Appendix we shall use without comment any symbols defined in the main text.) It is the elastic contribution whose calculation must be refined. Flory calculates it in terms of three partial processes: (i) breaking the network into separated polymer chains; (ii) stretching the separated polymer chains from their equilibrium length distribution to that characteristic of the swollen network; (iii) reconstructing the network (now swollen) from the stretched chains. The free energy change for processes (i) and (iii) is $(bkT/2)\ln(v/v_0)$, independently of the degree to which the chains are stretched in process (ii). This assumes the cross-linking is done by tetrafunctional groups, as is most frequently the case.

Process (ii) is affected by the finite chain lengths. We proceed as did Flory, but we will replace the Gaussian *a priori* probability distribution by an expression more appropriate to stretched chains. The free energy change for process (ii) is

$$\Delta A_{ii}(v) = -kT \int_0^\infty g(v, h)\ln[bW(h)/g(v, h)] \, dh \tag{9B.2}$$

where $g(v, h)$ is the number density of polymer chains at end-to-end distance h in a resin of volume v, and $W(h)$ is the *a priori* probability density that a chain will assume the end to end distance h. The basic assumption of Flory's method is that swelling proportionally stretches every chain, so that $g(v, h)v\,dh = g(v_0, \{v_0/v\}^{1/3}h)v_0 \, dh$, where the factors v and v_0 take into account the fact that the volume element does not stretch. On this basis, the formula for g must be

$$g(v, h) \, dh = b \left(\frac{3}{2\pi} \right)^{3/2} \exp\left\{ -\frac{3}{2} \left(\frac{h}{\langle h_0^2 \rangle^{1/2}} \right)^2 \left(\frac{v_0}{v} \right)^{2/3} \right\} \left(\frac{v_0}{v} \right) \frac{4\pi h^2 \, dh}{\langle h_0^2 \rangle^{3/2}}. \tag{9B.3}$$

Here $\langle h_0^2 \rangle^{1/2}$ is the root mean square length of a chain in a relaxed network (of volume v_0). Equation (9B.3) cannot be exactly correct at large swellings because it asserts that the Gaussian character of the distribution is not altered.

However, as g is normalized at the volume v, the free energy calculated from Eq. (9B.2) cannot be far from the proper value.

The *a priori* probability density $W(h)$, on the other hand, enters Eq. (9B.2) importantly for only certain values of h which may not be those for which $W(h)$ is large. Thus, even though $W(h)$ is also normalized, the calculated free energy can be quite sensitive to errors in the form of $W(h)$. We therefore introduce for $W(h)$ an expression more accurate than the Gaussian function for large h,

$$W(h)\, dh = \text{const} \times \exp\left[-\frac{h_{max}}{\langle h_0^2\rangle}\int_0^s L^*\left(\frac{h}{h_{max}}\right)dh\right]4\pi h^2\, dh \qquad (9B.4)$$

where h_{max} is the contour length of a chain, and L^* is the inverse Langevin function. We assume all the polymer chains have the same value of h_{max} In using Eq. (9B.4), we approximate L^* by an expression employed by Kuhn and Grün,

$$L^*(x) \approx 3x\left(1 + \frac{0.6x^2}{1-x^2}\right). \qquad (9B.5)$$

Inserting Eqs. (9B.3)–(9B.5) into (9B.2), and performing the indicated operations, one may find

$$\Delta A_{ii}(v) = \frac{3}{2}kTb\left\{0.4\left(\frac{v}{v_0}\right)^{2/3} - 1 - 0.6\left(\frac{v_m}{v_0}\right)^{2/3}\ln\left[1 - \left(\frac{v}{v_m}\right)^{2/3}\right]\right\} - bkT\ln\left(\frac{v}{v_0}\right).$$
$$(9B.6)$$

In the analysis leading to (9B.6), one additional approximation has been introduced. We have replaced

$$\int_0^\infty g(v,s)\ln\left(1 - \frac{h^2}{h^2_{max}}\right)ds \equiv b\left\langle \ln\left(1 - \frac{h^2}{h^2_{max}}\right)\right\rangle_v \text{ by } b\ln\left(1 - \frac{\langle h^2\rangle_v}{h^2_{max}}\right).$$

These expressions differ in terms of order $(\langle h^2\rangle_v/h^2_{max})^2$, but the nature of the proceedings suggests that this will not be the major source of uncertainty in the swelling theory. In (9B.6), we have also identified $\langle h^2\rangle_v/\langle h_0^2\rangle$ with $(v/v_0)^{2/3}$, and $\langle h^2\rangle_v/h^2_{max}$ with $(v/v_m)^{2/3}$ so as to cast the result entirely in terms of observable quantities.

Combining Eq. (9B.6) with the result quoted for processes (i) and (iii), one obtains

$$A_{str} = \frac{3}{2}kTb\left[0.4\left(\frac{v}{v_0}\right)^{2/3} - 1 - 0.6\left(\frac{v_m}{v_0}\right)^{2/3}\ln\left\{1 - \left(\frac{v}{v_m}\right)^{2/3}\right\}\right] - \frac{bkT}{2}\ln\left(\frac{v_0}{v}\right)$$
$$(9B.7)$$

which, together with Eq. (9B.1), leads to Eq. (9.4.2) of the main text. From the way in which Eq. (9B.7) was obtained, we see that it presupposes that the polymerization took place in such a way as to make the elastically relaxed state of volume v_0. This is presumably what would occur if the polymerization was carried out in the absence of solvent and was allowed to proceed essentially to completion. If these conditions are not met, then Eq. (9B.7) should be replaced by an equation which takes into account the initial elastic condition of the resin.

There are at least two complications inherent in the use of the Flory–Huggins formula for the mixing free energy which was displayed without comment in Eq. (9B.1). First, it has been assumed that the total volume of the system remains essentially constant even though the resin volume may change. Secondly, the Flory–Huggins formula has been used in the form appropriate to a binary solution even though the " solvent " actually consists of a solution of small particles. After writing the Flory–Huggins free energy expression, taking cognizance of all the solvent species, one may show that the mixing term of Eq. (9B.1) results, if v_s and χ_1 are suitable composites of the properties of the individual " solvent " species.

Appendix 9C.   Electrostatic and Distributional Free Energy

The free energy contribution A_{int} defined in the main text consists of the total free energy of a resin system of specified volume and numbers of bound ions of each kind, less free energy contributions A_{conf} and A_{chem} which are the same for all the different arrangements of the bound ions on the resin. A_{int} may not be computed, however, by breaking up the remaining contributions to the free energy into additive terms, because different arrangements of the same numbers of bound charges involve different electrostatic interaction energies. Instead, it is necessary to proceed from a somewhat more fundamental viewpoint, and (i) compute the " free energy " for each " microscopic " state characterized by an assignment of bound ions to specific sites; then (ii) combine these microscopic states, with weighting factors dependent on their " free energies," to form a partition function from which may be obtained the free energy of the desired macroscopically specified state of the resin.

The only contribution to the free energy of each microscopic state which need concern us here is the electrostatic free energy. If we number the n_0 charged sites in any microscopic state and refer to them by indices i or j, the electrostatic free energy may be expressed in the form

$$A_{\text{elec}} = \frac{1}{2} \sum_{\substack{i\,j \\ i \neq j}}^{n_0} \frac{q^2 e^{-\kappa' r_{ij}}}{D_E r_{ij}} - q^2 \frac{n_0 \kappa'}{D_0} \tag{9C.1}$$

where the zero of A_{elec} refers to a completely uncharged resin. Any symbols not defined in this Appendix have been defined in the main text. In Eq. (9C.1), r_{ij} is the distance between the ith and jth resin charges. The second term of Eq. (9C.1) is the free energy required to build ion atmospheres about the charged sites and unbound exchangeable ions within the resin. Equation (9C.1) is formulated subject to two restricting conditions, the first of which is that $\kappa' R \gg 1$ where R is of the order of a dimension of a resin particle. The second restriction is that the concentration of charged sites in the resin be much higher than the concentration of the external solution, so that all the counter ions to the resin charges may be expected to remain within the resin. It is the first restriction which allows the familiar Debye–Hückel term to be used, while the second restriction makes appropriate the use of the ion atmosphere terms characteristic of the interior of the resin.

It would be a formidable problem to evaluate Eq. (9C.1) for any one charge distribution, let alone for the enormous number of charge distributions occuring in any macroscopically specified resin state. We therefore restrict our consideration to only closely neighbouring charge groups. For the purpose of computing distances in the resin, we regard the sites as in some sort of average spatial configuration, thereby assuming that the electrostatic energy is insensitive to the polymer configuration at specified volume, just as was done when computing the configurational free energy. In typical resins the degree of cross-linking is sufficiently high that many sites which are not near neighbors along the polymer chains are spatially in close proximity. Therefore, consideration of interactions only between exchange sites which are nearest neighbors along the chains, as in the treatment of linear poly-electrolytes, will no longer in general be appropriate. We shall, instead, regard the exchange sites as being distributed throughout the volume of the resin with a liquid-like distribution, and use an approximate method for evaluating interactions between spatially closest neighboring sites irrespective of the manner of their connection along the chains.

In accordance with the plan outlined in the preceding paragraph, we place the Z exchange sites of the resin on a lattice of coordination number z and total volume v. We then wish to compute the partition function for the situation where n_0 of the sites remain charged, and $\{n_\sigma\}(\sigma \neq 0)$ are occupied by ions of types σ, when each charged site (plus counterion) contributes $-q^2 n_0 \kappa'/D_0$ to the free energy, and there is an interaction energy $q^2 \exp(-\kappa' r_0)/D_E r_0$ between each pair of nearest neighboring sites (at a distance apart r_0), both of which are charged. The problem just described is essentially that of a lattice liquid with " holes," with all the " molecules " fixed at their lattice points. The only complication is that the " holes " are of different kinds σ, with different arrangements of " kinds of holes " being energetically equivalent. We may therefore compute the free energy of an appropriate lattice liquid with but one kind of " holes." Using the quasi-chemical method to evaluate the partition function, one obtains directly all but the last term of Eq. (9.9.3). The first term of Eq. (9.4.3) is the ion atmosphere free energy, and the intermediate terms are just the usual lattice liquid result. The final term of (9.4.3) is the ideal free energy of mixing of the $\{n_\sigma\}(\sigma \neq 0)$ ions of types σ among the bound sites.

In the linear polyelectrolyte theory, the approximation equivalent to use of the ideal free energy in the last term of Eq. (9.4.3) was found not to be in exact accord with experimental observations on polyacids. Such a discrepancy could have resulted if the sites for ion-pair formation were not in one to one correspondence with the polymer anions, or if nonelectrostatic effects, such as, for example, those of a geometrical nature, cause interaction between the modes of occupation of neighboring sites. The deviation from experiment caused by assuming random mixing among the bound sites did not, however, markedly affect any important features of the linear polyelectrolyte theory, and it is therefore assumed that similar considerations apply to the resin systems.

It may be seen that the above treatment of A_{int} employs the combinatorial arguments and the lattice energy of a simple hole theory of liquids, but leaves the configurational entropy to be calculated in a manner appropriate to a cross-linked polymer. Thus, we have simplified the computation of the interaction energy without removing by our simplifications either the discreteness and distinguishability of the exchange sites of the resin, or the mechanical characteristics of the polymer chains.

Appendix 9D. Numerical Evaluation Methods

The simultaneous solution of Eqs. (9.4.7) and (9.4.9) is required to determined numerical values of $\{n_\sigma\}$ and v. We found it convenient to proceed by a partly numerical, partly analytical method. First, Eq. (9.4.7) was rearranged to the form

$$\frac{1 - \alpha'}{\beta + 1} = w(r_0) \tag{9D.1}$$

and a plot was made of the relatively complicated function $w(r_0)$. Then Eqs. (9D.1) and (9.4.4b) were solved simultaneously to obtain α' and β in terms of w and a quantity $T(r_0) = \exp(-q^2 e^{-\kappa' r_0}/D_E r_0 kT) - 1$. It is found that

$$\alpha' = (1 - 2w)/(1 + 4w^2 T) \tag{9D.2a}$$

$$\beta = (1 - 4w^2 T + 4wT)/(1 + 4w^2 T). \tag{9D.2b}$$

Substituting Eqs. (9D.2) into (9.4.9), that equation may be reduced to

$$\ln\left(\frac{2w}{1-2w}\right) - (z-1)\ln(1+2wT) = \ln\left(\sum_{\sigma \neq 0} \frac{c_\sigma}{\kappa_\sigma}\right) - \frac{q^2\kappa'}{D_0 kT} + \frac{q^2 z e^{-\kappa' r_0}}{D_{Er_0} kT}.$$

$$(9D.3)$$

Equation (9D.3), a single equation in r_0 alone, is readily solved by any of the usual successive approximation methods. From the value of r_0 thus determined v can be calculated, and $\alpha' \equiv n_0/Z$ can be obtained by direct evaluation of Eq. (9D.2a).

10. Expansion of Polyions

10.0 Introduction

As pointed out in Chapter 5 it is common to study the configurational properties of an isolated polyion in solution by separating the free energy of the system into two parts. These are, respectively, the entropic contribution due to the multiplicity of possible polymer configurations for a given end-to-end separation of the chain, and the electrostatic free energy. Both of these contributions are dependent upon the instantaneous chain configuration and their changes therefore depend upon the change in skeletal configuration arising from a change in polymeric extension. The condition that the free energy be a minimum at equilibrium clearly requires that

$$-T \frac{\partial S_{conf}}{\partial h} = \frac{\partial A_{elec}}{\partial h}; \qquad h = h_{equil}. \tag{10.0.1}$$

It is not easy to find good approximations to S_{conf} or A_{elec}. The configurational entropy is usually taken to be the same as that for a nonionic polymer of the same end-to-end extension,

$$S_{conf} = k \ln W_0(h) + \text{const.} \tag{10.0.2}$$

with $W_0(h)$ the probability density of finding an end-to-end distance h. We will defer to a later section any discussion of the physical adequacy of the approximation to $W_0(h)$ mentioned above, i.e.,

$$W_0(h) = \text{const. } h^2 e^{-3h^2/2NA^2}$$

$$\langle h^2 \rangle = NA^2 \tag{10.0.3}$$

$$h_{max} = NA$$

with N the number of Kuhn elements and A the length of a Kuhn element in the equivalent random chain chosen to duplicate the mean square end-to-end separation $\langle h^2 \rangle$ and the contour length h_{max} of the real chain.

We have already noted that the electrostatic free energy of the polyion depends upon the instantaneous skeletal configuration. It is clear that configurations which place large numbers of charges close to one another will have much smaller statistical weight (the Boltzmann factor is included in this usage of statistical weight) than extended configurations which tend to maximize the distances between charges. The probability density of finding an end to end extenstion h is, in fact, modified from the form given in Eq. (10.0.3) to the form

$$W(h) = \text{const.} \, h^2 \exp\left[-\left(\frac{3h^2}{2NA^2} - \frac{A_{\text{elec}}}{kT}\right)\right]. \qquad (10.0.4)$$

The usage of Eqs. (10.0.3) and (10.0.4) and the statement that the configurational entropy may be approximated by that of an uncharged polymer at the same end-to-end extension leads to the following explicit form for the configurational entropy:[1,2,3]

$$S_{\text{conf}} = k\left[\tfrac{3}{2}(\alpha_E^2 - 1) - \ln \alpha_E^2\right]$$

$$T\frac{\partial S_{\text{conf}}}{\partial \alpha_E} = 3kT\left(\alpha_E - \frac{2}{3\alpha_E}\right) \qquad (10.0.5)$$

$$\alpha_E^2 = \langle h^2 \rangle / \langle h_0^2 \rangle$$

with $\langle h_0^2 \rangle$ the mean square separation of the chain ends in the absence of electrostatic interactions. If the polyion is highly extended, the Gaussian distribution displayed in (10.0.3) and (10.0.4) is not accurate, and the distribution function becomes

$$W(h) = \text{const} \exp\left[-\frac{1}{A}\int_0^h L^*(y)\,dy\right]$$

with $L(x)$ the Langevin function and L^* its inverse,

$$L(x) = \coth x - \frac{1}{x} = y$$

$$\qquad (10.0.6)$$

$$x = L^*(y) = 3y + \frac{9}{5}y^3 + \frac{297}{175}y^5 + \cdots.$$

To quantitatively examine the several theories of polyion expansion it is necessary to have unambiguous measurements of the mean square end-to-end separation of the chain. Light scattering from solutions of polyelectrolytes provides such information. Viscosity data are often used, though the difficulties of the hydrodynamic theory render the derived dimensions somewhat uncertain. To determine the polymer dimensions by light scattering it is usual to use the extrapolation method introduced by Zimm. The relevant relation is

$$\left(\frac{Kc}{R_\theta}\right)_{c\to 0} = \frac{1}{M_W}\left[1 + \left(\frac{16\pi^2 n_0^2}{3\lambda^2}\right)\langle R_G^2\rangle \sin^2\frac{\theta}{2} + \cdots\right]$$

$$K = \frac{2\pi^2 n_0^2}{\lambda^4 N_0}\left(\frac{\partial n}{\partial c}\right)^2$$

(10.0.7)

with n_0 the refractive index of pure solvent and n that of the solution, λ the wavelength of light, $\langle R_G^2\rangle$ the mean square radius of gyration, and R_θ the Rayleigh ratio at angle θ. For a polydisperse system $\langle R_G^2\rangle$ does not correspond to the same average as does the molecular weight* and care must be exercised in its interpretation when the system is not monodisperse. The connection between $\langle R_G^2\rangle$ and $\langle h^2\rangle$ is easily established and for the case of randomly coiled polymers is

$$\langle h^2\rangle = 6\langle R_G^2\rangle. \tag{10.0.8}$$

Equation (10.0.7) is a rigorous relationship between the angular distribution of scattered light and the properties of the solute molecule. Although the effects of polydispersity may make interpretation difficult, this difficulty should not obscure the exactness of Eq. (10.0.7). In contrast, the determination of polymer dimensions by viscosity measurements is much less rigorous and the dimensions so obtained often differ when alternative theories are used for numerical calculation. The most widely used relationship between polymer dimension and intrinsic viscosity is the semiempirical equation of Flory and Fox[2],

$$[\eta] = \Phi\frac{\langle h^2\rangle^{3/2}}{M} \tag{10.0.9}$$

with Φ a universal constant having the value 2.1×10^{21} when $[\eta]$ is expressed in units of 100 cc/gm. One may question the applicability of Eq. (10.0.9) to polyelectrolytes because of the underlying assumption that the mass distribution is Gaussian. Recent determinations of Φ for polyelectrolytes by Orofino and Flory[4] give the values of 0.86×10^{21} for sodium polyacrylate and 0.90×10^{21} for potassium polystyrene p-sulfonate. Moreover, Φ does not approach the value 2.1×10^{21} as α tends to zero. This uncertainty alone leads to an uncertainty of 30% in values of $\langle h^2\rangle^{1/2}$ derived from Eq. (10.0.9).

The theories of Debye–Bueche[5] and Brinkman[6] may also be used to calculate an average polymeric dimension. In this case

$$[\eta] = \frac{\Omega_s}{100M}\phi(\sigma) \tag{10.0.10}$$

where $[\eta]$ is again measured in units of 100 cc/gm. Of the undefined symbols,

* In 10.0.7 the symbol M_W stands for the weight average molecular weight.

$$\Omega_s = \frac{4\pi R_s^3}{3}$$

$$\phi(\sigma) = \frac{5}{2} \frac{1 + \dfrac{3}{\sigma^2} - \dfrac{3}{\sigma}\coth\sigma}{1 + \dfrac{10}{\sigma^2}\left[1 + \dfrac{3}{\sigma^2} - \dfrac{3}{\sigma}\coth\sigma\right]}$$

$$\sigma = R_s\left(\frac{v_m\Upsilon}{\eta_0}\right)^{1/2} \tag{10.0.11}$$

$$v_m = \frac{3}{4\pi}\left(\frac{Z}{R_s^3}\right)$$

where R_s is the radius of an equivalent sphere assumed to have a uniform distribution of polymer segments, Υ the frictional coefficient of a segment, and the other symbols have been defined previously. In terms of more usual dimensions,

$$\langle h^2 \rangle = \frac{18}{5} R_s^2 \tag{10.0.12}$$

for random coils, while for polymers with a fixed end-to-end distance h,

$$R_s^2 = \frac{5}{36} NA^2 \left[1 + \frac{h^2}{NA^2}\right]. \tag{10.0.13}$$

The value of Υ is usually determined by fitting $[\eta]$ versus M data but this procedure is of limited use for the case of polyelectrolytes. The determination of Υ then provides a major stumbling block in the application of Eq. (10.0.10). Needless to say, the Kirkwood–Reisman theory may also be used to estimate coil dimensions provided that the proper precautions are taken.

10.1 Polymer Models Employing Uniform Spherical Charge Distributions

Calculations of the electrostatic free energy of a polyelectrolyte have been attempted by many investigators. The earliest theories were published by Hermans and Overbeek[1] and by Kuhn et al.[7] and it is interesting to note that the models employed by these groups, which were markedly different, set the pattern for two important directions in the later work of other investigators. Hermans and Overbeek represented the polyion by a sphere having the same number of uniformly distributed charges as the original polyion, while Kuhn, Künzle, and Katchalsky used a randomly coiled chain identical with the usual model for noncharged polymers. We shall call the model of Hermans and Overbeek a uniform sphere model and the model of Katchalsky et al. a random chain model.

Hermans and Overbeek considered in their calculation of the electrostatic free energy that the counterion distribution around a polyion must be generated automatically by the presence on the polyion of fixed charges, and, therefore, that the entropy change due to changes in the distribution of the simple ions could be neglected. That is, the entropy change due to the redistribution of simple ions from a place where the electrostatic potential is ψ_1, to a place where the electrostatic potential is ψ_2 is compensated by the change in the potential energy of the ions, so that

$$\Delta A = \Delta E + T\Delta S = q(\psi_2 - \psi_1) + kT \ln \frac{c_2}{c_1} = 0 \qquad (10.1.1)$$

since

$$c_2 = c_1 \exp[-q(\psi_2 - \psi_1)/kT]. \qquad (10.1.2)$$

Thus, the only contribution to the free energy computed by Hermans and Overbeek is the internal energy:

$$A_{\text{elec}} = \frac{1}{2} \int_0^\infty \rho_f \psi \, dv \qquad (10.1.3)$$

where ρ_f is the density of fixed charges on the polyion (as usual the counterions are not included in ρ_f)

$$\rho_f = \frac{Zq}{(4\pi/3)R_s^3}. \qquad (10.1.4)$$

As described in Chapters 3 and 5, for this model of the polymer the electrostatic potential is given by

$$\psi = \frac{4\pi\rho_f}{D\kappa^2} - \frac{2\pi\rho_f}{D\kappa^3} e^{-\kappa R_s}(1 + \kappa R_s) \frac{e^{\kappa r} - e^{-\kappa r}}{r} \qquad (r < R_s)$$

$$\psi = \frac{2\pi\rho_f}{D\kappa^3} \left[(1 + \kappa R_s) e^{-\kappa R_s} - (1 - \kappa R_s) e^{\kappa R_s}\right] \frac{e^{-\kappa r}}{r}. \qquad (r > R_s)$$

$$(10.1.5)$$

The substitution of Eq. (10.1.5) into Eq. (10.1.3) gives

$$A_{\text{elec}} = \frac{3}{5} \frac{Z^2 q^2}{DR_s} \left[\frac{5}{2p^2} - \frac{15}{4} \frac{p^2 - 1 + (1 + p)^2 e^{-2p}}{p^5} \right]$$

$$\cong \frac{3}{5} \frac{Z^2 q^2}{DR_s} (1 + 0.6p + 0.4p^2)^{-1} \qquad (10.1.6)$$

where

$$p = \kappa R_s$$

$$\kappa^2 = \frac{4\pi}{DkT} \sum_i c_i^0 q_i^2. \qquad (10.1.7)$$

Hermans and Overbeek showed that this result is but little changed by the details of the model employed for the segment distribution inside the polyion domain.

After substitution of Eq. (10.1.6) into Eq. (10.0.4), evaluation of $\langle h^2 \rangle$ and some simplification, the following relationship is obtained:

$$y^3 \frac{y^2 - 2}{y^2 - 1} = \frac{\beta}{3} \frac{1 + 1.2p + 1.2p^2}{(1 + 0.6p + 0.4p^2)^2} \tag{10.1.8}$$

where

$$\beta = \frac{18}{5\sqrt{5}} \frac{1}{A\sqrt{N}} \frac{Z^2q^2}{DkT}; \qquad y^2 = \frac{\langle h^2 \rangle}{\langle h_0^2 \rangle} + 1.$$

For large p, it follows that

$$y^2 = 1.55 + 0.53 \frac{a}{\kappa} \tag{10.1.9}$$

where

$$a^2 = \left(\frac{36}{5\langle h_0^2 \rangle}\right)^{3/2} \left(\frac{3Z^2q^2}{2DkT}\right)$$

and

$$\langle h_0^2 \rangle = NA^2.$$

Pals and Hermans[8] introduced a relationship between R_s and $[\eta]$ into Eq. (10.1.9) to obtain an equation for the intrinsic viscosity of a polyion in salt solution. The viscosity equation used by them is obtained from Eq. (10.0.10) in the limit $\sigma \to 0$, and therefore corresponds to the case when the hydrodynamic interaction between segments is negligible. The resultant limit is

$$[\eta] = \frac{R_s^2 \Upsilon}{1000 M_u \eta_0} \tag{10.1.10}$$

with $M_u = M/Z$ the molecular weight of a segment. Substitution of Eq. (10.1.10) into Eq. (10.1.9) gives

$$[\eta] = [\eta]_\infty + \frac{B}{(c_s^0)^{1/2}} \tag{10.1.11}$$

where the notation c_s^0 denotes the concentration of salt in the solvent, and $[\eta]_\infty$ and B are constants which can be expressed in terms of molecular parameters.

The linear relationship between $[\eta]$ and $\left(c_s^0\right)^{-\frac{1}{2}}$ has been demonstrated by many investigators to be in good agreement with experiment. Figure 10.1.1 shows an example of the experimental verification of Eq. (10.1.11).[8] However, this agreement is only qualitative. Quantitatively, there is a large disagreement between the calculated slopes [Eq. (10.1.11)] and the experimental slopes. In terms of molecular parameters, the dimensions of a segment which must be chosen so that agreement may be obtained between the theory and the experiments are much too large to be considered reasonable for polyions which are basically vinyl derivatives.[9] We should note that the dimensions deduced from similar measurements for cellulose derivatives are not so unreasonable.[8]

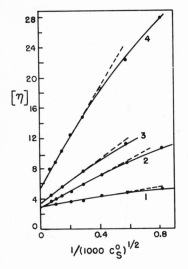

FIG. 10.1.1. The dependence of $[\eta]$ on the external salt concentration for a typical flexible polyelectrolyte [Reproduced from ref. 8.]

Several reasons can be advanced for this quantitative discrepancy. A most obvious source of difficulty is the failure of the assumption of negligible hydrodynamic interaction between segments used in Eq. (10.1.10). As was pointed out by Pals and Hermans, this assumption may be more accurate for the cellulose derivatives than for vinyl polymer derivatives inasmuch as the intrinsic viscosities of cellulose derivatives are proportional to the molecular weights of the polymers and the cellulose compounds may be closer to the nonexistent free draining limit. For vinyl compounds, however, this assumption cannot be correct. A second source of error may be the neglect of counterion binding. We defer a discussion of this until Section 10.4. Finally, it might be suspected that the linearization of the Poisson–Boltzmann equation may break down badly, especially in the region near the polyion. This latter approximation was studied by Nagasawa and Kagawa,[10] but was found to be unimportant as a factor in the calculated expansion.

Nagasawa and Kagawa have described a method of calculating the electrostatic free energy which is somewhat different from the method of Hermans and Overbeek. The Hermans–Overbeek procedure implies that the electrostatic free energy is equal to the work done during the time when all of the charged groups of the polyion are placed, one by one, on a noncharged skeleton of the polyion in a simple salt solution. That the Hermans–Overbeek charging procedure has this physical interpretation may be seen clearly by comparison of Eq. (10.1.3) and the following relation,

$$A_{\text{elec}} = \int_0^{R_s} \int_0^Z \frac{q\psi}{\frac{4}{3}\pi R_s^3} 4\pi r^2 \, dZ \, dr \qquad (10.1.12)$$

both of which give the same result for A_{elec}. Thus, the Hermans–Overbeek theory assumes that the counterions can be automatically supplied from the solvent while the fixed sites are being charged. It is doubtful whether this assumption can be justified for the case of electrolyte solutions, since electrical neutrality must be maintained throughout the solution.

To avoid the difficulty cited, Nagasawa and Kagawa[10] used the following relation to calculate A_{elec}:

$$A_{\text{elec}} = \int_0^\infty \int_0^1 \psi(\lambda) \frac{\rho(\lambda)}{\lambda} 4\pi r^2 \, d\lambda \, dr \qquad (10.1.13)$$

where λ denotes the fraction of the final charge which the ions have at a given stage of the integration. This equation implies that the electrostatic free energy is equal to the work which must be done when all the ions in the solution are charged simultaneously from the imaginary noncharged state to the real state, keeping the interior and the exterior of the polyion domain at equilibrium. From the potential and charge distribution functions presented by Nagasawa and Kagawa, A_{elec} becomes

$$A_{\text{elec}} = \frac{D(\kappa R_s^2)}{4\kappa} E_D^2 \int_0^{\kappa R_s} f(x) \, dx$$

$$= \frac{D(\kappa R_s)^2}{4\kappa} E_D^2 \hat{\sigma}(\kappa R_s) \qquad (10.1.14)$$

where

$$f(x) = \frac{(3 + 6x + 5x^2 + 2x^3)e^{-2x} + (x^2 - 3)}{x^3} \qquad (10.1.15)$$

$$\hat{\sigma}(\kappa R_s) = \int_0^{\kappa R_s} f(x) \, dx.$$

If $\hat{\sigma}(\kappa R_s) = 1$, Eq. (10.1.14) is the same as the electrical surface free energy of a sphere which has a Helmholtz double layer at its surface.[10]

Finally, from (10.1.14) Nagasawa and Kagawa obtained by procedures already described, the following relationship for the equivalent radius of a polyion,

$$kT \left[\frac{3}{R_s^2 - (5/36)NA^2} - \frac{108}{5} \frac{1}{NA^2} \right]$$

$$= \frac{D(\kappa R_s)E_D}{2} \left[E_D \left\{ \hat{\sigma}(\kappa R_s) + \frac{\kappa R_s}{2} \frac{d\hat{\sigma}}{d(\kappa R_s)} + R_s \frac{dE_D}{dR_s} \hat{\sigma}(\kappa R_s) \right\} \right] \qquad (10.1.16)$$

where

$$E_D = -\frac{kT}{q} \ln \frac{1 + \sqrt{1 + (8\pi R_s^3/3Z)c_s^0}}{(8\pi R_s^3/3Z)c_s^0} \qquad (10.1.17)$$

Z is the number of charges on one molecule, and c_s^0 is the concentration of added salt. The radius R_s can be determined from this equation by using a graphical method. Despite the elimination of the approximation mentioned, the values of R_s obtained from (10.1.16) and (10.1.17) are still too large to be considered reasonable. One example of the comparison between calculated values of R_s and experimental values obtained from viscosity data by using Eqs. (10.0.9) and (10.0.10) is shown in Table 10.1.1. The length of a segment was assumed to be 20 A, a value for vinyl polymers in agreement with much data in uncharged systems.

TABLE 10.1.1

COMPARISON OF THE CALCULATED RADIUS OF A POLYION WITH THE
EXPERIMENTAL RADIUS

$c_{NaCl}(N)$	$R_{s\eta}$(cm)	R_s (calc.) (cm)
1.00	1.73×10^{-6}	1.75×10^{-6}
5.00×10^{-1}	1.89	2.07
1.00	2.25	2.83
5.02×10^{-2}	2.41	3.34
2.01	2.76	4.14
1.00	2.99	4.65
5.04×10^{-3}	3.31	5.60
2.02	3.69	6.75
1.31	3.81	8.25

R_s (calc.): the values calculated from Eq. (10.1.16). $R_{s\eta}$: the values determined from viscosity data. Sample: Sodium polyvinyl alcohol sulfate and sodium chloride.

In view of the assumptions inherent in Eq. (10.1.16), the examination of the discrepancy illustrated in Table 10.1.1 has led Nagasawa and Kagawa to conclude that the neglect of site binding or equivalently of the marked nonideality of the counterions is the greatest source of error in the theory.

The polyelectrolyte theory advanced by Flory[11] bears some strong resemblances to those just discussed. Flory assumes that the net charge of a polyion in a simple salt solution is substantially zero throughout the greater part of the molecular domain. He therefore calculates the free energy of formation of the neutral polyion sphere by superposing the entropy of mixing of the small ions onto the ordinary terms describing the thermodynamic interactions of a nonionic polymer skeleton and solvent. That is, the free energy of formation of the system is taken to be

$$\Delta A = \sum_j \Delta A_{M_j} + \Delta A_{conf} \qquad (10.1.18)$$

where ΔA_{cont} is given by Eq. (10.0.4) and ΔA_{M_j} is the free energy of formation of the solution of n_j monomer units, solvent and mobile ions occurring in the jth shell of the polymeric domain (which is divided into virtual shells for the purpose of calculation). In this model, ΔA_{M_j} is given by

$$\sum_j \frac{\partial \Delta A_{M_j}}{\partial \alpha_E} = \sum_j \left(\frac{\partial \Delta A_{M_j}}{\partial x_{0j}}\right)\left(\frac{\partial x_{0j}}{\partial \alpha_E}\right)$$

$$= \sum_j (\mu_{0j} - \mu_0^*)\left(\frac{\partial x_{0j}}{\partial \alpha_E}\right) \tag{10.1.19}$$

where μ_{0j}, μ_0^* are the chemical potentials of water inside the jth shell and outside the polyion domain and x_{0j} is the mole fraction of water in the jth shell. By a simple combination of Flory's theory of nonionic polymer solutions and the Donnan theory of membrane equilibrium, $\mu_{0j} - \mu_0^*$ is seen to be

$$\mu_{0j} - \mu_0^* = kT\left[\psi_1\left(1 - \frac{\Theta}{T}\right)c_{pj}V_p^2 + (2c_{sj} + \alpha c_{pj} - 2c_s^0)\right] \tag{10.1.20}$$

for electrolytes of the 1–1 type. The symbols c_{pj} and c_{sj} represent the concentrations of polymer units and added electrolytes in the jth shell, c_s^0, the concentration of the added electrolyte outside the polymer domain. By the substitution of the equation of membrane equilibrium into the second term on the right-hand side of Eq. (10.1.20)

$$c_{sj}(c_{sj} + \alpha c_{pj}) = c_s^{02} \tag{10.1.21}$$

and summation over all j, we have the following relationship between the expansion of the polyion and the concentration of simple electrolyte in the medium.*

$$\alpha_E^5 - \alpha_E^3 = 2C_M\psi_1\left(1 - \frac{\Theta}{T}\right)M^{1/2} + 2C_I\frac{M^{1/2}}{c_s^0}. \tag{10.1.22}$$

The first term in Eq. (10.1.22) is the same as the corresponding term for nonionic polymers, and the second term is the term characteristic of polyions. The first term can be neglected relative to the second term under ordinary conditions. Moreover, when $T = \Theta$, the first term vanishes identically. C_I is given by

$$C_I = \frac{10^3}{2^{3/2}}\left(\frac{3}{\pi^{1/2}}\right)^3\frac{(M/\langle h_0^2\rangle)^{3/2}}{N_0 M_u^2} \tag{10.1.23}$$

where $M_u = M/Z$.

A comparison of Eq. (10.1.22) with experiment was carried out by Flory and Osterheld.[12] It was found that there is very good linearity between $1/c_S^0$ and $\alpha_E^5 - \alpha_E^3$ as predicted by Eq. (10.1.22) (see Fig. 10.1.2), but the slope

---

* Although a general equation is given in the original paper of Flory,[11] only the equation for electrolytes of 1–1 type is displayed herein. Some higher order terms are also neglected.

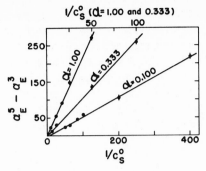

FIG. 10.1.2. The expansion factor α_E, as a function of external salt concentration. [Reproduced from ref. 12.]

of the curve is much smaller than the calculated slope (see Table 10.1.2).

TABLE 10.1.2[a]

COMPARISON OF OBSERVED AND CALCULATED SLOPES OF
$\alpha_E^5 - \alpha_E^3$ AGAINST $1/c_S^0$

Polymer fraction	α	Salt	Slope Obs.	Calc.	Obs./Calc.
A-7	1.00	NaCl	5.5	265	0.021
A-7	0.333	NaCl	2.6	29.4	0.089
A-6	0.333	NaCl	3.0	30.0	0.100
A-7	0.100	NaCl	0.54	2.65	0.20
A-6	0.100	NaCl	0.55	3.0	0.18
A-6	0.100	Na_2SO_4	0.55	3.0	0.18
A-6	0.100	$CaSO_4$	0.120	3.0	0.040
A-6	0.100	$CaCl_2$	0.088	3.0	0.029

[a] Reproduced from ref. 12.

In addition to the theories discussed, Oosawa et al.[16] have presented a calculation wherein the electrical free energy is determined by the addition of the internal energy and the entropy of mixing of the ions. Their theory is quite different in detail from those mentioned above, but is in the same general category.

10.2 Randomly Coiled Chain Models

The randomly coiled chain model considered by Kuhn et al.[7] is a much more realistic representation of the polyion than the models of Section 10.1. As already mentioned in Chapter 5, these authors calculate the average reciprocal distance between two fixed charges by using a Gaussian distribution of chain segments. With this result it is easy to obtain the total energy due to electrical interactions by summing the contributions of all possible pairs of interacting charges. In the earliest version of the theory, the contribution of the counterions was neglected. Katchalsky and Lifson[13] later improved the theory by providing for the shielding of the fixed ions by ionic atmospheres, with the shielding parameter as given by Debye and Hückel for simple electrolytes. Thus, the probability density of finding a distance r between two given charges is

$$W(r, \xi, h) = \left(\frac{2\pi\langle h_0^2\rangle\xi(1-\xi)}{3}\right)^{-\frac{1}{2}} \frac{r}{h\xi}$$
$$\times \left\{\exp\left[-\frac{2(r-\xi h)^2}{3\langle h_0^2\rangle\xi(1-\xi)}\right] - \exp\left[-\frac{2(r+\xi h)}{3\langle h_0^2\rangle\xi(1-\xi)}\right]\right\} \quad (10.2.1)$$

when the end-to-end distance of the polymer is kept fixed at h. We use the notation k for the number of segments between the two given charges, and $\xi = (k/N)$. The repulsive energy of a pair of ions is given by

$$u_{ij} = \frac{q^2}{Dr_{ij}} e^{-\kappa r_{ij}} \quad (10.2.2)$$

whereupon the average contribution to the repulsive energy of a pair of ions is

$$\langle u_{ij}\rangle = \int_0^\infty W(r, \xi, h)u_{ij}(r) \, dr. \quad (10.2.3)$$

After the introduction of (10.2.2) into (10.2.3) and simplification the following relation is obtained for the electrostatic free energy of a polyion having a constant end-to-end distance h:

$$A_3 = \frac{v^2q^2}{Dh} \ln\left(1 + \frac{6h}{\kappa\langle h_0^2\rangle}\right) \quad (10.2.4)$$

with v the number of ionized fixed charges and κ^2 is, as usual, $4\pi q^2 \sum c_i^0/DkT$.

In addition to the above electrostatic free energy, it is necessary to calculate the free energy, A_7 [Eq. (7.4.13) in Chapter 7] of building up the ionic atmosphere around all ions. This contribution to the free energy is conveniently approximated by the Debye–Hückel result. It should be remarked that although A_7 is unimportant as a driving potential for expansion, in Chapter 8 it was shown that the ion atmosphere has a marked effect on the activity coefficient of a simple ion.

By minimizing the free energy with respect to the end-to-end separation of the polyion, Katchalsky and Lifson show that

$$\frac{3h^*kT}{\langle h_0^2 \rangle} \lambda = \frac{v^2 q^2}{Dh^{*2}} \left[\ln\left(1 + \frac{6h^*}{\kappa\langle h_0^2 \rangle}\right) - \frac{6h^*/\kappa\langle h_0^2 \rangle}{1 + 6h^*/\kappa\langle h_0^2 \rangle} \right]. \quad (10.2.5)$$

If $\langle h_0^2 \rangle$ is calculated from viscosity data extrapolated to infinite concentration of added electrolyte, h^* can be calculated from (10.2.5) without the use of any adjustable parameters. Several comparisons between the theory and experiment have been carried out.[14] In no case is good agreement obtained. In Table 10.2.1 an example of one such comparison is given.

TABLE 10.2.1[a]

COMPARISON BETWEEN THE END-TO-END DISTANCE CALCULATED FROM
THE THEORY OF KATCHALSKY AND LIFSON h^*(K.L.) AND THE
EXPERIMENTAL END-TO-END DISTANCE CALCULATED USING FLORY'S THEORY,
h^*(F)[14] [b]

NaCl conc. (N)	0.1000	0.0500	0.0201	0.0101	0.0050	0.000991
h^*(K.L.) $\times 10^5$ (cm)	3.37	3.47	3.57	3.66	3.70	3.75
h^*(F) $\times 10^5$ (cm)	0.552	0.604	0.604	0.791	0.886	1.11

[a] Reproduced from ref. 14.
[b] Sample: Na–polyvinyl alcohol sulfate and sodium chloride. Polymer concentration is zero.

Notwithstanding the uncertainties in the polymer dimensions deduced from the Flory–Fox viscosity theory, it is clear that the calculated values are much too large to be considered reasonable.

There are several possible sources for the overlarge end-to-end distances calculated from the Katchslsky–Lifson theory. Lifson[15] has pointed out that the chain configurational entropy is considerably underestimated. However, in order that the entropic force $\partial A_{cont}/\partial h$ be equal to the electrostatic force $\partial A_3/\partial h$ at any reasonable value of h^*, the entropic force must be almost ten times larger than that given by the Kuhn entropy terms. This appears to be very unlikely. It is more likely that the electrostatic force is very greatly overestimated, although it may also be true that the Kuhn entropy is not adequate to describe a polyion. As one source of the overestimate, Nagasawa et al.[14] have suggested that the ionic atmospheres around the fixed charges must be compressed more closely to the groups than Katchalsky and Lifson calculated because of the interactions between the fixed charges. Such a compression has no effect on the mean activity coefficient because ions of one sign have their atmospheres dilated and of the opposite sign, compressed.

The net result is a cancellation of the effect when consideration is given to both positive and negative ions. The atmosphere compression cannot be neglected when computing the end-to-end distance since the ionic atmospheres around the fixed ions are composed of ions of one sign of charge.

10.3   Site-Binding Models

Aside from the obvious physical effects neglected in the preceding spherical and chain models, both also neglect the effects on the chain configuration of the extraordinarily large interaction between the polyion and a counterion. Such an interaction may be described in terms of an abnormally low counterion activity coefficient, or as ion binding. The utility of ion binding as a scheme for classifying the counterions was first suggested by Kern[18] and Kagawa,[19] but the first attempts to quantitatively analyze its implications were due to Harris and Rice.[17] This work has been described in Chapters 5 and 7 and no further comment will be made here. Subsequent to this work other investigators have made calculations of polyion configurational changes including the effects of ion binding.

Inagaki et al.[20] calculated the extent of ion binding by requiring that the experimental intrinsic viscosity agree with the theory of Hermans, Overbeek, and Pals [Eq. (10.1.11)]. The resultant net charge was then used to calculate the second virial coefficient. This procedure leads to fairly good agreement between the experimental second virial coefficients and calculations.

Nagasawa used the same procedure to improve Eq. (10.1.16). As stated in Chapter 8, the additivity rule for activity coefficients is an accurate description in systems with added electrolyte. Thus the activity coefficient of, say, the sodium counterion in a polyelectrolyte solution is not changed by the addition of simple salts without a common ion. This fact implies that the sodium ions originating from the polyelectrolyte can be classified and treated separately from the ions originating from the added salt. It is clear that the effective concentration of the counterions is much lower than the analytical concentration. It is also to be noted that the introduction of an effective concentration has the same effect as the explicit introduction of site binding. Both of the above descriptions also have the same physical effect, namely a decrease of the electrostatic potential inside the polymer domain.[10] The effective concentration (or the charge on the polyion) was calculated by Nagasawa from $Z_{\text{eff}} = (\gamma_{\text{Na}}^p Z)_{c_p \to 0}$ and the radius of the polymer domain R_s was calculated from the Debye–Bueche equation (10.0.10) using for the frictional constant of the monomer unit the value determined from electrophoresis experiments. (See Chapter 11.) The comparison between experiment and calculation is shown in Table 10.3.1. Good agreement is obtained if 40 A is assumed for the length of a segment.

TABLE 10.3.1

COMPARISON BETWEEN THE MAGNITUDE OF THE EXPANSION
CALCULATED FROM EQ. (10.1.16) USING AN EFFECTIVE CHARGE, AND THE
EXPERIMENTAL EXPANSION[a]

$c_{NaCl.}$	$R_{s\eta}$ (D.B.)	R_s (calc.)(A = 40 A)
0.2 N	187 A	216 A
0.1	205	220
0.05	232	233
0.02	265	262
0.01	295	291
0.005	335	330
0.002	395	387

[a]Sample: sodium polyvinyl alcohol sulfate; degree of polymerization = 1600; degree of esterification = 0.61. $(\gamma_{Na}^p)_{c_p \to 0} = 0.14$.

10.4 Critique of Chain Models

The calculations of Harris and Rice and Katchalsky and Lifson are not the only theories based on chain models. Krieger[22] has considered the problem of $N + 1$ freely linked beads interacting with a screened Coulomb potential. Assuming that the distance between successive beads in the chain had a spherically symmetric Gaussian distribution with mean square length NA^2, and using an approximation in the potential energy, Krieger developed a perturbation series for the mean square end-to-end distance of the chain with interacting charges. Because the perturbation series appears to converge very badly under the usual experimental conditions, we shall not present any of the details of Krieger's arguments.

There are familial resemblances between the several chain models discussed in this book. To examine these we follow the arguments of Fisher[23] and consider the canonical partition function for a polyion,

$$Q(h) = \sum_{\omega(h)} e^{-E_\omega(h)/kT} = g(h)e^{-\tilde{A}(h)/kT} \qquad (10.4.1)$$

where the summation is performed over all configurations $\omega(h)$ with the Boltzmann weight determined by the energy of that configuration, $E_\omega(h)$. In the second form of Eq. (10.4.1) $g(h)$ is the density of configurations of length h and $\tilde{A}(h)$ is then, by definition, the free energy of interaction.

Under the condition that $\tilde{A}(h)$ is small compared with the thermal energy kT, the right-hand side of Eq. (10.4.1) may be expanded and only the first two terms retained. A more general procedure is to use the high-temperature moment expansion

$$\tilde{A}(h) = \langle E \rangle - \frac{1}{2kT} (\langle E^2 \rangle - \langle E \rangle^2)$$

$$+ \frac{1}{6(kT)^2} (\langle E^3 \rangle - 3\langle E \rangle \langle E^2 \rangle + 2\langle E \rangle^3) - \cdots \qquad (10.4.2)$$

with

$$\langle E^n(h) \rangle = \frac{1}{g(h)} \sum_{\omega(h)} E_\omega^n(h) \qquad (10.4.3)$$

and the indicated averages are over the configurations of the uncharged polymer chain. This expansion was used in Chapter 3 in the discussion of fluctuation forces and has found extensive application in other problems of statistical mechanics. Equation (10.4.3) is conveniently rewritten as an integral over the coordinates of the chain segments,

$$\langle E^n(h) \rangle = \frac{1}{g(h)} \int d\mathbf{r}_1 \cdots \int d\mathbf{r}_{N-1} \int \frac{dh}{dh} p(\mathbf{r}_1 \cdots \mathbf{r}_{N-1}, \mathbf{h}) E^n(\mathbf{r}_1 \cdots \mathbf{r}_{N-1}, \mathbf{h})$$

$$(10.4.4)$$

where the chain is specified by the space coordinates $\mathbf{r}_1 \cdots \mathbf{r}_{N-1}$ and the angular coordinates of $\mathbf{r}_N = \mathbf{h}$. The function $p \, d\mathbf{r}_1 \cdots d\mathbf{r}_{N-1} \, d\mathbf{h}$ is the probability of a chain configuration in which the zeroth segment is at the origin, the first in the volume element $d\mathbf{r}_1$, and so forth. The function p refers to the chain without charge–charge interactions, in agreement with the averages defined in Eq. (10.4.3). For the most purposes it is sufficient to write the energy of the chain in the form

$$E(\mathbf{r}_1 \cdots \mathbf{r}_{N-1}, \mathbf{h}) = \sum_{i=0}^{N} \sum_{j=i+1}^{N} u(r_{ij}). \qquad (10.4.5)$$

Using (10.4.5) in (10.4.4) we find

$$\langle E(h) \rangle = \frac{1}{g(h)} \sum_{i=0}^{N} \sum_{j=i+1}^{N} \int d\mathbf{r}_i \int d\mathbf{r}_j \int \frac{dh}{dh} p_{ij}(\mathbf{r}_i, \mathbf{r}_j, \mathbf{h}) u_{ij}$$

$$p_{ij}(\mathbf{r}_i, \mathbf{r}_j, \mathbf{h}) = \int \cdots \int_{\substack{\text{excluding} \\ i,j}} d\mathbf{r}_1 \cdots d\mathbf{r}_{N-1} p(\mathbf{r}_1 \cdots \mathbf{r}_{N-1}, \mathbf{h}). \qquad (10.4.6)$$

Note the similarity between these reduced distribution functions and those defined in Chapter 1. By changing to relative coordinates and performing one more integration, Eq. (10.4.6) becomes

$$\langle E(h) \rangle = \frac{N(N+1)}{2} \int_0^\infty \mathscr{P}_1(\mathbf{r}, \mathbf{h}) u(r) \, dr$$

$$\mathscr{P}_1(\mathbf{r}, \mathbf{h}) = \sum_{i=0}^{N} \sum_{j=i+1}^{N} 2p_{ij}(\mathbf{r}, \mathbf{h})/N(N+1)g(h). \qquad (10.4.7)$$

Note that $p_{ij}(\mathbf{r}, \mathbf{h})\, d\mathbf{r}\, d\mathbf{h}$ is the probability that the ith and jth beads lie in the intervals $d\mathbf{r}$ about $\mathbf{r}$ and $d\mathbf{h}$ about $\mathbf{h}$. For the Gaussian chain (Kuhn element chain)

$$p(\mathbf{r}_1 \cdots \mathbf{r}_{N-1}, \mathbf{h}) = \left(\frac{3}{2\pi A^2}\right)^{3N/2} \prod_{i=1}^{N} \exp\left[-\left(\frac{3}{2A^2}(\mathbf{r}_i - \mathbf{r}_{i-1})^2\right)\right].$$

$$(10.4.8)$$

By similar procedures to those outlined above we may write

$$\langle E^2(h)\rangle = \left[\frac{N(N+1)}{2}\right]^2 \int_0^\infty \int_0^\infty \mathscr{P}_2(\mathbf{r}, \mathbf{r}', \mathbf{h})u(r)u(r')\, d\mathbf{r}\, d\mathbf{r}' \quad (10.4.9)$$

with the usual relation

$$\int_0^\infty \mathscr{P}_1(r, h)\, d\mathbf{r} = \int_0^\infty \int_0^\infty \mathscr{P}_2(\mathbf{r}, \mathbf{r}', \mathbf{h})\, d\mathbf{r}\, d\mathbf{r}' = 1. \quad (10.4.10)$$

The configurational distribution functions $\mathscr{P}_n$ contain all the structural information about the uncharged chain.

A complete calculation of $\tilde{A}(h)$, the free energy of interaction, must of course proceed by calculation of all the moments $\langle E^n \rangle$. The Katchalsky–Lifson theory represents $\tilde{A}(h)$ as equal to $\langle E \rangle$, and neglects any contributions from the higher moments. By retaining only the first term in the high-temperature series, their result is only valid when kT is very large relative to $\langle E \rangle$. Under ordinary experimental conditions this is not true.

In addition to the fundamental objection stated above Katchalsky and Lifson make further approximations when they use the approximate configurational distribution function

$$\mathscr{P}_1(r, h) = \frac{2}{h}[1 - \exp(-6hr/\langle h_0^2 \rangle)]. \quad (10.4.11)$$

Mazur[24] has made an exact calculation of the first-order configurational distribution function in the limit of a continuous Gaussian chain. The result is

$$\mathscr{P}_1(r, h) = \frac{2}{h}\left[1 - \exp\left[-\frac{6r(h+r)}{\langle h_0^2 \rangle}\right] - \frac{\sqrt{6r}}{\langle h_0^2 \rangle^{1/2}} \times \right.$$
$$\left. \exp\left(\frac{3h^2}{2\langle h_0^2 \rangle}\right)\left(\mathrm{erf}\left[\sqrt{\frac{3}{2}}\frac{h}{\langle h_0^2 \rangle^{1/2}}\right] - \mathrm{erf}\left[\sqrt{\frac{3}{2}}\frac{2r+h}{\langle h_0^2 \rangle^{1/2}}\right]\right)\right] \quad r > h$$

$$\mathscr{P}_1(r, h) = \frac{2}{h}\left[\exp\left(-\frac{6r(r-h)}{\langle h_0^2 \rangle}\right) - \exp\left[-\frac{6r(r+h)}{\langle h_0^2 \rangle}\right] - \frac{\sqrt{6r}}{\langle h_0^2 \rangle^{1/2}} \times \right.$$
$$\left. \exp\left(\frac{3h^2}{2\langle h_0^2 \rangle}\right)\left(\mathrm{erf}\left[\sqrt{\frac{3}{2}}\frac{2r-h}{\langle h_0^2 \rangle^{1/2}}\right] - \mathrm{erf}\left[\sqrt{\frac{3}{2}}\frac{2r+h}{\langle h_0^2 \rangle^{1/2}}\right]\right)\right]. \quad r > h$$

$$(10.4.12)$$

Katchalsky and Lifson have also used Eq. (10.4.12) to calculate $\tilde{A}$ in the approximation $\tilde{A} = \langle E \rangle$. As will be seen this is still subject to error. In Fig. 10.4.1 are depicted the exact values of $\mathscr{P}_1(r, h)$, as calculated from

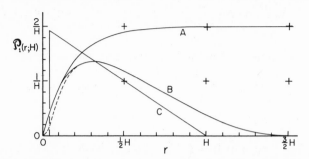

FIG. 10.4.1. The first-order configurational weighting function $\mathscr{P}_1(r; h)$ plotted as a function of r for $h = NA^2 = H$, according to the results of Katchalsky and Lifson (curve A and B), and according to Krieger (curve C). The dotted curve shows the exact behavior near the origin. [Reproduced from ref. 23.]

Eq. (10.4.12), as well as the approximate distribution function for weak interactions, $(h = h_0)$. The approximate distribution function, Eq. (10.4.11) is not normalizable and only agrees with the exact value for $r < \langle h_0^2 \rangle^{1/2}/6$. For a real chain there are also some differences from the formulas given in Eq. (10.4.12) near the origin. Since the potential is very large in this region it is not at all clear that the contribution due to this error is negligible. Katchalsky and Lifson calculated the most probable length h^* by using

$$\frac{\partial}{\partial h} \ln g(h) = \frac{1}{kT} \frac{\partial \tilde{A}}{\partial h} \tag{10.4.13}$$

and it was pointed out by Harris and Rice[25] that their result was in error due to an incorrect expression for $g(h)$. This error leads to the physical absurdity that $\langle h^2 \rangle$ tends to zero as the charge tends to zero whereas the obvious result must be that $\langle h^2 \rangle \to \langle h_0^2 \rangle$ as the charge tends to zero. Correction of the Katchalsky–Lifson result removes the incorrect limiting behavior but otherwise has little effect on the computed end-to-end separations.

Krieger has used the approximation

$$\mathscr{P}_1^K(r, h) = \frac{2}{h}\left(1 - \frac{r}{h}\right), \qquad \frac{h}{N} \leqslant r \leqslant h$$

$$\mathscr{P}_1^K(r, h) = 0 \qquad\qquad \text{otherwise} \tag{10.4.14}$$

which has the incorrect limiting value

$$\lim_{h \to 0} \mathscr{P}_1^K(r, h) = \delta\left(r - \frac{h}{3}\right). \tag{10.4.15}$$

The correct limiting form is obtained from Eq. (10.4.12) and is

$$\lim_{h \to 0} \mathscr{P}_1(r, h) = \frac{12r}{\langle h_0^2 \rangle} \exp\left(-\frac{6r^2}{\langle h_0^2 \rangle}\right) \tag{10.4.16}$$

which has a maximum at $r = 0.289 \langle h^2 \rangle^{1/2}$.

Consider the case of a potential with a cutoff such that $u = 0$ for $r \geqslant a$. Then it may be shown that the most probable end-to-end extension is

$$\frac{h^*}{\langle h_0^2 \rangle^{1/2}} = 1 + \left(\frac{3}{2}\right)^{1/2} \frac{u}{kT} \left(\frac{a}{A}\right)^3 \sqrt{N} + \cdots ; \qquad \text{Katchalsky–Lifson}$$

$$\frac{h^*}{\langle h_0^2 \rangle^{1/2}} = 1 + \frac{1}{6} \left(\frac{3}{2}\right)^{1/2} \frac{u}{kT} \left(\frac{a}{A}\right)(\sqrt{N})^3 + \cdots ; \qquad \text{Kreiger} \tag{10.4.17}$$

$$\frac{h^*}{\langle h_0^2 \rangle^{1/2}} = 1 + 2\left(\frac{3}{2\pi}\right)^{1/2} \text{erf}\left[\left(\frac{1.066}{2N}\right)^{1/2}\right] \frac{u}{kT} \left(\frac{a}{A}\right)^3 \sqrt{N} + \cdots ; \qquad \text{exact}$$

showing that the Katchalsky–Lifson calculation differs from the exact case by only the factor $1 + [(2 \times 1.066)/N]^{1/2} + \cdots$ whereas the Kreiger result gives the wrong functional dependence on N and on the cutoff distance a. It should be noted that the Katchalsky–Lifson result does not agree with higher terms in the exact expansion at all. For instance, in the exact series for $h^*/\langle h_0^2 \rangle^{1/2}$ the term in $(a/A)^4$ is absent, whereas in the Katchalsky–Lifson theory it appears with the coefficient $-3(3/2)^{1/2}(u/kT)$. Of course, all the expansions in Eq. (10.4.17) as well as the theories on which they are based are valid only for $(u/kT) \ll 1$, corresponding to very weak potentials. This situation cannot be assumed to prevail under the usual conditions in a polyelectrolyte solution.

It is clear that the theory of polyion expansion is far from quantitative and that much work, both experimental and theoretical, remains to be done. It is our opinion that a most valuable approach will be based upon the use of the Mayer cluster expansion somewhat similar in form to that detailed in Chapter 1. Mayer has extended his theory to electrolyte solutions by a diagram classification technique and calculated properties of the electrolyte solution to higher order than the Debye–Hückel theory. Similarly, the cluster technique has been applied with great success to uncharged polymeric systems. A program of this type would, we believe, yield a correct theory of expansion without being subject to the physical objections sketched in this section.

REFERENCES

1. J. J. HERMANS AND J. T. G. OVERBEEK, Rec. trav. chim. 67, 761 (1948).
2. P. J. FLORY, "The Principles of Polymer Chemistry." Cornell Univ. Press, Ithaca, New York, 1953.
3. F. E. HARRIS AND S. A. RICE, J. Polymer Sci. 15, 151 (1955).
4. T. A. OROFINO AND P. J. FLORY, J. Phys. Chem. 63, 283 (1959).
5. P. DEBYE AND A. M. BUECHE, J. Chem. Phys. 16, 573 (1948).
6. H. C. BRINKMAN, Koninkl. Ned. Akad. Wetenschap., Proc. B50, 618, 821 (1947).
7. W. KUHN, O. KÜNZLE, AND A. KATCHALSKY, Helv. Chim. Acta 31, 1994 (1948).
8. D. T. F. PALS AND J. J. HERMANS, Rec. trav. chim. 71, 433 (1952).
9. H. FUJITA, K. MITSUHASHI, AND T. HOMMA, J. Colloid. Sci. 9, 466 (1954).
10. M. NAGASAWA AND I. KAGAWA, Bull. Chem. Soc. Japan 30, 961 (1957).
11. P. J. FLORY, J. Chem. Phys. 21, 162 (1953).
12. P. J. FLORY AND J. E. OSTERHELD, J. Phys. Chem. 58, 653 (1954).
13. A. KATCHALSKY AND S. LIFSON, J. Polymer Sci. 11, 409 (1953).
14. M. NAGASAWA, M. IZUMI, AND I. KAGAWA, J. Polymer Sci. 37, 375 (1959).
15. S. LIFSON, J. Polymer Sci. 23, 131 (1957).
16. F. OSAWA, N. IMAI, AND I. KAGAWA, J. Polymer Sci. 13, 93 (1957).
17. F. E. HARRIS AND S. A. RICE, J. Phys. Chem. 58, 725, 733 (1954).
18. W. KERN, Z. physik. Chem. A181, 249, 283 (1938).
19. I. KAGAWA, J. Chem. Soc. Japan, Ind. Chem. Sect. 42, 574 (1944).
20. H. INAGAKI AND T. ODA, Makromol. Chem. 21, 1 (1956).
21. M. NAGASAWA, J. Am Chem. Soc. 83, 300 (1961).
22. I. M. KRIEGER, J. Chem. Phys. 26, 1 (1957).
23. M. E. FISHER, J. Chem. Phys. 28, 756 (1958).
24. J. MAZUR, Appendix in J. Polymer Sci. 11, 409 (1953).
25. F. E. HARRIS AND S. A. RICE, J. Polymer Sci. 15, 151 (1955).

11. Some Dissipative Processes in Solutions of Flexible Polyelectrolytes

11.0 Introduction

In this chapter, we shall briefly discuss the response to an applied electric field of both the polymeric ions and the counterions in a solution. It is apparent that said response will be entirely different from the motion induced by a shearing stress as well as from diffusion in response to a concentration gradient. The origin of the difference is in part due to the fact that the electric forces on a polyion and on the liquid immediately surrounding and inside the polyion have opposite directions. Moreover, these forces are largest within the polymeric domain. Thus, although the viscosity increment due to a polymer molecule in solution may be treated in terms of an equivalent sphere model in which the polyion and the counterions inside the polymer domain are assumed to move together, such a model is clearly inapplicable to the study of the electrophoretic motion of the polyion. Now it is possible to observe directly the mean motion of the polymeric species by optical methods. There is, therefore, little uncertainty in the experimental determination of the mobility of the polyion. On the other hand, conductivity measurements are sensitive to not only the movement of the polyion but also the movement of all the simple ions in the solution and it is in general quite difficult to analyze the contributions of different ions to the total conductivity without using auxiliary assumptions. For this reason we shall devote most of our attention to the study of the mobility of the polyion.

The subject matter of Chapter 4 dealt with the electrophoresis of rigid spherical particles. We shall now be concerned with the corresponding properties of solutions of flexible polyelectrolytes. It will be seen that the theory is in a much less satisfactory state than for the case of rigid molecules and that several fundamental questions have not even been attacked. The most important of these questions is related to the necessity of applying the hydrodynamics of continuous media to particles of molecular dimensions. For the case of globular proteins or colloids, this is not so serious a matter because the macroion dimensions are large relative to the dimensions of a solvent molecule. In the case of flexible molecules, however, we must consider the flow of fluid around obstacles of which two linear dimensions are of the same order of magnitude as the mean diameter of the solvent molecules.

Whether or not the usual formulation of the hydrodynamic equations of motion remains valid under these circumstances is unknown.

We shall proceed, as in Chapter 4, to examine first some simple models and some features of the hydrodynamical problem. This will then be followed by a study of the current theory of electrophoresis of flexible polyions and a survey of the available experimental data.

11.1 A Fluid Drop Model

We have seen in preceding chapters that the behavior of a polyelectrolyte molecule in solution can often be approximated by consideration of an equivalent sphere with almost no net charge. Since the small ions in the internal polyion domain are free to move about from point to point, and since the flexibility of the polymer skeleton also permits considerable motion of the fixed charges, we may anticipate that the flow of solvent about the equivalent sphere is quite different from the flow of solvent about a solid sphere. For a solid particle the fluid velocity with respect to the center of the sphere vanishes at the solid–fluid boundary. On the other hand, if the equivalent polymer sphere is representable as a fluid medium (analogous to a solvent entrapping coil) the normal component of fluid velocity must vanish at the boundary, but the tangential component is merely continuous. In the following we shall consider the analysis of the electrophoretic motion of a spherical fluid drop. In addition to serving as a simple model, our considerations will provide a link (via Chapter 4) with the analogous behavior of rigid polyions. The arguments to be used are due to Booth[1] and will be based on the following assumptions.

(1) The fluid sphere remains spherical when the external field is nonzero.

(2) The charge and potential distribution inside and outside the sphere are spherically symmetric in the steady state when there is no external field.

(3) The field distribution when an external field is applied is the vector sum of the field in the undisturbed state and the field which would obtain if there were no charge within the fluid drop. This assumption is equivalent to the neglect of relaxation effects in the electrolyte as well as any assymetry in the internal charge distribution caused by the fluid flow.

(4) The inertia terms in the hydrodynamic equations of motion may be disregarded.

(5) Both the solvent and the " fluid polyion domain " are incompressible.

(6) The dielectric constant, viscosity, and electrical conductance of both fluid phases retain their macroscopic values throughout the entire domain.

If ψ is the electrostatic potential outside the polyion, the equations of motion of the electrolyte are [see Eq. (4.1.4)]

$$\eta \nabla \times \nabla \times \mathbf{v} + \nabla p = -\rho \nabla \psi \tag{11.1.1}$$

$$\nabla \cdot \mathbf{v} = 0$$

where, due to assumption (3), it follows that

$$\psi = \psi(X = 0) - Xr \cos \theta \left(1 + \frac{\lambda b^3}{r^3}\right)$$

(11.1.2)

$$\lambda = \frac{\sigma - \sigma_0}{2\sigma + \sigma_0}.$$

The general solution to Eqs. (11.1.1) and (11.1.2) suitable to the problem at hand was obtained in Section 4.1, and displayed in Eqs. (4.1.34). The hydrodynamic equations of motion, Eqs. (11.1.1) also apply to the interior of the polyion domain in this model, with a viscosity η_0 and electrostatic potential ψ_0. In place of Eq. (11.1.2) the potential in the interior is

$$\psi_0 = \psi_0(X = 0) - Xr \cos \theta (1 + \lambda)$$

(11.1.3)

and the analogs of Eqs. (4.1.34) are

$$p_0 = -\int_0^r \rho_0(z) \frac{d\psi_0}{dz} \, dz + A_{00} + \cos \theta \left[A_{10}r + \frac{B_{10}}{r^2} - \frac{D_0 X(1 + \lambda)}{4\pi} \frac{d\psi_0}{dr} \right]$$

$$v_{0r} = \cos \theta \left[\frac{A_{10}r^2}{10\eta_0} + \frac{B_{10}}{\eta_0 r} + A_{20} + \frac{B_{20}}{r^3} \right.$$

$$\left. + \frac{D_0 X(1 + \lambda)}{6\pi\eta_0} \left(\int_0^r \frac{d\psi_0}{dz} \, dz - \frac{1}{r^3} \int_0^r z^3 \frac{d\psi_0}{dz} \, dz \right) \right]$$

$$v_{0\theta} = \sin \theta \left[-\frac{A_{10}r^2}{5\eta_0} - \frac{B_{10}}{2\eta_0 r} - A_{20} + \frac{B_{20}}{r^3} \right.$$

$$\left. - \frac{D_0 X(1 + \lambda)}{6\pi\eta_0} \left(\int_0^r \frac{d\psi_0}{dz} \, dz + \frac{1}{2r^3} \int_0^r z^3 \frac{d\psi_0}{dz} \, dz \right) \right]$$

$$v_{0\Omega} = 0.$$

(11.1.4)

As in the considerations of Section 4.1, we use the boundary conditions to evaluate the integration constants of Eq. (11.1.4). The condition that the pressure and velocity be finite at the origin and at infinity requires that B_{10}, B_{20}, and A_1 vanish. At the boundary between the model fluid polyion and the external solution, the normal components of the velocity vanish,

$$\left. \begin{array}{l} v_r(b, \theta) = 0 \\ v_{0r}(b, \theta) = 0 \end{array} \right\} \, 0 \leqslant \theta \leqslant 2\pi$$

(11.1.5)

whereas the tangential components become equal,

$$v_\theta(b, \theta) = v_{0\theta}(b, \theta); \quad 0 \leqslant \theta \leqslant 2\pi.$$

(11.1.6)

Finally, in order that the fluid drop be in a stable configuration we require continuity of the tangential and normal stress across the boundary. This is, in mathematical terms, a requirement that

$$\eta\left[\left(\frac{\partial v_\theta}{\partial r}\right)_{r=b} - \frac{v_\theta(b, \theta)}{b}\right] + \frac{DX(1 + \lambda)}{4\pi} \sin\theta\left(\frac{d\psi}{dr}\right)_{r=b}$$

$$= \eta\left[\left(\frac{\partial v_{0\theta}}{\partial r}\right)_{r=b} - \frac{v_{0\theta}(b, \theta)}{b}\right] + \frac{D_0 X(1 + \lambda)}{4\pi} \sin\theta\left(\frac{d\psi_0}{dr}\right)_{r=b}$$

$$\frac{2\Phi_{10}}{b} + p(b, \theta) - 2\eta\left(\frac{\partial v_r}{\partial\theta}\right)_{r=b} - \frac{D}{8\pi}\left(\frac{d\psi}{dr}\right)_{r=b}\left[\left(\frac{d\psi}{dr}\right)_{r=b} - 2X(1 - 2\lambda)\cos\theta\right]$$

$$= p_0(b, \theta) - 2\eta_0\left(\frac{\partial v_{0r}}{\partial\theta}\right)_{r=b} - \frac{D_0}{8\pi}\left(\frac{d\psi}{dr}\right)_{r=b}\left[\left(\frac{d\psi}{dr}\right)_{r=b} - 2X(1 + \lambda)\cos\theta\right]$$

$$(11.1.7)$$

where the last terms arise from the electric stresses across the interface, and are obtained from the Maxwell stress tensor introduced in Chapter 2 by the neglect of terms of order X^2. Φ_{10} is the interfacial tension between the polyion domain and the external solvent. By use of the boundary conditions to evaluate the integration constants, and the condition that the electrophoretic velocity U be the negative of the fluid velocity at infinity, Booth finds that[1]

$$U = -\frac{X}{6\pi\eta(3\eta_0 + 2\eta)}\left[3D\eta_0\int_b^\infty \xi(z)\,dz\right.$$

$$+ \eta\left\{D\left[2\int_b^\infty \xi(z)\,dz - b(1 + \lambda)\left(\frac{d\psi}{dr}\right)_{r=b} + b\xi(b)\right]\right.$$

$$+ \left.\left.\frac{5D_0(1 + \lambda)}{b^3}\int_0^b z^3\frac{d\psi_0}{dz}\,dz\right\}\right] \qquad (11.1.8)$$

which reduces to Henry's result when $\eta_0 \to \infty$ corresponding to the transition to a solid particle.

From Eq. (11.1.8) we may immediately observe that the charge distribution inside the polymeric domain markedly affects the electrophoretic velocity. If, for instance, there is a uniform charge distribution throughout the internal spherical region, with charge density ρ_0,

$$D_0\frac{d\psi_0}{dr} = -\frac{4\pi\rho_0 r}{3} \qquad 0 \leqslant r < b \qquad (11.1.9)$$

by Gauss' theorem as applied to a spherical region of radius r. If there is no charge on the surface

$$\rho_0 = \frac{3D\zeta(b + 1)}{4\pi b^2}. \qquad (11.1.10)$$

To utilize these considerations, the potential in the solvent must be specified. In the Debye–Hückel approximation

$$\psi = \frac{b\zeta}{r} e^{\kappa(b-r)} \tag{11.1.11}$$

whereupon

$$U = \frac{X}{6\pi\eta(3\eta_0 + 2\eta)} \left[D\zeta\{3\eta_0(1 + \lambda F_1(\kappa b)) \right.$$

$$\left. + 2\eta(1 - \lambda F_2(\kappa b))\} - \frac{5D_0\eta(1 + \lambda)}{b^3} \int_0^b z^3 \frac{d\psi_0}{dz} dz \tag{11.1.12}$$

with

$$F_1(\kappa b) = \frac{(\kappa b)^2}{8} - \frac{5(\kappa b)^3}{24} - \frac{(\kappa b)^4}{48} + \frac{(\kappa b)^5}{48} + \frac{(\kappa b)^4}{4} e^{\kappa b} Ei(\kappa b) \left(1 - \frac{(\kappa b)^2}{12}\right) \tag{11.1.13}$$

$$F_2(\kappa b) = \frac{1}{2} + \frac{(\kappa b)}{2} - \frac{(\kappa b)^2}{4} + \frac{(\kappa b)^3}{4} - \frac{(\kappa b)^4}{4} e^{\kappa b} Ei(\kappa b). \tag{11.1.14}$$

Note that $F_1(\kappa b)$ tends to zero in the limit as κb is very small relative to unity and tends to unity when κb is very large relative to unity. $F_2(\kappa b)$ tends to the values one half and two, respectively, in the same limiting cases. The substitution of Eqs. (11.1.9) and (11.1.10) into the general relation (11.1.12) gives

$$U = \frac{XD\zeta}{6\pi\eta(3\eta_0 + 2\eta)} [3\eta_0\{1 + \lambda F_1(\kappa b)\} + \eta\{3 + \kappa b + \lambda(1 + \kappa b - 2F_2(\kappa b))\}]. \tag{11.1.15}$$

If the conductances inside and outside the polyion domain are equal, $\lambda = 0$, and if κb is small, Eq. (11.1.15) reduces to

$$U = \frac{2\rho_0 X b^2(\eta + \eta_0)}{3\eta(2\eta + 3\eta_0)}. \tag{11.1.16}$$

A more interesting limit for our purposes is the case when κb is large and when the salt concentration in the polyion domain is large. Under these circumstances, $\lambda \approx 1$ and the electrophoretic velocity becomes

$$U = \frac{XD\zeta(3 + \kappa b)}{3\pi(3\eta_0 + 2\eta)}. \tag{11.1.17}$$

To what extent are considerations of a fluid drop model of the polyion useful? We note first that ζ decreases as κb increases at a rate such that U decreases as κb increases. Although it is of great importance that the hydrodynamic part of the problem can be handled with accuracy for the fluid drop model, this conclusion is not in agreement with the experimental data thus far published. Under most conditions, we know that the polyion behaves more nearly like a free draining coil than a solvent entrapping coil. Under these circumstances the definition of an equivalent fluid drop for the polymer domain becomes impossible. Although it may still remain possible to ascribe a spherical domain to the polyion, this domain must be regarded as permeable to the solvent, the extent of the permeability depending upon the local density of skeletal segments. Despite these facts, the considerations of this section retain value in the sense of providing a limit to be compared with the solvent entrapping polyion.

11.2 A Porous Sphere Model

In the development of the porous sphere model relaxation effects[2-4] will at first be ignored. Hermans and Fujita[5] have developed a theory of electrophoresis for permeable coils by combining the equilibrium theory of polyelectrolytes proposed by Hermans and Overbeek[6] and the viscosity–sedimentation theory of Brinkman,[7] and Debye and Bueche.[8] Overbeek and Stigter[9] and Hermans[10] have also reported similar simplified treatments of the problem, both of which lead to substantially the same results as those derived from the calculations of Hermans and Fujita. In all of these theories, the relaxation of the ionic atmosphere is neglected. In the following we shall examine the results of the above calculations. Since the variations between the several theories depend upon the techniques of calculation rather than the basic model, and since all the final results are in general concordance with one another, we shall present a simplified analysis due to Hermans.[10]

We consider, as in the preceding theories, that the polyion has a steady velocity U in the x direction. It will be assumed that the external field acting on the polyion has spherical symmetry and therefore depends only on the distance from the center of the molecule. The velocity field due to the force generated by the external field is not spherically symmetric. Moreover, due to the velocity field, the polymer skeleton exerts a force on the fluid and the resultant total force is not spherically symmetric. In Hermans's calculation the hydrodynamic interactions are computed from the Oseen equation

$$\mathbf{v} = \frac{1}{8\pi\eta r}\left[\mathbf{F} + \frac{\mathbf{F}\cdot\mathbf{rr}}{r^2}\right] \tag{11.2.1}$$

where the fluid velocity $\mathbf{v}$ is produced at the point $\mathbf{r}$ when the force $\mathbf{F}$ is applied at the origin. The fundamental approximation made by Hermans is that not only is the external force spherically symmetric, but also that the velocity generated by the force may be replaced in the Oseen relation by its average value over a spherical surface.

Consider, then, a spherical shell of radius γ subject to a force $4\pi\gamma^2 F_x(\gamma)\, d\gamma$. By hypothesis $F_x(\gamma)$ is a function only of γ. The x component of the flow generated at a point $Q(x, y, z)$ by the force density $F_x(\gamma)$ acting at the point P, with γ the distance between P and the center of the molecule, can from Eq. (11.2.1) be shown to be

$$\delta v_x = \frac{F_x}{8\pi\eta} \left(\frac{\partial^2 h}{\partial y^2} + \frac{\partial^2 h}{\partial z^2} \right) \tag{11.2.2}$$

with h the distance between P and Q. Equation (11.2.2) follows at once from Eq. (11.2.1) by replacement of r by h and explicit evaluation of the derivatives indicated. Note that we have also written δv_x in place of v_x but no differentiation is implied. If h is integrated over the surface of a sphere of radius γ, one finds that

$$4\pi\gamma^2\langle h \rangle = 4\pi\gamma^3 + \frac{4\pi}{3}\gamma r^2 \qquad \gamma > r$$

$$4\pi\gamma^2\langle h \rangle = 4\pi\gamma^2 r + \frac{4\pi}{3}\frac{\gamma^4}{r} \qquad \gamma < r \tag{11.2.3}$$

with r the distance from Q to the center of the molecule. By assumption, $\langle h \rangle$ can be used in place of h in the Oseen equation of motion. Straightforward differentiation of Eq. (11.2.3) and substitution into Eq. (11.2.2) then gives

$$\delta v_x = \frac{2\gamma F_x(\gamma)\, d\gamma}{3\eta} \qquad \gamma > r$$

$$\delta v_x = \frac{\gamma^2 F_x(\gamma)\, d\gamma}{2\eta} \left[\frac{1}{r} + \frac{x^2}{r^3} + \gamma^2 \left(\frac{1}{3r^3} - \frac{x^2}{r^5} \right) \right]. \tag{11.2.4}$$

Moreover, when the velocity δv_x is averaged over the surface of a sphere with radius r it is found that

$$\langle \delta v_x \rangle = \frac{2\gamma F_x(\gamma)\, d\gamma}{3\eta} \qquad \gamma > r$$

$$\langle \delta v_x \rangle = \frac{2\gamma^2 F_x(\gamma)\, d\gamma}{3\eta r} \qquad \gamma < r. \tag{11.2.5}$$

To obtain the total velocity generated by the force density we need only integrate Eq. (11.2.5) over the volume of the system. The result obtained is

$$\langle v_x \rangle = \frac{2}{3\eta} \left[\frac{1}{r} \int_0^r d\gamma\, \gamma^2 F_x(\gamma) + \int_r^\infty d\gamma\, \gamma F_x(\gamma) \right]. \tag{11.2.6}$$

By the device of averaging, the actual flow is approximated by a spherically symmetric flow. We shall later see that similar approximations are made in other hydrodynamic theories.

The result displayed in Eq. (11.2.6) may be used to find an explicit expression for the frictional coefficient Υ. Consider a force density which has the property that

$$F_x(r) = 0 \qquad\qquad\qquad r > R_s$$

$$F_x(r) = v_m \Upsilon [U - \langle v_x(r) \rangle] \; \Upsilon \qquad r < R_s \qquad (11.2.7)$$

$$v_m = \frac{3Z}{4\pi R_s^3}$$

where R_s the radius of the porous sphere, $\dot{Z}$ the degree of polymerization, and v_m the internal segment density. The substitution of Eq. (11.2.6) into Eq. (11.2.7) gives the relationship

$$F_x(r) = \frac{2v_m \Upsilon}{3\eta} \left[\frac{1}{r} \int_0^r \gamma^2 F_x(\gamma)\, d\gamma + \int_r^{R_s} \gamma F_x(\gamma)\, d\gamma \right]. \qquad (11.2.8)$$

By application of the operator d^2/dr^2 we find

$$\frac{d^2 F_x(r)}{dr^2} = \frac{2v_m \Upsilon}{3\eta} F_x(r) \qquad (11.2.9)$$

which has the solution

$$F_x(r) = \text{const.} \; \frac{\sinh(K_H r)}{r} \qquad (11.2.10)$$

$$K_H^2 = \frac{2v_m r}{3\eta}.$$

Note that the parameter K_H places a role analogous to the shielding parameter κ in ordinary electrolyte theory. To evaluate the constant of integration, Eq. (11.2.10) is substituted into Eq. (11.2.8) and the indicated quadratures carried out. The result of these operations is

$$F_x(r) = \frac{v_m \Upsilon U}{K_H \cosh(K_H R_s)} \; \frac{\sinh(K_H r)}{r}. \qquad (11.2.11)$$

The total frictional force experienced by the molecule is determined by integrating $F_x(r)$ over the volume of the porous sphere. If this resistance is denoted Ω, Hermans finds

$$\Omega = 6\pi\eta R_s U \Gamma(K_H R_s)$$

$$\Gamma(K_H R_s) = 1 - \frac{\tan(K_H R_s)}{K_H R_s} \tag{11.2.12}$$

In the limit as $K_H R_s$ tends to zero,

$$\lim_{K_H R_s \to 0} \Omega = ZYU \tag{11.2.13}$$

which is the case of the free draining coil. On the other hand, in the limit as $K_H R_s$ tends to infinity,

$$\lim_{K_H R_s \to \infty} \Omega = 6\pi\eta R_s U \tag{11.2.14}$$

which is Stokes law applied to the sphere of radius R_s. The frictional forces predicted by this formula clearly have the proper limiting values since $K_H R_s \to 0$ corresponds to complete permeability of the coil due to the vanishing local segment density and $K_H R_s \to \infty$ corresponds to an impenetrable sphere due to a very large local segment density. In the intermediate region, $\Gamma(K_H R_s)$ can be accurately approximated by

$$\Gamma(K_H R_s) = \frac{K_H^2 R_s^2}{3 + K_H^2 R_s^2}. \tag{11.2.15}$$

Having found the frictional coefficient it is now pertinent to proceed to a calculation of the electrophoretic velocity. It will be recalled that the three principal effects to be considered are the distortion of the lines of force of the external electric field due to the presence of the macromolecule, the retardation caused by ion atmosphere relaxation, and the electrophoretic counter flow of the solvent. As in the preceding section we shall neglect relaxation effects. The net force acting on the polyion depends upon the counterion density within the polymeric domain. Hermans uses for this density s the relation

$$s = -\rho_f + \rho_f(1 + p)e^{-p}\frac{\sinh\xi}{\xi}, \qquad r < R_s$$

$$s = -\rho_f p \cosh p\Gamma(p)\frac{e^{-\xi}}{\xi}, \qquad r > R_s \tag{11.2.16}$$

$$p = \kappa R_s$$

$$\xi = \kappa r$$

with $\Gamma(p)$ defined by Eq. (11.2.12) and ρ_f the fixed charge density due to the polyion. Consider now the application of an electric field X in the x direction. The force density becomes for this case

$$
\begin{aligned}
F_x(r) &= v_m \Upsilon[U - \langle v_x(r) \rangle] + Xs(r), & r < R_s \\
F_x(r) &= Xs(r) & r > R_s
\end{aligned}
\qquad (11.2.17)
$$

with the value of the velocity $\langle v_x(r) \rangle$ determined by Eq. (11.2.6). By substitution of Eq. (11.2.6) into Eq. (11.2.17) and use of the operator d^2/dr^2, it is found that the relation corresponding to Eq. (11.2.9) is

$$
\frac{d^2 F_x(r)}{dr^2} - \frac{2 v_m \Upsilon}{3\eta} F_x(r) = X \frac{d^2 s(r)}{dr^2}. \qquad r < R_s \qquad (11.2.18)
$$

Moreover, by use of Eq. (11.2.16) and solution of the resultant differential equation, it is found that the force density becomes

$$
F_x(r) = \frac{B}{r} \sinh(K_H r) + \frac{X \rho_f \kappa (1 + p) e^{-p}}{r(\kappa^2 - K_H^2)} \sinh(\kappa r) \qquad (11.2.19)
$$

with the constant B determined by

$$
BK_H \cosh(K_H R_s) = v_m \Upsilon U - X \rho_f
$$

$$
+ X \rho_f (K_H R_s)^2 e^{-p} \cosh p \left[\frac{\Gamma(p)}{p} - \frac{1+p}{p^2 - K_H^2 R_s^2} \right]. \qquad (11.2.20)
$$

As in the discussion of the frictional coefficient, the total force acting on the polyion is obtained from Eq. (11.2.19) by integration over the volume of the equivalent porous sphere. But the total electric force is just $(4\pi/3)R_s^2 \rho X$, whereupon

$$
U = \frac{\rho_f X}{v_m \Upsilon} \left[1 - \frac{K_H^2 R_s^2}{p} e^{-p} \Gamma(p) \cosh p \right.
$$

$$
\left. + \frac{(1 + p)K_H^2 R_s^2}{p^2 - K_H^2 R_s^2} e^{-p} \cosh p \left(1 - \frac{K_H^2 R_s^2}{p^2} \frac{\Gamma(p)}{\Gamma(K_H R_s)} \right) \right] \qquad (11.2.21)
$$

which, after use of Eq. (11.2.15) becomes

$$
U = \frac{\rho_f X}{v_m \Upsilon} \left[1 + \frac{K_H^2 R_s^2}{2} \frac{1 + e^{-p}}{3 + p^2} \right]. \qquad (11.2.22)
$$

There are several interesting limiting cases which deserve mention. In the limit of completely permeable coils,

$$\lim_{K_H R_s \to 0} U = \frac{\rho_f X}{v_m \Upsilon} \tag{11.2.23}$$

corresponding to the free draining coil, as anticipated. In the other extreme when the macromolecule behaves as a compact sphere,

$$\lim_{K_H R_s \to \infty} U = \frac{2\rho_f R_s^2}{3\eta} \frac{e^{-p}}{p^2} \Gamma(p) \cosh p \tag{11.2.24}$$

in agreement with Henry's theory. An interesting case is also posed by the high ionic strength limit for which

$$\lim_{p \to \infty} U = \frac{\rho_f X}{v_m \Upsilon} \tag{11.2.25}$$

and the molecule is once again freely drained. Finally, when $K_H R_s$ and p are both large,

$$U = \frac{\rho_f X}{v_m \Upsilon} \left[1 + \frac{K_H^2 R_s^2}{2p^2} \frac{2 + (K_H R_s / p)}{1 + (K_H R_s / p)} \right]. \tag{11.2.26}$$

Before discussing the physical nature of the limits presented it is pertinent to note that the relative velocity of the liquid with respect to the polyion is

$$u_r = \langle v_x \rangle - U$$

$$u_r = -\frac{1}{v_m \Upsilon} \left[X\rho_f + \frac{B}{r} \sinh(K_H r) \right.$$

$$\left. + \frac{X\rho_f \kappa (1 + p) e^{-p}}{\kappa^2 - K_H^2} \frac{K_H^2}{\kappa^2 r} \sinh(\kappa r) \right], \qquad r < R_s$$

$$u_r = -U + \left(\frac{2R_s^3}{3\eta r} \right) X\rho_f e^{-\kappa r} \Gamma(p) \frac{\cosh p}{p^2}, \qquad r > R_s \tag{11.2.27}$$

which relations are obtained from Eq. (11.2.6) by insertion of Eq. (11.2.19). By use of Eq. (11.2.27) the conductance σ of the solution may be calculated by use of the formula

$$\sigma = \sigma^0 + \frac{n}{X} \int u_r s(r) \, dv \tag{11.2.28}$$

with σ^0 the contribution to the specific conductance by the small ions, and n

the number of polyions per unit volume. The explicit formulas obtained by integration of (11.2.28) will not be displayed here.

It is also pertinent to note that Hermans and Fujita[5] were able to solve the hydrodynamic problem without use of the averaging assumption which gives rise to Eq. (11.2.26). The calculations are subject only to the approximations inherent in the use of the Hermans–Overbeek counterion distribution and the use of a representative porous sphere. Hermans and Fujita solved the following equations for the velocity of a polyion in an infinite volume of solvent containing a simple salt:

$$\eta \nabla^2 \mathbf{v} - \text{grad } p = -s\mathbf{X}, \qquad r > R_s$$

$$\eta \nabla^2 \mathbf{v} - v_m \Upsilon \mathbf{u}_r - \text{grad } p = -s\mathbf{X}, \qquad r < R_s \qquad (11.2.1')$$

$$\text{div } \mathbf{v} = 0$$

which relations are also the starting equations of Brinkman, Debye, and Bueche for sedimentation and viscosity when there is no $s\mathbf{X}$ term. As before the symbol Υ represents the frictional coefficient for the relative motion of a polymer segment and the solvent, and s is the charge density in the liquid due to small ions. They obtained as the result of a complicated analysis the following relation for the electrophoretic mobility of the polyion:

$$U_p^0 = U/X = \frac{\rho_f}{v_m \Upsilon} \left[1 + \frac{\phi(\tilde{\sigma}, p)}{1 - \tanh \tilde{\sigma}/\tilde{\sigma}} \right]$$

$$\phi(\tilde{\sigma}, p) = \frac{2}{3} \left(\frac{1 + p}{p^2 - \tilde{\sigma}^2} \right) e^{-p} [F(\tilde{\sigma}, p) + G(\tilde{\sigma}, p)]$$

$$+ \frac{2}{3} \left(\frac{e^{-p}}{p^2} \right) \left(\cosh p - \frac{\sinh p}{p} \right) H(\tilde{\sigma}, p)$$

$$F(\tilde{\sigma}, p) = \left(1 + \frac{2}{3} \tilde{\sigma}^2 \right) \left(1 - \frac{\tilde{\sigma}^2}{p^2} \right) \cosh p$$

$$+ \left(\frac{\tilde{\sigma}^2}{p^2} + \frac{2}{3} \frac{\tilde{\sigma}^4}{p^2} - 1 + \frac{1}{3} \tilde{\sigma}^2 \right) \frac{\sinh p}{p} \qquad (11.2.29)$$

$$G(\tilde{\sigma}, p) = \left[\left(\frac{\tilde{\sigma}^2}{p^2} - 1 - \tilde{\sigma}^2 \right) \cosh p + \left(1 - \frac{\tilde{\sigma}^2}{p^2} \right) \frac{\sinh p}{p} \right] \tanh \tilde{\sigma}/\tilde{\sigma}$$

$$H(\tilde{\sigma}, p) = \tfrac{1}{3} \tilde{\sigma}^2 (1 + p) + (\tilde{\sigma}^2 p + 1 + p) (1 - \tanh \tilde{\sigma}/\tilde{\sigma})$$

$$\tilde{\sigma} = R_s \left(\frac{v_m \Upsilon}{\eta} \right)^{1/2} = \left(\frac{3}{2} \right)^{1/2} K_H R_s.$$

Numerical calculations based on Eq. (11.2.29) are difficult unless use is made of a table of U_p^0 for various values of $\tilde{\sigma}$ and p published by Fujita.[12] Consider the limiting case when both p and $\tilde{\sigma}$ are large compared with unity, whereupon Eq. (11.2.29) can be simplified to:

$$U_p^0 = \left(\frac{\rho_f}{v_m \Upsilon}\right)\left(1 + \frac{\tilde{\sigma}^2}{3p^2} \frac{2 + (\tilde{\sigma}/p)}{1 + (\tilde{\sigma}/p)}\right) \qquad (11.2.30)$$

which result is substantially identical with Eq. (11.2.26). Equations (11.2.26) and (11.2.30) become identical when $K_H R_s \gg p$ and when $K_H R_s \ll p$. When $K_H R_s$ and p are of comparable magnitude they nowhere differ more than 15%. It is important to note that under ordinary experimental conditions both p and $\tilde{\sigma}$ are much larger than unity.

We now consider the nature and implications of the limiting cases cited previously.

(a) When the hydrodynamic shielding ratio $K_H R_s$ is zero, and the molecule is free draining, the mobility is determined by Eq. (11.2.23). In this limit the electrophoretic mobility of the polyion is equal to the mobility of a segment. Only in pure aqueous solutions at infinite dilution is it possible that the mobility may be adequately described by Eq. (11.2.23).

(b) When p tends to become infinite while $\tilde{\sigma}$ remains finite, we find for the mobility the same result as in the preceding case [Eq. (11.2.25)]. Now, p tends to infinity faster than $\tilde{\sigma}$ under ordinary experimental conditions, so that this result means that a polyion has the same mobility as a segment in highly concentrated salt solutions and hence the mobility is independent of the molecular weight and shape of the molecule.

(c) The mobility of a polyion must be larger than the mobility of a segment if p tends to infinity faster than $\tilde{\sigma}$. We thereby deduce that there must be a maximum in the mobility as a function of the concentration of added salt.

Conclusion (b) stands in marked contrast to the behavior expected under the assumption that the polyion behaves as an equivalent solid sphere. From the experimental point of view conclusion (b) is particularly important, in that it suggests the possibility that the charge density of a polyion could be determined from electrophoretic measurements. Although the total charge of a solid particle can be calculated from the electrophoretic velocity by using Henry's theory, prior to the paper published by Hermans and Fujita there had been no corresponding relation between mobility and charge for a flexible polyion.*

*An application of Henry's theory simply gives some unknown measure of the surface charge of the polyion coil.

To implement Eqs. (11.2.29) and (11.2.30), we must know the values of $\tilde{\sigma}$ and R_s as functions of the concentration of added salt. Since Hermans and Fujita employed the same model and mathematical procedures as did Brinkman, Debye, and Bueche for sedimentation and viscosity, it appears reasonable to calculate $\tilde{\sigma}$ and R_s from viscosity data by using the latter theory. According to this theory, valid for uncharged linear polymers in dilute solution, the intrinsic viscosity $[\eta]$ is given by the equations:

$$[\eta] = \Omega_s / M \phi(\tilde{\sigma}) \tag{11.2.31}$$

where

$$\Omega_s = \left(\frac{4\pi}{3}\right) R_s^3$$

$$\phi(\tilde{\sigma}) = \frac{5}{2} \frac{1 + (3/\tilde{\sigma}^2) - (3/\tilde{\sigma}) \coth \tilde{\sigma}}{1 + (10/\tilde{\sigma}^2)[1 + (3/\tilde{\sigma}^2) - (3/\tilde{\sigma}) \coth \tilde{\sigma}]} \tag{11.2.32}$$

$$\tilde{\sigma} = R_s (v_m \Upsilon / \eta)^{1/2}$$

with Υ again the frictional factor of a segment, and η here the viscosity of the solvent.

It is clear that a knowledge of Υ suffices to calculate $\tilde{\sigma}$ and R_s from Eq. (11.2.31) and the values of U_p^0 calculated from Eqs. (11.2.29) or (11.2.30) using these values of $\tilde{\sigma}$ and R_s can be compared with experiment. On the other hand, if we accept both of the theories as correct, we can fit the values of Υ (and R_s) so that both the viscosity and electrophoresis are correctly given by the theories. The value of Υ thus determined should be compared with Stokes law and the dimensions of a segment or a free monomer.

A combination of the two theories mentioned above probably has the following merit; it is well known that the values of Υ chosen to obtain agreement between the theory of Brinkman, Debye, and Bueche and the experimental intrinsic viscosities usually correspond to a much smaller radius for the segment than that corresponding to the known segment volume. In view of this fact, it should be recognized that either the model or the details of the calculation used must include some unrealistic approximations. In other words, the absolute values of $\tilde{\sigma}$ and R_s calculated from the Debye–Bueche theory cannot be considered to be correct, no matter what value of Υ is used. However, it may be anticipated that many of the ambiguities can be eliminated if $\tilde{\sigma}$ and R_s as derived from the viscosity theory are inserted into the electrophoresis theory. If this expectation is fulfilled, the values of U_p^0 thus calculated would have greater reliability than the values of $\tilde{\sigma}$ or R_s.

II.3 Experimental Studies of Electrophoresis

The basic method for measuring the electrophoretic velocity of a solid colloid particle was developed by Tiselius and incorporated by him in the well-known apparatus bearing his name. However, measurement of the electrophoretic mobility of a flexible polyion poses new difficulties for two reasons: the dependence of the mobility of the polyion upon the salt concentration and the strong electrostatic interactions between the polyions. Recently, it has been shown[13,14] that these difficulties can be eliminated by extrapolating the observed electrophoretic velocities to zero polymer concentration. Reliable electrophoretic mobilities have been obtained by this method, and Fig. 11.3.1 shows an example of the dependence of the elctrophoretic mobility upon the polymer concentration.

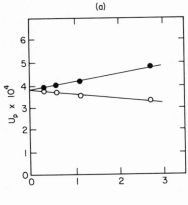

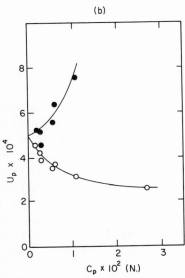

Despite the recent publication of several studies of electrophoretic mobility in solutions of polyelectrolytes[13-20] the data available for a quantitative comparison of theory and experiment are very limited. To examine the theory described in Section 11.2, it is necessary to know the value of Υ, or the mobility of a segment. Two methods for determining Υ have been reported.[13,14,17] Nagasawa, Soda, and Kagawa determined the mobility of the polyion in pure aqueous solution at infinite dilution by comparing the electrophoretic mobility with the mobility determined from measurements of the diffusion potential and conductivity. The frictional coefficient was computed from Eq. (11.2.23) by identifying a segment with a unit acid. (The determination of mobilities from diffusion potentials and

FIG. 11.3.1. Dependence of electrophoretic mobility upon polyion concentration. ●, calculated from ascending boundary; ○, calculated from descending boundary. Sample: Na-polyvinyl alcohol sulphate; solvent: (a) NaCl, 0.05N; (b) NaCl, 0.005N. [Reproduced from ref. 13.]

conductivity will be discussed in the following section.) In Fig. 11.3.2 the electrophoretic mobilities of sodium polyvinyl alcohol sulfate, U_p^0, are plotted against the added salt concentration c_{NaCl} and in the same figure the mobilities in salt-free solutions as calculated from the diffusion potential and conductivity are plotted against the polymer concentration c_p. Although the two curves may be expected to behave differently, both must tend to the same point on the ordinate, i.e., at $c_p = 0$ for the lower curve and at $c_{NaCl} = 0$ for the

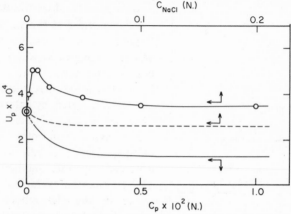

FIG. 11.3.2. Dependence of mobility of polyion in infinitely dilute solution U_p upon NaCl concentration. ——, mobility of polyion in salt free system; - - - -, hypothetical mobility of monomer acid. [Reproduced from ref. 13.]

upper curve. This requirement makes it possible to determine systematically the most consistent value of the mobility for a given polyion at $c_p = 0$ and $c_{NaCl} = 0$. Nagasawa, Soda, and Kagawa found the limiting value of U_p^0 to be about 3.2×10^{-4} cm²/sec volt for a sodium polyvinyl sulfate by this method. This value gives $\Upsilon = 5.0 \times 10^{-9}$ gm/sec, which corresponds to an equivalent sphere of radius 3.0×10^{-8} cm, in good agreement with the actual length of the unit acid, 4.1×10^{-8} cm. Nagasawa, Soda, and Kagawa also pointed out that the extrapolated mobility is in good agreement with the hypothetical mobility of the basic unit acid, which mobility can be estimated from an empirical relationship between the mobility of a univalent organic acid and the number of atoms in the molecule.[59]

With the frictional coefficient determined, it is easy to calculate $\tilde{\sigma}$ and R_s from the intrinsic viscosity by using Eq. (11.2.31). U_p^0 may then be computed from Eq. (11.2.29) and compared with experiment. A comparison of this type for the case of Na–polyvinyl sulfate is shown in Fig. 11.3.3, where the value of Υ was determined by the method of Nagasawa, Soda, and Kagawa. Very good agreement is found at high concentrations of NaCl but marked deviations between theory and experiment are evident at low concentrations.

In particular, the position of the maximum calculated from Eq. (11.2.29) is at a much lower salt concentration than the experimentally determined maximum. Thus, the qualitative features described in Section 11.2 are clearly satisfied. Further qualitative confirmation comes from several experiments which have shown that the electrophoretic mobility is independent of the molecular weight.[11]

The origins of the disagreement at low concentrations of NaCl may be numerous. Hermans[11] has pointed out that: (1) The basic segment required to describe the hydrodynamic interactions may increase to the dimer, trimer,

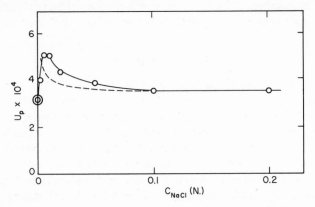

FIG. 11.3.3. Comparison between observed mobilities and mobilities calculated from Hermans–Fujita theory. ——, observed; - - - -, calculated. [Reproduced from ref. 13.]

etc. with decreasing ionic strength. (2) It is not necessary that $\tilde{\sigma}$ and R_s as calculated from the intrinsic viscosity with the use of the frictional coefficient determined by a method other than that based solely on the theory of Debye and Bueche be correct.

We believe these two criticisms to be untenable. For, since R_s is determined from a theory of intrinsic viscosity based on a spherical smeared density model, the actual hydrodynamics of a segment would seem to play no role. Moreover, it seems reasonable to assume a priori that the polyion sphere is composed of beads of monomer acids, and, as previously discussed, it is probable that the ambiguity in $\tilde{\sigma}$ and R_s is not serious for the purposes of the comparison described. There seems no logical reason why only the original procedure of determining Υ from the viscosity theory should be followed. It should be mentioned at this juncture that the values of R_s calculated from Υ as determined by Nagasawa, Soda, and Kagawa are in good agreement with the values calculated from the Flory theory of intrinsic viscosity. We believe that the disagreement at low concentrations of NaCl should be attributed to the failure of the uniform sphere model.

In addition to the studies cited, Napjus and Hermans,[14] and Van Geelen[17] examined the theory of Hermans, Fujita, Overbeek, and Stigter by calculating Υ from the electrophoretic velocity. To obtain values of R_s, Napjus and Hermans used viscosity data and Van Geelen used the potentiometric titration curve. The values of Υ determined by them appear to be constant, independent of the added salt concentration, but depend upon the degree of neutralization. A detailed verification of the theory, however, requires more data.

Even without detailed calculation, it is clear from the above discussion that the available experimental data indicate that the relaxation of the counterion atmosphere is likely to be an effect of secondary importance for the description of the electrophoretic behavior of linear polyelectrolytes.

11.4 Conductance

A discussion of conductance is much more complicated than that of electrophoresis, because a large part of the current carried is due to simple ions. In the treatment of the properties of the activity coefficient it was pointed out that the state of the simple ion and the electrostatic influence of the polyion upon the simple ion is very poorly understood in spite of the fact that the ionic distribution around the polyion has been studied in detail. The fly in the ointment is of course the very great difficulty in treating the properties of simple ions in the presence of a polyion. In calculating the equilibrium ionic distribution around a polyion, it is only necessary to consider the positions of all simple ions averaged over an ensemble or over a long period of time. Thereby, the equilibrium counterion distribution around the polyion is more or less symmetric, and the counterions are adequately described by this averaged distribution. However, it is immediately apparent that the ionic distribution around a simple ion in the bulk solution must be different from the distribution observed when the ion is assumed to be in the ionic atmosphere around the polyion. The difference arises from the restriction whereby the small ion position is specified relative to the polyion. The atmosphere is then determined by an average over the positions of all remaining small ions keeping the specified small ion and the polyion fixed in relative position. This inevitably gives rise to more positive charge near a negative ion and more negative charge near a positive ion than the distributions predicted from the assumption that all ions are in the averaged ionic distribution around the polyion. We shall denote this relative ionic distribution as the local ionic atmosphere. The local ionic atmosphere may play an important role in determining the mobilities of the simple ions as well as affecting the ionic activities. That is, although a spherical model of the polyion wherein all ions are smeared out uniformly may under some conditions be an adequate approximation to the behavior of the polyion, this model forces the counterions to be distributed almost uniformly inside the polyion sphere and, therefore, cannot be considered to be a good model for describing the motion of simple ions. In support of this argument we note that it is known from the experiments of Wall and co-workers that the movements of the counterions around a polyion vary markedly with the positions of the ions.

In the theory of Fujita and Hermans,[21] which is the only theoretical study of the electrical conductivity of linear polyelectrolytes thus far published, both the deviation of the local atmosphere from the averaged atmosphere and the relaxation of the counterion atmosphere are neglected. The influence of the polyion on the mobilities of the simple ions is introduced into the calculation only through the electro-osmotic flow of the liquid inside and outside the polyion sphere. In view of these approximations, it is instructive to see how well the theory agrees with experiment.

Fujita and Hermans used the same model and assumptions for the calculation of conductance as in their previous calculation of the electrophoretic velocity. Because the liquid around the polyion has a velocity $\mathbf{v}$ due to the electro-osmotic effect, the velocity of an anion in the neighborhood of a polyion must be $Z_i \omega_i \mathbf{X} + \mathbf{v}$. The total contribution of all the simple ions to the conductivity J' is given by summing the current due to a simple ion $q z_i^2 \omega_i \mathbf{X} + z_i q \mathbf{v}$ over all simple ions. The result is clearly

$$J' = qX \sum_i z_i^2 \omega_i c_i + n \int u_r s \, dv \qquad (11.4.1)$$

with ω_i and c_i the mobility and the stoichiometric concentration of the ith ion. The integration in Eq. (11.4.1) extends over a single polyion and its surrounding double layer. To perform the summation, it is assumed that the double layers of different macroions do not overlap. Further, if U is the velocity of the polyion, we may write $\mathbf{v} = \mathbf{U} + \mathbf{u}_r$ where $\mathbf{u}_r$ is the velocity of the liquid relative to the polyion. The result of integrating $Us \, dv$ will be $-UZq$ and this will cancel the contribution of the polyion to the conductivity. Therefore, the total current density will be

$$J' = J_0' + n \int u_r s \, dv \qquad (11.4.2)$$

and the specific conductance becomes

$$\sigma = \sigma_0 + \frac{n}{X} \int u_r s \, dv. \qquad (11.2.28)$$

The equation obtained by integration of the last term is very complicated and will not be displayed herein.

In Eq. (11.2.28) σ_0 is the specific conductance which would be observed if the polyion could be removed. Now, the characteristic feature of Eq. (11.2.28) is that the second term on the right hand side is always positive.

Therefore, if a polyelectrolyte solution is diluted keeping σ_0 constant, σ must decrease with dilution. However, according to Longworth and Hermans,[22] the observations do not always conform with this expectation. A direct comparison between Eq. (11.2.28) and experiment was carried out by Napjus and Hermans,[14] but the agreement was not satisfactory.

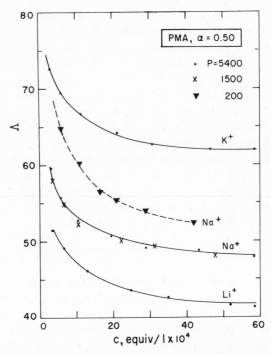

FIG. 11.4.1. Equivalent concentration dependence of equivalent conductance of poly-methacrylate solutions at 25°C. Degree of neutralization 0.50; ●, degree of polymerization 5400; ×—, degree of polymerization 1500; ▼, degree of polymerization 200. [Reproduced from ref. 29.]

In view of the present status of the theory of conductivity, it may be of some value to summarize the experimental data thus far reported. Experimental studies of the electrical conductivity of polyelectrolyte solutions have been carried out by many investigators.[3-35] One very characteristic feature of the observed conductivity is that the limiting equivalent conductance cannot be obtained by extrapolation. A typical example of the conductivity-concentration curve in the absence of added salt is depicted in Fig. 11.4.1. The curves show a very marked increase with dilution of the solutions, and deviate markedly from the linearity between Λ and $c^{1/2}$ predicted for simple salt solutions. It does not appear that the mobility of the polyion can be determined from conductivity measurements alone, whereas such determinations are often carried out for simple salts. However, it seems almost certain that the mobility of the polyion is independent of the nature of the counterion.[9-30] Eisenberg[29] studied salts of polymethacrylic acid and polyacrylic acid and showed that the differences $\Lambda_{COOK} - \Lambda_{COONa}$, $\Lambda_{COOK} - \Lambda_{COOLi}$, $\Lambda_{COONa} - \Lambda_{COOLi}$ are constant over the entire range of concentration investigated. Moreover, the following equality was found to hold:

$$\phi(\alpha) = (\Lambda_{COOK} - \Lambda_{COONa})/(\lambda_{K^+}^0 - \lambda_{Na^+}^0)$$

$$= (\Lambda_{COOK} - \Lambda_{COOLi})/(\lambda_{K^+}^0 - \lambda_{Li^+}^0) \tag{11.4.3}$$

$$= (\Lambda_{COONa} - \Lambda_{COOLi})/(\lambda_{Na^+}^0 - \lambda_{Li^+}^0)$$

where the Λ's and λ's are the equivalent conductances of the salts and ions respectively and $\phi(\alpha)$ is a constant depending upon the degree of neutralization α. It should be noted that this relationship does not seem valid for all polyelectrolytes.[35]

In an extensive series of studies, Fuoss and co-workers[25] have shown that the logarithm of the equivalent conductance of copolymers of vinyl pyridine in various organic solvents bear a linear relationship to the logarithm of the polyelectrolyte concentration with a slope of -0.13, and also that the logarithmic conductance of a polyelectrolyte solution is linear in the reciprocal dielectric constant, as shown in Fig. 11.4.2.[25]

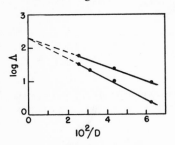

FIG. 11.4.2. The dependence of the conductance of a polyelectrolyte solution on the macroscopic dielectric constant. [Reproduced from ref. 25.]

The conductances of polyelectrolyte solutions are very low when compared with those of simple salt solutions. It has been shown[14,17] that the contribution of a polyelectrolyte to the conductance is of almost the same magnitude as the ionic conductance of the polyion, if it is assumed that the conductance of the simple salt in the solution is not changed by the presence of the polyion. Occasionally this assumption leads to the conclusion that the contribution of the counterion becomes negative.[17] This may be interpreted to show the presence of the same kind of ionic association as observed by Wall and co-workers in studies of the transference number of polyelectrolytes in pure water solution.

From light scattering studies as well as viscosity measurements, it is known that polyions when placed in pure water solution without added salts are considerably expanded relative to nonelectrolyte polymers of similar structure. These extended forms of the polyions are responsible for some interesting conductance phenomena. If the polyions are oriented along the direction of an applied force such as is obtained by streaming the solution or using an orientating electric field, the conductivity of the solution is different in different directions relative to the orienting field.[36-42] Figure 11.4.3 shows an example of such measurements.

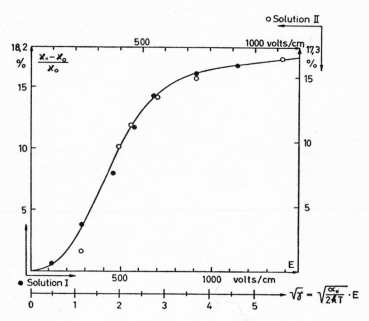

FIG. 11.4.3. Relative change of conductivity in stationary state parallel to the orienting field: theoretical curve (full line) and experimental values for two different samples. [Reproduced from ref. 40.]

Studies of the electrostatic interaction between a polyion and its atmosphere can be made by determination of the Wien effect and/or Debye–Falkenhagen effect but these measurements have been carried out in only a few instances.[43-47] In general it is observed that measurable high field conductance and measurable frequency dependence of the conductance occur at smaller values of the voltage and frequency than those required for the same effects in solutions of simple ions.

11.5   The Transference Number

In order to analyze the behavior of polyelectrolyte solutions under the influence of an applied electric field, it is necessary to know the mobilities of all the ions in the solution. Electrophoresis experiments give the mobility of a polyion without ambiguity, but there is no direct method for determining the mobility of the counterions and byions individually since conductivity measurements give only a sum of mobilities for all the ions in the solution. In ordinary electrolyte solutions, transference numbers for the ions can be combined with electrical conductivity measurements to obtain the mobilities of the individual ions in the solution. In a similar fashion, experimental studies of transference numbers have occupied an important position in the general study of transport phenomena in solutions of polyelectrolytes since the experiments of Wall and his co-workers[48-53] showed that some counterions migrate with the polyion against the applied electrical force.

The transference number of the ith ion, t_i, is defined as follows:

$$t_i = \frac{z_i \omega_i c_i}{\sum z_i \omega_i c_i}.$$ (11.5.1)

But, $\sum z_i \omega_i c_i$ is equal to the conductivity of the solution, so that the combination of electrical conductivity measurements and transference number measurements suffices to determine the mobilities of individual ions. There are several methods of determining a transference number. Wall and co-workers employed a modified Hittorf method and Figure 11.5.1 shows the apparatus used.[52] The cell is separated into two parts by a glass frit, with a platinum electrode fitted into each part. If a salt solution is placed in both compartments of the cell and electrolyzed with direct current, the transference number of the ionic species can be calculated from the relative increase in concentration in each compartment. The experiments of Huizenga et al.[48] show, however, that the transference number of the polyacrylate ion is larger than unity and that of Na+ becomes negative at a degree of neutralization higher than 60%. That is, Na+ did not move to the cathode but rather moved to the anode along with the polyion. This is interpreted as meaning that there exists an association of sodium ions with polyacrylate ions. With this assumption it is possible to calculate the amount of associated sodium from the change in concentration of the polyion and the counterion in the anode compartment, using the further assumption that unassociated counterions can migrate with the same mobility as sodium ions in sodium chloride solutions of the same ionic strength. The amount of sodium associated with a polyacrylate ion is plotted against the degree of

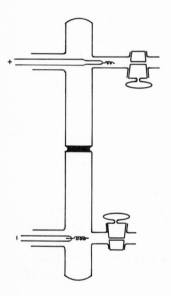

FIG. 11.5.1. The transference cell used by Wall and co-workers[52]

neutralization in Figs. 9.3.6 and 9.3.7. Despite the arbitrary nature of the assumption that the mobility of unassociated ions is the same as that in NaCl solutions, the values thus obtained are quite reasonable and agree with the results obtained from other independent measurements.[53,54] For a molecular interpretation of ionic association, Wall and his co-workers suggested two models: in one of their papers[56] they showed that the fraction of associated ions is almost equal to the fraction of ions occluded inside the polyion. In another paper[55] they considered the associated ions to be bound on the ionizable groups of the polyion. It is noteworthy that many electrical transport phenomena do not

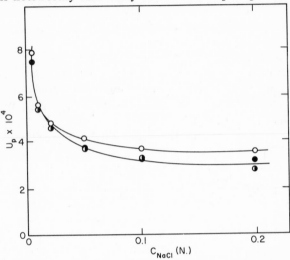

FIG. 11.5.2. Dependence of U upon degree of polymerization. ●—Na-PVS No. 1; ◐—No. 2; ○—No. 4, polyelectrolyte concentration $0.02N$. [Reproduced from ref. 57.]

depend on the degree of polymerization or the viscosity of the solution in a direct way, but only on the charge density of the polyion. In view of this, the latter model is superior to the former. The proposal that ion localization occurs at specific sites was introduced into theoretical calculations by Rice and Harris.

To determine the transport numbers of the polyion and the counterions, Nagasawa et al.[57] used a method different from that of Wall; the measurement of diffusion potentials. The diffusion potential of a salt (Na_ZP^{-Z}) is defined by the following Nernst equation:

$$-d\mathscr{E} = \frac{RT}{\mathscr{F}} \left[t_{Na^+} \, d \ln a_{Na^+} - \frac{t_{p-z}}{Z} d \ln a_{p-z} \right] \qquad (11.5.2)$$

where the a's are activities of the respective ions and the t's are transference numbers. The second term on the right-hand side can be neglected, since Z is a large number. Therefore, t_{Na^+} can be calculated from a relationship between the diffusion potential $\mathscr{E}$ and the activity of the sodium ion, and t_{p-z} is defined by $1 - t_{Na^+}$. The mobilities of the respective ions can be calculated from the products of the specific conductivity and the transference numbers by using Eq. (11.5.1). The mobilities obtained for a sodium poly-vinyl alcohol sulfate are shown in Fig. 11.5.2. The mobility of Na^+ thus obtained is an average value, neglecting the possible variation in mobility due to counterion configuration relative to the polyion. Also, the mobilities thus obtained can be profitably compared with the ionic activity coefficient

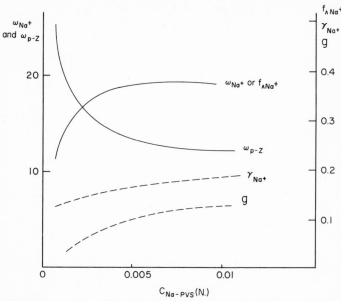

FIG. 11.5.3. Ionic conductances of Na-Polyvinyl alcohol sulphate (degree of esteri-fication $\simeq 0.73$). ω_{Na^+}—mobility; $f_{\Lambda \cdot Na^+} (= \omega_{Na^+}/\omega_{Na^+}^0)$—conductivity coefficient; γ_{Na^+} activity coefficient; g—osmotic coefficient.

of the sodium ion or the osmotic coefficient (see Fig. 8.1.1). It is observed that all three quantities decrease with dilution. In dilute solutions, the mobility of the polyion is much larger than that of the sodium ion.

In addition to electrophoresis, conductivity and transference number measurements, charge effects appear in the diffusion and sedimentation of a polyelectrolyte since differential motion of the polyion and the counterions gives rise to an electromotive force and this induced potential reacts upon the motion of the ions including the polyion. Unfortunately, few studies of these effects have been made. Howard and Jordan[58] measured the diffusion constant and sedimentation constant of sodium polymethacrylate with and without added salt, but they discussed their results mainly from the standpoint of the comparison with nonelectrolyte polymer solutions. One of their interesting results is that the molecular weight of the polyion calculated from the sedimentation constant and the diffusion constant is a function of the salt concentration and a reliable molecular weight can be obtained by extrapolation to the hypothetical infinitely concentrated simple salt solution.

Kedem and Katchalsky[60] measured the diffusion constants of polymethacrylates in salt free solutions and calculated the hydrodynamic resistance of the polymer by allowing for the osmotic behavior of the polyelectrolyte in solution. The results are interpreted to indicate a rodlike shape for the molecule at high degrees of ionization. This deduction, however, seems to represent an extreme interpretation of the data. It is unlikely that a flexible polyion has a rodlike shape, at least in a solution of finite concentration. Despite the fact that in diffusion the counterions and polyion move with the same mean velocity the neglect of the hydrodynamic interactions between the counterions and the polyion may be serious, and we believe the interpretation presented by Kedem and Katchalsky must be modified.

11.6   An Approximate Treatment of Relaxation Effects

Longworth and Hermans[22] have presented an approximate method of accounting for the previously neglected relaxation effects. The argument proceeds as follows. Consider a spherical polyion in the Donnan approximation, i.e., the interior domain will be assumed to be electrically neutral. It will be further assumed that the polyion is freely drained by the solvent so that

$$U = \frac{\rho_f X_2}{v_m \Upsilon} \tag{11.6.1}$$

with X_2 the local electric field. If the concentrations of simple ions inside and outside the polymeric domain are denoted c_j and c_j^*, then by hypothesis

$$q \sum_j c_j^* z_j = -\rho_f$$

$$\sum_j c_j^* z_j = 0. \tag{11.6.2}$$

In the steady state when the polyion velocity is U, the net effect of atmosphere deformation is the accumulation of charge on opposite sides of the spherical particle. It is assumed that the external potential ψ and the relaxation potential are linearly superimposable upon the Donnan potential of the resting polyion.

Consider now a spherical surface of radius r concentric with the polymeric domain. Let the vector $\mathbf{r}$ make an angle θ with the applied field. At an arbitrary point on this surface, $U \cos \theta$ is the velocity of the surface in a radial direction, and the velocity of a simple ion at this point will be $-z_j \omega_j (\partial \psi / \partial r)_s$. The velocity of these ions in a radial direction relative to the surface is then the algebraic sum $-z_j\omega_j(\partial\psi/\partial r)_s - U\cos\theta$, giving rise to the radial current density J_r of magnitude

$$J_r = -q\left(\frac{\partial\psi}{\partial r}\right)_s \sum_j z_j^2\omega_j m_j - qU\cos\theta \sum_j z_j m_j \qquad (11.6.3)$$

where m_j, the concentration of ions j, will be either c_j or c_j^* outside or inside the polyion. If we define

$$\sigma_0 = q \sum_j z_j^2\omega_j c_j$$

$$\sigma_m = q \sum_j z_j^2\omega_j c_j^* \qquad (11.6.4)$$

then by use of Eq. (11.6.2),

$$J_{0r} = -\sigma_0\left(\frac{\partial\psi_0}{\partial r}\right)_s + \rho_f U\cos\theta$$

$$J_{mr} = -\sigma_m\left(\frac{\partial\psi_m}{\partial r}\right)_s \qquad (11.6.5)$$

showing that Brownian motion is ignored. In the case of the conductance of small electrolytes, the neglect of Brownian motion leads to an error of $\sqrt{2}$ in the computed retarding forces. Accepting an uncertainty of this order, we note that the surface charge density satisfies the boundary condition

$$4\pi\rho_s = D_0\left(\frac{\partial\psi_0}{\partial r}\right)_s - D_m\left(\frac{\partial\psi_m}{\partial r}\right)_s \qquad (11.6.6)$$

$$\frac{\partial\rho_s}{\partial t} = J_{0r} - J_{mr}.$$

These boundary conditions, plus the Laplace equations

$$\nabla^2\psi_0 = 0, \qquad r < R_s$$

$$\nabla^2\psi_m = 0, \qquad r > R_s \qquad (11.6.7)$$

and the usual conditions that ψ be finite everywhere and continuous at the boundary give, in the steady state when $(\partial\rho_s/\partial t) = 0$,

$$3\sigma\psi_0 = [\rho_f U - 3\sigma_m X]r\cos\theta$$

$$3\sigma\psi_m = \left[-3\sigma X + (\sigma_0 - \sigma_m)\frac{XR_s^3}{r^3} + \frac{\rho_f UR_s^3}{r^3}\right]r\cos\theta \qquad (11.6.8)$$

$$3\sigma = \sigma_0 + 2\sigma_m.$$

Thus, from Eq. (11.6.8) we see that the electric field inside the polymeric domain is not X but rather

$$X_2 = \frac{3\sigma_m X - \rho_f U}{3\sigma} \qquad (11.6.9)$$

and thereupon

$$U = \frac{\rho_f X}{v_m \Upsilon} \frac{3\sigma_m}{3\sigma + (\rho_f^2/v_m \Upsilon)}. \tag{11.6.10}$$

Note that nonuniform liquid flow due to the forces exerted on the surface charges has been neglected in (11.6.10).

In the absence of relaxation effects the polyion was assumed to be freely drained. For this limit of the hydrodynamic interactions, U would be independent of the degree of polymerization. The inclusion of relaxation destroys this independence but slightly. Although ρ_f, σ, and σ_m depend upon the polymer dimensions, the fact that ρ_f and v_m are geometrically linked and that only the ratio ρ_f/v_m appears shows that the molecular weight dependence of U is small.

By use of the Donnan approximation to calculate the internal salt concentrations, Longworth and Hermans find for the specific conductance of the solution,

$$\sigma(\text{soln.}) = \sigma_m + \frac{ZqcU}{X} + \frac{3(\sigma_0 - \sigma_m)\sigma(\text{soln.})\phi}{\sigma_0 + 2\sigma(\text{soln.}) + (\rho_f^2/v_m \Upsilon)} \tag{11.6.11}$$

with ϕ the volume fraction of polyelectrolyte. This relation shows that the excess specific conductance $\sigma_0 - \sigma_m$, as a result of the excess small ion concentration inside the polymeric domain over that outside, does not contribute to the conductance, but only to the polarization. This behavior is a result of the approximation that the internal region is always electrically neutral.

The preceding crude discussion of relaxation phenomena has not been compared with experiment because of the difficulty of quantitatively estimating ϕ, ρ_f, and v_m. Nevertheless, it does show that the relaxation effect can in principle account for a decrease in specific conductance when polyelectrolyte is added to the solution under conditions such that the concentrations of simple ions are maintained constant. This observation may be of importance when changes in apparent mobility are interpreted in terms of ion binding. We do not believe that all apparent binding phenomena can be so explained, but the quantitative aspects of ion association might be somewhat altered from what they are now thought to be.

11.7 Some Comments about Hydrodynamic Interactions

All of the preceding comments are concerned with a particular model—the porous sphere. In this section we turn to an examination of the relationship between this model and a more realistic representation of the polymer in solution. We shall still make the very severe approximation of regarding the solvent as a continuous medium to which ordinary hydrodynamics is applicable. The more realistic representation to which we allude concerns itself with the detailed structure of the polymer skeleton.

The general formulation of the theory of irreversible processes in solutions containing polymer molecules has been studied by Kirkwood.[61] The system considered consists of a linear polymer of $Z = 2n + 1$ segments immersed in a solvent of viscosity η. Each segment has a monomer molecular weight M_u and friction constant Υ with respect to the solvent. The hydrodynamic resisting elements are subject to certain structural restraints; for example under ordinary conditions we require that the bond angles and bond lengths be fixed. The residual degrees of freedom are concerned with internal rotation about the valence bonds. Suppose then that there are μ actual degrees of freedom for motion subject to the restraints mentioned. As usual, three of the μ degrees of freedom are concerned with the translational motion of the center of mass and three more are concerned with the specification of the external orientation of the polymer.

Now the configuration space of the molecule has, of course, $6n + 3$ dimensions. The constraints mentioned above reduce the degrees of freedom to $2n + 4$ and these will be considered to form a new space—chain space. The $2n + 4$ coordinates consist of the three coordinates of the center of mass relative to some external coordinate system, the two orientation angles of the polymer molecule in the external system, and $2n - 1$ angles of internal rotation. The $4n - 1$ coordinates of configuration space complementary to chain space will be denoted $q^{2n+5} \cdots q^{6n+3}$. These bond angle and bond length coordinates can be chosen in a number of different ways.

The distribution function of the coordinates $f(q, t)$ may be used in the usual manner to define averages,

$$\bar{\phi} = \int \cdots \int \phi(q, t) f(q, t) \sqrt{g} \prod_\alpha dq^\alpha \tag{11.7.1}$$

where the volume element has been transformed from configuration space to chain space and its complement by use of the well-known transformations[62]

$$ds^2 = \sum_{\alpha,\beta} g_{\alpha\beta} \, dq^\alpha \, dq^\beta$$

$$g_{\alpha\beta} = \sum_{l=1}^{Z} \frac{\partial \mathbf{R}^l}{\partial q^\alpha} \frac{\partial \mathbf{R}^l}{\partial q^\beta} \tag{11.7.2}$$

$$\mathbf{R} = \sum_{l=1}^{Z} \mathbf{R}^l$$

with ds the element of length and $g_{\alpha\beta}$ an element of the metric tensor. As usual, g denotes the determinant of the elements of the metric tensor,

$$g = \|g_{\alpha\beta}\|. \tag{11.7.3}$$

In Eq. (11.7.2) $\mathbf{R}^l$ is the position vector of element l in the usual three-dimensional configuration space of element l. Now it is possible to choose the coordinates of the complementary space such that they are orthogonal to the coordinates of chain space. Under these circumstances, the components of the metric tensor connecting the chain space and the complementary space vanish. Thus one may write

$$g = g_{(1)}g_{(2)}$$
$$g_{(1)} = \|g_{\alpha\beta}\| \qquad \alpha, \beta \leqslant 2n + 4 \qquad (11.7.4)$$
$$g_{(2)} = \|g_{\alpha\beta}\|. \qquad \alpha, \beta \geqslant 2n + 5$$

To study the hydrodynamic interactions in ordinary configuration space it is convenient to use the covariant vectors

$$\mathbf{a}_\alpha = \sum_l \frac{\partial \mathbf{R}^l}{\partial q^\alpha} \qquad (11.7.5)$$

from which, as usual, the covariant components of any vector or tensor are[63]

$$F_\alpha = \mathbf{F} \cdot \mathbf{a}_\alpha \qquad (11.7.6)$$
$$T_{\alpha\beta} = \mathbf{a}_\alpha \cdot \mathbf{T} \cdot \mathbf{a}_\beta.$$

The mixed components (contravariant) will involve contravariant components of the basis vectors and therefore also the contravariant components of the metric tensor $g_{\alpha\beta}$. These are defined by [63]

$$g^{\alpha\beta} = \frac{\|g_{\alpha\beta}\|_{\alpha\beta}}{g}. \qquad (11.7.7)$$

The contravariant components of an arbitrary vector or tensor then become

$$F^\alpha = \sum_{v=1}^{2n+4} g^{\alpha v} \mathbf{a}_v \cdot \mathbf{F} \qquad (11.7.8)$$
$$T^{\alpha\beta} = \sum_{\mu,v}^{2n+4} g^{\alpha\mu} g^{\beta v} \mathbf{a}_\mu \cdot \mathbf{T} \cdot \mathbf{a}_v.$$

For the components in chain space, the sums need extend only over chain space since the other terms vanish. With the assumption that the frictional force acting on segment l is

$$\mathbf{F}^l = \Upsilon(\mathbf{v}^l - \mathbf{u}^l) \qquad (11.7.9)$$

with $\mathbf{v}^l$ the perturbed velocity of the fluid at $\mathbf{R}^l$ and $\mathbf{u}^l$ the velocity of the segment at $\mathbf{R}^l$, we are now ready to calculate the frictional resistance to flow encountered by the polymer molecule.

If $\mathbf{v}^{0l}$ is the velocity of the fluid at $\mathbf{R}^l$ prior to the introduction of the polymer then the Oseen equation becomes

$$\mathbf{v} = \mathbf{v}^0 - \mathbf{T} \cdot \mathbf{F}$$

$$\mathbf{T} = \sum_{l,s} T^{l,s} = \sum_{l,s} \frac{1}{8\pi\eta R_{ls}} \left[\mathbf{1}^{ls} + \frac{\mathbf{R}_{ls}^l \mathbf{R}_{ls}^s}{R_{ls}^2} \right]$$

$$\mathbf{F} = \sum_l \mathbf{F}^l$$

$$\mathbf{v}^0 = \sum_l \mathbf{v}^{0l} \tag{11.7.10}$$

$$\mathbf{u} = \sum_l \mathbf{u}^l$$

$$\mathbf{v} = \sum_l \mathbf{v}^l$$

where R_{ls} is the separation of segments l and s, $\mathbf{1}^{ls}$ is the unit tensor in the ordinary three-dimensional space common to segments l and s and all other symbols have been previously defined. Note that the frictional tensor ζ with components $\zeta_{\alpha\beta}$ is defined in configuration space by

$$F_\alpha = \sum_{\beta=1}^{6n+3} \zeta_{\alpha\beta}(v^{0\beta} - u^\beta) \tag{11.7.11}$$

and the sum extends over the complementary space as well as chain space. The components of ζ do not disappear between chain space and complementary space and therefore velocity differences in complementary space contribute to the frictional force in chain space.

If Eq. (11.7.10) is substituted into Eq. (11.7.9) and the resultant equation solved for F_α, one finds by comparison with Eq. (11.7.11) that

$$\zeta_{\alpha\beta} = \zeta g_{\alpha\beta} - \zeta \sum_{\nu=1}^{6n+3} T_{\alpha}^{\nu} \zeta_{\gamma\beta}. \tag{11.7.12}$$

The notation used is as follows: Greek superscripts and subscripts refer to components along the basis vectors $\mathbf{a}_\alpha$ and $\mathbf{a}^\alpha$; Latin superscripts refer to projections of $6n + 3$-dimensional vectors onto the three-dimensional subspace of the element indicated; Latin subscripts refer to the same projections when referred to the Cartesian system common to all segments. By inversion of (11.7.12), it is found that the elements of the inverse friction tensor satisfy the relation

$$(\zeta^{-1})^{\alpha\beta} = \frac{g^{\alpha\beta}}{\zeta} + \sum_{\mu,\nu=1}^{2n+4} g^{\alpha\mu} g^{\beta\nu} T_{\mu\nu} \tag{11.7.13}$$

and by use of the Einstein relation,

$$(\zeta^{-1})^{\alpha\beta} = \frac{D^{\alpha\beta}}{kT} \tag{11.7.14}$$

with $D^{\alpha\beta}$ the contravariant components of the diffusion tensor.

The determination of the general elements of $D^{\alpha\beta}$ requires detailed specification of the molecular structure. If we consider only the translation of the center of mass,

$$g_{\alpha\beta} = Z\delta_{\alpha\beta} \qquad \alpha = 1, 2, 3$$

$$g^{\alpha\beta} = \frac{\delta_{\alpha\beta}}{Z} \qquad \beta = 1, \cdots 2n + 4.$$

(11.7.15)

By substitution we find the simple and exact result,

$$D^{\alpha\beta} = kT\left[\frac{\delta_{\alpha\beta}}{Z\Upsilon} + \frac{1}{8\pi Z^2\eta} \sum_{\substack{l \neq s \\ =1}}^{Z} \left(\frac{\delta_{\alpha\beta}}{R_{ls}} + \frac{X_{ls}^\alpha X_{ls}^\beta}{R_{ls}^3}\right)\right] \qquad \alpha, \beta = 1, 2, 3 \quad (11.7.16)$$

with X_{ls}^α the components of distance between segments l and s along the rectangular axis α of the coordinate system to which the position of the center of mass is referred. The mean translational diffusion coefficient is defined as one-third the trace of $D^{\alpha\beta}$ averaged over the internal coordinates. Thus,

$$\zeta^{-1} = \frac{\langle D \rangle}{kT} = \left[\frac{1}{Z\Upsilon} + \frac{1}{6\pi Z^2\eta} \sum_{\substack{l \neq s \\ =1}}^{Z} \left\langle\frac{1}{R_{ls}}\right\rangle\right].$$

(11.7.17)

Assuming a Gaussian distribution of intersegment distances to be valid for all separations,

$$\left\langle\frac{1}{R_{ls}}\right\rangle = \left(\frac{6}{\pi}\right)^{\frac{1}{2}} b^{-1}|l - s|^{-\frac{1}{2}}$$

(11.7.18)

where b is the effective length of a monomer unit,

$$b^2 = b_0^2 \frac{(1 + \cos\theta)(1 + \langle\cos\phi\rangle)}{(1 - \cos\theta)(1 - \langle\cos\phi\rangle)}$$

(11.7.19)

θ is the valence angle, ϕ the azimuthal angle describing internal rotation, and b_0 is the bond length. Now if we define λ_0 by the relation

$$\frac{\Upsilon}{\lambda_0} = (6\pi^3)^{\frac{1}{2}}\eta b$$

(11.7.20)

then it may be shown that the use of Eq. (11.7.18) is Eq. (11.7.17) gives, finally,

$$\zeta = \frac{\Upsilon Z}{1 + (8\lambda_0/3)Z^{\frac{1}{2}}}.$$

(11.7.21)

In the limit as $Z \to \infty$, Eq. (11.7.21) gives the same frictional resistance as Stokes law if we assign to the polymer a radius of $0.26bZ^{\frac{1}{2}}$. The exact hydrodynamic treatment of the porous sphere gives a Stokes law limit with the sphere assigned a radius $0.53bZ^{1/2}$. The discrepancy between these two values is a measure of the difference in frictional resistance introduced by consideration of the details of the skeletal distribution of mass and by mathematical approximations.

It is not at all obvious whether the major source of the discrepancy cited is based in the mathematical approximations or in the difference in models. Isihara and Toda[64] have shown that the respective starting points of the two theories are equivalent. The argument starts from Eq. (11.7.10) rewritten in the form

$$v(R) = -\frac{\Upsilon}{8\pi\eta} \sum \left[\frac{v^{0s}}{|R - R^l|} + \frac{(R - R^l)(R - R^l) \cdot v^{0l}}{|R - R^l|^3} \right]$$

$$p = -\frac{\Upsilon}{4\pi} \sum \frac{(R - R^l) \cdot v^{0l}}{|R - R^l|^3}.$$

(11.7.22)

Suppose, now, that the distribution of segments about the center of mass $v_m(s)$ is known. Then by substitution, Eq. (11.7.22) can be transformed into

$$v(R) = -\frac{\Upsilon}{8\pi\eta} \left[\int \frac{v^0(s)v_m(s)}{|R - s|} ds - \int (R - s) \cdot v^0(s)\nabla_R\left(\frac{1}{|R - s|}\right) v_m(s) ds \right]$$

$$= -\frac{\Upsilon}{8\pi\eta} \left[2 \int \frac{v^0(s)v_m(s)}{|R - s|} ds - \nabla_R \int \frac{(R - s) \cdot v^0(s)}{|R - s|} v_m(s) ds \right].$$

(11.7.23)

By application of the Laplacian operator to both sides of Eq. (11.7.23), one finds that

$$\nabla^2 v(R) = -\frac{\Upsilon}{8\pi\eta} \left[-8\pi v^0(R)v_m(R) + 2\nabla_R \int \frac{(R - s) \cdot v^0(s)}{|R - s|^3} v_m(s) ds \right]$$

(11.7.24)

which, for a spherical particle in laminar flow becomes

$$\nabla^2 v^0(R) = \frac{\Upsilon v_m}{\eta} v^0(R) + \frac{1}{\eta} \nabla p.$$

(11.7.25)

After comparison with Eq. (11.2.1') it is seen that the starting points of Kirkwood and of Debye and Bueche are equivalent. By starting from Eq. (11.7.25) and using the converse arguments, Eq. (11.7.22) can be derived.

It is now clear that the differences in the theories arise from mathematical approximations which are closely related to the models used. Debye and Bueche assume that v_m is a constant, and Kirkwood assumes that Eq. (11.7.18) is valid for all segment distances. Both of these approximations correspond in part to a sphericalization of the mass distribution, but in very different ways. Although in principle the porous sphere approach can be improved by making the segment density a function of distance from the center of mass, the model is still subject to the ambiguity of definition inherent in the use of an equivalent sphere to replace the real polymer chain and there appears to be no unique way of defining $v_m(\mathbf{R})$ consistent with our knowledge of chain structure. On the other hand, the Gaussian approximation can in principle be improved but the calculations become very involved. Moreover, there does not appear to be, at present, any suitable manner to describe the counterion distribution about a real polymer skeleton. In the absence of knowledge of the counterion distribution, the Kirkwood method cannot be used to calculate the electrophoretic velocity of a linear polyion. Despite the preceding observations it is our opinion that an improved theory of transport in solutions of linear polyelectrolytes must be based on a realistic model of the chain, such as that proposed by Kirkwood.

In accordance with the preceding we shall close this chapter with a brief examination of the hydrodynamic interactions in a non-Gaussian coil. It can be shown[65] that the probability density characterizing the distribution of segments about the center of mass is altered by the presence of a velocity gradient. Even in the case of translational diffusion, where every particle moves with the same mean velocity there are effects which arise when the end to end separation of the chain becomes large. Katchalsky et al.[66] have shown that the distribution function for any intersegmental distance

$$\mathbf{r}_{kl} = \mathbf{r}_l - \mathbf{r}_k, \qquad l - k > 0 \tag{11.7.26}$$

with the end-to-end distance vector $\mathbf{h}$ held fixed, is

$$W_{kl}(\mathbf{r}_{kl}) = \left(\frac{3Z^2}{2h^2 j(Z-j)\pi}\right)^{1/2} \exp\left[-\left(\frac{3Z^2(\mathbf{r}_{kl} - (j/Z)\mathbf{h})^2}{2h^2 j(Z-j)}\right)\right]. \tag{11.7.27}$$

If terms which vanish in the averaging over all polymer configurations are omitted, then the Oseen formulation of the hydrodynamic interaction of two spheres of radius a becomes (in scalar form)

$$v^j = \frac{F^j}{6\pi\eta a} + \frac{F^k}{8\pi\eta}\left[\frac{1 + \cos^2\theta_{jk}}{r_{jk}} + \frac{a^2(1 - 3\cos^2\theta_{jk})}{r_{jk}^3}\right] \tag{11.7.28}$$

with v^j the velocity of the jth sphere, F^j and F^k the forces acting on spheres j and k, r_{jk} the distance between the spheres, and θ_{jk} the angle between $\mathbf{v}^j$ and $\mathbf{r}_{jk}$. In the usual treatment of the Gaussian coil it is assumed that the distribution of segments is spherically symmetric so that $\langle\cos^2\theta_{jk}\rangle = \frac{1}{3}$ whereupon the interaction term reduces to $\langle 1/r_{jk}\rangle = (6/\pi)^{1/2}(1/\langle r_{jk}^2\rangle)^{1/2}$ as used previously. However, with the use of the distribution function displayed in Eq. (11.7.27) it is found that the interaction term in the direction parallel to $\mathbf{h}$

$$T_{jh} = \frac{1}{8\pi\eta h}\left(\frac{2Z}{j}\right)\left[\left(1 - \frac{1}{2\xi^2}\right)\Psi(\xi) + \frac{e^{-\xi^2}}{\xi\pi^{1/2}}\right]; \quad j = k - l \quad (11.7.29)$$

differs from the interaction term for the direction normal to **h**,

$$T_{jn} = \frac{1}{8\pi\eta h}\left(\frac{Z}{j}\right)\left[\left(1 + \frac{1}{2\xi^2}\right)\Psi(\xi) - \frac{e^{-\xi^2}}{\xi\pi^{1/2}}\right]; \quad j = l - k \quad (11.7.30)$$

with

$$\xi^2 = \frac{3}{2}\left(\frac{h^2}{\langle h^2\rangle}\right)\frac{j}{Z - j}$$

$$\Psi = \frac{2}{\pi^{1/2}}\int_0^\xi e^{-t^2}\,dt. \tag{11.7.31}$$

When $h^2/\langle h^2\rangle$ is sufficiently small T_{jh} becomes equal to T_{jn}. However, when $h^2/\langle h^2\rangle$ is large, $T_{jh} = 2T_{jn}$. The preceding derivation has been carried through with the neglect of the r_{jk}^{-3} term. Peterlin and Copic[67] argue that when r_{jk} is small the distribution is almost isotropic and then $\langle 1 - 3\cos^2\theta_{jk}\rangle$ must almost vanish, and when r_{jk} is large the factor r_{jk}^{-3} decreases so rapidly that the contribution of this term is negligible. We concur with this argument for intermediate and large values of r_{jk}, but for small r_{jk} it seems most plausible that the distribution is anisotropic (see Chapter 6). Nevertheless, the importance of Eqs. (11.7.29) and (11.7.30) lies in the observation that if h is fixed, even a Gaussian coil gives rise to nonspherical hydrodynamic interactions.

Most linear polyelectrolyte molecules are randomly coiled in the discharged state. When fully charged we picture them as being distended coils (but not rods). Since in many instances the ratio of root mean square end-to-end separations of the charged to uncharged coils exceeds 0.2, the distribution function deviates from Gaussian behavior. Suppose that [68]

$$\langle h^2\rangle = \langle h_0^2\rangle f(\alpha Z) \tag{11.7.32}$$

where $\langle h_0^2\rangle$ is the unperturbed mean square end-to-end separation in the discharged state. In general, if $f(\alpha Z)$ is analytic,

$$f(\alpha Z) = f(0) + \left(\frac{\partial f}{\partial(\alpha Z)}\right)_0 \alpha Z + \cdots$$

$$= 1 + A_1\alpha Z + \cdots. \tag{11.7.33}$$

This expansion can be used to estimate the modification in the distribution function by writing

$$W(h) = C_0\left(\frac{3}{2\pi\langle h^2\rangle}\right)^{3/2}\exp\left(-\frac{3h^2}{2\langle h^2\rangle}\right)[1 + C_1 h + C_2 h^2 + \cdots]. \tag{11.7.34}$$

The use of Eq. (11.7.34) to determine $\langle h^2 \rangle$ leads to the relations

$$C_0 = [1 + \tfrac{3}{2}A_1\alpha Z]^{-1}$$
$$C_1 = 0 \qquad\qquad\qquad (11.7.35)$$
$$C_2 = 3A_1\alpha Z/2\langle h^2 \rangle.$$

The coefficient A_1 is related to the repulsive force which expands the chain. The entropic force tending to contract the polymer molecule is

$$\mathbf{F}_S = kT\nabla \ln W. \qquad\qquad (11.7.36)$$

By the substitution of Eqs. (11.7.34) and (11.7.35) into (11.7.36) one finds that

$$\Delta \mathbf{F}_S = \mathbf{F}_S - \mathbf{F}_S(\text{Gaussian}) = 2kT\mathbf{e}_h\frac{3A_1}{2\langle h^2\rangle}\alpha Zh$$

$$= 3kTA_1\alpha\frac{Zh}{\langle h^2\rangle}\mathbf{e}_h \qquad (11.7.37)$$

with $\mathbf{e}_h$ a unit vector pointing along the end-to-end separation vector $\mathbf{h}$. In the approximation carried through, the excess repulsive force is linear in the polyion charge.

To see what effect this non-Gaussian distribution has upon the hydrodynamic interactions, consider the perturbation to the velocity at the jth segment due to the other chain segments. This is obtained from Eq. (11.7.28) by summing the contributions from all segments. By use of the non-Gaussian distribution function displayed in Eq. (11.7.34), the value of $\langle 1/r_{jk}\rangle$ is found to be

$$\left\langle\frac{1}{r_{jk}}\right\rangle = \left(\frac{6}{\pi\langle h^2_{0,jk}\rangle}\right)^{1/2}\left(1 - \frac{A_1\alpha}{2}|j - k| + \cdots\right) \qquad (11.7.38)$$

and after introduction of Eq. (11.7.38) one finds from Eq. (11.7.17) the friction coefficient for the translational motion of the polymer,

$$\zeta = \frac{Z\Upsilon}{1 - \dfrac{3.20\Upsilon}{6\pi b}\left(\dfrac{6}{\pi}\right)^{1/2} + \left(\dfrac{6}{\pi}\right)^{1/2}\dfrac{\Upsilon}{6\pi b}Z^{1/2}\left[\dfrac{8}{3} - \dfrac{8}{30}A_1\alpha Z\right]} \qquad (11.7.39)$$

to the order of the terms retained in Eq. (11.7.38). This result is different in its functional dependence upon Z from the result displayed in Eq. (11.7.21)

It is clear that the theory of transport phenomena in solutions of polyelectrolytes is far from satisfactory. It is our opinion that a satisfactory theory will necessarily involve a detailed description of the hydrodynamics of non-Gaussian coils. Before this program can be carried through, however, the equilibrium distribution of counterions about a polyion must be calculated for a realistic model.

REFERENCES

1. F. BOOTH, *J. Chem. Phys.* **19**, 1331 (1951).
2. H. H. PAINE, *Proc. Cambridge Phil. Soc.* **28**, 83 (1932); J. J. BIKERMAN, *Kolloid-Z.* **72**, 104 (1935; J. J. HERMANS, *Phil. Mag.* (7) **26**, 650 (1938); *Trans. Faraday Soc.* **36**, 133 (1940).
3. J. T. G. OVERBEEK, *Kolloid-Beih.* **54**, 287 (1943).
4. F. BOOTH, *Proc. Roy. Soc.* **A203**, 514 (1950).
5. J. J. HERMANS AND H. FUJITA, *Koninkl. Ned. Akad. Wetenschap., Proc.* **B58**, 182 (1955).
6. J. J. HERMANS AND J. TH. G. OVERBEEK, *Rec. trav. chim.* **67**, 762 (1948).
7. H. C. BRINKMAN, *Koninkl. Ned. Akad. Wetenschap., Proc.* **50**, 618 (1947); *Physica* **13**, 447 (1947); *Appl. Sci. Research* **A7**, 27 (1947).
8. P. DEBYE AND A. M. BUECHE, *J. Chem. Phys.* **16**, 573 (1948).
9. J. TH. G. OVERBEEK AND D. STIGTER, *Rec. trav. chim.* **75**, 543 (1956).
10. J. J. HERMANS, *J. Polymer Sci.* **18**, 529 (1955).
11. J. J. HERMANS, *in* "The Structure of Electrolytic Solutions" (W. J. Hamer, ed.), p. 308. Wiley, New York. 1959.
12. H. FUJITA, *J. Phys. Soc. Japan* **12**, 968 (1957).
13. M. NAGASAWA, A. SODA, AND I. KAGAWA, *J. Polymer Sci.* **31**, 439 (1958).
14. P. J. NAPJUS AND J. J. HERMANS, *J. Colloid Sci.* **14**, 252 (1959).
15. E. B. FITZGERALD AND R. M. FUOSS, *J. Polymer Sci.* **14**, 329 (1954).
16. A. KATCHALSKY, N. SHAVIT, AND H. EISENBERG, *J. Polymer Sci.* **13**, 69 (1954).
17. B. VAN GEELEN, Thesis, Utrecht, 1958.
18. M. HOSONO, N. ISE, AND I. SAKURADA, *Mem. Fac. Eng., Kyoto Univ.* **19**, 408, 417 (1957).
19. N. ISE, M. HOSONO, AND I. SAKURADA, *Polymer Chem. Japan* **15**, 339 (1958) (in Japanese).
20. T. ISEMURA AND A. IMANISHI, *J. Polymer Sci.* **33**, 337 (1958).
21. H. FUJITA AND J. J. HERMANS, *Koninkl. Ned. Akad. Wetenschap., Proc.* **B58**, 188 (1955).
22. R. LONGWORTH AND J. J. HERMANS, *J. Polymer Sci.* **26**, 47 (1957).
23. D. EDELSON AND R. M. FUOSS, *J. Am. Chem. Soc.* **70**, 2832 (1954).
24. R. M. FUOSS AND D. EDELSON, *J. Polymer Sci.* **6**, 532 (1951).
25. G. I. CATHERS AND R. M. FUOSS, *J. Polymer Sci.* **4**, 121 (1949).
26. R. MACFARLANE, JR. AND R. M. FUOSS, *J. Polymer Sci.* **23**, 403 (1957).
27. R. M. FUOSS, *J. Polymer Sci.* **12**, 185 (1954).
28. A. OTH AND P. DOTY, *J. Phys. Chem.* **56**, 43 (1952).
29. H. EISENBERG, *J. Polymer Sci.* **30**, 47 (1958).
30. F. T. WALL AND R. H. DOREMUS, *J. Am. Chem. Soc.* **76**, 1557 (1954).
31. F. T. WALL AND S. J. GILL, *J. Phys. Chem.* **58**, 740 (1954).
32. T. SEIYAMA, *J. Chem. Soc. Japan, Ind. Chem. Sect.* **53**, 122 (1950) (in Japanese).
33. T. TAKAHASHI, K. KIMOTO, AND TAKANO, Y., *J. Chem. Soc. Japan, Pure Chem. Sect.* **72**, 292 (1951).
34. I. KAGAWA *J. Chem. Soc. Japan, Ind. Chem. Sect.* **54**, 394 (1951); *Ibid.* **56**, 857 (1953).
35. H. EISENBERG AND G. R. MOHAN, *J. Phys. Chem.* **63**, 671 (1959).
36. H. HECKMAN, *Naturwissenschaften* **40**, 478 (1953); *Kolloid-Z.* **136**, 67 (1954).
37. B. JACOBSON, *Nature* **172**, 666 (1953); *Rev. Sci. Instr.* **24**, 949 (1953).

38. U. SCHINDEWOLF, *Z. physik. Chem.* [N.F.] **1**, 129 (1954); *Naturwissenschaften* **40**, 478 (1953); *Z. Elektrochem.* **58**, 697 (1954).
39. K. G. GÖTZ AND K. HECKMANN, *J. Colloid Sci.* **13**, 266 (1958).
40. M. EIGEN AND G. SCHWARZ, *Z. physik. Chem.* [N.F.] **4**, 380 (1955); *J. Colloid Sci.* **12**, 181 (1957).
41. G. SCHWARZ, *Z. Physik* **145**, 563 (1956).
42. C. T. O'KONSKI AND A. J. HALTNER, *J. Am. Chem. Soc.* **79**, 5634 (1957).
43. H. SCHÄFER, *Z. Physik* **77**, 117 (1932).
44. H. FRICKE AND H. J. CURTIS, *Phys. Rev.* **48**, 775 (1935).
45. G. SCHMID AND A. V. ERKKILA, *Z. Elektrochem.* **42**, 737 (1936).
46. F. E. BAILEY, JR., A. PATTERSON, JR., AND R. M. FUOSS, *J. Polymer Sci.* **9**, 285 (1952); *J. Am. Chem. Soc.* **74**, 1845 (1952).
47. K. F. WISSBRUN AND A. PATTERSON, JR., *J. Polymer Sci.* **33**, 235 (1958).
48. J. R. HUIZENGA, P. F. GRIEGER, AND F. T. WALL, *J. Am. Chem. Soc.* **72**, 2636 (1950).
49. F. T. WALL, J. J. ONDREJICIN, AND M. PIKRAMENON, *J. Am. Chem. Soc.* **73**, 2821 (1951).
50. F. T. WALL AND P. F. GRIEGER, *J. Chem. Phys.* **20**, 1200 (1952).
51. F. T. WALL, P. F. GRIEGER, J. R. HUIZENGA, AND R. H. DOREMUS, *J. Chem. Phys.* **20**, 1206 (1952).
52. F. T. WALL AND R. H. DOREMUS, *J. Am. Chem. Soc.* **76**, 868, 1557 (1954).
53. F. T. WALL, H. TERAYAMA, AND S. TECHAKUMPUCK, *J. Polymer Sci.* **20**, 477 (1956).
54. J. R. HUIZENGA, P. F. GRIEGER, AND F. T. WALL, *J. Am. Chem. Soc.* **72**, 4228 (1950).
55. H. TERAYAMA AND F. T. WALL, *J. Polymer Sci.* **16**, 357 (1955).
56. F. T. WALL AND J. BERKOWITZ, *J. Chem. Phys.* **26**, 114 (1957).
57. M. NAGASAWA, S. OZAWA, AND K. KIMURA, *J. Chem. Soc. Japan, Ind. Chem. Sect.* **59**, 1201 (1956) (in Japanese); M. NAGASAWA, S. OZAWA, K. KIMURA, AND I. KAGAWA, *Mem. Fac. Eng., Nagoya Univ.* **8**, 50 (1956) (in English).
58. G. J. HOWARD AND D. O. JORDAN, *J. Polymer Sci.* **12**, 209 (1954).
59. R. LORENZ, " Raumerfüllung und Ionen Beweglichkeit." 1922. See N. KAMEYAMA, " Theory of Electrochemistry and its Application," p. 61. Maruzen, Tokyo, 1947 (in Japanese).
60. O. KEDEM AND A. KATCHASLKY, *J. Polymer Sci.* **15**, 321 (1955).
61. J. G. KIRKWOOD, *Rec. trav. chim.* **68**, 649 (1949); J. G. KIRKWOOD, *J. Polymer Sci.* **12**, 1 (1954); J. ERPENBECK AND J. G. KIRKWOOD, *J. Chem. Phys.* **29**, 909 (1958).
62. See, for example, A. TAYLOR, " Advanced Calculus." Ginn, Boston, 1952.
63. See H. WILLS, " Vector and Tensor Analysis." Dover, New York, 1958.
64. A. ISIHARA AND M. TODA, *J. Polymer Sci.* **7**, 277 (1951).
65. A. ISIHARA, *J. Polymer Sci.* **8**, 574 (1952).
66. A. KATCHALSKY, O. KÜNZLE, AND W. KUHN, *J. Polymer Sci.* **5**, 283 (1950).
67. A. PETERLIN AND M. COPIC, *J. Chem. Phys.* **27**, 434 (1956).
68. This technique was first used by A. PETERLIN, *Bull. sci., Conseil. acad. RPF Yougoslavie* **2**, 98 (1955).

AUTHOR INDEX

Numbers in parentheses are reference numbers. Numbers in italic show the page on which the complete reference is listed.

A

Accascina, F., 89(20b), *129*, 442(43), *487*
Ackermann, H., 429(10), *486*
Alexandrowicz, Z., 403
Alfrey, T., Jr., 221(28), 233(41), 235(41), 236(41), *247*, *248*
Ander, P., 436(27), 447(27), 458(27), 460(27), *487*
Anderegg, G., 244(61), *248*
Armstrong, S., 246(70), *248*
Arrhenius, S., 83(3), 85(3), *129*
Ascoli, F., 393(4), *425*

B

Bailey, F. E., Jr., 438(46), 454(64e), 461(77, 78), *488*, 538(46), *554* ⟍
Ballhausen, C. J., 242(52), *248*
Barkin, S., 246(73, 74), *248*, 436(31), *487*
Basolo, F., 428(7), 429(7), 444(7), *486*
Batchelder, A. C., 412(34), *426*
Beir, M., 187(2), 189(2), *205*
Bellemans, R., 47(16), 67(16), 71(16), *79*
Benoit, H., 212(14, 15), *247*
Berg, P. W., 233(41), 235(41), 236(41), *248*
Bergmann, M., 183(25), *184*
Berkowitz, J., 1(5), *79*, 159(14), *184*, 238(44), 239(44), *248*, 380(47), *390*, 447(54), 449(54), *487*, 540(56), *554*
Bikerman, J. J., 522(2), *553*
Bjerrum, J., 431(12), *486*
Bjerrum, N., 88(12), *129*, 241(47), *248*, 249(1), 269(23), *281*, 287(11), *389*, 437(33), *487*
Block, B. P., 429(9), *486*
Blout, E., 366(40), 378(44), *390*
Bluestone, S., 436(28), 447(28), 458(28), 459(28), 460(28), *487*

Bockris, J. O'M., 89(20a), *129*
Bonner, O. D., 472(87), 473(87), 483(87), *488*
Booth, F., 131(3), *184*, 188(5–8), 196(7), 197(7), 199(5, 7), 200(5), 204(5), *205*, 518(1), 522(4), 536(4), *553*
Botré, C., 393(4), *425*
Botty, M. C., 226(35b), *248*
Boyd, G. E., 461(80), 481(80), 482(92), *488*
Bradbury, A., 366(40), *390*
Bragg, J. K., 372(41), 377(41), 378(41), *390*
Branson, R., 374(42), *390*
Bregman, J. I., 476(88), 480(88), *488*
Brinkman, H. C., 499(6), *516*, 522(7), 536(7), *553*
Brody, O. V., 249(6), 255(6), *281*
Brout, R., 13(8), *79*
Brown, A., 412(34), *426*
Brown, W. B., 47(16), 67(16), 71(16), *79*
Bueche, A. M., 499(5), *516*, 522(8), 536(8), *553*
Buff, F. P., 1(2), 10(2), 13(8), *79*
Burbach, J. C., 246(72), *248*
Byrd, R., 36(10), 38(10), *79*

C

Calvin, M., 431(13), *486*
Cathers, G. I., 536(25), 537(25), *553*
Chu, V. H., 241(49), 242(49), *248*, 249(4), 255(4), *281*
Ciferri, A., 215(22), *247*
Clausius, R., 82(2), *129*
Cohn, E. J., 166(11), *184*, 341(34), *390*
Coleman, B., 181(24), 182(24), *184*
Collie, C. H., 224(33), *247*, 255(11), *281*, 318(22), 347(37), *389*, *390*, 471(86), *488*

SUBJECT INDEX

A

Absolute activity, 253, 314, 340, 353, 354, 367, 377, *see also* Fugacity, Grand Partition Function, Semi-Grand Partition Function
Absorption isotherm, 483
Activated state, 181
Activity coefficient, 16ff, 30, 98, 165, 180, 279–281, 391ff, 401, 418, 422–424, 436–439, 450, 457, 460, 481, 508–510, 534, 540, *see also* Excess Thermodynamic Functions, Osmotic coefficient, Osmotic pressure, Virial coefficient
    single ion, 392ff
Additivity of activities, 397–399
Adsorption, 131, 132, 135, 137, 140, 246
Anion binding, 246
Atactic polymer, 215

B

Bolaform electrolyte, 241, 249ff, 452
Born cycle, 440
Boyle temperature, 217
Branching, 207
Brownian motion, 87, 89, 91, 102, 200, 266, 543, 552, *see also* Diffusion constant, Distribution function
Butane, 209, 210

C

Canonical ensemble, 4, 27
Canonical partition function, 170, 355, 511, *see also* Canonical ensemble, Configurational partition function, Partition function
Capacity, of polyion-counterion system, 448ff, *see also* Double layer, Surface charge
Carboxymethyl cellulose, 328–331, 395, 396
Cation buffer, 431

Cell cluster, 54, 57, 72–74
Cell cluster integral, 56, 59, 66, *see also* Virial coefficient
Cell model, 1, 53ff, 88, 269, 441ff, 465, 493ff, *see also* Cell cluster, Cell cluster integral, Cell partition function, Ion-pair partition function
Cell partition function, 307
Cellulose sulfate, 395
Chain configuration, 207ff, *see also* Brownian motion, Distribution function, Stereochemical conformations
Chain energy, 218
Chain extension, 208, 210, 221, 222, 231, 232, 323ff, 330ff, 345ff, 349, 358ff, 401, 402, 415, 497ff, *see also* Brownian motion, Distribution function, Electrostatic free energy, Swelling
Chain length, 207
Chain model, 210ff, 257ff, 283ff, 401, 507ff, 544ff
    critique of, 511ff
Chain stiffness, 207, 214, 221
Characteristic function, 300
Charge correlations, 105, 114, 132, 335, 360ff, *see also* Charge fluctuations, Enzyme kinetics, Grand partition function, Partition function
Charge density, 90ff, 100ff, 109, 126, 127, 525
Charge distribution, 89–104, 125ff, 132, 170ff, 520
    discrete, 162ff, 284ff
Charge fluctuations, 116, 117, 132, 178ff, 273, 345ff, 359, 386, *see also* Charge correlations, Enzyme kinetics, Grand partition function
Charge transfer spectrum, 429, 444
Charging parameter, 107, 133, 157, 298, 504
Charging process, 113ff, 133, *see also* Charging parameter, Electrostatic free energy

563